Exegetical Crossroads

Judaism, Christianity, and Islam – Tension, Transmission, Transformation

Edited by Patrice Brodeur, Assaad Elias Kattan, and Georges Tamer

Volume 8

Exegetical Crossroads

Understanding Scripture in Judaism, Christianity and Islam in the Pre-Modern Orient

Edited by
Georges Tamer, Regina Grundmann, Assaad Elias Kattan, and Karl Pinggéra

DE GRUYTER

ISBN 978-3-11-056144-9
e-ISBN (PDF) 978-3-11-056434-1
e-ISBN (EPUB) 978-3-11-056293-4
ISSN 2196-405X

This volume is text- and page-identical with the hardback published in 2018.

Library of Congress Cataloging-in-Publication Data
A CIP catalog record for this book has been applied for at the Library of Congress.

Bibliographic information published by the Deutsche Nationalbibliothek
The Deutsche Nationalbibliothek lists this publication in the Deutsche Nationalbibliografie; detailed bibliographic data are available on the Internet at http://dnb.dnb.de.

Printing and binding: CPI books GmbH, Leck
♾ Printed on acid-free paper
Printed in Germany

www.degruyter.com

Table of Contents

Georges Tamer, Regina Grundmann, Assaad Elias Kattan and Karl Pinggéra
Exegetical Crossroads
Understanding Scripture in Judaism, Christianity and Islam in the Pre-Modern Orient — **1**

William A. Graham
"A Wandering Aramean was My Father"
An Abrahamic Theme in Jewish, Christian, and Muslim Scriptures and Interpretations — **9**

Cornelia Horn
Jesus, the Wondrous Infant, at the Exegetical Crossroads of Christian Late Antiquity and Early Islam — 27

Martin Heimgartner
The Letters of the East Syrian Patriarch Timothy I
Scriptural Exegesis between Judaism, Christianity and Islam — **47**

Mark N. Swanson
Scripture Interpreting the Church's Story
Biblical Allusions in the *History of the Patriarchs of Alexandria* — **61**

Sidney Griffith
Use and Interpretation of Scriptural Proof-Texts in Christian–Muslim Apologetic Literature in Arabic — 73

Najib George Awad
"Min al-'aql wa-laysa min al-kutub"
Scriptural Evidence, Rational Verification and Theodore Abū Qurra's Apologetic Epistemology — **95**

Juan Pedro Monferrer Sala
The Lyre of Exegesis
Ibn al-Ṭayyib's Analytical Patterns of the Account of the Destruction of Sodom — **119**

Alison Salvesen
“Christ has subjected us to the harsh yoke of the Arabs”
The Syriac Exegesis of Jacob of Edessa in the New World Order — **145**

Haggai Ben-Shammai
From Rabbinic Homilies to Geonic Doctrinal Exegesis
The Story of the Witch of En Dor as a Test Case — **163**

Lennart Lehmhaus
“Hidden Transcripts” in Late Midrash Made Visible
Hermeneutical and Literary Processes of Borrowing in a Multi-Cultural Context. — **199**

Martin Accad
Theological Deadlocks in the Muslim-Christian Exegetical Discourse of the Medieval Orient
Identifying a Historical *Meta-Dialogue* — **243**

Nicolai Sinai
Two Types of Inner-Qur’ānic Interpretation — 253

Gabriel Said Reynolds
Moses, Son of Pharaoh
A Study of Qurʾān 26 and Its Exegesis — **289**

Stefan Wild
Unity and Coherence in the Qur’ān — 303

Berenike Metzler
Qur’ānic Exegesis as an Exclusive Art – Diving for the Starting Point of Ṣūfī Tafsīr — 319

Reza Pourjavady
Ibn Kammūna’s Knowledge of, and Attitude toward, the Qur’ān — 329

Bibliography — 339

Authors — 379

The Editors — 381

Georges Tamer, Regina Grundmann, Assaad Elias Kattan and Karl Pinggéra

Exegetical Crossroads

Understanding Scripture in Judaism, Christianity and Islam in the Pre-Modern Orient

Introduction

Judaism, Christianity and Islam do not only share the broad geographic and multicultural context of their respective origins in the Orient, but also numerous characteristics intrinsic to their constitutions. Most specifically, they are religions of revelation, with revelation understood primarily as communication. Indeed, the foundational narrative of each of these three religions is characterized by an act of communication. In Judaism, God gives Moses, in conversation, the two Tablets of the Torah; in Christianity, the Word of God is incarnated as a communicating human being; in Islam, the Qur'ān, which includes God's words, is communicated orally. These three 'world religions' are, thus, word-religions. The God they proclaim does not eternally persist in dark seclusion. According to the three traditions, God shares some of himself with humans, speaks to them, and lets them recognize something of him at certain times in history. Captured in scriptures, God's communicative action incites further communication. His narrated communication with man is once again re-communicated among them in the context of community. That what is believed to be divine revelation takes its final literary shape through the activity of communication-based communities who ultimately canonize such interactions and transmit them in the form of a holy scripture from generation to generation. The result of such diverse, accumulative, multi-faceted and, as long as religion persists, never-ending communicative action builds the corpus of each of the three religions.

Although the original act of communication at the foundation of each of these religions occurs under specific historical conditions within particular social and cultural contexts, scriptures possess, in the context of their interpretive communities, normative universality due to their belief in a divine origin that transcends, for the faithful, the boundaries of human experience. The interpretation of Scriptures consists primarily in making God's message, which believers claim to be communicated to people at a certain point of history, accessible to later generations under changing historical conditions. Interpretation is, in

https://doi.org/10.1515/9783110564341-001

fact, a complex matrix of communication. The interpreter communicates with the transmitted Scriptures and attempts to penetrate their depths by engaging in dialogue with them. He/She unpacks these texts within their respective contexts and thus introduces them to new forms of communication.

Not only does monotheistic belief lie at the core of all three religions, but such belief is also based on core Scriptures that have normatively determined the relationship between man and God, and between man and his environment. Over the course of centuries, Judaism, Christianity and Islam have developed different methods of interpreting these Scriptures. Every generation of religious scholars that has attempted to disclose their 'true' meaning has faced the same challenge: that is linking their own interpretations to a specific exegetical tradition and, at the same time, finding answers to questions arising in their own particular era.

In this, the three exegetical traditions have exerted influence on one another, either through demarcation of boundaries or through appropriation. The exegetical developments unfolded ultimately by and large in a culturally heterogeneous environment marked by mutual influence. While public discourse today seems to be focusing on the differences rather than similarities between the three religious traditions, we tend to ignore the high degree of religious and cultural commonality that has characterized Judaism, Christianity and Islam. Most centrally, the position of revealed Scripture at the very core of every religious community in the pre-modern Orient is one of those commonalities that have provoked further cross-cultural entanglements.

The religious traditions considered in the present volume appear nowadays to be sources of dispute and conflict in some regions of the world, especially in the context of their origin. Nevertheless, religious scholars operating within culturally heterogeneous contexts such as the pre-modern Orient had to deal with each other as well as other traditions. They demonstrated cooperation in multifarious ways through mutual influence and the demarcation of boundaries. How productive were these interactions for the further development of their own respective traditions? Have there been blurred spaces of scholarly activity that transcended sectarian borders? What was the role played by mutual influence in how these scholars demarcated the boundaries of their own traditions? In what way did dynamic processes within particular traditions remain alive via discussion between younger exegetes and their past masters? These and other related questions have been dealt with by exegetes in all three religions who actually shared similar interests, similar worries and similar struggles for answers as some of the contributions in the present volume document.

The exact investigation of these questions as well as a critical assessment of the relationship between exegetical traditions in the pre-modern Orient gives us the opportunity to expand our understanding of these traditions and, subsequently, of our present time. This is necessary not in the least because the contemporary religious and cultural traditions of all three religions are based on exegetical methods and inner-religious discourses of that era. It is, therefore, an important task of research to illuminate this area of common heritage. The alignment of this volume with this particular focus seems to be all the more urgent, as this topic deserves more scholarly attention than it has received up to now.

This can be accomplished via interdisciplinary cooperation between scholars from relevant areas of research. Accordingly, most contributions in the present volume are devoted to the interrelationship of at least two of the three religious traditions. Interdisciplinary research remains invaluable for exploring the complex religious phenomena which developed in the Orient in Late Antiquity and the Middle Ages. We hope that this volume can offer a useful contribution to interdisciplinary scholarship related to these intertwined religions, particularly in the cultural realm of the pre-modern Orient, which witnessed their rise and early decisive theological developments.

This examination of the reciprocity and interdependency between the exegetical traditions in Christianity, Judaism and Islam is the outcome of a conference organized by the editors at the Friedrich-Alexander-University in Erlangen between February 20 – 22, 2014. The contributions chiefly address the exegetical understanding of Scripture in the three religions. They also tackle interpreting Scriptures in pluralistic religious contexts, taking into account apologetic and polemical tendencies intended to establish certain lines in the sand. Another topic addressed concerns how later interpreters assessed the approaches and results of earlier exegetes in order to determine the continuity or discontinuity of discourses in their respective traditions.

The journey to discover exegetical crossroads in Judaism, Christianity and Islam starts in this volume with Abraham, the prototype of the faithful, whose obedience to God still serves as a model for exemplary behavior in all three religions. William Graham offers a selective account of interpretive trajectories in the three traditions that regard Abraham as sojourner and founder of sacred sites. He points to ways in which, both in Scripture and particularly in exegesis, this particular Abrahamic theme underwent significant interpretive expansion. Yet such expansion occurred with very different points of emphasis and for different ends within each of the three traditions. Graham argues that all three traditions resonate in a variegated fashion to the paradigm of Abraham as the man who, on pure faith, abandons his homeland to wander and establish new places of worship at God's behest, thereby ultimately founding a new tradition of monotheistic faith in a new land.

Cornelia Horn engages with processes of exegetical and interpretative recreation of "Jesus the Wonderworker" in the more widespread, unofficial, or popular literature of Christianity and Islam. She considers aspects of the presentation of Jesus as wonderworker among others in Christian apocryphal texts in Syriac and Arabic, as well as parts of the rich body of Islamic works on the Lives of the Prophets. This comparative approach using para-Biblical material permits a reconstruction of certain aspects of theological, polemical, and exegetical settings which do not surface in other records from this period. The comparison illuminates an interdependence that can potentially form the starting point for a more in-depth investigation.

Similarly, Martin Heimgartner presents exegetical works of the East Syriac Patriarch Timotheos I. written in an Islamic context regarding Jesus. Against the background of the Qur'ānic image of Jesus, Timotheos addresses the question of whether Christians may call Jesus God's 'servant', a Christological title borrowed in the early church from Deutero-Isaiah. The paper shows how Timotheos develops a classification of such writings into four groups out of the conventional distinction between literary and metaphoric writing. He often adds explanatory and interpretative words or sentences in quotations, a device that, in extreme cases, results in rendering the meaning of certain statements into its veritable opposite.

Mark Swanson deals with the use of Biblical quotations and echoes in the *History of the Patriarchs of Alexandria* which was accomplished by a team led by the Alexandrian deacon Mawhūb ibn Manṣūr ibn Mufarrij. The paper examines how intertextual connections are made through quoting and alluding to the Bible and how these quotations illuminate and shape the presentation of events in the historiographical work. In doing so, the paper evaluates the hypothesis that the quotation of, or allusion to, the Bible opens to Christian readers possibilities for interpretation not immediately available to Muslim readers, thus allowing for deeper and even slightly subversive readings of what has normally been considered a 'semi-official' history.

It is in the realm of Arabic that Christians made use of Biblical texts in order to support their positions against their Muslim interlocutors. This is the subject matter of Sidney Griffith's contribution. He demonstrates that from the very beginning of Christian apologetic literature in Arabic, authors made abundant use of quotations from the Old and New Testaments, along with allusions to and echoes of their narratives, in an effort to provide the scriptural proof necessary to support the beliefs and practices which the Islamic Scripture criticized. Arabic-speaking apologists adapted Biblical testimonies widely deployed in earlier Christian literature in Greek and Syriac in order to meet the requirements of the new challenge. In response, Muslim apologists also assembled Biblical testimonies, particularly in support of the Qur'ān's contention that the Torah and the Gospel had announced the coming of Muḥammad and Islam. And right from the start, in addition to Biblical proof texts, Christian apologists were not slow to enlist quotations from the Qur'ān in support of their own apologetic arguments. Thereby a spiraling, interscriptural, interreligious controversy ensued in the course of the early Islamic centuries that arguably reached its apogee in the 13th century.

Najib Awad discusses Theodore Abū Qurrah's apologetic epistemology in his article, examining whether Abū Qurrah's apologetic thought, developed in an Islamic context, presents a Christian *mutakallim* arguing from religious scriptures in defense of Christian faith, or rather depicts a Christian *mutakallim* who defends his religious belief primarily on the basis of reason. This inquiry is made in conversation with Sidney Griffith's publications on this subject. Awad endeavors to shed light on central claims and views in Abū Qurrah's literature and thought that would disclose which rule of argument is more genuinely definitive of his approach: 'arguing from reason', or 'arguing from Scriptures'.

Juan Pedro Monferrer-Sala deals with Ibn al-Ṭayyib's analytical patterns regarding the account of the destruction of Sodom with the saga recounted in Gen 19, containing the story of Sodom and the daughters of Lot as part of a broader episode (Gen 18 – 19) narrating a whole day in the life of Abraham, in which his nephew Lot plays a major role. This Biblical narrative is a textual example of what might be termed a 'shared tradition' common to Judaism, Christianity and Islam. The paper discusses Ibn al-Ṭayyib's treatment of the Biblical account and comes to the conclusion that the analytical approach he adopts, common amongst Aristotelian Eastern Christian thinkers, results from his attempt to preserve the Syriac Christian heritage in Arabic.

Alison Salvesen depicts how Jacob of Edessa drew an explicit parallel between the captivity of Judah under the Babylonians and the subjugation of Christians under the Arabs, which he attributed to the sins of the people of God. She argues that his reasoning reflects the pervasive influence of the Deuteronomistic theology of the Old Testament, which had also led rabbinic Jews to see their own loss of sovereignty under the Roman and Byzantine empires as the result of their community's failure to keep their covenant with God by observing the Jewish Law. This paper assesses how far Jacob's Biblical exegesis tried to meet the challenge of the contemporary social, political and religious reality of late 7th century Syria, or whether his approach is fundamentally a conservative one that attempts to preserve Syrian Orthodox identity without innovation.

Focusing on the Witch of En-Dor story in 1 Sam 28, Haggai Ben-Shammai discusses the shift from rabbinic homilies to geonic exegesis, which occurred in a multi-religious environment. Three stages in the history of the meaning and interpretation of the story in Jewish sources are examined: The plain meaning and message of the story in the Hebrew Bible, its focus in rabbinic sources and finally its focus in Judaeo-Arabic Bible exegesis. Examination of the different attitudes towards the appearance of Samuel in the different stages reveals an interesting relationship between Jewish pre-modern sources on the one hand and Christian and Muslim ones on the other.

For his part, Lennart Lehmhaus investigates hermeneutical and literary appropriations in geonic era Midrash within a cultural and religious plurality of the formative Islamicate period. Contrary to earlier scholarship, beginning with the *Wissenschaft des Judentums*, which was primarily occupied with the adoption of Jewish motifs, narratives and literary elements, he draws attention to more subtle forms of exchange and processes of mutual cultural formation in early Islam. The contribution discusses adoptions and adaptations that mirror developments in Arab-Muslim and Syriac Christian traditions as well as shifts within broader Jewish culture, especially among grammarians, Scripturalists and pre-Karaite groups. In this context, an increased attentiveness to Hebrew and a 'return to Scripture' can be observed. Moreover, in contrast to the polyphonic discourses in classical Midrash, authorial voices emerge in later texts. Most likely, the literary and intellectual blooming among non-rabbinic Jews played a major role in linking Arab-Muslim culture with Midrashic appropriations.

Martin Accad draws our attention to the large amount of Biblical exegetical material that developed within the Islamic tradition. He argues that an exploration of the Islamic exegetical discourse on the Bible reveals a serious epistemological challenge: the traditional understanding of the Qur'ān is the core hermeneutical key to Muslim discourse on the Bible. With the Christian academic approach to the Bible today in mind, he emphasizes the importance of recognizing the Muslim exegetical discourse as a key hermeneutical context for Christians in their attempt to theologize in the presence of Islam. For this purpose, he proposes a three-step method to deal with Islamic discourses on the Bible.

Some of the articles included in this volume center around the Qur'ān. Nicolai Sinai investigates processes of interpretive engagement with Biblical passages in the Scripture of Islam. He presents some of the ways in which Biblical narratives are manifested there. While dealing with the Qur'ānic reception of Biblical stories, the paper distinguishes between interpretively motivated secondary expansion and revision of Qur'ānic passages, and interpretive back-referencing in the treatment of chronologically earlier narratives in later sūras. The Qur'ānic Adam narratives are presented as examples for this hermeneutical technique.

Gabriel Reynolds offers in his study a discussion of passages in Q 26 *The Poets* on Moses and his relationship to Pharaoh and how these passages are interpreted in several classical and modern Qur'ān commentaries. He shows how the Qur'ānic accounts dissociate from the Biblical narrative in order to make Moses the son of the Egyptian ruler.

Drawing on recent studies, Stefan Wild discusses the topic of textual unity and coherence in the Qur'ān. He observes a shift which occurred in scholarship on formal aspects of the Qur'ānic text and which consists in moving from earlier endeavors emphasizing the unity of the sūra towards a new trend considering the textual incoherence of the Qur'ān as a sign of its divine origin.

Berenike Metzler takes a different direction in her contribution. She presents the exegetical work of the Muslim Sufi al-Muḥāsibī on understanding the Qur'ān, thus shedding light on the formative era of Sufi Qur'ānic exegesis. In this work, traditional skepticism towards the human capacity to understand God's word collides with the author's own practice as well as with the emerging idea of Qur'ānic exegesis as an exclusive art.

Reza Pourjavady investigates Ibn Kammūna's discourse on the Qur'ān and its development in his writings. Obviously, the Jewish thinker wrote his treatise on Judaism, Christianity and Islam for Muslim circles. In the chapter on Islam, he challenges the orthodox views on some issues dealing with the Qur'ān mainly by drawing upon the works of the Muslim scholar Fakhr al-Dīn al-Rāzī. The divine origin of the Qur'ān, the history of its revelation and canonization, and some Qur'ānic interpretations were among the issues Ibn Kammūna dealt with in this chapter as well as in his earlier works.

The editors wish to thank Dr. Stephan Kokew, Ms. Dorothea Dietzel M.A., Mr. Martin Herholz, Ms. Katharina Linnemann and Ms. Saskia Pilgram for their help getting this volume edited. We finally thank the publisher Walter de Gruyter, especially Dr. Sophie Wagenhofer, for accepting the volume in the book series *Judaism, Christianity, and Islam – Tension, Transmission, Transformation* and for professional assistance during the production process.

The Editors

William A. Graham

"A Wandering Aramean was My Father"

An Abrahamic Theme in Jewish, Christian, and Muslim Scriptures and Interpretations

1 Introduction

This volume highlights the complexities of scriptural hermeneutics and scriptural communities, not least because the Jewish, Christian, and Islamic traditions whose interpretive activities are at issue here have been theologically as well as historically so complexly intertwined. To be sure, the intertwining has not been such as to make the three traditions very compatible through much of their long history of interactions, which arguably have focused more on divergences than on commonalities. This notwithstanding, one finds today the irenic notion of a shared "Abrahamic tradition" used frequently as the preferred entrée into interreligious dialogue among the three traditions. My several forays over the years into the vast literature on Abraham in the three traditions have left me, not unlike my colleague Jon Levenson, who has written extensively on the subject,[1] somewhat wary of using Abraham as the ideal meeting point of Jews, Christians, and Muslims. Even though he is arguably the most obvious figure whom all three venerate, typically Abraham is invoked by each group to support an exclusivist claim for its own members being his true successors.

Fortunately, the present volume is aimed not at interfaith dialogue among the three traditions, but at exploration of intersections and crossings of scriptural interpretation among them, and the varied contributions treat both similar and disparate modes and instances of scriptural interpretation in all three. I am well aware that to attempt to range across the three traditions even on a delimited topic is both to have to skim the surface of the material and, further, to risk missteps in interpretation of material on which one or more of the other contributors are specialists. Nevertheless, in what follows I make bold to sketch something of the remarkable range of interpretive possibilities in both intra- and extra-scriptural exegesis that the figure of Abraham specifically in his role as a sojourner and pilgrim has opened up in each of these traditions.

1 See especially his book, Levenson, Jon, *Inheriting Abraham: The Legacy of the Patriarch in Judaism, Christianity, and Islam*. Princeton: Princeton University Press, 2012.

https://doi.org/10.1515/9783110564341-002

The Abraham story-cycle—as redacted in *Genesis*, then referenced and elaborated in other parts of Tanakh, New Testament, and Qurʾān, and even more extensively in the massive exegetical traditions of all three—is what I like to call a *patterning narrative*. By this I mean a foundational story that has been so ramifying for a given tradition that it has become a wellspring of allusions, metaphors, and meanings widely accessed not only in religious and scholarly discourse, but in everyday life in the cultures permeated by that tradition. *Patterning narratives* are important touchstones for not only religion but art, literature, language, and culture; and the Abraham narrative has been a particularly resonant one in Jewish, Christian, and Muslim contexts for centuries.

Abraham plays obviously a prominent role in the divergent understandings of *Heilsgeschichte* that the three traditions developed over time. Post-exilic Jewish tradition looks back for a model of monotheistic faith, obedience, and piety to Abraham as the pre-Mosaic patriarch of Genesis who first made a covenant with God and was promised descendants who would be his special people and inherit a special land. In New Testament and later Christian interpretation, Abraham figures as patriarch and model of faith from long before Moses and Sinai. In the Qurʾān and later Muslim tradition, he plays an even more prominent role as the paradigmatic, pre-Mosaic and pre-Christian *muslim*, or monotheistic "submitter," as well as the progenitor of the Arabs through his son Ishmael. All three traditions revere Moses, yet both Christian and Muslim scriptures take Moses and the Exodus-Sinai event as emblematic of the Jewish tradition. Thus in their polemic both look to pre-Mosaic, pre-Torah history for authority for their own traditions. And while for all three traditions, Adam and Noah are prominent in the pre-Mosaic story of God's earliest dealings with humanity, it is the postdiluvian figure of Abraham who figures, however differently, as the physical and/or spiritual progenitor of each of the three. In Gen 17:5, God calls Abraham "the father of a multitude of nations," and elsewhere in the Tanakh, as well as in the New Testament and Qurʾān, he is termed "father Abraham."[2] Each of the three monotheistic communities looks to him as the symbolic founder of its faith and practice. While Isaac and Ishmael and Abraham's grandson Jacob carry special if differing patriarchal or prophetic status in the three traditions, it is Abraham who stands symbolically for them all as the emblem of mon-

2 E.g., in the Tanakh: Exod 3:6, "I am the God of thy father (*avik*), the God of Abraham"; Josh 24:3 and Isa 51:2, "your father Abraham (*Avraham avikum*)." On the centrality of "Our father Abraham" (*Avraham avinu*), see Levenson, *Inheriting Abraham*, p. 3. In the New Testament, examples of "our father Abraham" are found in Luke 1:73 (*Abraàm tòn patéra hēmōn*), Rom 4:1 (*Abraàm tòn propátora hēmōn*), and Jas 2:21 (*Abraàm ho patēr* hēmon). Cf. Q 22:78: "*millat abīkum Ibrāhīm*", "the religion of your father Abraham."

otheistic faith. And in their treatment of Abraham, the three have emphasized, if differently and to differing degree, three aspects of Abraham's story in particular: (1) Abraham's monotheistic faith and rejection of idolatry; (2) Abraham's testing by God in the *aqedah*, or "binding" of his son for sacrifice; and (3) Abraham as father of the faithful and model of faithfulness.

I want to pursue a fourth, less-frequently treated dimension of Abraham as a paradigmatic figure. Deut 26:5 begins, "A wandering Aramean was my father (*Arami oved avi*)." While most rabbinic commentators identify this father as Jacob, significant interpreters from Rashbam[3] in the 12th century to Martin Buber in the 20th identify Abram/Abraham as Deuteronomy's wandering Aramean. Buber terms him "a nomad of faith."[4] Arguably he is the wandering Aramean *par excellence* of Genesis, and it is the varied interpretations of this role that I shall briefly explore: Abraham as pilgrim or peregrine sojourner—as stranger and exile who travels and settles in foreign territory. This theme encompasses also his founding of altars or shrines consecrated to the worship of the one God as he moves about Canaan and, in the Muslim case, on to Mecca. This sojourner and altar- or sanctuary-builder subtheme is found in all three scriptural and later interpretive traditions; it is always linked to the central motif of Abraham as paradigm of faith, but it is treated quite differently in each tradition.

2 Abraham the Sojourner in the Tanakh and Jewish Interpretation

The Tanakh itself offers a prime example of intra-scriptural exegesis in what the foundational narrative of Gen 11–25 does with the image of Abraham as a wandering sojourner. In this account, he appears as one who, at God's command, leaves home and becomes a stranger traveling and dwelling in tents in strange lands, with the promise of an eventual homeland for his progeny. Until his death he remains a sojourner, not a property owner; at his death the only land he possesses is a burial plot in the Promised Land. This peregrine dimension of Abraham is closely tied to his obedience to God's command to abandon the land of his birth and go where God directs. It is also tied to his faith in, and dedicated worship of God, signaled in part in Genesis by references to his build-

3 Samuel ben Meir, "Rashbam" (d. c.1158), grandson of Shlomo Yitzhaki, "Rashi" (d. 1105).
4 Buber, Martin, *The Prophetic Faith*, trans. from the Hebrew by Carlyle Witton Davies. New York: Harper & Row, 1960, p. 35.

ing of at least six altars or shrines to his Lord during his movements as a sojourner in Canaan.

However one interprets the complex textual history of Genesis, its redactors present Abraham's story clearly as a fundament of the larger *Heilsgeschichte* of God's dealing with the Hebrews, who are destined to become the Children of Israel through the Exodus and Sinai events. This is most evident in the depiction in Genesis of Abraham's pilgrim journeying as an analog, even a prefiguration, of Israel's defining experience under Moses in the Exodus. God's drawing Israel out of Egypt onto its extended journey to the Promised Land is preceded and prefigured by Abraham's own going forth from Mesopotamia to Canaan, his descent into Egypt, his rescue by God, and his return to Canaan. Much of Genesis in its eventual scriptural form likely took shape only in the 7th and 6th centuries BCE, which makes the anachronistic allusions and parallels in the Abraham narrative to Israel's experience in Egypt and Sinai unsurprising. Some scholars consider especially chapter 15 to be a late portion of Genesis, from not long before codification of the Torah.[5] It can be seen as an effort to harmonize the Abraham story with that of Moses, the Exodus, and the wilderness experience, and to align Patriarchal traditions with the whole of the Pentateuch. Compare God's self-introduction in Gen 15:7, "I am YHWH who brought you out from Ur of the Chaldeans" with the beginning of the Decalogue in Exod 20:2, "I am YHWH [...], who brought you out from the land of Egypt." Here the earlier focus in Gen 11 on Abraham's father Terah as the one who abandons Ur for Haran has shifted completely to Abraham as the one God draws forth from Ur. This is what we then find elsewhere in the Tanakh e.g., Neh 9:7: "Thou art the Lord God who chose Abram and brought him forth out of Ur of the Chaldeans"; or Josh 24:3: "I took your father Abraham from beyond the river and led him through all the land of Canaan."

Furthermore, later in Gen 15, God tells Abraham that his progeny, the Children of Israel will live "as strangers in a land that is not theirs" and end up in Egyptian slavery, after which they will "depart with great possessions" eventually to inherit the promised land of Canaan from Egypt to the Euphrates (vv. 13, 16) —all clear allusions to statements concerning the Exodus and exile in Deut 10 and 11, Josh 1:4, and elsewhere in the Tanakh.[6] Thus the theme of Abraham as exile and sojourner in Canaan, looking forward to a promised homeland for his people, is mapped onto the Israelites' long servitude in Egypt and their sub-

5 E.g., Römer, Thomas, "Abraham and the 'Law and the Prophets'," in *The Reception and Remembrance of Abraham*, ed. by Pernille Carstens and Niels Peter Lemche, pp. 87–101. Piscataway, N.J.: Georgia Press, 2011.
6 Ibid.

sequent departure and wandering as exiles and sojourners under Abraham's descendant, Moses, until they reach the Promised Land. In the Tanakh, Abraham's going down as an alien into Egypt prefigures Israel's sojourn there generations later, just as his safe escape with new wealth mirrors the Israelites' escape "with great possessions" into the wilderness of Sinai.

Post-Biblical Jewish interpretation only expands on this alignment of the Abraham and Exodus sojourning narratives. In the 2nd century BCE *Book of Jubilees* there are only indirect references to the parallel between his sojourning in Canaan and Egypt and that of the Children of Israel in Egypt and then Babylonia, but there are repeated references to Abraham as a sojourner in the land of Canaan and to the many altars to God that he built throughout the land.[7] In the much later (likely 2nd century CE) *Apocalypse of Abraham*, the final chapter ends with a pointed reference to the Exodus: God assures Abraham that in the seventh generation after him, his progeny, like Abraham himself, "will go out into an alien land. And they will enslave them and oppress them as it were for one hour of the impious age [...]."[8]

Still later, no earlier than the 5th century CE, the rabbinic Midrash on Gen 12:10 ff. in *Genesis Rabbah* makes explicit the analogy between Abraham and the Children of Israel under Moses. It begins, "You find that whatever is written in regard to our father, Abraham, is written also with regard to his children," after which the text pairs passages from the story of Abraham's Egyptian sojourn in Gen 12 and 13 with ones from Exodus, Numbers, later chapters of *Genesis*, and the Psalms regarding the Israelites' sojourn in Egypt before the Exodus. Thus Gen 12:10, "So Abram went down to Egypt to sojourn there, for the famine was severe in the land" is matched with three texts, "our fathers went down into Egypt" (Num 20:15); "we have come to sojourn in the land" (Joseph's brothers speaking to Pharaoh in Gen 47:4); "for the famine was heavy in the land" (Gen 43:1). Another pairing is of Gen 13:3, "And [Abram] went on his journey," with Num 33:1, "These are the journeys of the Children of Israel." The phrase-by-phrase parallels in this Midrash identify the story of Abram explicitly with the story of Israel: both he and the Israelites were sojourners and exiles in

7 E.g., in *Jubilees*, chs. 13, 14, 15, 16, and 18, in *The Apocrypha and Pseudepigrapha of the Old Testament*, trans. by Robert H. Charles. Oxford: Clarendon Press, 1913.
8 "The Apocalypse of Abraham," last modified September 22, 2015. http://www.pseudepigrapha.com/pseudepigrapha/Apocalypse_of_Abraham.html.

Egypt and God brought them forth, Abram "very rich in cattle, in silver, and in gold" (Gen 13:3) and the Israelites "with silver and gold" (Ps 105:37).[9]

Nor does the linking of Abraham with Moses and the Children of Israel stop with rabbinic *midrashim*. The major 13th century kabbalistic work, the *Zohar*, in making refining and purification in exile a preparation for spiritual consummation, likens the sojourn of Abraham in Egypt to that of the Israelites and to that of the personified Holy Land under pagans before the coming of God's chosen people:

> Come and see the secret of the word:
> If Abram had not gone down into Egypt
> and been refined there first,
> he could not have partaken of the Blessed Holy One.
> Similarly with his children,
> when the Blessed Holy One wanted to make them unique,
> a perfect people,
> and to draw them near to Him:
> If they had not gone down to Egypt
> and been refined there first,
> they would not have become His special ones.
> So too the Holy Land:
> If she had not been given first to Canaan to control,
> she would not have become the portion, the share
> of the Blessed Holy One.
> It is all one mystery.[10]

What we can say finally is that in Jewish interpretation, growing out of the scriptural narrative in *Genesis* and extending for centuries, the *galut*, or exile—a major theme of post-Biblical Judaism altogether—is arguably already inscribed upon the figure of *Avraham 'avinu*, "Our Father Abraham," whose life as a "stranger and sojourner" becomes a foreshadowing or pre-enactment of Egyptian exile and slavery, Babylonian exile and captivity, and even the *galut* after the fall of the Second Temple.[11]

---

9 *Genesis Rabbah: The Judaic Commentary to the Book of Genesis*, trans. and ed. by Jacob Neusner, vol. 2. Atlanta: Scholars Press, 1985, pp. 77–85, 90–91.

10 "*Zohar* 1:83a," in *Zohar: The Book of Enlightenment*, trans. by, Daniel C. Matt. Toronto: Paulist Press, 1983, p. 64.

11 Also ultimately, even a foreshadowing of, or looking forward to the eschatological (or, third) Temple being rebuilt—even while a substantial Jewish community continued to flourish in Palestine, as we know from the texts produced there, such as the Mishna, the Jerusalem Talmud, and late-antique and medieval Midrashic collections (Levenson, personal communication, Feb. 2014).

3 Abraham the Sojourner in Christian Interpretation

The Christian New Testament unsurprisingly offers a very different interpretation of Abraham as patriarchal forefather—one central to the early Christian effort to distinguish the new preaching of the messianic Christ from Jewish tradition by using that very tradition against itself. Best known here is the Apostle Paul's polemic in Romans and Galatians where the Jewish convert takes Abraham and his faith over against Moses and the Law, as model and symbol of true faith in God, a faith that Christians find in fullness only in the Christ. The Jews become synonymous with "works alone" according to the Law of Moses and the Christian faithful with "salvation by faith" after the model of their spiritual father Abraham. Note Paul's claim in Gal 3:7 that "those who have faith" are the true "sons of Abraham," or his words in Rom 9:6 – 8:

> For not all who are descended from Israel belong to Israel, and not all are children of Abraham because they are his descendants [...]. It is not the children of the flesh who are the children of God, but the children of the promise are reckoned as his descendants.[12]

While the theme of Abraham as model of faith takes pride of place in both New Testament and later tradition, the Epistle to the Hebrews vividly moves the emphasis to the pilgrim/sojourner theme of Abraham as stranger and exile. The unknown author, steeped in priestly tradition, identifies in Heb 5:5 – 6 the enigmatic priest-king Melchizedek of Genesis with the Christ and then in Heb 7:1 – 17 makes Abraham blessed by Melchizedek (Gen 14) a prototype of the Christian blessed by the atonement and intercession of Christ. In Heb 11, the author calls even more strongly upon the model of the first patriarch by referencing the Gen 23 image of Abraham as "stranger and sojourner," *ger we thoshav* (LXX: *xénos kai parepídēmos*; Vulgate: *advena et peregrinus*). The text, Heb 11:8 – 16, bears citing in full:

> [8] By faith Abraham obeyed when he was called to go out to a place that he was to receive as an inheritance; and he went out, not knowing where he was to go. [9] By faith he sojourned in the land of promise, as in a foreign land, living in tents with Isaac and Jacob, heirs with

---

12 This polemic is of course at odds with the Letter of James 2:20 – 24, which also calls on Abraham but asks, "Was not Abraham our father justified by works, when he offered up his son Isaac upon the altar? You see that faith was active along with his works, and faith was completed by works, and he was called the friend of God. So you see that a man is justified by works and not by faith alone."

> him of the same promise. [10] For he looked forward to the city that has foundations, whose builder and maker is God. [11] By faith Sarah herself received power to conceive, even when she was past the age, since she considered him faithful who had promised. [12] Therefore from one man, and him as good as dead, were born descendants as many as the stars of heaven and as the innumerable grains of sand by the seashore. [13] These all died in faith, not having received what was promised, but having seen it and greeted it from afar, and having acknowledged that they were strangers and pilgrims (*xénoi kai parepídēmoí*) on the earth. [14] For people who speak thus make it clear that they are seeking a homeland. [15] If they had been thinking of that land from which they had gone out, they would have had opportunity to return. [16] But as it is, they desire a better country, that is, a heavenly one. Therefore God is not ashamed to be called their God, for he has prepared for them a city.

Here the writer not only builds on Abraham and Genesis but also Lev 25:13, Ps 39:12, and Ps 119:19, all of which stress that as humans we are *all* "strangers and sojourners" on God's earth. Abraham is the model of the pilgrim sojourner in foreign territory, the prototype of the faithful Christian who in this world is, like him, "living in tents" (and of course also like Moses and the Israelites "tenting in the wilderness"). The Christian is a pilgrim stranger and sojourner in "a foreign land"—without a home on this earth, but looking "forward to a city which has foundations": one prepared by God to receive the pilgrim exile because of his or her faith. Like Abraham, the Christian has on faith left home and family without looking back, to follow Christ, yet now under the new dispensation s/he desires not the earthly Jerusalem and the land promised to Israel, but "a better country, that is, a heavenly one." In the city God has prepared for them, the faithful will be with Him eternally, no longer strangers and pilgrims on the earth.

Not surprisingly, the writer of Hebrews goes on to sketch the trials, faith, and rescue of Moses with his people from the Red Sea and then the rest of the *Heilsgeschichte* of the Israelites according to Biblical chronology, after which he polemicizes that of course these wandering, suffering Children of Israel, "though well attested by their faith, did not receive what was promised, since God had foreseen something better for us, that apart from us they should not be made perfect" (Heb 11:39–40). Finally, the last two chapters of Hebrews bring the Christian faithful "to Mount Zion and the city of the living God, the heavenly Jerusalem," and the author rejoices that

> [...] we have an altar from which those who serve the tabernacle have no right to eat. For the bodies of those animals whose blood is brought into the sanctuary by the high priest as a sacrifice for sin are burned outside the camp. So Jesus also suffered outside the gate in order to sanctify the people through his own blood. Therefore let us go forth to him outside

> the camp, bearing abuse for him. For here we have no lasting city, but we seek the city which is to come.[13]

Here Abraham as stranger and sojourner in the land is recapitulated, first in the crucified Christ, and then in the Christian follower as a stranger and sojourner on this earth, seeking the city to come, not the earthly Jerusalem (which of course in this polemic stands for the land of Zion, the Law, and the Jews). A similar echo of "stranger and pilgrim" in the Tanakh is found in 1 Pet 2:11, where the Christians in Asia Minor are addressed thus: "Beloved, I beseech you as strangers and sojourners (*paroíkous kai parepidēmous*) to abstain from the passions of the flesh that wage war against your souls," reminding Christians that they are only passing through this earthly world; it is not their permanent home.

In Christian interpretation ever afterward, this powerful scriptural linkage of the Christian faithful to Abraham as stranger and sojourner in this world bound for a heavenly promised land remains a persistent theme.[14] We find the linkage in the 2nd century C.E. *Epistle to Diognetus*, where the writer speaks of Christians as strangers and sojourners whose "existence is on earth, but their citizenship is in Heaven."[15] In the early 3rd century, Origen (*Contra Celsum* 8:74–5[16]) speaks similarly of Christians as citizens in earthly cities but members of the higher community of the Church of God, and by living their faith in earthly cities coming ultimately into "a divine and heavenly city." This idea is similar to Augustine's words in the opening lines of *The City of God:* that the City of God "pursues its way as a stranger among unbelievers" in this life, all the while belonging to "the secure and eternal home beyond," which the faithful citizens of the City wait patiently to realize on Judgment Day.[17] He follows a similar line in *On Christian Doctrine* I.4, describing humans as "wanderers in a strange country" entranced by its ephemeral beauties:

---

13 Heb 13:10–14.

14 One possible indication of this is the use of *Perigrinus* and *Viator* as early Christian personal names. See, e.g., "Behind the Name." Accessed on January 29, 2014, http://www.behindthename.com/names/usage/late-roman.

15 Lightfoot, J.B., "Epistle to Diognetus," in *The Apostolic Fathers: Revised Texts with Short Introductions and English Translations*, ed. and completed by J. R. Harmer. London: Macmillan, 1891, pp. 490–511. Online: "Early Christian Writings – Diognetus." Accessed on February 4, 2014.
http://www.earlychristianwritings.com/ text/diognetus-lightfoot.html.

16 Origen, *Contra Celsum*, trans. by Henry Chadwick. Cambridge: Cambridge University Press, 1953, p. 510.

17 Augustine, *The City of God*, abridged and trans. by John W. C. Wand, vol. 1. London and New York: Oxford University Press, 1963, Preface p. 1.

> We have wandered far from God; and if we wish to return to our Father's home, this world must be used, not enjoyed, that so the invisible things of God may be clearly seen, being understood by the things that are made, that is, that by means of what is material and temporary we may lay hold upon that which is spiritual and eternal.[18]

Augustine's 4th century contemporary, Ambrose of Milan, interprets the Abraham story almost entirely allegorically: he says that Abraham departed from his homeland not for another country but for true religion, since the meaning of *Canaan* is "true religion."[19] Similarly, he takes Gen 15:13, "Thy seed shall be a sojourner," to mean that either "we must all be sojourners on this earth—for Abraham is the father of all men" or "[...] the true seed of Abraham will be a sojourner in this world [...]. Finally, Ambrose closes his argument with the declaration, "For whoso is a stranger here is a citizen in Heaven [...]."[20]

This theme resounds through Christian exegesis down the centuries in the books and sermons of writers in various languages, from Gregory the Great in the 6th century and St. Boniface in the 7th through the Middle Ages to Chaucer and Dante and well beyond, most famously perhaps in 17th century England in John Bunyan's *Pilgrim's Progress*. Still later, even Lord Byron picks it up: "Man is a pilgrim spirit cloth'd in flesh/ And tenting in the wilderness of Time."[21] Indeed, whatever the permutations, what begins in Genesis with Abraham as "stranger and sojourner" and is mapped onto both the Exodus story and the Christ event has perdured in the rhetoric of Christian religion and culture.

4 Abraham the Sojourner in the Qurʾān and Muslim Interpretation

In the Qurʾān, the particular theme of Abraham as stranger and sojourner is little developed in contrast to Tanakh or New Testament interpretation. There are only

18 Augustine, "On Christian Doctrine," in *A Select Library of the Nicene and Post-Nicene Fathers of the Christian Church*, ed. by Philip Schaff, vol. 2. Grand Rapids: Eerdmans, 1977–86. Online: "Philip Schaff, A Select Library of the Nicene and Post-Nicene Fathers of the Christian Church, vol. 2 (St. Augustin's City of God and Christian Doctrine)." Accessed on February 4, 2014. http://faculty.georgetown.edu/jod/augustine/ddc.html

19 In his small book, *De Abraham*. See Ambrose of Milan, *On Abraham*, trans. by Theodosia Tomkinson. Etna: Center for Traditionalist Orthodox Studies, 2000, p. 51.

20 Ibid., p. 81.

21 Lord Byron, *The Soul's Pilgrimage: A Poem*. Cambridge: Metcalfe, 1818, p. 12.

three clear Qur'ānic references to Abraham's departure from Mesopotamia. In Q 19:48–9, Abraham says to the idol worshippers around him:

> I shall draw apart from you and whatever you call upon instead of God and pray to my Lord only [...] And after he had drawn apart from them and all that they worshipped instead of God, We gave him Isaac and Jacob and made each a prophet [...].[22]

His separating himself from the idolaters of his land of birth is also referenced in Q 37:99, when he says to them, "I shall depart and go to my Lord, as he will guide me." In the third instance, in Q 21:71, the only one that refers specifically also to Abraham's being guided to the Promised Land, God says, "We rescued him and Lot and [led them] to the land that We blessed for all beings." These spare statements about Abraham are, of course, consonant with the thoroughly "referential" style that is an earmark of the Qur'ān: in almost every mention of Abraham, the text is clearly alluding to a more extensive Abraham story-cycle so well known to its audience as only to need barest reference to make clear what is being referenced.

Not surprisingly, the familiar theme in rabbinic and Christian exegesis of Abraham's stalwart monotheism and rejection of polytheism (a theme that ironically is not explicit in Genesis itself[23])—does loom large in the Qur'ān alongside that of Abraham the true "submitter" (*muslim*) as the paradigm of faith. Both of these themes correspondingly permeate later Muslim tradition as much as, or possibly even more than they do Jewish and Christian tradition. A few examples from the Qur'ānic text itself are worth citing: In Q 3:65, we see a polemical call upon Abraham as model of faith: "O people of Scripture! Do not quarrel about Abraham, for the Torah and the Gospel were only revealed after him!" This is picked up two verses later: "Abraham was neither a Jew nor a Christian, but a *ḥanīf muslim* [a righteous person submitting to God alone] and not a *mushrik* [idolater]." Q 6:161 speaks of the "straight path" and "right religion" that is "the way of Abraham," *ṣirāṭ Ibrāhīm*; and Q 3:95 and Q 4:125 urge listeners to follow *millat Ibrāhīm ḥanīfan*, "the religion of Abraham, a righteous man," with the addition in Q 4:125, "God chose Abraham as friend (*khalīl:* cf. Isa 41:8)." According to the Qur'ān, Abraham's faith and monotheistic creed are what God has presented anew to humankind through Muḥammad in the Qur'ān. Not surprisingly, Abraham in the Qur'ān, and still more so in later interpretation, is taken as the prefiguration of Muḥammad, much as in Genesis and rabbinic tradition he is the prefiguration of Moses.

---

22 All Qur'ānic translations are the author's own.
23 Levenson, *Inheriting Abraham*, pp. 3–5.

By contrast, the *explicit* Biblical theme of Abraham as stranger and sojourner is at best only *implicit* in the Qur'ān, even though the text apparently assumes the broad outlines of the ancient story of his migration southwestward from Mesopotamia to Canaan (or greater Syria), adding specifically that he also reached Mecca. What we do see in the Qur'ān that may be closest to the Genesis theme of Abraham the sojourner and altar-builder is the Abraham of the Medinan-period revelations who goes to Mecca, and at God's command, with his son Ishmael's help, purifies God's "House," the Kaʿba, and rebuilds it as a sacred place of pilgrimage. In Q 2:125, God says:

> [Remember] when We made the House a refuge and a sanctuary for the people, [saying], 'Take the place where Abraham stood as a place of worship.' And We charged Abraham and Ishmael with the purifying of Our House for those who circle [around it], those who hold fast [to it], and those who bow and prostrate themselves.

On the basis of his cleansing and building up of the Kaʿba for the worship of God, Abraham becomes for Muslims forever linked to the Kaʿba and the rites of the *ḥajj*, which include stopping to pray at "the place where Abraham stood," *maqām Ibrāhīm*, during the *ṭawāf* (circumambulation) of the Kaʿba; stoning the *jamarāt*, the three pillars representing Satan, to recall Abraham's rejection of idolatry; sacrificing an animal at Minā just as Abraham did; and the *saʿy*, or running back and forth between the two points, *aṣ-Ṣafā* and *al-Marwa*, to remember Hagar's distress at lack of water for her infant Ishmael (to which distress God responded with the miraculous appearance of the spring of Zamzam—a close parallel to the story of Hagar in the wilderness in Gen 21).

It is worth considering, even though there is no evidence of direct influence, that the Qur'ānic Abraham's building (or restoring) of the Kaʿba and institution of the pilgrimage rites in the sacred territory of Mecca can be seen as paralleling the Biblical Abraham's building of altars or sites of worship to God wherever he pitches his tent (six instances: two each in Gen 12 and 13, and one each in Gen 21 and 22).[24] In the text of the Qur'ān, as in Tanakh and NT, we see reinterpretation and reconfiguration of older material in the *Gestaltung* of Abraham as not only the first *muslim*, but also the founder of the most sacred site in the world. And even if Abraham's going forth from his homeland to sojourn and to build altars

---

24 Gen 12:5–7; 12:8; 13:3–4; 13:17–18; 21:33; 22:9. Note that Firestone argues that the association of Abraham with establishing sacred sites in Genesis "was probably the source for his pre-Islamic connection with the founding of the Kaʿba". See Firestone, Reuven, *Journeys in Holy Lands: The Evolution of the Abraham-Ishmael Legends in Islamic Exegesis.* Albany: State University of New York Press, 1990, p. 82.

to God in new places is not overtly linked in the Qur'ān to his institution of the rites of the *ḥajj*, it is hard not to find there an echo of the pilgrim of faith whom God extracts from his home and sends to Canaan and Mecca—something that has been mirrored in every performance of the *ḥajj* down to the present day when pilgrims specifically remember Abraham at multiple points during the *manāsik*, or ritual acts of the *ḥajj*.

Finally, Muslim interpretation (*tafsīr*) of the Qur'ānic Abraham narrative similarly places primary emphasis on Abraham as pre-Mosaic paradigm of faith and the first *muslim*. He is, moreover, also understood as Muḥammad's forefather and his paradigmatic prophetic forebear. Thus, even in interpreting a Qur'ānic passage with no explicit mention of Abraham, the famous "Light Verse" of Q 24:35,[25] early and classical exegetes take its long metaphor of "the likeness of his light" to refer to the light as that of Muḥammad and "kindled from a blessed olive tree" to be a metaphor for Muḥammad's descent from Abraham, with "a tree [...] neither of the East nor of the West" taken to mean that Abraham was neither a Jew nor a Christian. Even the words "light upon light" are interpreted as meaning that Muḥammad was a prophet descended from another prophet, namely Abraham—in one exegete's words, "a prophet of prophetic descent."[26]

This kind of interpretive placement of Muḥammad and his prophetic vocation in a lineage of prophecy going back to Abraham is only one instance of the abiding emphasis in Muslim interpretation upon Abraham as paradigm of faith and the prophetic model for Muḥammad. Similarly, Muslim exegesis treats the sacrifice of Abraham's son extensively, with much attention given in early

25 The verse reads: "God is the light of the heavens and the earth. The likeness of His [his?] light is a niche (*mishkāt*), in which is a lamp (*miṣbāḥ*), the lamp in a glass (*zujājah*), the glass as it were a shining star (*kawkab durrī*), kindled from a blessed olive tree neither of the east nor of the west, the oil of which would almost light up (*yuḍī'u*) even though no fire touched it. Light upon light (*nūr 'ala nūrin*)! God guides to His light (*yahdī li-nūrihi*) whom He will. God coins similitudes (*al-amthāl*) for humankind, and God knows everything."

26 The early Sunnī interpreter Muqātil b. Sulaymān (d. 767) and the early Shī'ī exegete 'Alī b. Ibrāhīm al-Qummī (fl. early 10th century), as well as the pseudo-Ibn 'Abbās (d. 687) and the great aṭ-Ṭabarī (d. 923) all raise the possibility that the "light" in Q 24:35 refers to Muḥammad and go on to read Abraham into the metaphor as I have indicated. The phrase, "a prophet of prophetic descent", is from Muqātil b. Sulaymān, *Al-Wujūh wa-n-naẓā'ir fī l-qur'ān al-'aẓīm*, ed. by Ḥātim Ṣāliḥ aḍ-Ḍāmin. Riyad: Maktabat ar-Rushd Nāshirūn, 2010, p. 160. For a fuller summary of these readings of the text and further *tafsīr* references, see my forthcoming article, "Light as Image and Concept in the Qur'ān and Other Early Islamic Sources," in *God is the Light of the Heavens and the Earth: Light in Islamic Art and Culture*, ed. by Sheila Blair and Jonathan Bloom. New Haven: Yale University Press, 2015, pp. 45–59.

tafsīr works to Abraham's readiness to sacrifice his son (and much discussion of whether it was Isaac or Ishmael who was the son in question). However, among the remaining Abrahamic themes, it is that of Abraham as sojourner in Mecca and founder of God's holy house there that most clearly looms much larger in Muslim exegesis than in the Qur'ān itself.[27] This is a topic that Reuven Firestone has treated well in his survey of the treatment of Abraham in Muslim exegesis,[28] so I will not rehearse his findings. However, two main points based largely on his work are in order.

First, as already noted, there is nothing specific in the Qur'ān about Abraham's peregrinations through greater Syria, Egypt, and the western Arabian Hijaz. Consequently, this silence offered Muslim exegetes a wide scope for elaboration. They used accordingly the riches of Talmudic and likely other regional legendary lore to fill in the story of Abraham's time in Canaan, Egypt, and Mecca. The varying strands in their accounts of the sojourning forefather reflect the fact that the basic story is tacitly assumed in the Qur'ān and was probably already circulating in different versions in the Near East of the 7th century C.E. The exegetes elaborate on moments and halting places (Haran, Syria, Jordan, Egypt)[29] in Abraham's peripatetic career between Mesopotamia and Mecca, and flesh out the picture of him as a nomadic prophet-patriarch figure who travels to greater Syria, *ash-Shām* (Canaan), which they identify readily as the land "blessed for all beings" of Q 21:71. Where the Qur'ān is silent on how Abraham got to Mecca, the exegetes fill in the gap, generally having him take Hagar and Ishmael there after Sarah asks that they both be banished, something also not found in the Qur'ān itself. The commentators identify the Paran desert of Gen 21:21 (where Ishmael went) with the environs of Mecca, and after retelling the story of the miraculous appearance of the Zamzam well, most accounts have Abraham assure Hagar that he will come again to build or restore God's House there.[30]

Second, it is also evident from Firestone's survey that it is the Kaʿba's rebuilding and the institution of the pilgrimage thence by Abraham (both at God's command) that loom large in Muslim exegetes' treatment of Abraham's journey south from Canaan to Mecca. Only the binding of Isaac gets more attention in Muslim exegesis than does the (re-)building of the Kaʿba (on the foundations of Adam's original structure) and the associated institution of the rites of

---

27 Albeit still very much subordinate to emphasis on Abraham as paradigm of prophethood and faith, and forebear of Muḥammad.

28 Firestone, *Journeys*, esp. chs, pp. 3–12.

29 Ibid., pp. 25–30.

30 On Abraham's movements and time in Mecca in particular, see Firestone, *Journeys*, pp. 8–10.

the *ḥajj*.[31] Whether seen as the sole builder or having his son Ishmael as helper, Abraham becomes for Muslim tradition the effective founder (or post-dilivuian re-founder) of the Kaʿba and the associated pilgrimage rites of the *ḥajj*. As such, he is also a prefiguration of the Prophet Muḥammad, who cleanses the Kaʿba and establishes (or re-establishes) the proper *ḥajj* observances as recounted in *Sīra*, Hadith, and other traditional sources. The exegetes build their interpretations especially on Q 22:26–7,[32] "Remember when We prepared for Abraham the place of the House, [saying]: 'Do not ascribe any partner to me and purify my House for those who circumambulate, stand, bow, and prostrate. And proclaim to humanity the *ḥajj*.'" Although the sequence of events and the details vary in different exegetical accounts, most have Abraham proclaim the pilgrimage just as Q 22:27 enjoins him to. A lesser number describe him as making the first *ḥajj* himself, sometimes with Gabriel's help, then calling others to do the same.[33]

Overall, as Firestone's survey shows, Islamic exegetical tradition regarding Abraham focuses on (1) the Qurʾānic allusions to his departure from his homeland out of revulsion at its idolatry, (2) his trials as a monotheist in an idolatrous world, (3) his demonstration of faith by offering to sacrifice his son, and (4) his establishment of the Kaʿba and *ḥajj* rituals at Mecca where he settled Hagar and Ishmael and by extension his Arab progeny. With respect to our sojourner and altar-builder motifs, we can say that, unlike Jewish and Christian scripture and exegesis, both the Qurʾān and later exegesis emphasize less the wandering of Abraham and much more his rejection of his idolatrous homeland (no. 1 above). Further, instead of the Biblical portrayal of Abraham as a regular builder of altars to God wherever he goes, the Qurʾān and its interpreters focus on his building up of the Kaʿba, God's most holy House, and his attendant institution of the rites, or *manāsik*, of the *ḥajj*.

Abraham's move from the idolatrous land of his birth to Canaan, however emphasized, is clearly consonant with the older Biblical story of the patriarch first told in Genesis. And while there is no clear influence on the Kaʿba-building of Abraham from the altar-building of Genesis, it is hardly far-fetched to say that both are of a piece in their emphasis on Abraham's role in the three traditions as the postdiluvian man of faith chosen by God to be founder of monotheistic faith and practice before Moses and Sinai, before Jesus and his crucifixion, and before Muḥammad and his cleansing of the Kaʿba and institution of the *ḥajj*. Where

31 Firestone, *Journeys*, pp. 88.
32 Ibid., pp. 76–103.
33 Ibid., pp. 96–102.

Abraham in both Tanakh and Jewish tradition is the type and prefiguration of Moses and Israel in his travel to and emergence from Egypt, or in the New Testament and Christian tradition the stranger and sojourner who is the *type* of the Christian traveling through this world with eyes fixed on the next, in the Qur'ān and Muslim interpretation Abraham is not only the type of the faithful *muslim*, but also the father who settles his son Ishmael and his mother Hagar at Mecca and later returns to build with his son the holy House of the Kaʿba, which has always been seen as the earthly holy of holies for Muslims. Still more, even though it is not developed in classical *tafsīr*, because Abraham is also clearly a type and prefiguration of the prophet Muḥammad in the Qur'ān and later tradition, his abandonment of Mesopotamia and its idolatry might also be seen typologically as a type or prefiguration of the migration or *hijra* that Muḥammad makes to escape the persecution of the idolaters in Mecca and to found his new community of faith in Medina.

5 Conclusion

The foregoing has been a rapid review of interpretive trajectories in the three traditions regarding Abraham as sojourner and founder of sacred sites. In looking at his role as Buber's "nomad of faith," a tent dweller who is also a builder of altars or sanctuaries to the God whom he follows, I have tried to point to ways in which, in the scriptures and even more in the exegetical traditions of all three monotheisms, this particular Abrahamic theme underwent significant interpretive expansion, yet with very different emphases and for very different purposes in each of the three traditions. Apart from, but related to Abraham as the prototype of the person of pure faith and obedience, all three monotheistic traditions resonate variously to the paradigm of Abraham as the man who on faith abandons his homeland to wander, to establish new places of worship at God's behest, and ultimately to found a new tradition of monotheistic faith in a new land. Abraham has been a prolific source of interpretation and reinterpretation across the three traditions,—whether he becomes in Tanakh and Jewish tradition the prototype of Moses and the Children of Israel bound for a promised land, or in New Testament and Christian tradition the prototype of the faithful Christian seeking a heavenly promised land, or in Qur'ān and Muslim interpretation the prototype or prefiguration of Muḥammad, his prophetic successor, by establishing God's holiest sanctuary and instituting the pilgrimage rites. Abraham's story in all three traditions, for all its divergences, portrays him as a sojourner whose unshakable faith in God and dedication to His worship are paradigmatic for all who consider themselves his physical or spiritual progeny.

I would suggest that the rich variation in Abrahamic traditions, interpretations, and extensions of interpretations, one segment of which I have touched on briefly here, simply reminds us that interpretation is always inventive as well as conservative, always taking up new, often polemical, agendas as well as trying to clarify what has gone before. Interpreters are always capable of latching onto different elements, however small or secondary, of any sacred history and making these the basis for new hermeneutical trajectories. Nor can we forget that every scriptural text itself is already replete with ongoing interpretation in its own pages, even though all of its meanings rest ultimately in the hands of later interpreters of those pages.

In that regard, let me close by paraphrasing the likely spurious Hadith cited by al-Dārimī: *as-sunna qāḍiya 'alā l-qur'ān, wa-lā l-qur'ān bi-qāḍin 'alā s-sunna:* "tradition controls the Qur'ān, not the Qur'ān tradition."[34] Here we could easily substitute "interpretation" for "tradition" and "scripture" for "Qur'ān" to form a general axiom: "interpretation controls scripture, not scripture interpretation," for it is interpretation both within and beyond scriptural texts that always develops the meanings of those texts for communities of faith, which are of course always communities of interpretation.

---

34 Al-Dārimī, 'Abd Allāh b. 'Abd al-Rahmān, *Kitāb as-sunan*, ed. by 'Abdullāh al-Yamanī al-Madanī. 2 vols. Cairo 1386/1966, *Muqaddimah* [Introduction], section 49.

Cornelia Horn

Jesus, the Wondrous Infant, at the Exegetical Crossroads of Christian Late Antiquity and Early Islam

1 Introduction

One of the prominent features of Jesus' presentation in the canonical gospels highlights his role as a healer of illness and disability.[1] Healing the sick was an essential aspect of Jesus' ministry and message as well as of that of his disciples (see for example Luke 10:9 and Acts 2:22, 3:1–10). Yet the New Testament featured him as someone who worked also many miracles that did not pertain immediately to the realm of healing sicknesses. The gospels showed him to have turned water into wine (John 2:1–11), to have multiplied five loaves and two fish (Matt 14:13–21; Mark 6:30–44), to have walked on water (Matt 14:22–33; Mark 6:45–52), and at the very end of his life to have risen from the dead (Matt 28:1–10; Mark 16; Luke 24:1–12; John 20:1–18). For many Christian readers of the gospels in the ancient world, the narratives of Jesus' miracles supported his special character as a wonderworker. Yet the signs and wonders he was thought to have worked likewise strengthened belief in him as the Messiah and functioned as proof of his divinity. The representations of Jesus as a wonderworker supported and promoted the faith and self-identity of many ancient Christians as believers in God having become man in Jesus.[2]

1 The research and writing of this article for publication was supported through a Heisenberg Fellowship (GZ HO 5221/1–1), for which the author wishes to express her gratitude to the Deutsche Forschungsgemeinschaft (DFG).

2 See for instance Zeilinger, Franz, *Die sieben Zeichenhandlungen Jesu im Johannesevangelium*. Stuttgart: Kohlhammer, 2011. Main elements of the thematic trajectory of the image of Jesus as a miracle-worker in support of claims to Jesus' divinity in the ancient Syriac-speaking realm are discussed in Horn, Cornelia, "Jesus' Healing Miracles as Proof of Divine Agency and Identity: The Trajectory of Early Syriac Literature." in *The Bible, the Qur'ān, and Their Interpretation: Syriac Perspectives*, ed. by Cornelia Horn, Eastern Mediterranean Texts and Contexts 1, pp. 69–97. Warwick, RI: Abelian Academic, 2013.

https://doi.org/10.1515/9783110564341-003

Jesus' identity as a worker of miracles was a significant part of early Christians' interest in his presentation.[3] Polemical and theological writings offered exegetical and interpretive approaches to the New Testament witnesses on the topic. In the process, authors expanded the repertoire of miracles they thought could have been part of the story of Jesus' life. Especially early Christian apocrypha that were composed from the second century onward filled in periods of Jesus' life with wondrous activities that were not thoroughly or even not at all part of the texts that eventually came to be regarded as canonical.[4] Along with this, one perceives shifts in the representation of Jesus as a wonderworker that emphasized more strongly Jesus extraordinary qualities and powers. Responses of non-Christians to the New Testament portrait as well as to the representation of Jesus reflected in other Christian texts that established claims to the identity of Jesus as a wonderworker are in evidence in late ancient Jewish sources, the Qur'ān, and subsequent medieval Islamic writings. The present article aims to contribute to the ongoing project of tracing the shifting perceptions of Jesus as a wonderworker, healer, and man of miracles in developing Christian apocrypha, Jewish literature, and early and medieval Islamic texts. It focuses on exegetical intersections between infancy-of-Jesus episodes in Greek and Syro-Arabic apocrypha from the Christian realm and the Qur'ān.

3 For a helpful collection of some of the early evidence see Zimmermann, Ruben, in collaboration with Detlev Dormeyer and Susanne Luther, *Die Wunder Jesu*, Kompendium der frühchristlichen Wundererzählungen 1. Gütersloh: Gütersloher Verlags-Haus, 2013.

4 Accessible recent collections of Christian apocrypha in modern translations with scholarly introductions include Bovon, François; Pierre Geoltrain and Sever Voicu, et al., eds., *Les écrits apocryphes chrétiens*, Bibliothèque de la Pléiade 442 and 516. Paris: Gallimard, 1997–2005, and Markschies, Christoph and Jens Schröter, in collaboration with Andreas Heiser, eds., *Antike christliche Apokryphen in deutscher Übersetzung. I. Band in zwei Teilbänden: Evangelien und Verwandtes. 7. Auflage der von Edgar Hennecke begründeten und von Wilhelm Schneemelcher fortgeführten Sammlung der neutestamentlichen Apokryphen.* Tübingen: Mohr Siebeck, 2012. Further volumes are to appear in due course. Until then, the reader may still consult with benefit the two volumes of the English translation of the sixth edition of the Hennecke-Schneemelcher. See Hennecke, Edgar and William Schneemelcher, eds., *New Testament Apocrypha*, trans. by Robert McLachlan Wilson and Angus John Brockhurst Higgins. Philadelphia: Westminster Press, 1963–1966.

2 Interpretations of Jesus as a Wonderworker in Early Christian Apocrypha

Questions that consider which qualities enabled Jesus to work miracles were already part of the concerns that motivated the interactions, even if one-sided, between patristic writers and authors of other religious literature in the earliest Christian centuries. In some circles, particularly among so-called Gnostic writers, one seems to have thought that Jesus was only able to work miracles once Christ had descended upon him. Irenaeus of Lyons for instance adduced information concerning Cerinthus' opinion that "it was only after his baptism that Christ, [descending] from the highest power, which is above everything, had come down upon him in the form of a dove, and from then on, he [Jesus] had proclaimed the unknown father and had worked miracles."[5] From other Gnostic teachers, Irenaeus had gathered that "only when Christ descended upon Jesus, [Jesus] began to work miracles, heal, proclaim the unknown Father and reveal himself openly as Son of the First Anthropos."[6] Implicit in this perspective is

---

5 Rousseau, Adelin and Louis Doutreleau, eds., *Irénée de Lyon. Contre les hérésies. Livre I. Tome II*, Sources chrétiennes 264. Paris: Les Éditions du Cerf, 1979, pp. 344–346: "Et Cerinthus autem quidam in Asia non a primo Deo factum esse mundum docuit, sed a Virtute quadam valde separata et distante ab ea Principalitate quae est super universa et ignorante eum qui est super omnia Deum. Iesum autem subiecit non ex Virgine natum, impossibile enim hoc ei visum est, fuisse autem eum Ioseph et Mariae filium similiter ut reliqui omnes homines, et plus potuisse iustitia et prudentia et sapientia ab omnibus. Et post baptismum descendisse in eum ab ea Principalitate quae est super omnia Christum figura columbae, et tunc adnuntiasse incognitum Patrum et virtutes perfecisse; in fine autem revolasse iterum Christum de Iesu, et Iesum passum esse et resurrexisse, Christum autem impassibilem perseverasse, existentem spiritalem." In later Christian theology, various ideas about Christ's birth expressed themselves in different, but not unrelated ways. Consider for instance, as late as the eighteenth century, the Unctionist/Sost Ledat ("three Births") Christology prominent in Ethiopia/Eritrea. According to this theological perspective, Christ was born three times: once from the Father, once in the incarnation from the Virgin Mary, and once through the Holy Spirit. For comments, see for instance Kaplan, Steven, "Dominance and Diversity: Kingship, Ethnicity, and Christianity in Orthodox Ethiopia." *Church History* 89, 1–3 (2009): pp. 291–305, here 302–303.

6 Rousseau and Doutreleau, *Irénée de Lyon*, pp. 380–382: "Multos ergo ex discipulis eius non cognovisse Christi descensionem in eum dicunt; descendente autem Christo in Iesum, tunc coepisse virtutes perficere et curare et adnuntiare incognitum Patrem et se manifeste Filium Primi Hominis confiteri. In quibus irascentes Principes et Patrem Iesu, operatos ad occidendum eum; et in eo cum adduceretur, ipsum Christum quidem cum Sophia abstitisse in incorruptibilem Aeonem dicunt, Iesum autem crucifixum." See also Bauer, Johannes B., "Wunder Jesu in den Apokryphen." In *Heilungen und Wunder: Theologische, historische und medizinische Zugänge*, ed. by

the view that Jesus' miracles expressed that his nature and identity were not limited to being human, but that divine power dwelled within him. Simply to label this understanding as Gnostic and dismiss it may not be justified, since it had too strong a basis in mainstream Christian thinking. Christian art moreover readily promoted and continued the idea that Jesus' baptism was the event that empowered him to be able to work miracles.[7] Nevertheless, for groups within the emerging early mainline Church, this perspective was defective since it did not extend Jesus' possession of divine identity to the whole of his life. When early Christian apocryphal literature ascribed also to Jesus' childhood and youth the ability to work miracles, it became possible to show that Jesus' divine powers were present in him throughout his life.

In the canonical New Testament, Jesus began to work miracles as an adult. The miracle of turning water into wine at the wedding feast at Cana was said to have been the beginning (ἀρχή) of his signs (John 2:11).[8] Early Christian writers regularly highlighted this event and emphasized its Christological importance[9] and this scene was depicted in Early Christian art as well.[10] As early as the 5th

Josef Pichler and Christoph Heil, in Zusammenarbeit mit Thomas Klampfl, pp. 203–214. Darmstadt: Wissenschaftliche Buchgesellschaft, 2007, here p. 203.

7 See for instance the depiction of Biblical scenes on an early-fifth-century ivory plaque, preserved at the Staatliche Museen, Preussischer Kulturbesitz, in Berlin-Dahlem, that shows a row of three scenes, from top to bottom, of the massacre of the children at Bethlehem, Jesus' baptism in the Jordan, and the miracle of turning water into wine at the wedding feast at Cana. For a discussion of the dating and a depiction of the ivory plaque, see Kitzinger, Ernst, *Byzantine Art in the Making: Main Lines of Stylistic Development in Mediterranean Art, 3rd to 7th Century*. London: Faber and Faber, 1977, p. 47 and plate 84. Scholars have argued that this plaque was once part of a five-part diptych, which was intended to be used as a book cover. See for instance Schnitzler, Hermann, "Kästchen oder fünfteiliges Buchdeckelpaar?" in *Festschrift für Gert von der Osten*, ed. by Horst Keller, Rainer Budde, Brigitte Klesse, et al., pp. 24–32. Köln: DuMont 1970. For a full-page depiction of the Berlin ivory plaque, accompanied by a plaque now kept in Paris that may also have been part of this five-part diptych, see Schnitzler, "Kästchen", p. 25.

8 For refocusing the discussion of the wedding feast at Cana as the beginning of Jesus' miracles, see more recently the contribution by Förster, Hans, "Die johanneischen Zeichen und Joh 2:11 als möglicher hermeneutischer Schlüssel." *Novum Testamentum* 56 (2014): pp. 1–23.

9 See for instance the study of patristic exegesis of the wedding feast at Cana that is offered in Smitmans, Adolf, *Das Weinwunder von Kana. Die Auslegung von Jo 2,1–11 bei den Vätern und heute*, Beiträge zur Geschichte der biblischen Exegese 6. Tübingen: Mohr Siebeck, 1966. Smitmans shows a greater concern with Christology, and a lesser interest in Mariology, in patristic exegesis of the passage, when compared to the interests of Roman Catholic exegetes of his own time.

10 In addition to the above-mentioned five-part diptych, the scene is found for instance on a relief on the rear side of the archepiscopal ivory throne of Maximianus (545–553 CE) and on

century early Christian art may have integrated the scene within a cycle of depictions of scenes from Jesus' childhood.[11] The witness of non-Christian texts offers some evidence that the miracle of the wedding feast at Cana was indeed an important step in characterizing Jesus as a wonderworker. Early Islamic *qiṣaṣ al-anbiyā'* literature reworked this miracle and its context multiple times. The work of one or several Islamic redactor(s) may then have assembled material from different strands of traditions from within and outside of early Islam.[12] The Islamic variants on Jesus' presence at a wedding feast placed such a miracle within the context of the not-yet-twelve-year-old boy's sojourn in Egypt. This suggests that when stories of Jesus' miracles moved from one religious tradition to another, the overall framework, place, and role of the miracle within the context of Jesus' life story could be transformed. The more prominent and articulate emphasis on the reception of this particular miracle into the context of Jesus' childhood stories may serve as a pointer to the strong resonance and impact that the developing literature of apocryphal infancy gospels had at the exegetical intersections of Christian literature with the literatures of other religions. Firmly established elements of the stories of the Christian tradition became subject to change and transformation and entered the discussions at the interreligious crossroads in a new garb.

Different from the canonical record, early Christian apocrypha continued to increase the details of the image of Jesus displaying exceptional features already during his childhood. The present discussion examines one complex example of the *Traditionsgeschichte* of infancy miracles of the Christ-child, highlighting the special circumstances that accompanied Jesus' birth and the manifestations of wondrous powers of the newborn infant at the intersection of the Christian and Islamic traditions.

---

a 6th-century ivory fragment from Egypt (see MacLagan, Eric, "An Early Christian Ivory Relief of the Miracle of Cana." *The Burlington Magazine for Connoisseurs* 38, 217 (1921): pp. 178–195.

11 Schnitzler, "Kästchen," pp. 29–30, argues convincingly that the Berlin ivory plaque depicting the three scenes of the massacre of the Bethlehemite children, Jesus' baptism, and the wedding feast at Cana were part of a cycle of infancy-of-Jesus depictions, complemented by and paired with a separate cycle of miracles of Jesus as part of the decorations of a set of book covers.

12 See for instance Brinner, William M., trans., *'Arā'is al-majālis fī qiṣaṣ al-anbiyā' or "Lives of the Prophets" as Recounted by Abū Isḥāq Aḥmad ibn Muḥammad ibn Ibrāhīm al-Tha'labī*, Studies in Arabic Literature, Supplements to the Journal of Arabic Literature 24. Leiden: Brill, 2002, pp. 650–651.

3 The Wondrous Infant Working Miracles

The mid-to late-second-century apocryphal infancy gospel known as the *Protoevangelium of James* tells the story of Mary's childhood, youth, and of her conception and giving birth to Jesus.[13] In this account, the scene of the birth of Jesus is shrouded in mystery. Two marvelous events frame comments on how Joseph searched for a midwife while Mary was resting in the cave before the delivery of her child. In a scene that is preceding Jesus' birth, but could be construed to be simultaneous to it, time seemed to stand still and the characters came to be caught in the midst of their actions.[14] A second vignette tells how Joseph and the midwife beheld "a bright cloud overshadowing the cave." When the cloud withdrew, their "eyes could not endure" the light that appeared there. Only when "little by little that light withdrew," "the young child appeared," being already sufficiently developed physically to go and take "the breast of his mother Mary" on his own accord.[15] In the *Protoevangelium of James*, these two scenes suggested the special nature of the newborn child. They supported the midwife's claim that "a virgin ha[d] brought forth."[16] Yet these descriptions also revealed to the reader that the true nature of the child was not limited to the human realm.

Ideas about the special nature of the newly born Jesus were more widespread. This is illustrated in the second-century *Ascension of Isaiah*, a composite pseudepigraphical text of Jewish origins, which nevertheless is to be regarded as a Christian work in the form it assumes in its final redaction.[17] Building on the interpretation of Isa 7:14 as a prophecy of Jesus' birth, the second half of the *Ascension of Isaiah*, known as the *Vision of Isaiah*, featured a report of the Savior's miraculous birth as a child that both Mary and Joseph witnessed. *Ascension of*

---

13 See Pelegrini, Silvia, "Das Protevangelium des Jakobus." in *Antike christliche Apokryphen in deutscher Übersetzung. I. Band in zwei Teilbänden: Evangelien und Verwandtes. 7. Auflage der von Edgar Hennecke begründeten und von Wilhelm Schneemelcher fortgeführten Sammlung der neutestamentlichen Apokryphen*, ed. by Christoph Markschies and Jens Schröter, in collaboration with Andreas Heiser, I.2, pp. 903–929. Tübingen: Mohr Siebeck, 2012, here p. 907

14 De Strycker, Émile, ed., *La forme la plus ancienne du Protévangile de Jacques. Recherches sur le Papyrus Bodmer 5 avec une édition critique du texte grec et une traduction annotée*, Subsidia Hagiographica 33. Bruxelles: Société des Bollandistes, 1961, pp. 148–151.

15 De Strycker, *La forme*, pp. 154–157.

16 Ibid., pp. 158–159.

17 For the text of *Ascension of Isaiah*, together with extensive commentary, see Bettiolo, Paolo; Alda Giambelluca Kossova and Claudio Leonardi et al., eds., *Ascensio Isaiae: Textus*, Corpus Christianorum, Series Apocryphorum 7. Turnhout: Brepols, 1995; and Norelli, Enrico, *Ascensio Isaiae: Commentarius*, Corpus Christianorum Series Apocryphorum 8. Turnhout: Brepols, 1995.

Isaiah 11.7–11 tells of how one day, while Mary was pregnant, she "straightaway looked with her eyes and saw a small babe, and she was astonished."[18] Yet any signs of pregnancy on her body had disappeared immediately after parturition. Likewise, Joseph experienced that "his eyes were opened and he saw the infant."[19] The description of the appearance of the child in the *Protoevangelium of James* at the point when the bright light had subsided somewhat from the cave may have built on earlier traditions preserved in the *Ascension of Isaiah.* Or, this latter text may reflect revisions of the Christian redactor that had their origins in the *Protoevangelium of James.*[20] The birth of Jesus occurring without any labor pains and the child manifesting himself prominently through visual experiences serve as evidence that in those layers of the Christian society, in which apocryphal and pseudepigraphical texts circulated, ideas originated and found propagation that readily ascribed to Jesus' birth wondrous circumstances that collaborated and revealed the child's exceptional nature, a nature not limited to the restrictions of the physical human body.

Within the apocryphal Christian record, ideas about the special nature of the newborn child quickly translated into descriptions of manifestations of the wondrous powers of that child and his body. In the *Protoevangelium of James*, Salome did not intend to accept the virgin birth upon its mere proclamation.[21] When she physically examined Mary to test her virginity, Salome's hand withered and was said to have felt as if consumed by fire.[22] Yet when an angel instructed Salome to stretch out her hand and lift the baby up, Salome confessed the child to be the king of Israel, and she was healed.[23] In the literary construction of the scene of Salome's reaction, her healing was interpreted as resulting from the intertwining of her physical contact with the baby and her recognition or confession of faith in the child as Lord. Yet it was understood that had she not come into direct contact with the newborn Jesus' body, she would not have received healing. Miracles worked through immediate contact with Jesus, even while he was still a newborn

18 Bettiolo et al., *Ascensio Isaiae*, pp. 120–121.

19 Ibid.

20 Norelli, *Ascensio Isaiae*, pp. 65–66, argues that *Ascension of Isaiah* 6–11 is a product of the end of the first century and *Ascension of Isaiah* 1–5 likely had its origins in the first half of the second century. In his commentary on the passage that is relevant here, Norelli highlights the connections between the Gospel of Matthew and *Ascension of Isaiah*, and emphasizes the difference between *Ascension of Isaiah* and the *Protoevangelium of James.*

21 De Strycker, *La forme*, pp. 156–159.

22 Ibid., pp. 158–161.

23 Ibid., pp. 162–167.

baby, began to receive space and gain relevance in early Christian discourse that is revealed in apocryphal Christian texts.

In subsequent apocryphal traditions that built on this story, one observes various contractions and expansions of the interpretation of the newborn Jesus and his role in this context. The evidence of the Syriac and Arabic traditions is informative here. In some lines of transmission of the *Protoevangelium of James*, the story of Salome's testing of Mary's virginity was excised. The *Syriac Life of the Blessed Virgin Mary*, for instance, which Ernest A. Wallis Budge edited in 1899 on the basis of a 19th-century copy of a manuscript from the 13th or 14th century and one additional manuscript for which no date or provenance were provided, simply had the midwife-turned-into-old-woman praise God for the birth of the Redeemer of all.[24] Yet in other texts that are available as evidence for developments of this tradition within the Syro-Arabic realm, the scene was substantially expanded to include an interpretation of the newborn Jesus and his miraculous powers that either focused on the efficacy of entering into contact with his body through the mediation of his mother or that conveyed the idea that the miraculous powers of the newborn Jesus witnessed to his prophetic character. Thus, an investigation focusing on developments that emphasized how baby Jesus' miracles were interpreted as showing him forth as a prophet-in-the-making is of special relevance for a study that seeks to explore the intersections of traditions at the interreligious exegetical crossroads between Christianity and early Islam.

First, it pays off to consider briefly a text that speaks to interpretations of the infant Jesus as son of God. In the *Arabic Infancy Gospel* there occurs the following passage:

> We find what follows in the book of Josephus the high priest, who lived in the time of Christ. Some say that he is Caiaphas. He has said that Jesus spoke, and, indeed, when He was lying in His cradle He said to Mary His mother: "I am Jesus, the Son of God, the Logos, whom thou hast brought forth, as the Angel Gabriel announced to you; and my Father has sent me for the salvation of the world."[25]

24 Budge, Ernest A. Wallis, ed., *The History of the Blessed Virgin Mary and The History of the Likeness of Christ Which the Jews of Tiberias Made to Mock at*, 2 vols., Luzac's Semitic Text and Translation Series. London: Luzac and Co., 1899, vol. 1, p. 30 [Syriac] & vol. 2, p. 34 [English]. For information on the manuscripts underlying this edition, see Budge, *The History*, vol. 1, pp. v–vi.

25 Provera, Mario E., ed., *Il vangelo arabo dell'infanzia secondo il ms. Laurenziano Orientale (n. 387)*, Quaderni de "La Terra Santa". Jerusalem: Franciscan Printing Press, 1973, pp. 112–113.

The claim of the text clearly is that the speaking infant is the Son of God and the Logos. This Jesus is seen neither as only a human being, nor merely as a prophet.

How this text relates to the evidence of a speaking infant Jesus in the Qurʾān though (Q 3:46, Q 5:110, and Q 19:29–33) is less evident. Scholarship on the *Arabic Infancy Gospel* that has discerned the intimate connections between its Syriac *Vorlage* and ideas and images that derive from the Sasanian realm argues convincingly that the origins of the work are to be dated not later than the 7th century.[26] Yet the Arabic versions that have been studied thus far may suggest a process of transformation, potentially in conversation with the Qurʾān.

In both published editions of the *Arabic Infancy Gospel* one finds the description of a scene in which the infant Jesus speaks from a position of lying in his cradle. In the manuscript edited by Henricus Sike in 1697, this scene features as the first episode of the story, whereas the Laurenziana manuscript, which Mario Provera edited in 1973, placed the scene towards the middle of the text.[27] At present, the critical study of the complete manuscript evidence for the *Arabic Infancy Gospel* and its transmission is still in progress and so it is not yet possible to draw any firm conclusions about what the original location of that scene may have been.

Textual fluidity is an inherent feature of early Christian apocryphal texts. On the basis of the observation of the relatively unstable position of the scene in question in the *Arabic Infancy Gospel*, one could readily imagine that this scene was an addition to an earlier text which a redactor of the *Arabic Infancy Gospel* may have created upon receiving inspiration from or in reaction to a well-known scene, which occurs in the Qur'ān and which will be considered in more detail below. Yet one may also adduce this very same characteristic of the fluidity of apocryphal texts and argue that the variation in the position of this scene within the text does not have to have been motivated through contact with external ideas. It could have simply occurred in the process of a combina-

---

26 See Mescherskaya, Elena N., "'L'Adoration des Mages' dans l'apocryphe syriaque *Histoire de la Vierge Marie*." in *Sur les pas des Araméens chrétiens: mélanges offerts à Alain Desreumaux*, ed. by Françoise Briquel-Chatonnet and Muriel Debié, pp. 95–100, Cahiers d'Études Syriaques 1. Paris: Paul Geuthner, 2010, here p. 100. See also Horn, Cornelia, "Arabic Infancy Gospel." in *Encyclopedia of the Bible and Its Reception Vol. 2*, ed. by Hans-Josef Klauck et al., pp. 589–592. Berlin and New York: Walter de Gruyter, 2009.

27 For the 17th-century edition, see Sike, Henricus, *Evangelium infantiae vel liber apocryphus de infantia Salvatoris*. Trajecti ad Rhenum, 1697, pp. 2–5. Just two years later, the *Arabic Infancy Gospel* circulated in a German translation. See the publication "Evangelium infantiae: oder ein so genantes apocryphisches Büchlein, worinnen die Wunder-Geschichte unseres Herrn Jesu Christi, welche sich in seiner Kindheit ... begeben, beschrieben werden. Aus d. Arab. ins Latein durch H. Sike u. nun ins Hochteutsche übersetzet von H. A. v. R. Bi."

tion of lines of oral and written transmissions of the text. The scene in question, therefore, may quite as well already have been a part of a pre-Qur'ānic Syriac original or forerunner of the *Arabic Infancy Gospel*. The question is not an easy one to decide, but the second possibility gains some probability to the extent that one can identify other Christian texts that witness to the spread of ideas about the newborn Jesus speaking at an age that was prior to the age at which infants normally begin to speak.

When one considers the motif of the infant Jesus speaking up as a prophet-in-the-making, the *Arabic Apocryphal Gospel of John* is of special relevance.[28] This rather long, apocryphal work, its origins being in the Syriac tradition, is preserved in two Arabic manuscripts from the 12th and 14th centuries: the first kept at the Monastery of St. Catherine on Sinai and the second preserved in Milan.[29] Having been hailed by Oscar Löfgren as "the only, almost completely preserved ... New Testament apocryphal text, which in fact also represents the [literary] genre of a gospel,"[30] it is a composite apocryphal work that includes a retelling of Christ's birth that is composed among others from reminiscences of canonical traditions and sections of the *Protoevangelium of James* as well.[31] Elements that connect its account of Christ's birth with that of the *Protoevangelium of James* include the reference to the cave in which the birth took place, the presence of Salome, and her physical examination of Mary's virginity. Yet the story that is told in the *Arabic Apocryphal Gospel of John* also bridges gaps to non-Christian tradi-

28 See also Horn, Cornelia, "Syriac and Arabic Perspectives on Structural and Motif Parallels regarding Jesus' Childhood in Christian Apocrypha and Early Islamic Literature: the 'Book of Mary,' the *Arabic Apocryphal Gospel of John*, and the Qur'ān." *Apocrypha* 19 (2008): pp. 267–291; and Horn, Cornelia, "Editing a Witness to Early Interactions between Christian Literature and the Qur'ān: *Status Quaestionis* and Relevance of the *Arabic Apocryphal Gospel of John*." in *Actes du 8e Congrès International des Études Arabes Chrétiennes (Granada, septembre 2008)*, published in *Parole de l'Orient* 37 (2012): pp. 87–103, esp. pp. 100–103.
29 For the discussion of the manuscript evidence and preliminary results on the way towards creating a critical edition of this important apocryphal work, see Horn, "Editing a Witness".
30 Löfgren, Oscar, "Ein unbeachtetes apokryphes Evangelium." *Orientalische Literaturzeitung* 4 (1943): pp. 153–159, here p. 159.
31 For an edition of the *Arabic Apocryphal Gospel of John* on the basis of the Milan manuscript, together with a Latin translation, see Galbiati, Giovanni, ed., *Iohannis evangelium apocryphum arabice*. Milan: In aedibus Mondadorianis, 1957. Löfgren, Oscar, *Det Apokryfiska Johannesevangeliet: I översättning från den ende kända arabiska handskriften i Ambrosiana*. Stockholm: Natur och kultur, 1967, offered a Swedish translation of that same manuscript, which Per Prytz rendered into Norwegian just short of thirty years later. An Italian translation, also of the Ambrosiana manuscript, appeared in 1991. See Moraldi, Luigi, *Vangelo Arabo apocrifo dell'Apostolo Giovanni da un Manoscritto della Biblioteca Ambrosiana*, Biblioteca di Cultura Medievale. Milan: Editoriale Jaca Book, 1991.

tions about Jesus' birth, particularly the representation of the newborn baby in the early Islamic tradition that is manifested in the Qur'ān. One is alerted to this exegetical intersection by two interconnected motifs in the *Arabic Apocryphal Gospel of John*. The first motif is a reference to a cradle in which the newborn baby is placed. The second motif, closely connected with the first one, is found in that part of the storyline that develops the image of Jesus as a baby speaking pretty much right after his birth.

In the scene of Jesus' birth in the *Arabic Apocryphal Gospel of John*, Mary delivered her child, then wrapped him up "and prepared that manger for him like a cradle (*wa-ja'alat dhālika l-midhwada lahu mahdan*)."[32] In this context, one notes that the Arabic word *mahd* (cradle) is identical with the term the Qur'ān uses in Q 3:46, Q 5:110, and Q 19:29 for the resting place at which the newborn Jesus was put down and from which the child spoke in two of the three passages. One also observes that *Arabic Apocryphal Gospel of John* 5.2 employed the word *mahd* as a term that explained more directly the purpose for which the *midhwad* (manger), a feeding trough for animals, was intended to serve at this instance. Perhaps one might consider *mahd* as a gloss on *midhwad* that was used in order to aid a reader who might have been less familiar with Jesus' birth narrative in the gospels or one who would not have readily understood the Arabic *midhwad*. Yet it seems that the function of *mahd* could be explained well as being sufficiently motivated from within the literary context of the narrative itself. Since a manger normally serves as a feeding place for animals, its reuse as a bed for an infant was noteworthy. The emphasis provided through the special explanation of the purposes of the manger certainly attracts the reader's attention to this object.

The *Arabic Apocryphal Gospel of John* integrated the test of Mary's virginity at the midwife Salome's hands with details of Salome's subsequent illness as well as her healing through contact with the body of the newborn child that was mediated and directed through Mary. These interactions played out between Salome and Mary. The infant Jesus was only present through his healing and miracle-working body. This apocryphal text also introduced the concern of the value of Isa 7:14 as a prophecy of the birth of the Messiah into the conversations between Mary and Salome. Here the *Arabic Apocryphal Gospel of John* offered that Mary "took Salome's hand, which had dried up, and laid it on the [newborn]-child, and it was loosened [from its disease] at that moment and was restored to its [natural] condition completely."[33] Having experienced miraculous

32 Ibid., p. 31.
33 Ibid., p. 33.

healing in this way, Salome broke out in praise, saying "Truly, O [newborn]-child, you are the son of God, of whom the prophet Isaiah prophesied."[34] The acceptance of Isaiah's prophecy as a testimony to Jesus' manifesting God's presence among God's people is as old as Matt 1:22–23. Presenting Salome as a doubting midwife who was turned into a prophetess proclaiming the birth of the son of the Virgin was reminiscent of Luke's portrait of the figure of Anna, the aged prophetess (Luke 2:36–38).

Into this immediate context in response to Salome proclaiming the newborn Jesus as the fulfillment of Isaiah's prophecy, the newborn child, whose presence had thus far only been featured as an instrument in the text, broke out into speech, confirmed the truth value of Salome's interpretation, and revealed himself through the fact and content of his miraculous speech, not simply as a prophet, but as the Lord of the prophets. Thus, the *Arabic Apocryphal Gospel of John* narrated:

> The child opened his mouth and said, "Truly, this is the prophecy of Isaiah: since he did not rebuke King Uzziah or prevent him from entering into the altar of God, wherefore I hindered him from prophesying and made his lips leprous (2 Chron 26:19); but when he turned to me from his lapse, I sent to him an angel [having] with him tongs to carry the pure offering from the base of the fire. Then it approached him and he became cleansed of his leprosy and I reinstated [Isaiah] in his status as prophet and sent him to prophesy concerning my incarnation from this pure Mary, whom I have selected and taken from among all the women of the world, and from whom I desired to be incarnated for the salvation of Adam and his children from sin and [from] servitude to Satan. For a miracle does not arise from the strong one when he overcomes the weak, a miracle arises from the weak when he overcomes the strong one."[35]

The text identified this speech, in which the newborn one in the cradle positioned himself as the one who "reinstated [the prophet Isaiah] in his status as prophet and sent him to prophesy" concerning Jesus' incarnation and birth from Mary as the second "miracle ... that our Lord Christ performed as a child."[36] In counting miracles, this text continued a Johannine tradition.[37] Moreover, in the text's perspective, the newborn one was not merely a prophet, but revealed his divinity in his speech. With the child speaking in the first person singular and claiming that he was the one who was to come for the salvation of sinners, that he had sent the angel that purified the prophet's lips and empowered him with the gift of prophecy, and that he had commissioned the prophecy

34 Ibid., p. 33.
35 Ibid., pp. 33–34.
36 Ibid., p. 34.
37 Horn, "Editing a Witness", pp. 89–90.

of the birth of the Christ child, the author of the *Arabic Apocryphal Gospel of John* established that the newborn Jesus not only represented or offered God's perspective, but was and acted as God. Such claims either responded to, or invited, challenges.

The Qur'ān offered its readers three opportunities to encounter references to Jesus as an infant speaking prophetically. It is instructive to compare the *Arabic Apocryphal Gospel of John*'s scene of the newborn child in the cradle speaking from the position of one who is presented as being the Lord of the prophets with the presentation of Jesus as infant speaking prophetically as it is found in the early Islamic tradition, especially in Q 3:46, 5:110, and 19:29, both for the parallels and for the differences it reveals.

The reader of the Islamic material readily notices that in all three instances the miracle of the speaking infant is embedded in a shorter or longer list of other wondrous signs. Q 3:42–51 concatenated the miracle of the infant speaking in the cradle with a listing of other miracles of Jesus, namely his receiving exceptional knowledge from God in the form of the Torah and the Gospel, of the child Jesus breathing life into birds of clay, of the manifestation of Jesus' miraculous knowledge about what people eat and store at home, of the miracles of Jesus healing the blind and the leper, and of the miracle of Jesus raising the dead. Q 5:109–120 embedded such a list of Jesus' miracles in a narrative that envisioned conversations at the final Day of Judgment. The list of miracles here is almost the same and in the same order as in Q 3:46, merely omitting Jesus' knowledge of what people eat and store at home. In addition, this list of miracles is then expanded by an account of Jesus' prayer bringing about the miracle of the descent of heavenly food for a feast for the disciples. Finally, Q 19:27–36 presents the miracle of the infant speaking shortly after his birth in connection with a reference to the child's receipt of special knowledge from God in the form of the Scriptures. In the immediate context of the Qur'ān, the concatenation of these different miracles supports that the infant who is speaking while still being a child in the cradle is to be regarded as a prophet. Yet he also is a prophet, who was clearly subordinate to God.

While it is not possible to discuss exhaustively the Qur'ānic view of Jesus as a prophet here, one notes the emphasis of the relevant passages on the child speaking while still being in the cradle, that is, as a young, newborn infant. One also observes that Q 3:46, for instance, formulated that the infant spoke like an adult. With such a comment, the Qur'ān did not merely articulate the miraculous character of Jesus' speech here. It also entered into conversation with contemporary and preceding ancient Christian traditions of miracles, especially in Syriac martyrdom literature, in which young children were presented as having spoken with power and the rhetorical ability of adults long before the normal age at which such discourse usually is possible.

Ancient Christian literature developed a motif that showed young children who were able to engage in informed and rhetorically developed discourse beyond what is normally possible at their age, specifically in situations in which the circumstances required it. When featuring the martyr Romanus in one of the hymns that were part of the collection known as the *Crown of the Martyrs* (*Peristephanon*), the 4th-century Roman Christian poet Prudentius included a scene that described the brutal beating and death of a little boy, who was hardly weaned.[38] Given that children were weaned in antiquity considerably later than what is customary in many modern Western cultures, the child in question here may have possessed sufficient skills for regular, everyday-life communications. Nevertheless, Prudentius created a scene in which the ensuing dialogue between the persecutor and the young child showed the child to be speaking with theological wisdom and insight and with adult-like courage that transcended by far what any young child could deliver.

Prudentius' hymn does not constitute a singular example. Late ancient Syriac Christian hagiography was particularly rich in featuring very young children facing martyrdom and speaking out against their persecutors with greatest courage and defiance. The story of the two-year-old Mar Ṭalyā' of Cyrrhus and his encounter with the local governor serves as a particularly good example.[39] While the governor addressed him as a mere infant, Mar Ṭalyā' mocked and destroyed about a third of the idols in the governor's temple with ease and responded to the ruler's questions with daring and insulting comments that went beyond what even adults would have attempted to get away with. One notices that both the little boy who was featured in the hymn about the martyr Romanus and Mar Ṭalyā' addressed outside opponents in situations in which the children themselves, but also the parent who accompanied them, were confronted with adversities and aggressions against their well-being. These examples attest that a noteworthy, sufficiently wide-spread motif existed, that featured very young children who were speaking up in situations of being threatened when they were still younger than what is considered to be the normal age for rational

38 Prudentius, *Peristephanon 10:651–845*, ed. by Maurice P. Cunningham, Aurelii Prudentii Clementis Carmina, Corpus Christianorum Series Latina 126. Turnhout: Brepols, 1966, pp. 251–389, here pp. 352–359; text and trans. by H. J. Thomson, *Prudentius. Works*, The Loeb Classical Library, 2 vols. Cambridge: Harvard University Press, 1953, vol. 2, pp. 98–345, here pp. 272–285.
39 On the *Martyrdom of Mar Ṭalyā' of Cyrrhus*, see Horn, Cornelia, "Children and Violence in Syriac Sources: The *Martyrdom of Mar Ṭalyā' of Cyrrhus* in the Light of Literary and Theological Implications." *Parole de l'Orient* 31 (2006): pp. 309–326.

speech. This motif was identifiable in its main contours in the Eastern world and more widely in the ancient Christian realm.

These stories of speaking infants in Christian martyrdom stories throw a helpful light on certain aspects of the story of Jesus' birth from Mary as it is told in *Sūrat Maryam* (Q 19). It is worth considering how the miracle of the speaking infant Jesus relates to this Islamic birth narrative and its consequences.

Q 19:27 presented the mother at a point in time when she had already given birth to her child, had recovered from that experience, and was now returning to her people, carrying her child on her arms. Yet instead of receiving Mary with open arms and a warm welcome, those who belonged to "her people" and whom the audience of the Qur'ān would readily identify as Jews, accused her, the "sister of Aaron" as they called her, of having committed an illicit act, declaring, "your father was not a wicked man (*imra'a saw'in*), and your mother was not a prostitute (*ummuki baghiyyan*)" (Q 19:28). With a reference to her father's and her mother's honorable lives spent without evil or disgrace, Mary's fellow Jews were presented as having called into doubt her reputation and as suggesting that the child was the product of Mary's lack of chastity. The scene may readily be understood as one in which a mother suffered from aggression or persecution while she was holding her little newborn baby, who had not yet been weaned, on her arms. This scene offered the framework for Jesus' first miracle. The close physical connection between mother and child was emphasized, when the text formulated that the mother pointed at the child (*fa-ashārat ilayhi*). Such a gesture would not necessarily have led a reader or hearer to expect that in response to this prompting the baby might speak of his own initiative. Yet within the context of the story, verse 19:26 had stated that Mary had vowed not to speak to anyone on this day. When confronted with the implied accusation of her lack of chastity that the Jews brought against her, she opted not to break her vow and defend herself but instead to expect that through her child a direct resolution of her difficult situation might occur. Through this gesture, Mary appeared as a woman who was truthful in her relationship to God and who possessed superior faith and trust, expecting a miraculous intervention and in a way describing through her gesture ahead of time what the nature of that miracle would have to be. A Christian or once-Christian hearer among the Qur'ān's audience may have been reminded of Matt 10:19, "When they deliver you up, do not be anxious how you are to speak or what you are to say, for what you are to say will be given to you in that hour." Mary's fellow Jews in the Qur'ān, who did not possess any knowledge of her vow, may have expected her to defend herself. Yet instead they became witnesses to a physical and visually perceptible gesture, which they interpreted as Mary encouraging the child to speak. Thus they expressed their own astonishment when they asked, "How can we speak to one who is in the cradle, an infant" (*kayfa nukallimu man kāna fī l-mahdi ṣabiyyan*). Into this dramatic setting, in which the participants interacted with one another at different levels of awareness and with insight into what happened and what was expected to happen in the future, the child on Mary's arms began to speak and offered a rational discourse.

With this guided, but still surprising turn of events the Qur'ān used the opportunity of the miracle of Jesus speaking as an infant in order to pursue at least two goals: it clarified its position regarding Jesus' nature and character and it employed Jesus' speech as a tool with which to handle the presumed Jewish criticism of Mary. The infant Jesus declared himself to be "a servant" or "slave" of God (Q 19:30). Any hearer of the Qur'ān was to understand this as a statement that Jesus was neither God nor God's son. The second and third elements of Jesus' self-characterization in Q 19:30 pertained to him being a recipient of *al-kitāb*, the (sacred) book, and a prophet (*nabī*). The announcement of Jesus speaking as an infant in Q 19:29 served as a revelation of his prophetic quality. It placed this verse in a noteworthy parallel to verses Q 3:45–46 that indirectly spoke of Jesus' prophetic character by presenting his speaking as an infant as a consequence of his being a word from God and to verse Q 5:110 that presented the miracle of Jesus speaking in the cradle as a consequence of his reception of strength from the Holy Spirit.

It is quite obvious that the Qur'ān's presentation of the miracle of the speaking infant in *Sūrat Maryam* was criticizing Christian and Jewish positions. The intended correction of showing Jesus as a prophet and servant of God has already been addressed. Yet the text also dealt with Jewish polemic against Jesus' birth. The Qur'ān's statements on the topic gain in force when one considers that they also addressed an audience that was familiar with Jewish traditions that accused Mary of adultery. While one finds traces of such a discourse in early Christian authors like Origen of Alexandria and his interactions with Greek philosophers, or in rabbinic statements, perhaps the most prominent witness to such claims is found in the traditions of the *Toledot Yeshu*, a popular text and tradition, for which Aramaic fragments are extant as far back as the 4th or 5th centuries.[40] In the overall presentation of the matter in the Qur'ān we encounter

40 See Origen, *Contra Celsum*, trans. by Henry Chadwick. Cambridge: Cambridge University Press, 1980, ch. 1.28–32 and 1.69, pp. 27–32, and the discussion of Jesus' family in chapter one in Schäfer, Peter, *Jesus in the Talmud*. Princeton: Princeton University Press, 2007. One may access the Aramaic fragments of *Toledot Yeshu* in one place in Deutsch, Yaacov, "New Evidence of Early Versions of *Toldot Yeshu*." *Tarbiz* 69 (2000): pp. 177–197 [in Hebrew]. On the origins of this Aramaic material in the first half of the first millennium, see Smelik, Willem F., "The Aramaic Dialect(s) of the Toledot Yeshu Fragments." *Aramaic Studies* 7 (2009): pp. 39–73; and

here a combination of traditions concerning attacks on Mary and her defense through the speaking infant with traditions which emphasize Jesus in his role as a worker of miracles and a prophet, but with an emphasis on the idea that Jesus received from God that which enabled him to work miracles and prophesy.

In pursuit of its goal of defending Jesus' mother against Jewish accusations of a lack of chastity in Q 19, the Qur'ān presented Jesus through his speech as one who prayed, who practiced giving alms, and who respected his parent, that is, his mother. Jesus was one who did not join the side of those who accused Mary. Instead, he was shown to relate to God through actions that revealed his awareness of his dependence on God and his willingness to accept this. His respect for and appreciative actions towards his mother fulfilled basic commandments for the proper treatment of others within the human community, both within the wider society through almsgiving and the narrower context of the family.

In Late Antiquity and the Middle Ages, the lines between developing hagiographies, para-Scriptural literatures, and even Scriptures and rewritten Scriptures were not rigid, quite independent of the extent to which modern scholarship has been able to recognize and trace such correlations. Oriental Christian literature in the realm of apocryphal texts provides evidence from more than one source for the development and circulation of a motif of newborn children, specifically Jesus, being able to speak right from the cradle. The 8th-century apocryphal text of the Syriac *Revelation of the Magi*, for instance, presented baby Jesus as delivering several speeches of considerable length.[41] The content of these speeches qualifies best as revelatory discourse and as such could be considered as fitting for a young prophet. Although the immediate context in which these speeches occurred did not feature any further miracles, the extent to which the tradition of the speaking infant that is embedded in the Syriac *Revelation of the Magi* could serve as an intertext for the Qur'ānic miracle of Jesus speaking as an infant in the cradle or could itself have developed in dialogue with the Qur'ānic account as an intertext may repay further study.

Sokoloff, Michael, "The Date and Provenance of the Aramaic Toledot Yeshu on the Basis of Aramaic Dialectology." in *Toledot Yeshu ("The Life Story of Jesus") Revisited. A Princeton Conference*, ed. by Peter Schäfer, Michael Meerson and Yaacov Deutsch, Texts and Studies in Ancient Judaism 143, pp. 13–26. Tübingen: Mohr Siebeck, 2011.

41 Landau, Brent Christopher, "The Sages and the Star-Child: An Introduction to the *Revelation of the Magi*, An Ancient Christian Apocryphon." Ph.D. thesis Harvard University, 2008, pp. 41–44, 52–58, 60–61 [Syriac] and pp. 99–102, 113–120, and pp. 123–124 [English]. See also Horn, "Syriac and Arabic Perspectives", p. 289.

4 Conclusion

The examples of very young children speaking with wisdom and insight in adult-like fashion and often in the face of adversities derive from martyrdom texts, hagiographical writings, and apocryphal works, all of which envisioned wider circles of readers and listeners among their audiences. The motif of the infant speaking prematurely therefore is to be recognized as one with which late ancient Christian audiences were familiar and comfortable. The evidence one can assemble is not conclusive for addressing the question of the precise direction of the reception of the motif by one tradition from another one. The solution to this type of question is not obvious for a variety of reasons, including that the textual stage at which we catch a hold of this vignette in a given text, for instance in the *Arabic Apocryphal Gospel of John*, is not necessarily identical with the stage at which this story was produced in writing, and we know little to nothing about the stage in which it was when it began to circulate orally. When the infant Jesus is portrayed as speaking in the Syriac *Revelation of the Magi*, the text in which one encounters the scene represents 8th-century material. Yet given that the evidence we have occurs in a Syriac text and also in a Christian-Arabic text that itself has its roots in the Syriac tradition, it is possible that the story began to take shape prior to the transition of stories from their Syriac garb into an Arabic one.

One does not have to insist on identifying any one of the lines of transmission with final precision. Perhaps it makes more sense to acknowledge the shared occurrence of the motif of the speaking infant and to note that it contributes to our understanding of a reality of intersecting crossroads of traditions that appear to have been grounded in a common milieu of literary and cultural traditions. In such a shared milieu, tracing the reinterpretations of motifs along with the differences and similarities that mark their specific shapes in the respective traditions allows us to see more clearly the theological, polemical, or ideological emphases in pursuit of which the respective religious traditions availed themselves of a given motif. The motif of the newborn Jesus speaking in the cradle as a prophet as found in the Islamic tradition, or speaking as a divine being and especially as the Lord of the prophets in the Christian tradition, fits well into the theological expectations of the traditions' respective religious systems. It may not easily be decided whether the Christian tradition responded to the Qur'ān's claim that Jesus was only a prophet when speaking as an infant, or whether the Qur'ān reacted to an earlier narrative which presented Jesus speaking as an infant as an act in which Jesus revealed himself as one greater than the prophets. Yet one finds in this material a witness to a lively debate and to interactions between members of the respective communities in the course of which the different sides laid claims to possessing the authority to decide, at least for their own community, but more likely beyond its borders as well, which image of Jesus was the one to follow.

Martin Heimgartner

The Letters of the East Syrian Patriarch Timothy I

Scriptural Exegesis between Judaism, Christianity and Islam

Three letters constitute the main contributions of the East Syrian Catholicos-Patriarch Timothy I (780 – 823)[1] to scriptural hermeneutics.[2] Letters 34 – 36 were not written as free treatises but are a reply to a specific situation. A discussion arose among the Christians in Southern Iraq concerning the usage of the term "servant" for Christ. The early Christians already transferred this term to Christ as a title expressing his highness, mainly in theological reflection on the so-called *Gottesknechtslieder* in Deutero-Isaiah. However, in its confrontation with Islam, particularly in the spiritual centre of Baṣrā, the old title became problematic because Christ is referred to as a "servant" in the Qur'ān (Q 19:30; 43:59). Moreover, in this context, the title emphasises a created Christ, not "the Son of God." Q 19:30 demonstrates this clearly when Christ designates himself a "servant," and Q 5:116 – 118 when Christ denies being the Son of God. Therefore, the burning question was whether the New Testament title confirmed the Islamic understanding of Christ.

Naṣr, who was an intellectual Christian from Baṣrā, requested that Timothy provides an explanation of how to handle the traditional title. The problem was not merely intellectual because Naṣr explicitly asked him to write to the entire Christian community of Baṣrā and Huballat. Thus, Timothy wrote letter 34,

1 For Timothy, see the fundamental study: Berti, Vittorio, *Vita e studi di Timoteo I (†823) patriarca cristiano di Baghdad: Ricerche sull'epistolario e sulle fonti contigue*, StIr.C 41. Paris: Association pour l'avancement des Études iraniennes, 2009. Besides see Heimgartner, Martin and Roggema, Barbara, "Timothy I," in *Christian-Muslim Relations: A Bibliographical History*, ed. by David Thomas and Barbara Roggema, pp. 515 – 531, History of Christian-Muslim Relations 11. Leiden: Brill, 2009 (both publications with further secondary literature).—Thanks go to Rebecca Giselbrecht, PhD, for her language support.

2 All quotes are according to my edition and translation: Heimgartner, Martin, *Die Briefe 30 – 39 des ostsyrischen Patriarchen Timotheos I.* Corpus Scriptorum Christianorum Orientalium 662. Leuven: Peeters, 2016 (*Einleitung, Übersetzung und Anmerkungen*, CSCO, Leuven: Peeters, 2015). For more details, see the introduction and the notes to my translation. Also see: Griffith, Sidney H., "The Syriac Letters of Patriarch Timothy I and the Birth of Christian Kalām in the Mu'tazilite Milieu of Baghdad and Baṣrah in Early Islamic Times," in *Syriac Polemics: Studies in Honour of Gerrit Jan Reinink*, ed. by Wout Jac van Bekkum, Jan Willem Drijvers and Alexander Cornelis Klugkist, pp. 103 – 132, Orientalia Lovaniensia analecta 170. Louvain: Peeters, 2007.

https://doi.org/10.1515/9783110564341-004

which is a kind of sermon to be read to the Christian community. The doctrine for understanding Scripture that he drafted was therefore of general interest, composed to address a specific situation, and resulted in a scriptural hermeneutics for facing Islam.

While letter 34 is a 'letter sermon' to the whole community, Timothy addresses letter 35 to Naṣr the intellectual, who made the request. Timothy writes "in a rather brief manner" but more profoundly and in greater detail. In addition, letter 36 followed as a 'scientific appendix' for Naṣr including two detailed exegesis on Ps 110:1–3 and Ps 8, one study on the question of the one energy and the one will, and the other on the veneration of martyrs.

In the following, I will explore the Biblical hermeneutics of Timothy by highlighting the important statements in letters 34–36. I will include relevant materials from other letters where necessary.[3]

1 Systematizations of Scriptural Sense

The simplest possibility to escape from the narrow literal sense of a text and to integrate it into an overall theological concept is the use of a metaphorical sense. The letters of the Patriarch are also subject to this conventional distinction between a literal and metaphorical sense. In letter 34, however, we find two examples of real systematic differentiations.

1.1 Four Classes of Predicates

The first classification follows Aristotle's fundamental reflection in *The Categories* concerning what things are and what things are called. Timothy begins with

3 Given the current state of research, this theme cannot be dealt with in a comprehensive manner. Notably, there is a lack of studies of letter 1. As for letter 2, a primary fundamental study has been published in 2015: Berti, Vittorio, *L'au-delà de l'âme et l'en-deça du corps: Approches d'anthropologie chrétienne de la mort dans l'eglise syro-orientale.* Paradosis 57. Fribourg: Academic Press, 2015. For editions of the letters see: Heimgartner, Martin, *Die Briefe 42–58 des ostsyrischen Patriarchen Timotheos I.* Corpus Scriptorum Christianorum Orientalium 644. Leuven: Peeters, 2012 (*Einleitung, Übersetzung und Anmerkungen*, CSCO 645, Leuven: Peeters, 2012); quoted from now on as "Heimgartner, CSCO 644" and "Heimgartner, CSCO 645". Braun, Oskar, ed., *Timothei Patriarchae I Epistulae I.* Corpus Scriptorum Christianorum Orientalium Syri 2,67 Textus. Paris: J. Gabalda, 1914 (Reprint: CSCO 74/75 Syri 30/31, 1953).

two questions: Is Christ the predicate in question or not? Does Scripture use the predicates in question for Christ or not?

Using this double *diairesis*, the Patriarch arrives at a fourfold classification whose formal structure corresponds to the four predicables in Aristotle, *Topics* 1,8. This is no great surprise, considering that Timothy translated the *Topics* into Arabic by order of the caliph al-Mahdī.

	What Christ is:	What Christ is not:
What Scripture predicates on Christ:	a) statements to be understood literally	d) statements to be understood metaphorically
What Scripture does not predicate on Christ:	c) dogmatic formulas of the Church	b) contradictions in terms

a) What Scripture says about Christ and what he really is, including statements to be understood literally, for example, "God," "Man," "Son," and "Christ" (34,3,5).[4]
b) What Scripture does not say about Christ and what he is not. The second point encompasses predicates like "darkness." Timothy not only names the predicates as he did before but combines them with common predicates for Christ in order to form contradictions in terms. Using the predicate "true light" from John 1:9, he frames the sentence: "The light of truth is darkness" (34,3,15). The Biblical background can also be found in the following sentences; however, the statements become more and more absurd: "The square is a circle" (34,3,9) and "Two and two are fourteen" (34,3,20).[5]
c) What Scripture does not say about Christ, but what he still is. Here Timothy mentions key dogmatic formulas valid for Christ, but not to be found in the Bible. These include him being consubstantial with the Father, in so far as he is the Word (Logos); him being consubstantial with human beings, in so far as he is man; the two substances, respectively, natures of Christ (divinity and humanity); the one potency, the one will, and the one energy (34,3,22). The last example shows the monotheletic-monenergistic position of the East-Syrian Church.
d) What Scripture does say about Christ, but what he is not. Finally, Timothy mentions the predicates of Christ to be understood in a metaphorical sense (34,3,24–36) "stone" (Ps 118:22 = Matt 21:42; Dan 2:34); "sun" (Mal 3:20); "star" (Num 24:17); "sprout" (Isa 11:1); "shoot" (Isa 11:1); "door"

4 Examples: Bar 3:38; Jn 1:1; Isa 9:5; 1 Tim 2:5; 1 Cor 15:47; Isa 7:14; Dan 9:25f (34,3,6–13).
5 So "truthfully", "imperishable", "being" (34,4,16–18).

(John 10:7); "way" (John 14:6); "sin" (2 Cor 5:21); "curse" (Gal 3:13); and, of course, the key term of the entire letter—"servant."

In 34,4,16 f., the term "servant" in the context of John 8:34 is brought into an intimate correlation with sin.[6] It is inconceivable that Christ is equal to men in regard to sin.[7] That Christ is equivalent to men in regard to being a servant is also unimaginable. To speak of Christ being a servant while facing Islam is possible only as a peak statement, for instance, that Christ became sin and that he became a curse (compare 2 Cor 5:21 and Gal 3:13). According to Timothy, this form of speech should no longer be used. Thus, the 'four class pattern' also enables the classification of dogmatic expressions that are unfamiliar to the Bible. On the other hand, the metaphorical sense in this pattern turns out to be one of four special cases juxtaposed on the background of "being" and "being named."

1.2 The "Five Reasons"

Subsequent to this fourfold pattern, we find a second systematization. Timothy provides "five reasons" for metaphorical language usage (34,3,37)—to some extent furnishing copious examples.[8] These still address the prevailing topic of letters 34–36: why Christ may be called a "servant." The term "servant" fashions the example in all five groups.

The first reason is that something is named according to the "nature" wherefrom it comes and not according to its "individuality." Thus, the human is called a "seed" (Gen 22:18 in 34,3,49), the pearl is "water," and gold and silver are "dust" (34,3,39).[9] If such predicates are applied to Christ, they refer to the "nature" common to all of humankind and not to the "individuality" of Christ, that is, to his specific humanity.[10]

6 "Now, the defect of a servant of sin infects the whole beauty of human nature and covers it so that [human] nature is thereby called a servant as well – as our Lord also taught: "'Everyone who commits sin is the servant of sin.'" (John 8:34) 4,17 Hence, our Lord accepts the peculiarity of [human] nature together with [human] nature, 'yet without sin' (Heb 4:15), and he connects them [both] with his human individuality. At the same time, he accepts [human] nature together with [human] individuality without any disturbance. This is why our Lord assumed human individuality and human nature without servanthood." (1 Tim, Letter 34,4,16 f.)

7 "Whoever commits the sin, is a servant of the sin." (John 8:4)

8 Survey of the "five reasons" in 34,3,38. See detailed presentation in 34,3,39–80.

9 The passage is augmented with thoughts on the two natures (34,3,41–56) left apart here.

10 This differentiation of "according nature" and "according individuality" is found previously in letter 34,3,22–23 as an example of a possible form of non-verbal language usage.

Second, the prototype and image are often hastily mixed in relation to prophetic words in the Old Testament (34,3,57 f.). Isa 42:1 is the paramount example: "Behold, my servant, with whom I am well pleased." In a strict sense, the term "servant" applies to Zerubbable to whom Timothy attributes the quotation. However, if the saying is transferred to Christ, the term "servant" must be changed whenever it is to be understood in a strict sense. Therefore, Matt 3:17[11] reads: "Behold, my beloved son, with whom I am well pleased" (34,3,59 f.). Consequently, Ps 72:1, Isa 52:13 and Dan 7:13 have to be interpreted in a similar manner. Timothy provides a hermeneutical tenet for this purpose. He claims that sayings declaring Christ's highness—such as "king" in Ps 72:1—apply to Christ in the strict sense; while, sayings about his lowliness have a metaphorical sense as does the term "servant" in Isa 52:13. In the case of the Maccabees, it is the other way around; sayings of lowliness apply to them in a strict sense, and sayings of highness in a metaphorical sense (34,3,61–64). The tenet should not be mixed with a later one concerning the interpretations of expressions of highness and lowliness in relation to the two natures of Christ mentioned below.

Three further reasons for metaphorical predication include that for the sake of humankind, Christ has fulfilled the law voluntarily (Gal 4:4 in 34,3,65–67); he wanted to teach humility and readiness to help men through the example of his own speech and actions, depicted paradigmatically by the foot-washing (John 13:4 f. in 34,3,68–72); he voluntarily took upon himself suffering, crucifixion, death and the grave for the sake of humankind.

The "five reasons" give more of the impression of a parenetical-rhetorical intensification than of a real systematization. Notably, reasons 3–5 show a certain uniformity. For Christ bows to a certain level for the sake of humankind, although he does not need to; he accepts the law (reason 3). Furthermore, he performs deeds of humility (reason 4) and takes suffering and death upon himself (reason 5). Here the various stages of his earthly life are attributed to the different "reasons" for the sake of rhetoric. The same technique appears in other letters in which Timothy repeatedly presents the same argument in different wrapping.[12]

---

11 See also Matt 17:5; Mark 1,11; 9:7; Luke 3:22; 9:35.

12 So for instance: ep. 42,3. Compare Heimgartner, CSCO 645, XXX–XXXI. Also see: Heimgartner, Martin, *Timotheos I., ostsyrischer Patriarch: Disputation mit dem Kalifen al-Mahdī. Textedition*, Corpus Scriptorum Christianorum Orientalium 631, Syri 244. Leuven: Peeters, 2011, 7,25–30 (quoted from now on as "Heimgartner, CSCO 631". Compare: Heimgartner, Martin, *Einleitung, Übersetzung und Anmerkungen*, Corpus Scriptorum Christianorum Orientalium, p. 35 n. 128 (quoted from now on as "Heimgartner, CSCO 632"; Syri 245).

2 The Method of Interpretation "By Addition"

Timothy often uses a method of interpretation by adding words and parts of sentences to explain and clarify the sense of a Bible quote.[13] Berti characterizes this with *tosefta,*[14] an appropriate term as Timothy uses the verb *tausep* (to add). Nonetheless, the term *tosefta* used here has to be distinguished from the specific Jewish term *tosefta.*[15] Timothy not only uses this method for Bible exegesis but also for interpreting patristic texts, as letters 38 and 39[16] paradigmatically demonstrate.[17] Moreover, he does so again in letter 42,4 in his explanation of a paragraph of Aristotle's *Categories.*[18] We find two prime examples in letter 36,3,20 where Timothy specifies the Bible verse Luke 22:42, "Not my will, but thy will shall be done!" as "'Not my will shall be done', for I don't have a will, separated from thine, 'but' this one 'will' shall be done, which is common to thee and me, which is life and salvation for all!"[19] Likewise he explains John 6:38 "'I descended from heaven not to do my will' separately,—for I don't have a particular and separate will—'but to fulfil the will of him, who sent me', for his will and my will are one, for whose sake I accepted to become man, to suffer and to die for the

13 Already delineated by Isebaert-Cauuet, Isabelle, "Les pères dans les commentaires syriaques," in *Les Pères grecs dans la tradition syriaque*, ed. by Andrea Schmidt and Dominique Gonnet, pp. 77–88. Études syriaques 4. Paris: Geuthner, 2007, here p. 80.

14 Berti, *Vita*, pp. 351–353. Once, he uses the noun *tāwseptā*, but not as a *terminus technicus.* In 38,3 he speaks about an "illuminating addition".

15 Berti, *Vita*, p. 353.

16 "Hereby, we explain the chapter itself first with an illuminating addition of those words, which are absolutely necessary for understanding, and then we speak briefly about the meaning of the chapter. 4 "'If someone says that Christ perfected himself by deeds' and is not God's son by nature and according to nature, 'or that after baptism or after his resurrection from the dead he was acknowledged as the Son, as those' [teach], who being 'pagans', respectively Arians, 'introduce the [heros] illegally listed [as Gods]', and does not agree that Christ has been God's Son forever and eternally, 'then he be cursed. 5 For he, who neither began in virtue nor developed nor perfected himself is not' by nature God, 'even when he is named so because of his gradual manifestation.' (Gregor Nazianzene ep. 101,23 f) 6 But he who, neither begins nor perfected himself in virtue, is not human by nature even though he is named so because of the incarnation and his becoming a human. 7 All this words must thereby be added in order to clarify the teacher's thoughts." (Timothy, Letter 38,3–7).

17 See: Isebaert-Cauuet, *Pères*, p. 80, and Berti, *Vita*, pp. 351–357. Also consider: Heimgartner, Introduction to *Briefe 30–39*, chapter 1.3.

18 See: Heimgartner, CSCO 645, pp. 13–17, esp. the explanations on pp. XXXI–XXXVIII.

19 I use single quotation marks for the quotation.

salvation and the life of all" (36,3,20). The same method of interpretation "by addition" is to be found in Qur'ān exegesis.[20]

3 Scientific Bible Studies

Timothy reflects on the disagreements between the various translations of the Old Testament, notably the Greek and the Peshitta, in an extraordinary manner for the end of the 8th century. Yet, he perceives the topic to be reserved for experts; those who listen to his letter in Baṣrā and Huballat are not expected to realize the textual differences in Ps 110:3. However, in letter 36 to Naṣr, the well-educated spokesman of the community, Timothy gives a detailed exegesis on the divergent readings. East-Syriac Christology is the principle of interpretation. In any case, the eternal birth from the Father and the temporal birth through Mary must be juxtaposed.

The same is to be found in the *Disputation with al-Mahdī.*[21] He denies the caliph's accusation that the Christians should have adulterated Scripture. He says that there is no single point of difference between the Christian and the Jewish Bible text. [22] However, this is untrue as he is caught citing Ps 110:3, once according to the Greek and in another case following the Syriac text.[23] The background to this are Timothy's studies of the Syro-Hexapla;[24] we know that the court official Gabriel Būḥtīshū' provided him with a copy of the Syro-Hexapla.[25] In letter 47, Timothy describes the adverse conditions of copying the manuscript

---

20 See: Sinai, Nicolai, *Fortschreibung und Auslegung. Studien zur frühen Koraninterpretation.* Diskurse der Arabistik 16. Wiesbaden: Harassowitz, 2009, pp. 201–206.

21 Critical edition and German translation: Heimgartner, CSCO 631. Cf. also the older edition and English translation: Mingana, Alphonse, *Woodbroke Studies: Christian Documents in Syriac, Arabic, and Garshūni. Volume 2: 1. Timothy's Apology for Christianity, 2. The Lament of the Virgin, 3. The Martyrdom of Pilate.* Cambridge: W. Heffer & Sons Limited, 1928 (reprint in: *The Early Christian-Muslim Dialogue: A Collection of Documents from the First Three Islamic Centuries (632–900 A.D.). Translations with Commentary*, ed. by N. A. Newman. Hatfield: Interdisciplinary Biblical Research Institute, 1993, pp. 169–267).

22 Heimgartner, CSCO 631, 13,47.

23 Heimgartner, CSCO 631, 13,30 (Greek textual form) and 19,13 (Syriac textual form). The relevance of the textual differences for the apologetic context was already mentioned by R. Bas ter Haar Romeny, "Biblical Studies in the Church of the East: The Case of Catholicos Timothy I," in *Papers presented at the Thirteenth International Conference on Patristic Studies held in Oxford 1999: Historica, Biblica, Theologica et Philosophica*, ed. by Maurice F. Wiles and Edward J. Yarnold, pp. 503–510. Studia Patristica 34. Leuven: Peeters, 2001, here pp. 508–510.

24 See: Heimgartner, CSCO 645, p. 63 n. 299.

25 Ep. 47,2.

in detail.[26] Likewise, Timothy lends insight into his reflections on the divergence between the various Bible revisions.[27] In so doing, he also tells about the discovery of Bible manuscripts near the Dead Sea from which he had heard.[28] Timothy hopes that these manuscripts would contain the original text, which, in his opinion, would be represented by the Hebrew, Greek and Syriac text in three different variations.[29] The problem of the textual differences occupied Timothy in an existential manner; indeed, it tortured him. "This has become like a fire in my heart, burning, and glowing in my bones" (letter 47,29). Only well-educated Christians like Naṣr are expected to endure the difficult topic of the textual differences. *Vis-à-vis* to the Muslim accusations that Christians had adulterated Scripture, Timothy found no alternative in the religious dispute than to systematically deny the differences between the Christian and Jewish textual traditions.

4 Christian Qur'ān Exegesis[30]

John of Damascus and Theodor Abū Qurra spoke about Muḥammad and the Qur'ān with disdainful words.[31] Thus, it is surprising to find a Christian Timothy interpreting Qur'ān verses. Most of the examples are in his *Disputation*,[32] but Letter 34 also provides remarkable evidence therefore.[33] Timothy thought that

---

26 Ep. 47,3–12.

27 Ep. 47,5.13.24–29.

28 Ep. 47,16–20. For the interest in this text after the discoveries in Qumran see: Heimgartner, CSCO 645, p. 67 n. 324.

29 See also ter Haar Romeny, "Studies", p. 510.

30 I outlined these thoughts earlier in: Heimgartner, Martin, "Die Disputatio des ostsyrischen Patriarchen Timotheos (780–823) mit dem Kalifen al-Mahdī," in *Christians and Muslims in Dialogue in the Islamic Orient of the Middle Ages: Christlich-muslimische Gespräche im Mittelalter*, ed. by Martin Tamcke, pp. 41–56, particularly 48–50. Beiruter Texte und Studien 117. Beirut: Ergon Verlag Würzburg in Kommission, 2007.

31 Compare: John of Damascus, haer 100, in *Die Schriften des Johannes von Damaskos*, ed. by Bonifatius Kotter, vol. 4. Patristische Texte und Studien 22. Berlin: De Gruyter 1981, pp. 60–67, particularly 11.16.32.48.152; Abū Qurra, Theodor, "Opusculum 20," in *Johannes Damaskenos und Theodor Abū Qurra, Schriften zum Islam*, ed. by Reinhold Glei und Adel-Théodore Khoury. Corpus Islamo-Christianum, Series Latina 3. Würzburg: Echter, 1995.

32 Compare: Q 46:9 in Heimgartner, CSCO 631, p. 7,33 f.; Q 4,157, 19,33 and 3,55 in Heimgartner, CSCO 631, p. 9,17–25; Q 4:157 in Heimgartner, CSCO 631, p. 9,35–48 and others.

33 Compare: Q 4:171 in Letter 34,6,15–20: "15 If Jesus is God's Word and Spirit (cf. Q 4:171) and God is Lord, then it follows that Jesus is Lord. 16 No Lord is a servant. Then Jesus is also no servant. 17 Moreover, if Jesus is the Word and Spirit of God, and if every Word and Spirit are equal to the essence of the one they belong to, then Jesus is of the same essence as God. How can he be a

Muḥammad had not received revelation, but that his teaching is in harmony with the Old Testament prophets. Thus, the doctrine of the Qur'ān refers to Christ's revelation in the New Testament in the same way as the Old Testament does. Therefore, the Qur'ān has the character of truth but not revelatory character.[34] The status of the Qur'ān in Timothy seems to amount to the rank of an Old Testament Apocrypha, or as Luther once said in another context, they "are not equal to Holy Script," but for the Arabs, "still useful and good to read."[35] The concept of a Christian exegesis of the Qur'ān is a reply to a Muslim Bible exegesis that is already rudimentary to the Qur'ān,[36] in the Sīra of Ibn Ishāq,[37] in Timothy in the votes of the Caliphe al-Mahdī[38], and after that in Hārūn's letter to Emperor Constantine IV.[39] By no later than 'Alī ibn Rabban in 855, the examples are abundant.[40] Timothy recognises traces of the Trinity in several passages of the

---

servant when he is the same essence as God! 18 Moreover, if Jesus is God's Word, and David said: 'I want to praise God's word!' (Ps 56:10), thus Jesus deserves to be praised. 19 If Jesus is furthermore the Word and Spirit of God, as it says in your writings (cf. Q 4:171), and 'by the word of the Lord the heavens were made, and by the spirit of his mouth all the powers of heaven' (Ps 33:6), as the Prophet David said, then Jesus created the heavens and all the powers of heaven. 20 If he is creator of all, he is also Lord, and if he is Lord of all, he is also creator of all: How then should he be called a servant!" Moreover, compare Q 3:55 in ep 34,6,21–23: "21 Furthermore, if Jesus is in heaven, and heaven is God's throne, according to Scripture (compare Is 66:1), Jesus thereby sits on God's throne. 22 For they say, that God says to him: 'Behold, I will make thee die and raise thee to Myself.' (Q 3:55). He says not only: 'I will make thee die and raise thee', but he adds 'to Myself' because God has lifted up and lifts up many angels and human beings, who are worthy of this, to himself, but only of Jesus does he say that he lifts him up 'unto him' in a special, that is, particular way. 23 Now they mention seven heavens. Which of these [three possibilities apply]: Is God in all seven heavens or in one –namely, the highest and most high or is he infinite in all things and beyond everything, in so far as he is infinite?"Cf. also Q 4:171 in 34,6,33–38 as well as Q 3:49 and Q 5:110 in 34,6,42–47.

34 Heimgartner, "Disputatio", pp. 48–50, particularly 48.

35 Heimgartner, "Disputatio", pp. 48 f., the quotation 49.

36 Compare the quotation of Jesus in Q 61:6.

37 Wüstenfeld, Ferdinand, ed., *Das Leben Muhammed's nach Muhammed ibn Ishâk bearbeitet von Abd el-Malik ibn Hischâm. Aus den Handschriften zu Berlin, Leipzig, Gotha und Leyden*, vol. 1. Göttingen: Dieterichs, 1864, pp.149–150 (Arab text); *The life of Muhammad: A Translation of Isḥāq's Sīrat Rasūl Allāh*, with introduction and notes by A. Guillaume. London: Oxford University Press, 1955, pp. 103–104 (English translation).

38 Heimgartner, CSCO 631, p. 7,24 (the Paracletus is Muḥammad); 8,39 (interpreting Isa 21:7 on Muḥammad).

39 See: Eid, Hadi, *Lettre du calife Hârûn Al-Rašîd à l'empereur Constantin VI: Texte présenté, commenté et traduit par Hadi Eid, préface de Gérard Troupeau*. Études chrétiennes arabes. Paris: Cariscript, 1992, 46.51–52.72–76, p. 46, pp. 51–52 and pp. 72–76.

40 Fritsch, Erdmann, *Islam und Christentum im Mittelalter: Beiträge zur Geschichte der muslimischen Polemik gegen das Christentum in arabischer Sprache*. Breslauer Studien zur historischen

Qur'ān referring to God in plural like "we made." They are likewise to be found in the enigmatic three letters in the headlines of some sūras. Timothy withholds the fact, that there are also sūras with two, four, or five letters. The most important verse for him is Q 4:171, where he believes to find the second and third person of the Trinity named as word and spirit, "Christ Jesus, the son of Mary, is the ambassador of God; he is his Word, sent to Mary, and spirit of him." Timothy's main idea is that in the Qur'ān God's unity is pronounced openly, and God's trinity only in a veiled form because the Arabs would have misunderstood it as polytheism.

5 References to Jewish Exegesis

Muslims are Timothy's main interlocutors in the question of Bible exegesis. We find him conversing with Judaism only marginally. The Patriarch actually had a lively interest in the differences between the Christian Bible and the Hebrew Bible, which he consistently denied in front of the Muslims. Only one explicit controversy with Jewish exegesis is to be found; it is the explanation of Ps 110:1 in Letter 36,1,8 – 31. The long passage is riddled with problems that cannot be examined in detail here.[41] The following point must suffice. Timothy assumes נאם־יהיה to be the original Biblical text. He understands the double יה of the tetragram in this formula to be a double word: "Lord." "The Lord spoke to the Lord." Thus, the tetragram elementally mirrors the inner-Trinitarian face-to-face relation of God and the Logos, both being Lord equally. Timothy, therefore, assumes that the Jews mutated the text by complementing לאדני because,

---

Theologie 17. Breslau: Müller & Seiffert, 1930, pp. 6 – 12 and 77 – 94; Thomas, David, "'Alī l-Ṭabarī," in *Christian-Muslim Relations: A Bibliographical History*, ed. by David Thomas and Barbara Roggema, pp. 669 – 674. History of Christian-Muslim Relations 11. Leiden: Brill, 2009.

41 See my explanations in the footnotes to the translation of this text. As an example, I give here only ep. 36,14 – 18. Please note that here small capitals are used to distinguish Syriac words not in the emphatic case (i.e., *casus constructus* or *absolutus* or with possessive suffix signifying the vocative case): "14 In the same way 'Lord' appears rather [as a title] to elevate than to humiliate. 15 For behold: also in Hebrew, in which this verses were first written, they read the same word and the same characters differently and with altered recitation ... 17 and [they change the recitation] although in their hostility and bitter attitude they replaced 'Lord' through 'JH' and 'Lord' through 'Adon(y)', while in their ignorance and dull wittedness they ascribed the greater glory to the nature of God and the lesser honor to Christ or to any other person to whom they interpreted the Psalm. 18 For they believed—as they still believe—that it is impossible to ascribe the word 'JH' to any person, for they have a Jewish understanding of God as one nature and as one individuation."

according to Jewish belief, God exists as one single divine nature in one single individuality (hypostasis). Thereby, an interpretation of the prepositional object of a non-divine person became possible. Thereafter, when the Jews recited the text, they always read “adonai” instead of the Holy Name of God. According to Timothy, they thereby rescinded their mutation of the text inasmuch as the verse again corresponded to the original structure; however, other vocabulary was used—namely, the previous usage of two יה became twice אדונ(י). So, instead of נאם־יהיה they read נאם־אדוני לאדונ(י), “The Lord spoke to the Lord.” Thus, according to Timothy, the Hebrew text conforms to the Christian version, contrary to the intentions of the interpolators.[42]

6 Hermeneutic Patterns: The Interpretation of Sayings About Christ’s Highness and Lowliness

Space does not allow to situate Timothy’s position in the broader context of East-Syriac Biblical exegesis. A few final remarks must suffice. In the context of their editions of the commentaries of Īshū‘dād of Merv, Diettrich (1902) and van den Eynde (1955) indicate a methodological turnaround.[43] In the middle of the 9th century, Īshū‘dād favours an allegoric understanding of Scripture in contrast to the verbal interpretation method of Theodor of Mopsuestia. Both Diettrich and van den Eynde ascribe this to East-Syriac exegetical impulses from Ḥenānā the Adiabenite.[44] The author, who was condemned under Sabrīshū‘, is cited abundantly in Īshū‘dād’s commentaries. Indeed, the latter is a rich source of fragments of Ḥenānā’s many lost writings.

42 Compare: ep. 36,1,25: “I have said these things in order to show that one and the same Lordship appears in the flesh and in the Word in Hebrew too: in the former according to the union and the kingdom, in the latter according to nature and the essence; not in the sense of progress and development, but in the sense of the nature and union beginning with the first] movement of becoming and of incarnation onwards.”

43 Diettrich, Gustav, *Išô‘dâdh’s Stellung in der Auslegungsgeschichte des Alten Testamentes an seinen Commentaren zu Hosea, Joel, Jona, Sacharia 9–14 und einigen angehängten Psalmen veranschaulicht*. Beihefte zur Zeitschrift für die alttestamentliche Wissenschaft 6. Giessen: Ricker’sche Verlagsbuchhandlung, 1902; van den Eynde, Ceslas and Jacques-Marie Vosté, eds., *Commentaire d’Išo‘dad de Merv sur l’Ancien Testament: I. Genèse*, traduit par Jacques-Marie Vosté and Ceslas van den Eynde. Corpus Scriptorum Christianorum Orientalium 156, Syri 75. Louvain: Durbecq, 1955.

44 See: Diettrich , *Stellung*, pp. LXII–LXV, and Vosté/ van den Eynde, *Commentaire*, pp. XI–XII.

Timothy cannot be defined according to the alternatives of literal and metaphoric sense. The reason for this is not only the fact that we do not have actual commentaries from his own pen, except for the detailed exegesis in 36,1,2. Moreover, his explanations are strongly concentrated on terminology—something that reveals his Aristotelian background again. Despite his vivid interest in Biblical exegesis and textual research, he is ultimately probably less an exegete than a systematic theologian. His interpretations are always dedicated to the dogmatic statements of East-Syriac theology. He is guided by his hermeneutic concept: howsoever the text is interpreted, it must conform to East-Syriac theology. The term "servant" refers to Christ in an improper sense; the term "Lord" in a proper sense. His textual variants show the outcome to be the same. The unity of God is uttered openly in the Qur'ān, the Trinity cryptically. Sayings about the two wills of God and Christ are harmonised with the East-Syrian monotheletism by the method of addition.

Timothy uses a particularly elaborate hermeneutical concept to exegete the sayings concerning Christ's highness and lowliness. The method was previously applied by Diodore of Tarsos, who, to some extent, comes up with a pattern of interpretation which refers sayings of highness to the earthly Jesus only "according to grace." By the same token, Diodore refused the application of earthly predicates to the Logos as inappropriate abusive use of language, καταχρηστικῶς.[45] Only Mary's traditional title of "God bearer" is tolerated by Diodor; it can be used because of the "union" with the Logos.[46]

Timothy, however, employs a fully developed hermeneutical pattern of interpretation that distinguishes the two relations "according to nature" and "according to union."[47] The sayings of highness apply to the Logos "according to nature"

45 L 4, B 27, S 3, the numeration of the fragments follows Behr, John, *The Case against Diodore and Theodore: Texts and their Contexts.* Oxford Early Christian Texts. Oxford: Oxford University Press, 2011, there pp. 168–195 (B), 234–243 (S) and 310–313 (L).

46 The fragment in Tetz, Martin, *Eine Antilogie des Eutherios von Tyana*, Patristische Texte und Studien 1. Berlin: Walter de Gruyter, 1964, p. 62, likewise in Behr, *Case*, p. 160. For Diodor cf. also Heimgartner, Martin, "Neue Fragmente Diodors von Tarsus aus den Schriften 'Gegen Apollinarius','Gegen die Manichäer' und 'Über den heiligen Geist'," in *Apollinarius und seine Folgen*, ed. by Silke-Petra Bergian, Benjamin Gleede and Martin Heimgartner, pp. 211–232. Tübingen: Mohr Siebeck, 2015.

47 Essentially, it has to do with the Aristotelian principle of non-contradiction ("It is not possible that the same in the same way both become and not become at the same time." (τὸ γὰρ αὐτὸ ἅμα ὑπάρχειν τε καὶ μὴ ὑπάρχειν ἀδυνατὸν τῷ αὐτῷ καὶ κατὰ τὸ αὐτόν [Aristotle, *Metaphysics Γ3*, 1005b,19 f.]), W. D. Ross, *Aristotelis Metaphysica*, (SCBO) Oxford University Press, Oxford, 1957 (reprinted) Übersetzung M.H., nearly verbal quotation in Timothy, letter 35,8,14. The sayings

and to the earthly Jesus only "according to union." The formula is reversed concerning the sayings of lowliness. To the earthly Jesus, they apply "according to nature," but, to the Logos, only "according to union." Thereby, Timothy has an effective hermeneutical instrument to harmonise any problematic Biblical verse, to clarify many of the Father's allegedly vague quotations, and to bring them into unison with East-Syriac Christology.

In conclusion, Timothy's essential hermeneutical interpretive principles are always the East-Syriac dogmatics. The main elements of these are the Trinity, the Christology of the two natures, the one will, and the one energy. The examples show us that in accord with the Old Church, Timothy is not a Biblicist or even a fundamentalist. Rather, he understands the Bible and the patristic texts as a "visualisation" of all the truth approved by his community.[48] In light of Timothy's faith, which was confirmed by his community, the Bible and the Fathers' writings are read, interpreted, and, if necessary, adjusted.

of highness and of lowliness, categorically incompatible, are reconcilable with each other, if they are not valid "in the same relation".

48 My former teacher Hans-Dietrich Altendorf used to call this: "den gegenwärtigen Glaubensstand der Gemeinde," (the present state of the faith of the community).

Mark N. Swanson

Scripture Interpreting the Church's Story

Biblical Allusions in the *History of the Patriarchs of Alexandria*

1 The History of the Patriarchs of Alexandria as a Literary Composition

For students of the medieval history of the Coptic Orthodox Church in Egypt, an indispensable source is the Arabic-language compilation usually known as *The History of the Patriarchs* (*tārīkh al-baṭārikah*).[1] Although both among Copts and in the wider scholarly world it is still a reflexive habit to attribute the core of the work to the 10th-century Coptic bishop and theologian Severus or Sāwīrus ibn al-Muqaffaʿ,[2] the studies of scholars such as Kāmil Ṣāliḥ Nakhlah, David W. Johnson, and Johannes Den Heijer have not only sorted out the compositional history of the work but have shown the attribution to Sāwīrus to be false.[3] The work (in

---

1 Abbreviated as *HP* in the following. For an initial orientation to this text, see Den Heijer, Johannes, "History of the Patriarchs of Alexandria," in *The Coptic Encyclopedia*, ed. by Aziz Suryal Atiya, Vol. 4, pp. 1238–1242. New York and Toronto: Macmillan, 1991. The most widely used edition of the text is Evetts, Basil Thomas Alfred, ed., *History of the Patriarchs of the Coptic Church of Alexandria*, Patrologia Orientalis (PO) vol. 1, fasc. 2; vol. 1, fasc. 4; vol. 5, fasc. 1; and vol. 10, fasc. 5. Paris: Firmin-Didot, 1904–1915, abbreviated as *HPE* in the following. Continued by Yassā ʿAbd al-Masīḥ, Oswald Hugh Edward Burmester, Aziz Suryal Atiya et al., *History of the Patriarchs of the Egyptian Church, Known as the History of the Holy Church, by Sawīrus ibn al-Mukaffaʿ, Bishop of al-Ašmūnīn*, vol. 2–4. Cairo: Société d'Archéologie Copte, 1943–1974, abbreviated as *HPC* in the following. Perhaps because of the accompanying English translation, Evetts' edition has been more widely used than the competing edition of Seybold, Christian Friedrich, *Severus ben el Moqaffaʿ: Historia patriarcharum Alexandrinorum*, Corpus Scriptorum Christianorum Orientalium 52. Paris: Poussielgue, 1904. For the one printed edition of the "primitive" recension of *HP*, see below, note 15.

2 E.g., Atiya, Aziz Suryal, "Sāwīrus ibn al-Muqaffaʿ," in *Coptic Encyclopedia*, vol. 7, ed. by Aziz Suryal Atiya, pp. 2100–2102. New York: Macmillan, 1991, according to whom the *HP* "stands as a permanent monument to [Sāwīrus'] erudition and critical sense," p. 2101; or Trombley, Frank R., "Sawīrus ibn al-Muqaffaʿ and the Christians of Umayyad Egypt: War and Society in Documentary Context," in *Papyrology and the History of Early Islamic Egypt*, ed. by Petra M. Sijpesteijn and Lennart Sundelin, pp. 199–226. Leiden: Brill, 2004. Despite the title, Trombley is aware of modern discussions of authorship and throughout the essay refers to "ps.-Sawīrus."

3 The decisive study is Den Heijer, Johannes, *Mawhūb ibn Manṣūr ibn Mufarriğ et l'historiographie copto-arabe: Étude sur la composition de l'Histoire des Patriarches d'Alexandrie*, Corpus

https://doi.org/10.1515/9783110564341-005

its initial form) was, in fact, the accomplishment of a team led by the Alexandrian deacon Mawhūb ibn Manṣūr ibn Mufarrij, who in the year 1088 set out, with patriarchal blessing, to collect the relevant Coptic-language histories and then to translate and edit them into a single Arabic text containing the biographies of the first 65 patriarchs.[4] That accomplished, Mawhūb then composed, in Arabic, biographies of his contemporaries Christodoulos (the 66th patriarch, 1046–1077) and Cyril II (the 67th patriarch, 1078–1092).[5] And that was not the end of it; the work was regularly updated by continuators who added Arabic-language biographies to what had, in effect, become the semi-official history of the Coptic Orthodox Church.[6] The process continues down to the present day: during a recent visit to Cairo I came across a new three-volume printing of *The History of the Patriarchs* prepared by the Syrian Monastery in the Wādī al-Naṭrūn, which begins by reprinting an edition of the medieval *HP* (the 1st through the 74th patriarchs, up to 1216 CE), continues with Kāmil Ṣāliḥ Nakhlah's classic study of the 75th through the 111th patriarchs (1235–1906 CE), and continues on, concluding with a biography of Coptic Pope Kīrillos VI (the 116th patriarch, 1959–1971).[7] I am confident that we shall soon see an edition that includes a biography of Pope Shenouda III (the 117th patriarch, 1971–2012).

A great contribution of the scholars mentioned earlier is that they have enabled us to see, at least to a certain extent, behind the redacted Arabic text to its sources. We can assign names and fragments of biography to the authors of *HP*'s major sources, including those of the Coptic histories that Mawhūb and his team discovered.[8] Just so, we can now study *HP* both as a whole and in its component

---

Scriptorum Christianorum Orientalium 513. Leuven: Peeters, 1989, with full bibliography of earlier studies.

4 It is worth noting that this literary form, a series of biographies of successive patriarchs, is reflected in a title regularly found in the manuscripts: *Sīyar al-bīʿah al-muqaddasah* or *Lives of the Holy Church.*

5 Den Heijer, *Mawhūb*, pp. 81–116.

6 Den Heijer, *Mawhūb*, pp. 9–13, and see note 8 below.

7 *Silsilat tārīkh al-ābāʾ al-baṭārikah*, 3 vols., 2nd printing. Wādī n-Naṭrūn: Maktabat Dayr as-Suryān, 2011. Volume 1 reproduces the text of the Evetts–Cairo edition (see above, note 1); volume 2 reprints Kāmil Ṣāliḥ Nakhlah, *Silsilat Tārīkh al-bābāwāt baṭārikat al-kursī al-iskandarī*, 5 fascicles, 2nd printing. Wādī an-Naṭrūn: Dayr as-Suryān, 2001 (1951–1954); and volume 3 contains the work of more recent continuators.

8 In addition to the works of Den Heijer mentioned above, see now individual entries for sources of the *History of the Patriarchs* in Thomas, David et al., eds., *Christian-Muslim Relations: A Bibliographical History*, vols. 1–5. Leiden: Brill, 2009–2013, abbreviated as *CMR* in the following. See the entries of Swanson, Mark N., "George the Archdeacon," in *CMR* vol. 1, pp. 234–238; "John the Deacon," in *CMR* vol. 1, pp. 317–321; "John the Writer," in *CMR* vol. 1, pp. 702–705; "Michael of Damrū," in *CMR* vol. 3, pp. 84–88; "Mawhūb ibn Manṣūr ibn Mufarrij," in *CMR*

sources, with attention to the particular voice, characteristic emphases, and specific historical-theological assumptions of each individual author.[9] At the same time, there are unifying elements within the work as a whole. Most of the authors of patriarchal biographies were aware of the contributions of their predecessors; some had, in fact, copied out the entire *History* before bringing it up to date with new biographies.[10] And so, themes developed earlier in the compilation are often carried forward in later contributions.

I began by saying that the *History of the Patriarchs* is an indispensable source for the medieval history of the Coptic Orthodox Church, and indeed, much modern writing about that history consists largely in paraphrases of the work; this is true not only in the Arabic-language histories written in Egypt but also in Western writing since 1713, when Eusèbe Renaudot published his *Historia patriarcharum Alexandrinorum Jacobitarum* in Paris.[11] But while it has long been tempting to regard the work primarily as a reservoir of facts (many not available elsewhere), regularly to be repeated in various configurations, it is useful to examine *HP* as a literary composition, with attention to the characteristics of each component part and to how these component parts work together as a whole. Within the compilation as a whole we find what Johannes Den Heijer has recently termed "internal intertextuality," as certain themes, motifs, and patterns carry forward from earlier contributions and play their role in shaping later ones.[12] For example, I have argued that the well-known account of the initial encounter between the Arab conqueror of Egypt, ʿAmr ibn al-ʿĀṣ, and the just-returned-from-exile patriarch Benjamin, provided a paradigm which has echoes

vol. 3, pp. 217–222; "Ibn al-Qulzumī," in *CMR* vol. 3, pp. 409–413; "Marqus ibn Zurʿa," in *CMR* vol. 3, pp. 643–647; "Maʿānī ibn Abī l-Makārim," in *CMR* vol. 5, pp. 679–683, and "The Life of Patriarch Matthew I," in *CMR* vol. 5, pp. 396–401, and Moawad, Samuel, "Yūḥannā ibn Wahb," in *CMR* vol. 4, pp. 317–319.

9 One attempt to take this particularity seriously is Swanson, Mark N., *The Coptic Papacy in Islamic Egypt (641–1517)*, The Popes of Egypt 2. Cairo and New York: American University in Cairo Press, 2010, where the chapters largely correspond to *HP*, taken source by source.

10 For example, the first continuator after Mawhūb, Yūḥannā ibn Ṣāʿid ibn Yaḥyā ibn Mīnā known as Ibn al-Qulzumī, explicitly states that he copied out the entire work before adding his own contribution. See *HPC* vol. 2, pp. 232–233 (Arabic text), pp. 369–370 (English translation).

11 For more on this point, see Swanson, Mark N., "Reading the Church's Story: The ʿAmr–Benjamin paradigm' and its echoes in *The History of the Patriarchs*," in *Copts in Contexts: Negotiating Identity, Tradition, and Modernity*, ed. by Nelly van Doorn-Harder, pp. 157–168. Columbia: University of South Carolina Press, 2017.

12 See Den Heijer, Johannes, "*The Martyrdom of Bifām Ibn Baqūra al-Ṣawwāf* by Mawhūb ibn Manṣūr ibn Mufarrij and its Fatimid Background." *Medieval Encounters* 21 (2015): pp. 452–484, here p. 466.

in accounts of Muslim ruler – Coptic patriarch encounters throughout *HP*;[13] I will return to this argument in what follows. But there are other intertextual elements in *HP*, in particular, quotations from and allusions to the Bible and to other Christian literature, hagiographical texts in particular. A good example here comes in *HP*'s account of the tribulations of Patriarch Zacharias (64th patriarch, 1004–1032) during the persecution unleashed by the Fatimid caliph al-Ḥākim (r. 996–1020): at one point in the account, the patriarch was thrown to the lions, in a scene that not only echoes the Biblical story of *Daniel in the Lions' Den* (Dan 6) but also has verbal echoes of an account of St. Thecla being thrown to the beasts.[14]

2 Archdeacon George and his Use of the Bible

In the present essay, I would like to make a small contribution to a study of the relationship between *HP* and the Bible. It will, in a sense, not so much be an essay about the history of the Coptic Church's exegesis of the Bible as it will be about the Bible's exegesis of the Coptic Church's history—which, of course, for the medieval Copts, is the Church's history within the *dār al-Islām*. To make the essay manageable, I will restrict myself to a single one of *HP*'s sources: the contribution of one Archdeacon Jirjah or George, who was the scribe of Patriarch Simon I (the 42nd patriarch, 692–700).[15] His history began with lives of patriarchs around the time of the Council of Chalcedon in the mid-5th century, and continued through to the biography of Patriarch Simon at the very end of the 7th

13 Swanson, "Reading the Church's Story."

14 Swanson, Mark N., "Sainthood Achieved: Coptic Patriarch Zacharias according to The History of the Patriarchs," in *Writing 'True Stories': Historians and Hagiographers in the Late-Antique and Medieval Near East*, ed. by Arietta Papaconstantinou, Muriel Debié and Hugh Kennedy, Cultural Encounters in Late Antiquity and the Middle Ages 9, pp. 219–230. Turnhout: Brepols, 2010, here pp. 228–229.

15 On this George, see Den Heijer, *Mawhūb*, pp. 7–8; Swanson, "George the Archdeacon," in *CMR* vol. 1, pp. 234–238. The most convenient edition (with English translation) of the part of the *HP* for which George was the principal source is Evetts, *History*, vol. 2, *Peter I to Benjamin I (661)*, Patrologia Orientalis 1.2 (1904): pp. 445–518; and vol. 3, *Agathon to Michael I (766)*, Patrologia Orientalis 5.1 (1909): pp. 1–58. However, in this paper I will work primarily with the sole published edition of the older or "primitive" recension of *HP*: Seybold, Christian Friedrich, *Severus ibn al-Muqaffa': Alexandrinische Patriarchengeschichte von S. Marcus bis Michael I 61–767, nach der ältesten 1266 geschriebenen Hamburger Handschrift*. Hamburg: Lucas Gräfe, 1912, abbreviated as *HPH* in the following. On the distinction between the "primitive" and "Vulgate" recensions, see Den Heijer, *Mawhūb*, pp. 14–80; Den Heijer, "The Martyrdom of Bifām," 454–464.

century. He thus wrote about critical periods in the history of the Egyptian Church, including the struggle within Egypt and the Byzantine Empire more widely over the Chalcedonian definition, and, of course, about the Arab conquest and the establishment of new patterns of Christian existence in Egypt during the first half-century of Arab rule.

Important themes are established early on in Archdeacon George's history (and will be important in later parts of *HP* as well). The patriarch is a heroic confessor of the faith, who at times speaks the language of the martyrs, and who may be called upon to endure periods of exile.[16] If the 4th-century patriarch Athanasius the Great (the 20th patriarch, 328–373) was an early example of these themes of confession, near-martyrdom, and exile, they are re-emphasized in the stories of patriarch Severus of Antioch (512–538) and the Coptic patriarch Theodosius I (the 33rd patriarch, 536–567) in the 6th century, and carry over to the dawn of the Islamic period in the story of Coptic patriarch Benjamin (the 38th patriarch, 623–662), who spent ten years in hiding from the Byzantine authorities before eventually being recalled to Alexandria after the Arab conquest. Archdeacon George sees the steadfastness of these patriarchs as a model for all the faithful; at the end of his *Life of Patriarch Theodosius*, he exhorts his readers to stand firm and preserve the orthodox faith, as did Theodosius who "confessed before kings, rulers, and authorities in that transgressive generation, and the heretics."[17] One may perhaps hear in George's exhortation overtones of the Synoptic apocalypse, where Jesus tells his followers, "you will stand before governors and kings because of me, as a testimony to them" (Mark 13:9, cf. Luke 21:12);[18] the Biblical echo might add a note of eschatological urgency to George's exhortation.

But at the same time that Archdeacon George was well aware of the capacity of governors and kings to be oppressive, when he comes to the Arab period he becomes quite cautious in his portrayal of rulers—at least of the chief authority (however dastardly some of his subordinates might be). The pattern is set early on in the well-known story of the meeting between the Arab conqueror of Egypt, ʿAmr ibn al-ʿĀṣ, and Patriarch Benjamin after ʿAmr had recalled him from his exile.[19] As Archdeacon George tells the story, a Coptic notable arranged a meeting, which turned out to be one of mutual respect: ʿAmr admired the appearance and bearing of the patriarch, and promised him a certain autonomy in church

16 See Swanson, *Coptic Papacy*, chapter 1.
17 *HPH* 88/5–6; *HPE*, PO 1:468/10–11.
18 English Bible translations are taken from the New Revised Standard Version Bible. New York: National Council of Churches of Christ in the U.S.A., 1989.
19 *HPH* 100–101; *HPE*, PO 1:495–97.

affairs (and, implicitly, support over his Chalcedonian rivals)—in exchange for the patriarch's prayers for ʿAmr's continuing military success. Then the patriarch, we are told, preached a sermon in ʿAmr's presence and, at least in one recension of the work, prayed for him on the spot. This story provided a paradigm for patriarch-ruler relations that is revisited throughout *HP*, and which indeed has had echoes in "national unity" discourse in Egypt down to the present day.[20]

In addition to providing this paradigm for governor–patriarch relations, it is in Archdeacon George's account that a role of the Islamic governing authorities in major church appointments is recognized,[21] beginning with the Umayyad governor ʿAbd al-ʿAzīz ibn Marwān (685–705, the son, brother, and father of Umayyad caliphs). The election of Patriarch Simon's predecessor Isaac (the 41st patriarch, 690–692), as well as the election of Simon himself, were backed by the authority of ʿAbd al-ʿAzīz. It was the governor's intervention that secured the patriarchate for Isaac over a candidate who had been put forward without the governor's knowledge or approval—and Archdeacon George assures his readers that this occurred "by a command from God."[22] And in the case of the surprising choice of Simon rather than the bishops' first candidate, who happened to be Simon's spiritual father, a monk named John, the author has the bishops address the governor with the words, "command belongs to God *and to you*."[23] Archdeacon George is convinced that it is indeed *God* who chooses the patriarch, and backs this up with scriptural passages about God's electing activity (1 Tim 2:19; Heb 5:4, Ps 65:4);[24] but God may work through the Muslim governor. This measuredly positive portrayal of the Muslim authorities—which is fairly standard throughout the *History of the Patriarchs*—seems to be of a piece with the policy of Patriarch Simon whom George served, who exercised great care in relating to the governing authorities. He would not, for example, appoint a bishop for the "Indians" (the Nubians or the Ethiopians) without the governor's permission,[25] and, in general (according to Archdeacon George) "was striving all his life not to present a stumbling block between the Christians and the Muslims, so that no one would incur loss on his account."[26]

---

20 Swanson, "Reading the Church's Story."

21 Swanson, *Coptic Papacy*, pp. 6–11.

22 *HPH* 121/12; *HPE*, PO 5:24/1.

23 *HPH* 123/16. *HPE*, PO 5:29/1 (representing the later "Vulgate" recension of *HP*) adds the word thumma, to give the (less scandalous?) reading, "command belongs to God, and *then* to you."

24 *HPH* 120/17–19; *HPE*, PO 5:22/1–3.

25 *HPH* 127/4–5; *HPE*, PO 5:36/8–9.

26 Reading here with *HPE*, PO 5:43/2–3, rather than the unclear text at *HPH* 130/5–7.

Archdeacon George is careful in his portrayals of the Muslim authorities, at least as far as explicit description is concerned. We should note, however, that as an educated churchman he had available to him the resources of Scripture to describe and comment upon personalities and events: a well–placed simile, a partial quotation, or an allusion could provide a Biblical background against which to ponder and explore the meaning of late 7th-century events—that is, for Biblically competent readers who could recognize the cues and mentally access the appropriate passages. George could be fairly demanding in this regard. For example, when he reports on how, contrary to expectations, the monk Simon rather than his spiritual father John was made patriarch, he mentions—almost in passing—two pairs of names as Biblical precedents: Perez and Zerah, and Adonijah and Solomon.[27] The mention of Perez and Zerah reminds Biblically competent readers of the birth of twin boys to Judah and Tamar in Gen 38: while it was Zerah who first stuck out a hand (and had a crimson thread tied to it), it was in fact Perez who turned out to be the first-born. The second pair of names remind Bible readers of the succession to King David; that while Adonijah was for a time the heir-apparent, it was the younger brother Solomon who actually received the throne (1 Kgs 1). *Just so*, readers of George's history are left to conclude, the holy monk John may have been expected to become patriarch, but it was younger monk Simon who was actually chosen by God and who would be elevated to that office. Archdeacon George's brief allusions to the Biblical stories assume Biblically well-informed readers who could make sense of and be edified and/or entertained by the comparisons.

Nowhere is the use of the Bible more important than when Archdeacon George wants to characterize the patriarchs themselves or their opponents, including the ruling authorities. In describing the patriarchs, he often mentions various heroes of the Bible. While the young John of Samannūd might have been made bishop for a diocese, "God was preserving him for his flock like David"[28]—that is, God was preserving him to be patriarch of the whole church (as the 40th patriarch, 680 – 689), rather than to be a local bishop. As patriarch, John's face shone with light like that of Moses;[29] and all that he did was God-pleasing, like St. John the Evangelist.[30] That the future patriarch Benjamin

27 *HPH* 123/7–8; *HPE*, PO 5:28/4–5.
28 *HPH* 114/17–18; *HPE*, PO 5:8/11–9/1 (here reading li-ra'iyyatihi, "for his flock," with HPH rather than li-da'atihi, "for his gentleness," with *HPE*). On the preservation of David for the sake of his flock, see 1 Sam 17:34–37.
29 *HPH* 116/14–16; *HPE*, PO 5:12/10–13/1. On the shining of Moses' face, see Exod 34:29–35.
30 *HPH* 119/21; *HPE*, PO 5:20/2–3. This probably refers to the traditional identification of St. John the Evangelist with the disciple "whom Jesus loved" in St. John's Gospel.

should outstrip his spiritual father Theonas in knowledge had its Biblical precedent in Paul and his teacher Gamaliel (Acts 22:3).[31] And the late-6th century patriarch Damian (the 35th patriarch, 569–605), who faced a range of heretical opponents, made them, says Archdeacon George, "like Ahab in the days of our father Elijah the prophet" (1 Kgs 16:29–22).[32] In this last case, these few words are enough to remind the Biblically competent reader of the long contention between Elijah and Ahab, perhaps including Ahab's eventual repentance as found in 1 Kgs 21:27–29; certainly, the *History of the Patriarchs* regularly lauds the efforts of the patriarchs to reconcile their theological opponents to the orthodox faith. But I wonder whether the allusion also might quietly enable a comparison between the Meletians and Julianists of Damian's day with the priests of Baal in those of Elijah (1 Kgs 19).

This last example shows that Biblical allusions may also be used to cast an unflattering light on opponents. Other examples, briefly, include Archdeacon George's account of Severus of Antioch before the Byzantine emperor Justinian (r. 527–565), where we are told that when Justinian received Severus with honor, Severus chose to regard this good treatment as a kind of bribe—and quoted from St. Peter's speech to Simon Magus (Acts 8:20–23), who had offered money for the power to bestow the Holy Spirit.[33] A few pages later, George reports Justinian's death as follows: "the angel of the Lord smote him: he destroyed him and he died, like the death of Herod"[34]—leaving the reader to recall the 12th chapter of the Acts of the Apostles, in which Herod Agrippa unleashed a persecution against the church, but who, after having been proclaimed as a god and not a mortal, was brought low, as we read in verse 23: "And immediately, because he had not given the glory to God, an angel of the Lord struck him down, and he was eaten by worms and died." And yet another example: when Archdeacon George mentions that the Umayyad governor ʿAbd al-ʿAzīz ibn Marwān had recourse to corvée labor, he is quick to say that "he used people as did Pharaoh in his time"[35]—without further elaboration. But the simile alone sufficed to overlay the story of the account of Christians in Egypt under governor ʿAbd al-ʿAzīz with the story of the Israelites in Egypt, whose lives were made bitter under the pharaoh who did not know Joseph (Exod 1:8–14).

The examples just given mostly come in the form of similes—"like David," "like Moses," "like St. John the Evangelist," "like Ahab," "like the death of

---

31 *HPH* 97/2–4; *HPE*, PO 1:487/9–488/1.
32 *HPH* 90/22–24 (where the word "our father" is misplaced); *HPE*, PO 1:474/7–9.
33 *HPH* 80/13–15; *HPE*, PO 1:452/8–10.
34 *HPH* 89/11; *HPE*, PO 1:471/5.
35 *HPH* 130/4; *HPE*, PO 5:42/10–43/1.

Herod," "like Pharaoh"—but as simple as the allusions are, they are enough to take events or personalities from the history of the church and overlay them with Biblical accounts, thereby opening up a wide range of possibilities for interpretation and understanding. However, the use of Biblical material in Archdeacon George's account is not limited to these simple figures. For example, let us examine Archdeacon George's account of the initial encounter between figures already mentioned above, Patriarch John III of Samannūd and the Umayyad governor 'Abd al-'Azīz ibn Marwān.[36]

As Archdeacon George tells the story, their relationship did not begin well: the patriarch was not on hand to meet the governor when he (unexpectedly) arrived in Alexandria, and the patriarch's Chalcedonian rivals loudly interpreted his absence as a deliberate snub from a haughty and wealthy prelate. 'Abd al-'Azīz reacted with fury, had Patriarch John arrested, spoke severely to him, and turned him over to his officers with instructions to exact payment of a fine of 100,000 dinars. The patriarch was taken into custody, we are told, by a particularly merciless official named Samad, who intended to torture the patriarch into surrendering his supposed wealth. When Samad demanded payment, Patriarch John responded that he had no wealth, and added: "Whatever you desire to do [to me], do it! My body is in your hand, but my soul and body together are in the hand of my Lord Jesus Christ."[37] These words are reminiscent of scores of Christian martyrdom accounts,[38] giving the reader a context within which to understand Patriarch John's ordeal. But Christian martyrdoms are regularly interpreted in the light of Christ's passion—and that is certainly the case in the account of Archdeacon George, who relates that the jailer's cruel designs were only stayed by a message that came from the governor's wife: "Beware lest you do any evil to the man of God who has been handed over to you! This night I have endured great trials on account of him."[39]

The Biblically literate reader immediately recognizes here an echo of the passion narrative according to St. Matthew, in which Roman procurator Pontius Pilate, sitting in judgment over Jesus, received a message from his wife: "Have nothing to do with that innocent man, for today I have suffered a great deal because of a dream about him" (Matt 27:19). And just so, here by allusion rather than by explicit analogy, the account of patriarch and governor is overlaid

---

36 *HPH* 116–18; *HPE*, PO 5:13–18. I have also examined this account in Swanson, "Reading the Church's Story."

37 *HPH* 117/14–15; *HPE*, PO 5:14/10–11.

38 For a very early example of the "do what you will" language, see *The Martyrdom of St. Polycarp* 11.

39 *HPH* 117/17–19; *HPE*, PO 5:15/2–5.

with and interpreted by that of Jesus Christ before Pontius Pilate.[40] Let me emphasize the subtlety of this: Archdeacon George did not mention the *name* Pontius Pilate—and yet his narrative allows his readers to hold ʿAbd al-ʿAzīz and Pontius Pilate together in a single thought.

According to Archdeacon George, this intervention by the governor's wife changed the story-line from one of physical threat to one of negotiation, through which the fine imposed on the patriarch was gradually reduced from 100,000 to 10,000 dinars—at which point a group of Coptic notables stepped in to urge a settlement, arranging payment in return for the patriarch's release. They arranged a meeting between governor and patriarch that, the reader of Archdeacon George's history will recognize, in many ways echoes his earlier account of the encounter between ʿAmr ibn al-ʿĀṣ and Patriarch Benjamin. Archdeacon George tells us that ʿAbd al-ʿAzīz was awed by the appearance of the patriarch, seeing him "as if he were an angel of God."[41] He received him with courtesy and engaged him in a sharp but edifying conversation:

> [The governor] said to him: "Don't you know that the sultan is not to be opposed?" The saint responded and said to him: "The command of the sultan is to be obeyed as it is necessary and when it does not provoke the wrath of God! Beyond this, God has bid us, as it is said to us [in Scripture]: 'Do not fear the one who kills your bodies, but has no authority over your souls' [Matt 10:28]." The emir said to him: "Your God loves faithfulness and truth". The patriarch said to him: "My God is wholly Truth, and in Him there is no falsehood [John 1:5], but He destroys everyone who pronounces falsehood."[42]

Shortly afterwards, the governor released the patriarch "with honor,"[43] to the delight of his followers and the distress and shame of his enemies.[44]

What sort of evaluation of governor ʿAbd al-ʿAzīz does this story imply? The echoes of martyrdom accounts and especially the echoes of St. Matthew's passion narrative would suggest quite a negative evaluation: the governor plays the role of the Christian-persecuting ruler of classic martyrdom accounts; and if he is not directly involved in the worst elements of the patriarch's ordeal, that is perhaps reminiscent of Pontius Pilate, who worked through subordinates

40 The same motif of the governor's wife is found in the story of the 10th-century martyr Jirjis (Muzāḥim), and is briefly reported in the *Synaxarion* entry for Jirjis on 19 Ba'ūnah: Basset, René, *Le synaxaire arabe jacobite (rédaction copte), V: Les mois de Baouneh, Abib, Mesoré et jours complémentaires*, Patrologia Orientalis 17. Paris, 1923, pp. 578–580, here p. 579/3.

41 *HPH* 118/12; *HPE*, PO 5:16/10.

42 *HPH* 118/13–17; *HPE*, PO 5:17/1–6.

43 Lit. "glory" (*majd*); *HPH* 118/19; *HPE*, PO 5:17/8.

44 *HPH* 118/19–21; *HPE*, PO 5:17/8–10.

and, in the end, washed his hands of responsibility in the case of Jesus (Matt 27:24). On the other hand, as the story comes to echo Archdeacon George's earlier account of the meeting of Patriarch Benjamin and as ibn al-ʿĀṣ, the relationship of patriarch and governor appears to be restored to what, in Archdeacon George's view, it should be, one characterized by boldness of speech and by mutual giving of honor.

Perhaps the ambiguity that we perceive here is contained within the echo of the story of Pilate's wife. While this echo does, as we have said, allow the Biblically-literate reader to hold Pilate and ʿAbd al-ʿAzīz in a single thought, for the Copts this thought was not unambiguously negative—as they were familiar with stories according to which Pontius Pilate eventually repented, became a Christian, and was even martyred for his faith.[45] That is, the Pontius Pilate who supervised the proceedings that led to Christ's death later became Pontius Pilate, the martyr for the sake of Christ! And thus we see that while Archdeacon George's allusion to a detail in St. Matthew's passion narrative is suggestive and opens up new possibilities for the interpretation of key figures in the story he relates, it does not unambiguously guide that interpretation.

3 Observations

So where does this leave us? A number of points can be made (and eventually compared with the results of studies of the use of the Bible in other component parts of *HP*):

a) George the Archdeacon sees the Bible as providing a treasury of passages that illustrate the ways God works in human history, raising up leaders and bringing tyrants low; as well as a wide range of illustrations of virtue and belief and of wickedness and unbelief. These passages do not simply speak of particular events and individuals, but provide models that regularly recur in human history.
b) In particular, these biblical passages are useful for describing the patriarch and his opponents, including the ruling authorities. Even when the author is careful not explicitly to criticize the ruler, a word or two can evoke the figures of Pharaoh, Ahab, Pontius Pilate, or Herod Antipas. Not all of these evocations would be obvious to every reader,

45 See Luisier, Philippe, "De Pilate chez les Coptes." *Orientalia Christiana Periodica* 62 (1996): pp. 411–425. Two of the important Pilate-texts, *The Lament of the Virgin* and *The Martyrdom of Pilate*, are found in Syriac with English translations in Mingana, Alphonse, *Woodbrooke Studies: Christian Documents in Syriac, Arabic, and Garshūni*, vol. 2. Cambridge: Heffer & Sons, 1928, pp. 163–240 and 241–332 respectively.

and so could serve as a slightly subversive element in a text otherwise quite careful and restrained in its portrayal of Muslim rulers.

c) And still, these evocations may themselves contain an element of ambiguity. Ahab repented. Pontius Pilate was rehabilitated to the point that the *History of the Churches and Monasteries*, a Copto-Arabic text the core of which is from the late 12th century, reports the existence of a Church of Pontius Pilate in the village of Abyar in the Delta.[46]

d) In any event, the biblical allusions (and allusions to Christian apocrypha and hagiography) are not the only intertextual element at work in shaping these texts. In the passage just studied, it is the paradigm for patriarch–ruler relations, established in the story of the encounter between ʿAmr ibn al-ʿĀṣ and Benjamin, which is decisive in shaping the final outcome of the story of the encounter of ʿAbd al-ʿAzīz ibn Marwān and John of Samannūd. At the same time, the texture of the story has been greatly enriched through the overlay of the narrative of Christ's passion on that of the patriarch's arrest and brush with torture.

I hope these examples, from just one part of the *History of the Patriarchs*, are sufficient to remind us of the complex ways in which the Bible can function. For George the Archdeacon, the Bible provides universally-relevant patterns of God's work in history, of inter-human relationships, of lives lived faithfully and faithlessly—and these can be brought to bear, not in any neat or mechanical way but allusively, through suggested correspondences and analogies, on concrete historical cases from the early decades of Christian existence within the *dar al-Islām*. During this period, Christians in Egypt were remaking their identity in a world in which the old order that included a Christian emperor gave way to a new order of Muslim caliphs and governors. The writing of the histories that went into the *History of the Patriarchs* were themselves exercises in articulating that new identity; and in the writing of those histories, Sacred Scripture played a role.

46 Luisier, "De Pilate chez les Coptes", pp. 414–415.

Sidney Griffith

Use and Interpretation of Scriptural Proof-Texts in Christian–Muslim Apologetic Literature in Arabic

1

From the very beginning of Christian apologetic literature in Arabic (ca. 755–775), authors made abundant use of quotations from the Old and New Testaments, along with allusions to and echoes of their narratives, in an effort to provide the scriptural proof (*burhān*) that the Qur'ān demanded of the *Scripture People* (*ahl al-kitāb*) in support particularly of the Jewish and Christian beliefs and practices that the Islamic scripture critiqued.[1] Beginning with an arsenal of Biblical testimonies widely deployed in earlier Christian literature in Greek and Syriac, Arabic-speaking Christian apologists adapted them to the requirements of the new challenge and soon developed standard Testimony Lists of their own. In response, Muslim apologists also assembled Biblical testimonies, particularly in support of the Qur'ān's contention that the Torah and the Gospel had announced the coming of Muḥammad and Islam. And right from the start, in addition to Biblical proof-texts, Christian apologists were not slow to enlist quotations from the Qur'ān itself in support of their own apologetic arguments. There ensued in the course of the early Islamic centuries a spiraling, inter-scriptural, interreligious controversy that arguably reached its apogee in the thirteenth century.

After briefly describing this crescendo of inter-scriptural reasoning that often accompanied the more reason-based apologetic and polemic arguments between Christian and Muslim *mutakallimūn* in early Islamic times, this contribution highlights the arguments from Scripture in the works of the Muslim jurist Abū l-Baqā' Taqī al-Dīn Ṣāliḥ ibn Ḥusayn al-Ja'farī (1185–1270) and the contemporary Christian scholar al-Ṣafī ibn al-'Assāl (ca. 1205–ca. 1265) in controversy with one another in 13th century Cairo.

1 Q 2:111: "They Say, 'Only those who are Jews or Christians will enter the Garden.' These are their own vague ideas. 'Bring your proof, if you tell the truth.'" Quoted from Jones, Alan, *The Qur'ān Translated into English*. Cambridge: Gibb Memorial Trust, 2007, p. 37.

https://doi.org/10.1515/9783110564341-006

2

The earliest surviving apologetic text written in Arabic by a Christian in response to the Qur'ān's critique of Christian beliefs about Jesus the Messiah, Mary's son, and about the one God, whose word and spirit the Qur'ān identifies with Jesus (Q 4:171), seems on the testimony of its author to have been written just after the mid-8th century CE.[2] The now unknown author of this work that has somewhat inappropriately come to be called *On the Triune Nature of God*, wrote at one point about the endurance of Christianity against all odds up until his time. "If this religion were not truly from God," he said, "it would not have stood so unshakably for seven hundred and forty-six years."[3] Based on this remark, scholars have variously assigned a date between the years 750 and 775 CE as the year of composition for this "Apology for the Christian Faith," as Mark Swanson insightfully prefers to call it.[4]

Among the many interesting features of the treatise *On the Triune Nature of God* is the author's use of numerous Biblical *testimonia*, fifty and more of them, quoted along with several verses from the Qur'ān, all deployed as proof-texts intended to support the credibility of Christianity's truth claims regarding the Trinity, the Incarnation, the true religion, and the right understanding of the ancient Biblical prophecies. But the influence of the Qur'ān is not limited to quotations used as proof-texts; the whole treatise in its diction and style is suffused with echoes of the Qur'ān's language.

In the poetical introduction to the treatise, by allusion and choice of words and phrases the author already echoes the diction and style of the Qur'ān.[5] As

---

2 See the text published in Dunlop Gibson, Margaret, ed., "An Arabic Version of the Acts of the Apostles and the Seven Catholic Epistles from an Eighth or Ninth Century MS in the Convent of St Catharine on Mount Sinai, with a Treatise on *The Triune Nature of God*, with Translation, from the Same Codex." *Studia Sinaitica* VII (1899): pp. 1–36 [English], pp. 74–107 [Arabic]; Gallo, Maria, trans., *Palestinese anonimo; Omelia arabo-cristiana dell'VIII secolo.* Roma: Città Nuova Editrice, 1994. See Samir, Samir Khalil, "The Earliest Arab Apology for Christianity (ca.750)," in *Christian Arabic Apologetics during the Abbasid Period (750–1258)*, ed. by Samir Khalil Samir and Jørgen S. Nielsen, pp. 57–114, Studies in the History of Religions 63. Leiden: Brill, 1994.

3 Sinai Arabic MS 154, f. 100v. Margaret Gibson unaccountably left this leaf out of her edition of the text and so she took no notice of the internal indication of the date of its composition.

4 See Samir, "The Earliest Arab Apology", pp. 69–70; Swanson, Mark, "Beyond Prooftexting: Approaches to the Qur'ān in Some Early Arabic Christian Apologies." *The Muslim World* 88 (1998): pp. 297–319, esp. 305–308.

5 See Samir, "The Earliest Arab Apology", pp. 69–70; Swanson, "Beyond Prooftexting", p. 305.

Mark Swanson has rightly remarked, "The text simply *is* profoundly Qur'ānic."[6] One can see it even in English translation, as in this brief passage from the opening prayer:

> We ask you, O God, by your mercy and your power to put us among those who know your truth, follow your will, and avoid your wrath,[who] praise your beautiful names, (Q VII:180) and speak of your exalted similes. (cf. Q XXX 27) You are the compassionate One, the merciful, the most compassionate; You are seated on the throne, (Q VII:54) You are higher than creatures; You fill up all things.[7]

Shortly after this prayer, the author makes a statement that may well serve as an expression of his purpose in composing his work. Again, the attentive reader can hear the Qur'ānic overtones clearly. The author says:

> We praise you, O God, and we adore you and we glorify you in your creative Word and your holy, life-giving Spirit, one God, and one Lord, and one Creator. We do not separate God from his Word and his Spirit. God showed his power and his light in the Law and the Prophets, and the Psalms and the Gospel, that God and his Word and his Spirit are one God and one Lord. We will show this, God willing, in those revealed scriptures, to anyone who wants insight, understands things, recognizes the truth, and opens his breast to believe in God and his scriptures.[8]

One notices straightaway the author's intention to make his case for Christian teaching from the scriptures; he names the Law, the Prophets, the Psalms, and the Gospel, scriptures that are named as they are named in the Qur'ān. Moreover, in emphasizing God, his Word, and his Spirit, the author recalls the Qur'ān's own mention of these three names in the often quoted phrase, "The Messiah, Jesus, Son of Mary, was nothing more than a messenger of God, his word that he imparted to Mary, and a spirit from him." (Q 4:171) What is more, the author is willing to include explicit citations from the Qur'ān among the scripture passages he quotes in testimony to the credibility of the Christian doctrine. On the one hand, addressing the Arabic-speaking, Christian readers who were his primary audience, the author speaks of what "we find in the Law and the Prophets and the Psalms and the Gospel," in support of the Christian doctrines of the Trinity and the Incarnation. On the other hand, several times he rhetorically addresses Muslims; he speaks of what "you will find [...] in the Qur'ān," and he goes on to cite a passage or a pastiche of quotations from several sūras, in support of the

6 Swanson, "Beyond Prooftexting", p. 308.
7 Adapted from the text and translation in Samir, "The Earliest Arab Apology", pp. 67–68.
8 Gibson, *An Arabic Version*, p. 3 [English], p. 75 [Arabic]. Here the English translation has been adapted from Gibson's version.

doctrines, in behalf of the veracity of which he has been quoting or alluding to scriptural evidence from passages and narratives from the Old or New Testaments.[9] For example, at one point in the argument, in search of testimonies to a certain plurality in the Godhead, the author turns to the scriptures for citations of passages in which God speaks in the first person plural. Having quoted a number of such passages, he goes on to say:

> You will find it also in the Qur'ān that "We created man in misery" (Q XC:4), and "We have opened the gates of heaven with water pouring down" (Q LIV:11), and have said, "And now you come unto Us alone, as We created you at first." (Q VI:94) It also says, "Believe in God, and in his Word; and also in the Holy Spirit." (cf. Q IV:171) The Holy Spirit is even the one who brings it down (i.e., the Qur'ān) as "a mercy and a guidance from thy Lord." (Q XVI:64, 102) But why should I prove it from this (i.e., the Qur'ān) and bring enlightenment, when we find in the Torah, the Prophets, the Psalms, and the Gospel, and you find it in the Qur'ān, that God and his Word and his Spirit are one God and one Lord? You have said that you believe in God and his Word and the Holy Spirit, so do not reproach us, O men, that we believe in God and his Word and his Spirit: we worship God in his Word and his Spirit, one God and one Lord and one Creator. God has made it clear in all of the scriptures that this is the way it is in right guidance and true religion.[10]

Evidently in this passage the Christian author is addressing himself directly, at least in part, to readers of the Qur'ān as well as to the devotees of the Christian Bible. He speaks of what "we find in the Torah, the Prophets, the Psalms, and the Gospel," and of what "you find ... in the Qur'ān." One also notices in this passage the prominence of the author's references to God, his Word, and his Spirit, and how they provide a continual evocation of Q 4:171. Like almost every Arab Christian apologetic writer after him, the author of *On the Triune Nature of God* takes this verse as Qur'ānic testimony to the reality that the one God is in fact possessor of Word and Spirit and that they are He, the Son of God, and the Holy Spirit, three persons, one God, as the Christians say.

In a further passage, the author of *On the Triune Nature of God* takes advantage of another verse in the Qur'ān to explain how it came about that by the action of the Holy Spirit, God's Word, the Son of God, became incarnate and was clothed, even veiled (*iḥtajaba*),[11] in Mary's human nature. "Thus," he says, "God was veiled (*iḥtajaba*) in a man without sin."[12] The veiling language here once

9 See, e.g., Gibson, *An Arabic Version*, pp. 5–6 [English]; pp. 77–78 [Arabic]. See the passage quoted and discussed in Griffith, Sidney H., *The Church in the Shadow of the Mosque: Christians and Muslims in the World of Islam*. Princeton, NY: Princeton University Press, 2008, p. 55.

10 Translation adapted from Gibson, *An Arabic Version*, pp. 5–6 [English], pp. 77–78 [Arabic].

11 See Gibson, *An Arabic Version*, p.11 [English], p. 83 [Arabic].

12 Gibson, *An Arabic Version*, p. 13 [English], p. 85 [Arabic].

again evokes a particular passage in the Qur'ān: "God speaks with man only by way of revelation (*waḥy*), or from behind a veil (*ḥijāb*), or he sends a messenger (*rasūl*) and he reveals by His permission what He wishes." (Q 42:51) The author of our treatise likens Jesus' humanity to the veil, from behind which the Qur'ān says God might choose to speak to humankind and in this way he once again evokes Qur'ānic language in a bid to make use of its probative potential.

On the face of it, since the apology is addressed to the Qur'ān's strictures against Christian dogmas and since it opens with a prayer "profoundly influenced by the language of the Qur'ān," and since the text includes a number of direct quotations from the Qur'ān, one might reasonably suppose that the author of the apology was mindful of the Qur'ān's challenge to those who claimed that none would enter Paradise save those who had become Jewish or Christian, to wit, "Produce your proof (*burhānakum*) if you are telling the truth." (Q 2:111). And it is clear in the context that the proof required would be scriptural proof. For the Qur'ān itself regularly recalls the scriptural witness of the Biblical prophets and messengers, among others, in testimony to the truth of its own call for submission to its message. In taking up the Qur'ān's challenge, the author of this apology for the Christian faith was doubtless also aware of the Qur'ān's further admonition to Muḥammad himself in the course of his controversies with the *Scripture People*, "If you (2ms) are in doubt about what We have sent down to you, ask those who were reading scripture before you." (Q 10:94) So the author composed his apology with a selection of carefully chosen proof-texts drawn from the very scripture in question and presented them within the framework of what he viewed as their proper interpretation, supporting the very doctrines critiqued in the Qur'ān. It is as if one of his purposes was to correct what he viewed as the Qur'ān's flawed interpretation of the very scriptures to which it refers Muḥammad for the solution of his doubts.

An interesting fact about the Biblical passages quoted in Arabic in *On the Triune Nature of God* is that they are among the earliest texts so far known of the Bible in Arabic writing. They are of an age with the earliest known Arabic translations of the Gospels for which we have textual evidence, which is to say the mid-to-second half of the 8th century according to the most probable hypothesis. The same may be said of the Psalms, the most often quoted single book of the Bible cited in the treatise *On the Triune Nature of God*; the earliest textual evidence for their translation into Arabic dates from the late 8th or early 9th century.[13] So the question arises, from what source did the author of this very orig-

13 See in this connection the discussion in Griffith, Sidney H., *The Bible in Arabic: The Scrip-*

inal apology for the Christian religion in Arabic take the numerous Biblical quotations he cited in the text? So far no one has been able to connect them with any known Arabic translation of any Biblical book. The supposition must be that the author of *On the Triune Nature of God* translated them himself from Greek or Syriac texts he had in hand.

Scholars who have studied the Biblical quotations in *On the Triune Nature of God* have long ago concluded that in the ensemble they reflect the lists of scriptural *testimonia* typically found in early Christian "Testimony Collections" that circulated in the churches for centuries before the rise of Islam.[14] This finding would suggest that the Christian apologist in early Islamic times drew on this traditional resource in constructing his defense of Christian doctrines against the challenge of the Qur'ān's critique of Christian doctrines and practices and that he simply put these traditional *testimonia* forward as the proof (*al-burhān*) demanded by the Qur'ān. But it would be a mistake to think that in the composition of his text the Arab Christian apologist worked just from a simple list or catena of quotations from the Old Testament, interpreted and explained in reference to passages in the New Testament. Rather, the presentation in *On the Triune Nature of God* literarily speaking resembles no other genre in its mode of discourse more readily in terms of its contents and mode of scriptural reasoning than it does the Syriac doctrinal *mêmrê* that circulated so widely among Aramaic/Syriac-speaking Christians in the very centuries of the rise of Islam and the first Christian responses to the challenges of the Arabic Qur'ān.[15] It would not in fact be far-fetched to think that the Arabophone author of *On the Triune Nature of God* had a Syriac-speaking background.

Unlike the Syriac *mêmrâ*, which was typically composed in isosyllabic lines of verse, the treatise *On the Triune Nature of God* is a prose work artfully phrased especially in its opening paragraphs in the Arabic cadences of the Qur'ān. But very much in its contents and mode of scriptural reasoning, not to mention its homiletic character, *On the Triune Nature of God* mirrors the *mêmrâ*'s character-

---

tures of the 'People of the Book' in the Language of Islam. Princeton, NJ: Princeton University Press, 2013, esp. pp. 97–125.

14 See Bertaina, David, "The Development of Testimony Collections in Early Christian Apologetics with Islam," in *The Bible in Arab Christianity*, ed. by David Thomas, pp. 151–173, The History of Christian-Muslim Relations 6. Leiden: Brill, 2007.

15 On this important genre of Syriac exegetical literature, look for Griffith, Sidney H., "Disclosing the Mystery: The Hermeneutics of Typology in Syriac Exegesis; Jacob of Serūg on Genesis XXII," in *Interpreting Scriptures in Judaism, Christianity and Islam: Overlapping Inquiries*, ed. by Mordechai Z. Cohen and Adele Berlin, pp. 46–64. Cambridge, UK: Cambridge University Press, 2016; and Griffith, Sidney H., "The Poetics of Scriptural Reasoning: Syriac *Mêmrê* at Work." Forthcoming.

istic, typological presentation of the Christian creed in a tissue of quotations, allusions, images, and evocations of Old Testament narratives and *testimonia* (including several passages from the Qur'ān), all recollected in reference to the Gospels and other New Testament presentations of Jesus' words and actions in such a way that the New is seen to fulfill and complete the Old, affirming the humanity and divinity of Jesus the Messiah and the treble reality of the one God.

Clearly one can see the influence of the traditional Christian Testimony Collections in the background of their deployment in *On the Triune Nature of God.* In all likelihood the standard testimonies the author employed were readily available to him most immediately through the medium of earlier homiletic and apologetic compositions such as those embodied in such compositions as the Syriac *mêmrê* mentioned above. In his hands these passages and proof-texts were then brought into service *mutatis mutandis* to meet the Qur'ān's demand for scriptural proof for the credibility of the Christian doctrines that the Islamic scripture critiqued. By the 9th and 10th centuries, Christian apologists writing in Arabic had abandoned the homiletic style of *On the Triune Nature of God* in favor of the style and methods of the contemporary Muslim *mutakallimūn,*[16] but they did not altogether abandon the scriptural Testimony Collections. Rather, they developed and adapted them to the new challenges posed by Muslim controversialists. Methodologically, apologists now regularly argued both *from scriptural tradition* (*min al-naql*) and *from reason* (*min al-'aql*), and even the most rigorous of the rationalists seldom neglected to cite the standard proof-texts if only in passing. But the Christians were not the only participants in the *'ilm al-kalām* of early Islamic times to employ Biblical testimonies; Muslim *mutakallimūn* readily employed passages from the New and Old Testaments in their own arguments *min al-naql.* What is more, like their Christian counterparts, Muslim apologists also soon had their own testimony lists of Biblical proof-texts.

3

In texts produced from the 9th century onward, one observes a double appropriation of Biblical proof-texts in the works of Muslim scholars. On the one hand, in the circumstances that John Wansbrough so aptly termed "the Sectarian Milieu" of the early Islamic period,[17] many writers seem to have been concerned to claim

---

16 See in this connection the discussion in Griffith, *The Church,* esp. pp. 75–105.

17 See Wansbrough, John, *The Sectarian Milieu: Content and Composition of Islamic Salvation History.* Oxford: Oxford University Press, 1978. See also Wansbrough, John, *Qur'anic Studies: Sources and Methods of Scriptural Interpretation.* Oxford: Oxford University Press, 1977.

the authority of the Bible to warrant the scriptural authenticity of Muḥammad, the Qur'ān, and Islamic teaching more generally. One might call it the process of 'Biblicizing' the Islamic prophetic claims. On the other hand, given the concomitant, and growing Islamic concern about the corruption of the earlier scriptures, and observing the consequent, divergent cast of many of the Islamic presentations of Biblical texts, one may also speak of a simultaneous process of 'Islamicizing' Biblical material.[18] In the earlier period, when the concern was generally more to 'Biblicize' Islamic prophetology, some writers showed a keen interest in the Biblical text. In this connection one might mention quotations, albeit often corrected (and therefore 'Islamicized'), allusions and paraphrases to be found in the work of Wahb ibn Munabbih (d. 732), the *Sīrah* of Abū 'Abd Allāh Muḥammad ibn Isḥāq (d.ca. 767), as transmitted by Abū Muḥammad 'Abd al-Mālik ibn Hishām (d. 834), and the copious quotations in the works of Ibn Qutaybah (d. 889) and al-Ya'qūbī (d. 897).[19] Similarly, in the apologetic and polemical texts composed by both Christian and Muslim *mutakallimūn* from the 9th and 10th centuries onward, the Bible was an authority to which the disputants on both sides could readily appeal in their arguments *min al-naql. Arguing from scripture*, long a staple element in the Jewish-Christian encounter, also claimed a place early on in Christian-Muslim controversies. The earliest Muslim authors of refutations of the Christians, for example, quoted not only from the Qur'ān, but they readily cited passages from the Old Testament and the New Testament in testimony to Islam as the true religion and to the authenticity of Muḥammad's claim to genuine prophethood. One might mention in this connection one of the earliest Muslim *kalām* texts explicitly to counter Christian doctrines, al-Qāsim ibn Ibrāhīm ar-Rassī's (d. 860) *Kitāb ar-radd 'alā n-naṣārā*, in which the author quotes abundantly from the Gospels according to Mat-

18 See this matter discussed in Griffith, Sidney H., "Arguing from Scripture: The Bible in the Christian/Muslim Encounter in the Middle Ages," in *Scripture and Pluralism: Reading the Bible in the Religiously Plural Worlds of the Middle Ages and Renaissance*, ed. by T. J. Heffernan and T. E. Burman, pp. 29–58, Studies in the History of Christian Traditions 123. Leiden: Brill, 2005.

19 See Adang, Camilla, *Muslim Writers on Judaism and the Hebrew Bible: From Ibn Rabban to Ibn Ḥazm*, Islamic Philosophy, Theology, and Science 22. Leiden: E.J. Brill, 1996; Accad, Martin, "The Gospels in the Muslim Discourse of the Ninth to the Fourteenth Centuries: An Exegetical Inventorial Table." *Islam and Christian-Muslim Relations* 41 (2003): pp. 67–91, 205–220, 337–352, 459–479; Griffith, Sidney H., "The Gospel in Arabic: An Inquiry into Its Appearance in the First Abbasid Century." *Oriens Christianus* 69 (1985): pp. 126–167; Griffith, Sidney H., "The Gospel, the Qur'ān, and the Presentation of Jesus in al-Ya'qūbī's *Ta'rīkh*," in *Bible and Qur'ān: Essays in Scriptural Intertextuality*, ed. by John C. Reeves, pp. 133–160, Symposium Series 24. Atlanta, GA: Society of Biblical Literature, 2003.

thew and John, interpreting the passages in such a way that they support the Islamic positions he defends against the doctrines of the Melkite, Jacobite, and the Nestorian Christians. The final portion of his work includes a long list of quotations from the Gospel according to Matthew, particularly from the Sermon on the Mount.[20] Similarly, al-Qāsim's contemporary, the septuagenarian 'Nestorian' convert to Islam, 'Alī ibn Sahl Rabbān al-Ṭabarī (d. ca. 860), cited abundant proof-texts from the Bible, together with Islamic interpretations of them, in both his *Kitāb ar-radd 'alā n-naṣārā*, a polemic against the verity of Christian doctrines, and in his *Kitāb ad-dīn wa-d-dawlah*, a defense of Muḥammad's claims to genuine prophetic status.[21] As we shall see, Rabbān al-Ṭabarī's *Radd 'alā n-naṣārā*, composed ca. 850 CE, was still being read and commented upon by both Muslim and Christian writers four centuries after its composition, in 13th century Cairo. But in the meantime, Christian *mutakallimūn* themselves were not slow to continue the practice, inaugurated by the author of *On the Triune Nature of God*, of drawing proof-texts in support of Christian doctrines from the Qur'ān.

4

Just as Muslim, anti-Christian polemicists in the early Islamic period soon made use of Biblical proof-texts in their arguments against the truthfulness of the Christian doctrines and practices that the Qur'ān critiqued, so too did Christian apologists from the very beginning regularly draw proof-texts from the Qur'ān in support of these same Christian doctrines and practices. As we have seen, the author of *On the Triune Nature of God* began this practice and it continued right up to the 13th century and beyond. Almost every Christian apologist writing in Arabic whom one can name, at some point in his apology for Christianity quoted the passage from Q 4:171, in which Jesus the Messiah, Mary's son, is said to be God's word (*kalimat Allāh*) and a spirit from Him (*rūḥun minhu*). Christian writers not only regularly quoted verses from the Qur'ān for the probative value of their scriptural authority but in their polemical arguments they often simultaneously claimed the Qur'ān to have been originally a Christian composition

---

20 See di Matteo, I., "Confutazione contro I Christiani dello Zaydita al-Qāsim b. Ibrāhīm." *Rivista degli Studi Orientali* 9 (1921–23): pp. 301–364.

21 On these works, see Thomas, David, "'Alī ibn Rabbān al-Ṭabarī: A Convert's Assessment of his Former Faith," in *Christians and Muslims in Dialogue in the Islamic Orient of the Middle Ages*, ed. by Martin Tamcke, pp. 137–156, Beiruter Texte und Studien 117. Würzburg: Ergon Verlag, 2007; Thomas, David, "Ṭabarī's Book of Religion and Empire." *Bulletin of the John Ryland's Library* 69 (1986): pp. 1–7.

subsequently distorted by Jews, while at the same time they sometimes explicitly disparaged the claims made by Muslims in behalf of the Qur'ān's miraculous quality as a sign from God, its inimitability, and even its credibility.[22] Perhaps the apogee of the Christian attempt to make use of passages from the Qur'ān for apologetic purposes came in the early 13th century when Paul of Antioch, the Melkite bishop of Sidon, wrote his well-known *Letter to a Muslim Friend.*

It is unclear just when Paul of Antioch composed his *Letter to a Muslim Friend*; it has to have been sometime after the death in 1046 of Elias of Nisibis, whose work Paul quoted in his other treatises, and before the year 1232, when the earliest surviving copy of Paul's *Letter* was made.[23] The *Letter* opens with an account of the bishop's sojourn in what he calls "the homelands of the Romans, Constantinople, the country of Amalfi, some Frankish provinces, and Rome," where he says he came into conversation with "the most important people ... their most eminent and learned men." Paul says that his Muslim friend wanted to know what these learned Byzantines thought of Muḥammad, and in the course of his *Letter* he proceeds to recount what they had to say about the Muslim prophet, the Qur'ān, and Islam, along with what Paul presents as their defense of basic Christian teachings by way of a Christian interpretation of selected passages from the Qur'ān. He concludes his *Letter* by expressing his wish that if the report pleases his friend, God be praised, "since He will have made quarreling cease between His servants the Christians and the Muslims." Otherwise, says Paul, he would be willing as a mediator to convey his Muslim friend's objections to the learned Byzantines he had met abroad in expectation of a suitable reply.

Surely the most striking feature of the *Letter* is Paul's heavy use of selected quotations from the Qur'ān to bolster his argument that Muḥammad was a prophet sent with the Arabic Qur'ān to the pagan Arabs and not to other peoples; that Christians are believers in the one God and they are not polytheists, nor are they called by God to accept Islam or the Qur'ān. Moreover, in addition

---

22 See Griffith, Sidney H., "The Qur'ān in Arab Christian Texts: The Development of an Apological Argument: Abū Qurrah in the *Mağlis* of al-Ma'mūn." *Parole de l'Orient* 24 (1999): pp. 203–233; Griffith, Sidney H., "Christians and the Arabic Qur'ān: Prooftexting, Polemics, and Intertwined Scriptures," i*ntellectual History of the Islamicate World* 2 (2014): pp. 243–266.

23 For the Arabic text and a French translation, see Khoury, Paul, ed. and trans., *Paul d'Antioche, évêque Melkite de Sidon (XIIe. S.)*, Université Saint-Joseph, Institut de lettres orientales, Recherches 24. Beirut: Imprimerie Catholique, 1964, pp. 5983 (Arabic), pp. 169–187. For an English translation, see Griffith, Sidney H., "Paul of Antioch," in *The Orthodox Church in the Arab World 700–1700: An Anthology of Sources*, ed. by Samuel Noble and Alexander Treiger, chap. 10, pp. 216–235, 327–331. DeKalb, IL: Northern Illinois University Press, 2014; for a Spanish translation, see Sarrio Cucarella, Diego R., "Carta a un amigo musulmán de Sidón de Pablo de Antioquia." *Collectanea Christiana Orientalia* 4 (2007): pp. 189–215.

to the conventional lines of reasoning customarily used by Christian apologists in the Islamic milieu, he also used Qur'ān quotations to support the reasonableness of Christian doctrines, including the doctrines of the Trinity and the Incarnation, the Orthodox doctrine of the union of divinity and humanity in Christ and even the integrity of the Christian scriptures and the superiority of Christianity as a religion over Judaism and Islam. Overall Paul's language is polite, peaceful, and respectful of Muḥammad, the Qur'ān, Islam, and of his Muslim correspondent, albeit that he commends the superiority of Christianity. It is nevertheless true that in his use of passages from the Qur'ān, while the quotations are most often textually accurate, Paul nevertheless quotes verses out of context, sometimes distorting them, and he ignores other verses which could not be reconciled with his Christian exegesis of his Qur'ānic proof-texts. It is no wonder that Paul's *Letter* elicited strong responses from Muslim scholars.

Unlike many other Christian apologetic texts written in Arabic, Paul's *Letter*, with its citation of numerous proof-texts from the Qur'ān came almost immediately after its composition and dissemination to the attention of Muslim scholars in Cairo. Straightaway, the famous Muslim jurist, Shihāb al-Dīn Aḥmad ibn Idrīs al-Qarāfī (1228 – 1285), without mentioning Paul's name, included a refutation of his arguments point by point in his notably polemically titled, anti-Christian work, *al-Ajwibah al-fākhirah 'an al-as'ilah al-fājirah fī r-radd 'alā l-millah al-kāfirah*. Meanwhile, in the wake of the *Letter's* initial impact, a now anonymous, Arabic-speaking Christian apologist in Cyprus in the early years of the 14th century reworked and expanded Paul's *Letter* and promptly sent copies to two prominent Muslim scholars of the day,[24]—to Ibn Taymiyyah (1263 – 1328) in 1316 and to Ibn Abī Ṭālib al-Dimashqī (d. 1327) in 1321, both of whom wrote strong rejoinders to the Cypriot's considerably revised edition of Paul's original *Letter.*[25] In his well-known book, *al-Jawāb aṣ-ṣaḥīḥ li-man baddala dīn al-masīḥ*, Ibn Taymiyyah spoke explicitly of the *Letter.* He wrote:

> One of the reasons we treat this religion and its appearance is that a letter arrived from Cyprus in which there is an argument for the religion of the Christians. [...] Thus it demands that we mention by way of answer the final conclusions to which [their arguments] lead,

---

24 In all likelihood, the reviser of Paul's *Letter* was a Nestorian and not a Melkite like Paul himself. See the convincing study by Treiger, Alexander, "The Christology of the *Letter from the People of Cyprus.*" *The Journal of Eastern Christian Studies* 65 (2013): pp. 21 – 48.

25 See Ebied, Rifaat and Thomas, David, *Muslim-Christian Polemic during the Crusades: The Letter from the People of Cyprus and Ibn Abī Ṭālib al-Dimashqī's Response*. Leiden: Brill, 2005. For a comprehensive study of the relationships between the texts mentioned here see Sarrió Cucarella, Diego R., *Muslim-Christian Polemics across the Mediterranean: The Splendid Replies of Shihāb al-Dīn al-Qarāfī (d. 684/1285)*, History of Christian-Muslim Relations 23. Leiden: Brill, 2015.

> and that we make clear their straying from what is correct. [...] We have found them making use of this treatise before now; their scholars hand it down among themselves, and old copies of it still exist. It is attributed to Paul of Antioch, the monk, bishop of Sidon. He wrote it to one of his friends, and had previously written works about the supremacy of Christianity [...] We will show—to God be praise and Strength—that all which they adduce as religious argument, whether from the Qur'ān or from the books preceding the Qur'ān, as well as reason itself is an argument, not for them, but against them.[26]

As a matter of fact, albeit that Paul's *Letter* and its Cypriot expanded revision caused a considerable stir among Cairene Muslim jurists at the turn of the 13th and 14th centuries due in large part to its Qur'ānic, scriptural reasoning, it was not actually the only precipitating factor in the surge in Christian-Muslim altercations in 13th century Cairo. Already prior to al-Qarāfī's refutation of Paul's *Letter* in his *al-Ajwibah al-fākhirah*, local Christian and Muslim intellectuals in controversy with one another had already begun a new phase in their exegesis of one another's scriptures in their efforts the more effectively to argue on behalf of the truth claims of their opposing doctrines *min al-naql*. One supposes that political and social changes were among the precipitating factors behind this new and more confrontational phase in Christian-Muslim interactions, factors such as the victories over the Crusaders from the late 12th century onward, the change from Fatimid (909 – 1161) to Ayyubid (1169 – 1260) rule in Cairo and Damascus, and from the mid-13th century onward, the intellectual and cultural shift from Baghdad to Damascus and Cairo and the subsequent change from Ayyubid to Mamlūk (1250 – 1517) rule, not to mention the new challenge from the Mongols from the mid-13th century onward.[27]

5

From the 9th century onward, Christian and Muslim controversialists continued to engage one another in arguments about the proper interpretations of passages from the Bible. Muslims found passages that according to their readings of them foretold the coming of Muḥammad and supported Qur'ānic Christology, while their Christian interlocutors interpreted the same passages differently.

---

26 Michel, Thomas F., *A Muslim Theologian's Response to Christianity: Ibn Taymiyya's al-Jawab al-sahih*. Delmar, NY: Caravan Books, 1984, pp. 140 – 142.

27 See the discussion in Sarrió Cucarella, *Muslim-Christian Polemics*, pp. 28 – 35. See also Werthmuller, Kurt J., *Coptic Identity and Ayyubid Politics in Egypt, 1218 – 1250*. Cairo and New York: The American University in Cairo Press, 2010; Mikhail, Maged S. A., *From Byzantine to Islamic Egypt: Religion, Identity and Politics after the Arab Conquest*. London and New York: I. B. Tauris, 2014.

One has only to recall in this connection al-Qāsim ibn Ibrāhīm's and Rabbān al-Ṭabari's lists of Biblical proof-texts and even earlier, Ibn Isḥāq's focus on an Islamic reading of John 15:23 – 16:1.[28] In due course, in tandem with their search for Biblical attestations to Muḥammad's prophetic status, there was the concomitant development among Muslim scholars of the doctrine of the inimitability (*al-i'jāz*) of the Qur'ān, which renewed interest in the topic of the corruption of the scriptures prior to the Qur'ān at the hands of Jews and Christians. By the 11th and 12th centuries, major writers, such as al-Ghazālī (1058 – 1111) and Ibn Ḥazm (994 – 1064), were concerned to refute the scriptural reasoning of Jews and Christians alike by demonstrating in great detail the unreliability of their scriptures.[29] But in Cairo in the 13th century something different was beginning to happen. Christian and Muslim scholars began arguing with one another about the details of Biblical passages and weaving their exegeses into the fabric of their apologies for their respective doctrinal positions. Whereas in the past, Christian and Muslim controversialists each deployed their own list of proof-texts within their own interpretive frameworks,[30] without much interaction with one another about the details, in Cairo in the 13th century the protagonists argued in response to one another more closely about the proper quotation and interpretation of individual proof-texts. Muslim polemicists of the period, such as Ṣāliḥ ibn al-Ḥusayn al-Ja'farī (d. 1270) and Shihāb al-Dīn al-Qarāfī (d. 1285) actually recycled the scripture passages already put forward by the aforementioned, 9th century Muslim polemicist, Rabbān aṭ-Ṭabarī, to which the Christian apologist, al-Ṣafī ibn al-'Assāl (d. ca. 1265/75) replied in some detail, an example of which we shall see below. Then by the beginning of the 14th century, Muslim polemicists such as Najm al-Dīn aṭ-Ṭūfī (d. 1316), and subsequently Ibrāhīm ibn 'Umar al-Biqā'ī (d. 1480), began to take a keener interest in the texts of the Bible as the Christians actually had them, in particular the so-called *Alexandrian Vulgate* used by the Copts of Egypt.[31] On the Christian side of the conversation the posture became generally one of reaction; apologists offered detailed critiques of Muslim interpretations of particular proof-texts and attempted to show that they could not be rightfully

---

28 Regarding Ibn Isḥāq's discussion of the passage from the Gospel according to John, see Griffith, "The Gospel in Arabic".

29 See in this connection, Adang, *Muslim Writers*, sub nomine.

30 See Accad, "The Gospels".

31 See Saleh, Walid, ed., *In Defense of the Bible: A Critical Edition and an Introduction to al-Biqā'ī's Bible Treatise*, Islamic History and Civilization 73. Leiden: Brill, 2008; Demiri, Lejla, *Muslim Exegesis of the Bible in Medieval Cairo: Najm al-Dīn al-Ṭūfī's (d. 716/1316) Commentary on the Christian Scriptures—a Critical Edition and Annotated Translation with an Introduction*, History of Christian-Muslim Relations 19. Leiden: Brill, 2013.

used to contravene Christian teaching. For this undertaking a new generation of Arabophone, Christian scholars arose in Cairo, one which consciously built on the achievements of their forbears not only in Egypt but also in Syria and Mesopotamia in Abbasid times.[32]

The now most well-known of the 13th century Christian apologists of Cairo were the so called Awlād al-'Assāl, a family of Coptic Christian scholars, one of the most productive of whom was the aforementioned al-Ṣafī ibn al-'Assāl (ca. 1205 – ca. 1265). He was in all probability a layman who served as a secretary for the Coptic patriarch. In addition to translating Greek patristic texts into Arabic and publishing an important Arabic edition of the *Nomocanon*, al-Ṣafī collected and published summaries of and extracts from the works of earlier, Arab Christian writers of the Abbasid era, such as Yaḥyā ibn 'Adī, 'Ammār al-Baṣrī, Elias of Nisibis, and Patriarch Timothy I, thereby preserving their works for posterity.[33] But for our purposes his most pertinent works were those in which he set out to rebut the anti-Christian polemics of Muslim scholars. The first of them in which he responded in detail was his *Jawāb an-Nāshi' al-akbar*, a response to the ninth century Mu'tazilī scholar of Baghdad of that name, who had moved to Egypt just prior to his death in 906 CE. Al-Ṣafī reproduced and responded to large sections of an-Nāshi's *ar-Radd 'alā n-naṣārā*, originally excerpted by the Jacobite Yaḥyā ibn 'Adī from al-Nāshi's larger work, *Fī l-maqālāt.*[34] Al-Ṣafī's next and arguably most important response to a Muslim writer's anti-Christian polemic was a work he entitled, *aṣ-Ṣaḥā'iḥ fī jawāb an-naṣā'iḥ*;[35] a detailed refutation of the aforementioned *Kitāb ar-radd 'alā n-*

32 See Sidarus, Adel Y., "Essai sur l'âge d'or de la litterature copte arabe (XIIIe-XIVe siècles)," in *Acts of the Fifth International Congress of Coptic Studies, Washington, 12 – 15 August 1992*, ed. by David W. Johnson, pp. 443 – 462, Papers from the Sections, Part 2, vol. 2. Rome: C.I.M., 1993; Sidarus, Adel Y., "La pré-renaissance copte arabe du Moyen Âge (deuxième moitié du XIIe/début du XIIIe siècle)," in *Eastern Crossroads: Essays on Medieval Christian Legacy*, ed. by Juan Pedro Monferrer-Sala, pp. 191 – 216. Piscataway, NJ: Gorgias Press, 2007; Rubenson, Samuel, "Translating the Tradition: Some Remarks on the Arabization of the Patristic Heritage in Egypt." *Medieval Encounters* 2 (1996): pp. 4 – 14; Parker, Kenneth S., "Coptic Language and Identity in Ayyūbid Egypt." *Al-Masāq* 25 (2013): pp. 222 – 239.

33 See Awad, Wadi, "Al-Ṣafī ibn al-'Assāl," in *Christian-Muslim Relations: A Bibliographical History, vol. 4 1200 – 1350*, ed. by David Thomas and Alex Mallett, pp. 538 – 551. Leiden: Brill, 2012.

34 See this work discussed, re-edited, and translated into English in Thomas, David, *Christian Doctrines in Islamic Theology*, History of Christian-Muslim Relations 10. Leiden: Brill, 2008, pp. 19 – 77.

35 Jirjis, Murqus, ed., *Aṣ-Ṣafī ibn al-'Assāl: Kitāb aṣ-Ṣaḥā'iḥ fī jawāb an-naṣā'iḥ & kitāb nahj as-sabīl fī takhjīl muḥarrifī l-injīl*. Cairo: Maṭba'at 'Ayn Shams, 1926; two works of aṣ-Ṣafī published together.

naṣārā by the 9th century writer, Rabbān al-Ṭabarī,[36] which seems to have still been in circulation among Cairene Muslim scholars in the 13th century, for a local Muslim jurist relied upon al-Ṭabarī's work in the composition of his own anti-Christian polemic. This jurist was none other than al-Ṣafī's contemporary, Abū l-Baqā' Taqī al-Dīn Ṣāliḥ ibn al-Ḥusayn al-Ja'farī (1185–1270), the author of one of the earliest substantial, Cairene Muslim anti-Christian polemics entitled, *Takhjīl man ḥarrafa t-tawrāt wa-l-injīl*, a long text which the author subsequently distilled into two smaller works, *al-Bayān al-wāḍiḥ al-mashhūd min faḍā'iḥ an-naṣārā wa-l-yahūd*, and the compendious *Kitāb ar-radd 'alā n-naṣārā*.[37] The Coptic Orthodox patriarch of the time, probably Patriarch Cyril III (1235–1243), ordered aṣ-Ṣafī to write a refutation of the *Takhjīl*, a task which he reluctantly undertook, complaining that he had effectively accomplished the task in his earlier response to the work of Rabbān al-Ṭabarī, on which al-Ja'farī had relied heavily. Nevertheless the patriarch prevailed and al-Ṣafī dutifully composed the response called, *Nahj as-sabīl fī jawāb takhjīl muḥarrifī l-injīl.*[38] In both his works against the anti-Christian polemics of Rabbān al-Ṭabarī and al-Ja'farī, al-Ṣafī engaged in detailed exegeses of the Biblical proof-texts they cited in behalf of their positions.

Before looking at some of the disputed proof-texts, one should note that in his *al-Ajwibah al-fākhirah*, in which he refuted Paul of Antioch's arguments in behalf of Christianity based on proof-texts quoted from the Qur'ān, al-Qarāfī included large portions of the work of al-Ja'farī, albeit that al-Ja'farī makes no mention of Paul's *Letter.* Al-Ṣafī however, probably did know of the *Letter* for he seems to have incorporated a substantial portion of it, without naming Paul or the *Letter* explicitly, in his *Ṣaḥā'iḥ fī jawāb an-naṣā'iḥ*, indicating thereby that the *Letter* was already known in Cairo early in the 13th century.[39]

6

The Muslim, anti-Christian polemicists of Cairo in the 13th and 14th centuries included numerous Biblical proof-texts in their works. For example, as Diego

36 See Khalil Samir, Samir, "La réponse d'al-Ṣafī ibn al-'Assāl à la refutation des chrétiens de 'Alī al-Ṭabarī." *Parole de l'Orient* 11 (1983): pp. 281–328.

37 See Demiri, Lejla, "Al-Ja'farī," in *Christian-Muslim Relations: a Bibliographical History*, vol. IV, ed. by David Thomas and Alex Mallett, pp. 480–485. Leiden: Brill 2013. See also the discussion and further bibliographical details in Sarrió Cucarella, *Muslim-Christian Polemics*, pp. 74–82.

38 See the edition by Murqus Jirjis cited above, n. 35.

39 See the discussion in Sarrió Cucarella, *Muslim-Christian Polemics*, p. 71.

Sarrió Cucarella has recently shown, in al-Jaʿfarī's *Takhjīl* and al-Qarāfī's *al-Ajwibah al-fākhirah* there are fifty-one Biblical proof-texts cited in support of just the one claim that the coming of Muḥammad as a prophet was foretold in the Bible![40] At other points in their works they similarly adduce Biblical proof-texts in support of the Qur'ān's and Islamic theology's positions regarding other matters as well and particularly concerning the truth about Jesus of Nazareth from an Islamic perspective, in response to what they regard as Christian exaggerations in religion. Time and space allow for only one or two examples here, so two extended passages, one from the Muslim al-Jaʿfarī and one from the Christian aṣ-Ṣafī should be sufficient to demonstrate the detail in which the Muslim authors cite the scripture texts and the corresponding thoroughness with which al-Ṣafī responds to them.

6.1 Al-Jaʿfarī: Invalidating the Allegation of Christ's Death on the Cross

In his work, *ar-Radd ʿalā n-naṣārā*,[41] al-Jaʿfarī begins his discussion of the Christian affirmation of Christ's death by crucifixion with a brief survey of the Christological formulae of the Melkites, the Jacobites, and the Nestorians. And he argues that on the basis of their mutual affirmation of the union of divinity and humanity in Christ, belief in his death by crucifixion should logically be ruled out to begin with. Then, from the perspective of a Muslim jurist, he asks if reports of the crucifixion and death have been transmitted by way of a due process of continuous transmission (*tawātur*), or merely by way of disjointed, random accounts, such as would yield no certain knowledge. Arguing from his Islamic perspective that due to their discrepancies the Christian accounts of the crucifixion and death of Jesus on the cross do not measure up to the strict requirements for the transmission of authentic traditions (*tawātur*), al-Jaʿfarī concludes that they therefore cannot yield necessary knowledge. By way of proof he recounts in a free-wheeling style, without any quotation or exactitude, the disparate Gospel stories of Christ's arrest, concentrating on the behavior of his disciples, their confusions and denials. As for the Jews involved in the affair, al-Jaʿfarī claims that the number of witnesses is fewer than would be required for an authentic, continuous transmission, remarking along the way that even the Christian eye-wit-

---

40 See Sarrió Cucarella, *Muslim-Christian Polemics*, p. 218.

41 Abū l-Baqā' Ṣāliḥ ibn al-Husayn al-Jaʿfarī, *ar-Radd ʿalā n-naṣārā*, ed. by M. Muḥammad Ḥasanayn. Cairo: Maktabat Wahbah, 1409/1988, pp. 71–76.

nesses were just weak women. Moreover, he says, subsequent Jewish testimonies were reliant on the reports of the originally insufficient number of witnesses. Therefore, in his view there exists no reliable, continuous transmission of eyewitness reports of Christ's crucifixion and death sufficient to yield certain knowledge of the event. And it is at this point in his discussion that al-Jaʿfarī says, "So let us follow up with some obvious referents (*ẓawāhir*) from the Gospel that will impugn the reliability of the crucifixion and death of the Messiah, and they will refer it to someone else."[42] He then proposes eight arguments:

1) Jesus the Messiah was well known among the Jews as an important person from Galilee, who bested many in argument. So what would require them to hire one of his twelve disciples to identify his countenance to them if no look-alike were available?

2) The one who was crucified and killed was someone other than the Messiah. The Gospel of Matthew reports, the High Priest adjured the one who was arrested by the living God to tell us if you are the Messiah, the son of the living God, and he said to him, "You say so." (cf. Matt 26:63–64) He did not say I am the Messiah. And the Gospel of Luke concurs. (Luke 22:67–68) Were he the Messiah he would not have disguised his answer or used a trick to put the priest off. How could the Messiah, made to swear by the living God, not have said, "I am the Messiah"?

3) Luke says in his Gospel, "Jesus went up to a mountain of Galilee and Peter, James, and John were with him. While he was praying his visage was changed, his garments became white and shone like lightening. And they saw that Moses ibn ʿImrān and Elijah had appeared to him and a cloud overshadowed them. As for those who were with him, sleep fell upon them and they went sound asleep." (cf. Luke 22:67–68) "This passage that Luke transmitted is proof of the 'taking up' (*rafʿ*) of the Messiah and his protection from his enemies, the Jews. God, exalted be He, left them in the lurch."

4) Concerning God's protection of His prophet, the Messiah. The Gospel speaks of the horrible treatment of the one who was arrested, his scourging, his disfigurement, his carrying of the wooden beam, and his crucifixion. (cf. Matt 27–31) But at the beginning of his Gospel, Luke testifies that Gabriel announced to Mary, the mother of the Messiah, that God would seat her son on the throne of David and that He would give him rule over the house of Jacob forever. (Luke 1:21–22) Gabriel spoke the truth. And God's account is true. So if we were to say that the wretch who was crucified was the Messiah, it would require the invalidation of the veritable Good News.

5) The disciple John reports in his Gospel, "The one arrested went out to those who apprehended him from the garden in the Jordan valley and, his person being hidden from them, he said to them, 'Whom do you seek?' And they said, 'Jesus'. They put the question many times and he gave the same answer." (cf. John 18:4–8) That is proof of the production of a look-alike (*at-tashbīh*). For while he grew up among them and was raised with them, they did not recognize his face.

6) The Gospel of Luke says, "The Messiah companioned two men from Jerusalem who were heading for a village called Emmaus after his resurrection. He was walking along fol-

42 Al-Jaʿfarī, *ar-Radd*, p. 74.

> lowing them, but their eyes were withheld from recognizing him, but when he spoke with them they recognized him." (cf. Luke 24:13–32) This is proof of a change of state (*ḥāl*). How could his state have changed for his companions on the road and for his disciples in Galilee if there were no look-alike? Luke also said, "While the disciples were in the Upper Room, the Messiah was all of a sudden present in their midst and they did not recognize him. He asked them for something to eat and they gave him some fish and some honey." (cf. Luke 24:42–46) If his identity were concealed from his own disciples, how would the Jews have recognized him to have killed him?
>
> 7) John said, "The Messiah came upon his disciples while they were fishing and he said to them, 'Boys, do you have anything to eat?' But they did not recognize him and they said, 'no'. So he said, 'Cast the net on the right side.' They did it and it came to the breaking point with fish and then they recognized him." (John 21:1–8) There is much of that sort of thing in the Gospel.
>
> 8) Luke says, "Gabriel came to Mary in Nazareth and gave her the news that her son the Messiah would be a king for the Israelites, and that he would sit on the throne of his father, David." (cf. Luke 1:21–22) How then can the Christians differ with this promise and give the lie to Gabriel in his report, without a work of it being carried out? Rather, the opposite happened. He was mistreated and mocked. How could this tradition (*an-naql*) be sound, while God, exalted be He, was reporting on the tongue of Gabriel that he would be on the highest levels? How wrong can you be! There is no doubt about it, anyone who alleges the killing of the Messiah has told a lie.[43]

At the conclusion of these allusions to Gospel proof-texts that al-Jaʿfarī cites according to his own Islamic interpretation of them to support the Islamic view that the Messiah was neither crucified nor killed, he says that the very texts the Christians proffer to prove the contrary actually refute their own view, due to their insufficiency in providing the evidence of credible witnesses to the actuality of the alleged events. Al-Jaʿfarī then concludes that the Gospel texts he has cited actually go a step further and make clear God's protection of his prophet the Messiah from his enemies and they also make clear the existence of a look-alike, he says, by means of whom God distracted the Messiah's enemies from him. He says that people who uncritically accept and follow false reports without proof should carefully consider the passages from the Gospel he has cited.[44]

6.2 Al-Ṣafī's Explanation of Matt 24:36 and Mark 13:32

In their polemical texts against Christian beliefs, Rabbān al-Ṭabarī, al-Jaʿfarī, and al-Qarāfī all cited, as Biblical testimony to the fact that the Messiah is not God's

43 Ibid., pp. 74–76.
44 See ibid., p. 76.

divine son, Jesus' saying in the Gospel regarding knowledge of the last day, "Of that day and hour no one knows, neither the angels of heaven, nor the son, but the father alone." (Matt 24:36 and Mark 13:32) Al-Ṣafī responded to the claim most completely in his *Kitāb aṣ-ṣaḥā'iḥ*. He does so in response to Rabbān al-Ṭabarī's citation of the passage in his *Kitāb ar-radd 'alā n-naṣārā* as one in a list of Biblical testimonies that al-Ṭabarī, along with al-Ja'farī and al-Qarāfī, had argued should disprove the Christian doctrine of Christ's divinity.[45] 'Alī al-Ṭabarī had written:

> In the eleventh chapter of his Gospel, Mark said that when his disciples asked him about the hour of the resurrection, the Messiah said to them that no one knows that day nor that hour, not the angels who are in the heavens, not even does the son know it, but the father alone knows it, and this is an acknowledgment on his part that he was lacking in knowledge, whereas God is the most knowing and most knowledgeable about the matter, and that he was not He.[46]

The first thing al-Ṣafī did in response was to quote the Gospel passage correctly; he says, "Here is the text: 'Of that hour, no one knows it, neither the angels who are in the heavens nor the son, but only the father'." And he says there are a number of possible and plausible interpretations (*ta'wīlāt*) of the verse.

One interpretation is that considering only his humanity, the son does not know the hour; he knows it only by means of his divinity, in which he and the father are one. Al-Ṣafī points out that the attributes of his humanity pertain to the Messiah and that attributing sonship to him in virtue of his humanity is as sound as attributing sonship to him in virtue of his divinity.

Another interpretation is that by "the son," the saying in question intends the community of believers, like what God is reported in the Torah to have said to Pharaoh on the tongue of Moses: "Send out my first born son to worship me, otherwise I will kill your first born son." He meant by "His son" the community of the Israelites and by "the son of Pharaoh" all the first-born sons of the Egyptian people, because in fact He did kill them all.

Another interpretation is that it means that no one knows the hour, not the angels and not the son, but only the father, because someone had asked about knowing the hour, and someone was longing for the knowledge of the angels, and someone else preferred to know the son.

---

45 See the discussion in aṣ-Ṣafī, *Kitāb aṣ-ṣaḥā'iḥ fī jawāb an-naṣā'iḥ*, pp. 48–50.

46 Khalifé, I.-A. and W. Kutsch, "Ar-Radd 'alā-n-naṣārā de 'Alī aṭ-Ṭabarī." *Mélanges de l'Université Saint Joseph* 36 (1959): pp. 3–36 (115–148), p. 15 (27).

Another interpretation is that in denying the son's knowledge of the hour the intention was economy (*as-siyāsa*), because revealing the moment would bring harm to people, especially those in whose time the hour would come. For the people of every religion (*sharīʿa*) believe that God does not reveal the knowledge of the hour due to the wisdom (*al-ḥikma*) involved in keeping it secret. Economy is not lying. With regard to speech that is at variance with the mind (*aḍ-ḍamīr*), if by it harm is intended and it entails moral evil (*qubḥ*) and blame (*dhamm*), it is named a lie (*kidhb*) and an evil (*sharr*). But if a benefit (*an-nafʿ*) is intended by it, and it entails something good (*khayr*) and praiseworthy (*madḥ*), it is named "economy." What is named "economy" among philosophers (*al-ḥukamāʾ*), the learned men of the Arabs call "dissemblance" (*tawriyah*). It is the situation in which they tell someone who asks about a matter something that does not indicate what they actually know about that matter. All intelligent people would tell the questioner the same thing, on the condition that by means of this economy a greater wrong than a wrong against the truth is avoided in that particular matter.[47]

Al-Ṣafī goes on at some length to explicate this last interpretation more fully, along the way citing appropriate Biblical proof-texts to support his position. He concludes his discussion of this verse by saying, "It is perfectly clear that on the basis of many things in the Gospel that the son knows the hour of the resurrection," and he goes on to quote passages that speak of the son's role in the unfolding of the event. Finally, quoting passages that he argues indicate that what the father knows, the son also knows, al-Ṣafī remarks, "So any doubt about the son's knowledge of the hour is dispelled, and one of the interpretations of the denial of his knowledge as listed above must have been the case."[48]

7

Looking back from the vantage point of the controversies between Christian and Muslim controversialists in Cairo in the 13th century one can see how it came about over the years that the practice of mining one another's scriptures for proof-texts to support the beliefs and practices of each community involved both the selective choice of the passages and the interpretation of them within the hermeneutical horizon of the alien community. Paul of Antioch's deployment of verses from the Qur'ān and al-Jaʿfarī's use of Gospel citations well illustrate

47 See aṣ-Ṣafī, *Kitāb aṣ-ṣaḥā'iḥ*, pp. 48–50.
48 Ibid., p. 50.

the phenomenon. Similarly, al-Ṣafī's counter exegesis of Gospel passages cited by Muslim polemicists and Ibn Taymiyyah's re-interpretation of the Qur'ān verses cited in such works as the Cypriot recension of Paul's of Antioch's *Letter* well illustrate how the arguments from one another's scriptures in the 'Exegetical Crossroads' of Cairo in Mamlūk times, just as surely as the arguments from reason, issued eventually in alienating Christian and Muslim systems of thought from one another rather than providing any measure of rapprochement between them.

From another perspective one might notice that the abundant use of Biblical proof-texts by Cairene Muslim scholars such as al-Jaʿfarī and al-Qarāfī in their anti-Christian polemical works marks the high-point in the 13th century of the centuries-long process of the 'Islamicization' of the Bible and the 'Biblicization' of Islamic prophetology. Furthermore, the familiarity with the text of the so-called *Alexandrian Vulgate*, or the *Egyptian Vulgate* of the Gospels in Arabic on the part of these writers testifies to their interest in quoting the proof-texts in the form in which the local, Egyptian Christians actually had them, albeit that they often quoted somewhat freely and inexactly. In this regard their work paved the way for that of the younger Muslim commentators on the Christian scriptures, texts such as that of the aforementioned Najm al-Dīn al-Ṭūfī (d. 1316), whose *at-Taʿlīq ʿalā l-anājīl al-arbaʿah wa-t-taʿlīq ʿalā at-tawrāh wa-ʿalā ghayrihā min kutub al-anbiyā'* we now have in a new critical edition thanks to the recent work of Lejla Demiri. And one must also include the work of the later, 15th century Qur'ān commentator, Abū l-Ḥasan Ibrāhīm ibn ʿAmr ibn Ḥasan al-Biqāʿī (d. 1469),[49] who not only consulted Arabic translations of the Jewish and Christian scriptures in his interpretations of Qur'ānic passages, but also wrote a separate monograph in which he made the case, as Walid Saleh has pointed out, that it was not just legitimate, but actually necessary for Muslim scholars to make use of the canonical Bible in their work. Clearly by Mamlūk times the Bible had in some sense become an Islamic scripture, one almost literally beyond the ken of contemporary Christians, whose reading of the same texts and whose scriptural reasoning took place within the shaping framework of a very different hermeneutical paradigm.

As for the Christianization of the Qur'ān, and the quotation of Qur'ānic proof-texts in Christian, apologetic and polemical tracts, one hears little of it after Paul of Antioch's *Letter to a Muslim Friend.* Nevertheless it remains true that Christian writers in Arabic frequently quoted from the Qur'ān, sometimes imitated its diction and style, wrote polemics against it, used its words and

---

49 See the bibliographical information for the work of Demiri and Saleh quoted in n. 31 above.

phrases as proof-texts in their own works, and appealed to the religious authority of the Islamic scripture for its probative value. In many ways the Qur'ān effectively structured Christian religious discourse in Arabic. The Qur'ān's vocabulary seeped into the Arabic-speaking Christians' religious lexicon, subtly Islamicizing the ways in which they spoke of themselves for a time as *an-naṣārā*, who received the Gospel of Jesus the Messiah from the apostles, the messengers (*ar-rusul*), who properly purveyed and interpreted the messages of the prophets who came before them, whose witness was recorded in the Torah, the Psalms, and the Wisdom, just as the Qur'ān speaks of them. But the Qur'ān did not thereby become a Christian scripture.

Ironically it turns out that one might seriously entertain the idea that in their controversies with one another, the Christian and Muslim practice of proof-texting from one another's scriptures in their apologetic and polemical tracts in early Islamic times, the interlocutors achieved neither rapprochement nor reconciliation but rather a permanent divorce.

Najib George Awad

"Min al-ʿaql wa-laysa min al-kutub"

Scriptural Evidence, Rational Verification and Theodore Abū Qurra's Apologetic Epistemology

1 Introduction

During the early Abbasid era, Muslim and Christian *mutakallimūn* alike used the religious books in their debates and polemics. They did not only use at some occasions these books' verses, sūras and chapters in a literal manner that is close to proof-texting.[1] They further endeavored to develop exegetical and philosophical explanations and methods out of a serious interaction with either the religious texts that were followed by the *mutakallimūn*, or with the texts that were followed by these *mutakallimūn's* opponents. The outcome of this was a development of what Sidney Griffith calls an "arguing from scriptures" apologetic methodology[2].

In this essay, I attempt to examine whether or not Theodore Abū Qurra's (ca. 755 – ca. 830) apologetic arguments illustrate before us a Christian *mutakallim* who follows this *arguing from scriptures* method. I pursue this inquiry in conversation with an interesting nuance, which I spot in some of Sidney Griffith's publications on this subject. In his reading of Theodore Abū Qurra's *Mayāmir*, Griffith seems to be shifting from an early interpretation of Theodore as a follower of *arguing from reason* apologetic method into a more recently alternative in-

1 On how Muslim *mutakallimūn* used biblical texts after a proof-text method, especially the Gospel of John, see Accad, Martin, "The Ultimate Proof-Text: the Interpretation of John 20.17 in Muslim-Christian Dialogue (Second/Eighth-Eighth/Fourteenth Centuries)," in *Christians at the Heart of Islamic Rule: Church Life and Scholarship in 'Abbasid Iraq*, ed. by David Thomas, pp. 199 – 214. Leiden: Brill, 2003. In his turn, Mark Swanson traces a similar approach in the Christian apologists use of the Qur'ān, yet he tends to argue that the Christian Arab *mutakallimūn's* use of the text is so varied and subtle and interesting so as to be merely described as "proof-texting": Swanson, Mark N., "Beyond Prooftexting: Approaches to the Qur'ān in Some Early Arabic Christian Apologies." *The Muslim World* 88 (1998): pp. 297 – 319.
2 Griffith, Sidney, "Arguing from Scripture: the Bible in Christian/Muslim Encounter in the Middle Ages," in *Scripture and Pluralism: Reading the Bible in the Religiously Plural Worlds of the Middle Ages and Renaissance*, ed. by Thomas J. Heffernan and Thomas E. Burman, pp. 29 – 58. Leiden: Brill, 2005.

https://doi.org/10.1515/9783110564341-007

terpretation of him as an exemplary follower of an *arguing from scriptures* strategy. This essay endeavors to shed lights on central claims and views in Theodore Abū Qurra's literature and thought that would disclose which rule of argument is more genuinely definitive of his approach: *arguing from reason*, or *arguing from scriptures*.

2 *Arguing from Scriptures* in Early *Kalām*

It does not take the reader of the extant Christian-Muslim *kalām's* apologetic texts—be they Arabic, Syriac or Greek—much effort to realize that the Christian and Muslim *mutakallimūn* alike consistently include religious scriptural attestations (*Qur'ān, Injīl, Tawrāt*) in the storage of their polemics' and apologetics' arsenal. Both the Christian and the Muslim *mutakallimūn*, from at least the late 7th century A.D. onwards, either frequently used verses from the Gospel, the Old Testament and the qur'ānic sūras in defending their faith (from within their religious scriptures as such, or via the attestations of the other religions' scriptures). Or, they implemented such verses in their nullification of the authenticity and truthfulness of the other religions by means of the textual verses of their scriptures or the scriptures of the antagonized others.[3]

3 On the use of the religious scriptures, Bible or Qur'ān, in some recent scholarship, see for example Griffith, Sidney, "The Qur'ān in Arab Christian Texts: The Development of an Apologetical Argument; Abū Qurrah in the *Mağlis* of al-Ma'mūn." *Parole de l'Orient* 24 (1999): pp. 203–233; Cook, David, "New Testament Citations in the Ḥadīth Literature and the Question of Early Gospel Translations into Arabic," in *The Encounter of Eastern Christianity with Early Islam*, ed. by Emmanouela Grypeou, Mark Swanson and David Thomas, pp. 185–224. Leiden: Brill, 2006; Roggema, Barbara, "A Christian Reading of the Qur'ān: the Legend of Sergius-Baḥīrā and its Use of Qur'ān and Sīra," in *Syrian Christians under Islam, the First Thousand Years*, ed. by David Thomas, pp. 57–74. Leiden: Brill, 2001; Beaumont, Mark Ivor, "Early Christian Interpretation of the Qur'ān." *Transformation* 22 (2005): pp. 195–203; Accad, Martin, "The Gospels in the Muslim Discourse of the Ninth to the Fourteenth Centuries: An Exegetical Inventorial table (Part I)." *Islam and Christian-Muslim Relations* 14 (2003): pp. 67–91; Beaumont, Mark Ivor, "Muslim Readings of John's Gospel in the 'Abbasid Period." *Islam and Christian-Muslim Relations* 19 (2008): pp. 179–197; Monferrer-Sala, Juan Pedro, "An Arabic-Muslim Quotation of a Biblical Text: Ibn Kathīr's *al-Bidāya wa-l-Nihāya* and the Construction of the Ark of the Covenant," in *Studies on the Christian Heritage: In Honour of Father Prof. Dr. Samir Khalil Samir, S.I. at the Occasion of His Sixty-Fifth Birthday*, ed. by Rifaat Ebied and Herman Teule, pp. 263–278. Leuven: Peeters, 2004; and most recently Griffith, Sidney, *The Bible in Arabic: The Scriptures of the 'People of the Book' in the Language of Islam* (Jews, Christians and Muslims from the Ancient to the Modern World). Princeton, NJ: Princeton University Press, 2013, pp. 175–203.

In an essay on the use of religious scriptures in early Christian-Muslim polemics, David Thomas traces the implementation of religious scriptures according to the above-mentioned twofold strategy. Thomas notices that religious texts were used in the polemic writings "as illustrations or supporting examples of points which have already been established on other grounds."[4] Thomas further conjectures that these religious texts were merely used for the purpose of pure demonstration and only demonstration, for there was no common agreement between the two debating sides on the cogency of the arguments or the claims of these religious texts' content. What the Christian and Muslim interlocutors resorted to was a pragmatically oriented strategy: they snatched from the religious texts of their opponents "verses which would support their own case"[5] and would defend their faith via nullifying the belief of the other. Many Christian apologetic texts from the 9th and the 10th centuries, as Thomas notices, do show us how many Muslim *mutakallimūn* started relying on Biblical and Qur'ānic verses alike as weapons against their opponents, and how Christian *mutakallimūn* followed the same strategy to launch a counter-attack against Islam and in defense of Christianity.[6] Despite each side's serious suspicions on the other's religious scripture, both polemicists evinced readiness to employ the opponent's religious scripture in their polemical discourse.[7] The logic behind such a pragmatic strategy, Thomas surmises, is that, despite its in-authenticity and irreliability as a whole, the religious scripture of the opponent contains examples and ideas that efficiently back up the polemics against this opponent's faith. Both Muslim and Christian interlocutors opted for isolating scriptural verses (from the texts of the other's religious books) and exploiting them as they saw it fit.[8] In his turn, Mark Beaumont concurs in principle with Thomas' analysis, as he demonstrates that the Christian *mutakallimūn* opted for such a method because they believed that it will basically show that 1) the Qur'ān contains materials that are unworthy of being called "divine revelation," 2) the Qur'ān also presents Muḥammad as less than the prophet that the Muslims claim he is;

---

4 Thomas, David, "The Bible in Early Muslim Anti-Christian Polemic." *Islam and Christian-Muslim Relations* 7 (1996): pp. 29–38, here 29. See also Albayrak, Ismail, "Reading the Bible in the Light of Muslim Sources: From *Isrā'īliyyāt* to *Islāmiyyāt*." *Islam and Christian-Muslim Relations* 23 (2012): pp. 113–128; and Schmidtke, Sabine, "The Muslim Reception of Biblical Materials: Ibn Qutayba and His *A'lām al-Nubuwwa*." *Islam and Christian-Muslim Relations* 22 (2011): pp. 249–274.

5 Thomas, "The Bible", p. 30.

6 Ibid.

7 Ibid., p. 31.

8 Ibid.

and 3) the Qur'ān sometimes even supports basic elements in Christian faith, namely that Jesus is God's divine Word, that Jesus was a miracle worker and that Jesus has risen from the dead and was raised up to heaven.[9]

On the other hand, in an essay published in 1998, Mark Swanson tackles the same subject and proposes describing the strategy of using scriptural verses in polemic discourses as a rather bad example of a "prooftexting" argument.[10] Swanson states that

> the sort of inter-textual policy often labeled 'prooftexting' does indeed play a major role in the Arabic Christian apologetic literature, especially in popular texts that, passed from Christian hand to Christian hand, answered readers of the truth of Christian faith and the certainty of its vindication in fair debate.[11]

This notwithstanding, after visiting some early Arabic Christian apologetics (e.g., the papyrus fragments catalogued as *Heidelberg Papyrus Schott-Reinhardt 438*; *Fī tathlīth Allah al-wāḥid*; and Abū Qurra's *Mayāmir*) and comparing their qur'ānic citations with the same verses in the Qur'ān itself, Swanson concludes, that what we have there is an evident case of "exploitative imitation"; that lies in "an exploitative use of the Qur'ānic subtext."[12] Such exploitation, Swanson states, offers us a scarcely little indication "that the Christian apologist was paying any particular attention to the original context of the [Qur'ānic] phrases."[13] Instead of exegeting the religious scriptures in their own terms and from their sub-contextual framework, the Christian (and the Muslim) *mutakallimūn* decided to exploit the scriptural text in an eisegetical "prooftexting" apologies that aimed only at granting each interlocutor victory over his opponents. From the Christian *mutkallimūn's* apologies, Swanson concludes that "the Qur'ānic subtext is 'updated' by identifying in it trajectories which reach their goal only when pulled into a Christian gravitational field."[14]

Among the contemporary leading historians and interpreters of Arabic Christianity, none more than Sidney Griffith has so far written on the use of religious scriptures in relation to philosophical reasoning in early Christian-Muslim *kalām*. In his latest publication, *The Bible in Arabic* (2013), and in his

9 Beaumont, "Interpretation of the Qur'ān", pp. 195–203.
10 Swanson, "Beyond Prooftexting", pp. 297–319.
11 Ibid., p. 303.
12 Ibid., p. 317.
13 Ibid., p. 316.
14 Ibid., p. 217.

essay he published in 2005,[15] Griffith opines that the early *mutakallimūn* of the Abbasid era (Muslims, Jews and Christians alike) used the twin approach of "arguing from scripture" and "arguing from reason" in parallel to "comment on the truth of the confessions of faith they championed."[16] Griffith argues in his essay of 2005 that, influenced by the intellectual context of the Islamic milieu which they lived in, the Christian *mutakallimūn* not only learned from Muslim *kalām* to rely on reason in defending Christian faith. They also, and more importantly, adopted from the Muslim religious writers of the early Islamic period the specific method of "arguing from scripture."[17] In order to back up his proposal with some evidences of the existence of the method of *arguing from scripture* in Christian *kalām*, Griffith pays in his relevant publications a particular attention to the extant Arabic *Mayāmir* of Theodore Abū Qurra. In his recent monograph, *The Bible in Arabic*, Griffith brings as an example of this *arguing from scripture* method some lines from Theodore Abū Qurra's *Maymar fī taḥqīq nāmūs Mūsā l-muqaddas wa-l-anbiyā'* (Maymar on the Verification of the Holy Law of Moses and the Prophets).[18] From this *Maymar*, Griffith cites in English the following lines, deeming them an exemplary evidence of a Christian *mutakallim's* appeal to an *arguing from scriptures* apologetic methodology:

> Christianity is simply faith in the Gospel and its appendices, the Law of Moses, and the books of the prophets in between. Every intelligent person must believe in what these books we have mentioned say, and acknowledge its truth and act on it, whether his own understanding attains to it or not.[19]

The same approach characterizes Griffith's argument in his essay of 2005, as he also considers Abū Qurra's discourse one of the main examples of the Christian

---

15 Griffith, "Arguing from Scripture".
16 Griffith, *The Bible in Arabic*, p. 143.
17 Griffith, "Arguing from Scripture", p. 45.
18 Abū Qurra, Theodore, "Maymar fī taḥqīq nāmūs Mūsā l-muqaddas wa-l-anbiyā' alladhīna tanabba'ū 'alā l-Masīḥ wa-l-Injīl al-ṭāhir alladhī naqalahu ilā l-umam talāmīdh al-Masīḥ al-mawlūd min Maryam al-'adhrā' wa-taḥqīq al-urthūdhuksiyya allatī yansubuhā n-nās ilā l-Khalkīdūniyya wa-ibṭāl kull milla tattakhidhu n-Naṣrāniyya siwā hādhī l-milla" (Maymar on the Verification of the Holy Law of Moses and the Prophets who prophesized about the Messiah, and on the Gospel, was conveyed to the nations by the disciples of the Messiah who is born from the Virgin Maryam, and on verifying the orthodoxy, which people attribute to Chalcedonianism and nullifying every religious group that arrogates Christianity other than this one). in *Mayāmir Thāwūdūrūs Abī Qurra, usquf Ḥarrān, aqdam ta'līf 'arabī*, ed. by Constantine Bacha, pp. 140–179. Beirut: Al-Fawā'id, 1904.
19 Griffith, *The Bible in Arabic*, p. 143. Griffith quotes this citation from Bacha, Constantine, *Les Œevres arabes de Théodore Abou-Kurra évêque d'Haran*. Beirut: Al-Fawā'id, 1904, p. 27.

mutakallimūn's reliance on an *arguing from scriptures* apologetic method. This time, Griffith depends on another Abū Qurran *Maymar* to demonstrate this conviction. He claims that the reliance of the Melkite theologian on *arguing from scriptures* rule "was clearly the structure in Theodore Abū Qurra's [...] Arabic treatise, *On the Existence of the Creator and the True Religion*, as well as in his treatise *On the Authority of the Mosaic Law, and the Gospel, and on the Orthodox Faith*."[20] He cites first the same saying of Theodore in the *Maymar on Moses's Law and Chalcedonian Orthodoxy*, which I just jotted down from Griffith's text of 2013. Yet this time, he adds to it the following citation from Abū Qurra's *Maymar fī wujūd al-khāliq wa-d-dīn al-qawīm* (*Maymar on the Existence of the Creator and the Right Religion*)[21]:

> The verification of the truth of our position is that in the pure Gospel Christ said to his disciples: 'as my father has sent me' (John 20.21) to you, 'go to all peoples, make them disciples and baptize them in the name of the Father and the Son and the Holy Spirit and teach them to keep everything which I have commanded you. And behold, I am with you all days to the end of the world' (Matt 28.19 – 20) 'whoever believes shall live (John 11.25) and whoever does not believe is defeated and abandoned'. (John 3.18) And after Jesus spoke to them he ascended into the sky and took his seat at the right hand of the Father, and they went forth and preached in every place, while the Lord helped them and confirmed their preaching with the signs and wonders which they worked (Mark 16.19 – 20) and because of these all the nations (cf. Matt 28.19) accepted them.[22]

From these two quotations, Griffith extracts what he deems to be a sufficient evidence, not only of Abū Qurra's adoption of an *arguing from scriptures* rule, but also as a sufficient demonstrator in general of "the importance of 'arguing from the scriptures', and specifically from the Gospel, for Christians engaged in dialogue with Muslims in the early Islamic period."[23] Before he then concludes that "in a way, the whole Christian/Muslim encounter in the early Islamic period could be characterized as a conflict over the proper understanding of the narratives in the scriptures."[24]

20 Griffith, "Arguing from Scripture", p. 46.

21 Abū Qurra, Theodore, *Maymar fī wujūd al-khāliq wa-d-dīn al-qawīm* (Maymar on the Existence of the Creator and the Right Religion), ed. by Ignace Dick. Jounieh: Libraire St. Paul, 1982.

22 Griffith, "Arguing from Scripture", p. 47. Griffith cites this from Cheikho, Louis, "Mīmar li-Tādurus Abī Qurra fī wujūd al-khāliq wa-d-dīn al-qawīm." *Al-Mashriq* 15 (1912): pp. 825 – 842, here 837.

23 Griffith, "Arguing from Scripture", p. 47.

24 Ibid., p. 56.

Sidney Griffith is undoubtedly one of the most seminal contributors on Arabic Christianity in the shadow of Islam whom we luckily have in the field. Surfing through his writings during my research on the subject of this essay, drove my reading track years back into yet another, much older, essay he wrote on Theodore Abū Qurra's apologetic methodology in relation to a particular dimension that is relevant to my inquiry in this paper. The essay I mean here was published in 1994, and is titled *Faith and Reason in Christian Kalām: Theodore Abū Qurra on Discerning the True Religion.*[25] In this essay, Griffith presents a systematic hermeneutic of Theodore Abū Qurra's argument for the truthfulness of Christian religion in his *Maymar on the Existence of The Creator and the Right Religion.* Griffith summarizes the aim of Theodore Abū Qurra's defense in this *Maymar* by stating that this text endeavors "to argue in favor of the adequacy of Christianity's description of God when it is compared with the descriptions proposed by other religious groups, particularly the Muslims."[26]

When the reader tries to deduce the question to which Griffith seems to be offering the above-mentioned answer, one comes out with, more or less, the following inquiry: from which standpoint, and upon which arguing rule, Theodore defends the adequacy of Christianity in this specific *Maymar:* from the standpoint of scriptures, or from the standpoint of reason? Instead of echoing the same appraisal we read in his essay of 2005, namely that this *Maymar* is an evidence of Abū Qurra's dependence on *arguing from scriptures* rule, Griffith elaborates in length on Abū Qurra's rather particular, reliance on an *arguing from reason* apologetic method. He suggests that "the modern reader will immediately recognize the essentially rationalist, even the neo-platonic character" of Abū Qurra's schemes in this *Maymar,* before he adds that this use of rationalism "[owes its] inspiration to the Islamic intellectual milieu in which Abū Qurra functioned as a *mutakallim.*"[27] For Griffith, Abū Qurra's arguments in *Maymar on the*

---

25 Griffith, Sidney, "Faith and Reason in Christian *Kalām:* Theodore Abū Qurrah on Discerning the True Religion," in *Christian Arabic Apologetics During the Abbasid Period (750 – 1258),* ed. by Samir Khalil Samir and Jørgen S. Nielsen, pp. 1 – 43. Leiden: Brill, 1994.

26 Griffith, "Faith and Reason", p. 17.

27 Ibid., pp. 10 – 11. Scholars do also agree that early Christian-Muslim *kalām* was not only influenced by Platonic and neo-Platonic views, but most dominantly with Aristotelian ones. On the influence of Aristotelianism on Christian and Muslim *kalām,* see Gutas, Dimitri, *Greek Thought, Arabic Culture: the Graeco-Arabic Translation Movement in Baghdad and Early 'Abbasid Society (2nd – 4th/8th – 10th Centuries).* Abingdon: Routledge, 2005; D'Ancona, Cristina, "Greek into Arabic: Neoplatonism in Translation," in *The Cambridge Companion to Arabic Philosophy,* ed. by Peter Adamson and Richard C. Taylor, pp. 10 – 31. Cambridge: Cambridge University Press, 2005; Ziakas, Gregorios D., "Islamic Aristotelian Philosophy," in *Two Traditions, One Space: Orthodox Christians and Muslims in Dialogue,* ed. by George C. Papademetriou, pp. 77 – 108. Boston, MA:

Existence of the Creator and the Right Religion may not actually depict before us a necessarily subtle philosophical thinker. Nevertheless, they sufficiently reveal a Christian *mutakallim* attempting to prove *from reason* "what he and the Muslims ... already believe about God from revelation."[28] Why would Theodore argue from reason instead of from revealed truths of the religious text in this *Maymar?* Griffith indirectly answers this question few pages below in the essay, when he states that "it is clear in the very diction of the tract that the author [Theodore Abū Qurra] *assumes* biblical, even Qur'ānic, faith in his readers, whom he expects to be convinced of the coincidence of meanings in the rational and scriptural languages about God which he has been investigating."[29]

As an avidly serious reader of Sidney Griffith's contributions, I cannot but realize a change in view between his 1994's reading of Theodore Abū Qurra's *Maymar on the Existence of the Creator and the Right Religion* as a sample of a Christian *kalām* that is shaped after *arguing from reason* rule, on one hand, and his 2005's and 2013's consideration of Theodore Abū Qurra's *Maymar on the Existence of the Creator and the Right Religion*, and *Maymar on Moses's Law and Chalcedonian Orthodoxy* as two evidential examples of the Christian *kalām's* reliance on an *arguing from scriptures* apologetic method. My essay is not primarily on Sidney Griffith's scholarship *per se*. I am, rather, developing a brief study of a particular methodological aspect in Theodore Abū Qurra's apologetic thinking. Griffith's stimulating studies on this subject invite me, specifically and thankfully, to ask the following question: which method of argument is more dominantly characteristic of Theodore Abū Qurra's thought? Was he primarily an *arguing from scriptures* or more noticeably an *arguing from reason mutakallim?* Which of these two methods of arguments is comparatively more implemented by this Melkite theologian, and which among them is more foundational?

In the remaining of this essay, I search for possible answers to this inquiry in some of Theodore Abū Qurra's *Mayāmir.* I do not have here enough space to visit every single writ of Theodore's complete extant Arabic collection of *Mayāmirs.* Since I launch this explanation in conversation with Sidney Griffith's views,

---

Somerset Hall Press, 2011; Peters, Francis E., *Aristotle and the Arabs: The Aristotelian Tradition in Islam.* London: University of London Press, 1968; Watt, John W., "The Strategy of the Baghdad Philosophers: the Aristotelian Tradition as a Common Motif in Christian and Islamic thought," in *Redefining Christian Identity: Cultural Interaction in the Middle East since the Rise of Islam*, ed. by Jan J. van Ginkel, Heleen L. Murre-van den Berg and Theo M. van Lint, pp. 151–165. Leuven: Peeters, 2005.

28 Griffith, "Faith and Reason", p. 13.

29 Ibid., p. 27. Italic is mine.

I will follow him in narrowing my work by looking in particular at Theodore's *Maymar on the Existence of the Creator and the Right Religion* and *Maymar on Moses's Law and the Chalcedonian Orthodoxy*, allowing myself to point at some elements in other Arabic texts from Theodore's legacy whenever this is needed.

3 Reason and Scriptures in Abū Qurra's Apologies

3.1 In *Maymar on Moses's Law and Chalcedonian Orthodoxy*

The first *Maymar* from Theodore Abū Qurra's writings, which Sidney Griffith invokes as an example of dependence on *arguing from scriptures* rule, is *Maymar on Moses's Law and Chalcedonian Orthodoxy.* As I mentioned earlier, Griffith cites a quotation from this *Maymar* to show such a dependence on scripture in it in both his book of 2013 and his essay of 2005.[30] In the ensuing paragraphs, I shall also start my reading of Theodore's texts from this same *Maymar* to see if this text truly corroborates that Theodore Abū Qurra is one of the primary employers of *arguing from scriptures* apologetic method or not. I do look in my study to Constantine Bacha's Arabic edition of this *Maymar.*[31]

The content of *Maymar on Moses's Law and Chalcedonian Orthodoxy* is basically divided into two main arguments: the first is Abū Qurra's explanation of the uniqueness of Jesus Christ's message, which made the nations eventually accept his Gospel; while, the second argument is about Abū Qurra's justification of the authenticity of the Orthodox faith that is followed by those Christians who adhere to Chalcedonianism. It is very crucial to perceive the identity of those who are actually the imagined opponents, before whom Abū Qurra launches his defense of Christianity. The first option the reader may spontaneously think of is Muslim interlocutors. The fact that the *Maymar* is composed in Arabic may be the first inspirer of a possible Muslim reader to whom Abū Qurra is ad-

30 Griffith, *The Bible in Arabic*, p. 143; Griffith, "Arguing from Scripture", p. 46: "Christianity is simply faith in the Gospel and its appendices, the Law of Moses, and the books of the prophets in between. Every intelligent person must believe in what these books we have mentioned say, and acknowledge its truth and act on it, whether his own understanding attains to it or not." In my search for this quotation in Constantine Bacha's Arabic edition of this *Maymar*, I could not actually find these lines as Griffith cites them. Griffith might be here translating into English Bacha's words, rather than Theodore's words in the Arabic text itself.

31 Abū Qurra, Theodore, "Maymar fī taḥqīq nāmūs Mūsā".

dressing the writ. But, actually, the Arabic language of the text does not automatically entail that Abū Qurra is writing the text to argue with no other than Muslims. It is now known by scholars that, starting from the 8th century CE onwards, Arabic was not just the *lingua sacra* of Islam, but also the *lingua franca* of all those who belonged to, and lived in, the territories of the Islamic caliphate, Muslims, Jews and Christians alike.[32] It is quite normal to see an 8th-9th-centuries' thinker like Abū Qurra composing his *Maymar* in Arabic. It is equally perfectly normal to expect him opting for writing in Arabic upon the hope that Muslims as well would also have an access to his *Maymar* if they wanted to. This conceded, the linguistic textile of the text does not sufficiently corroborate that the arguments which Theodore develops in the text are addressed as apologies against Muslim antagonists.

The actual content of the text conspicuously shows that, in this *Maymar*, Theodore is not actually debating with Muslims, but rather with Jews and other non-Chalcedonian, or non-Melkite, Christians. The first argument on Jesus' Gospel is developed in comparison and contrast with the Jewish Torah, or the Law of Moses.[33] Here, Abū Qurra argues against the Jews that the authenticity of the Christian Messiah and His Gospel is anticipated by Moses's Torah, and that the uniqueness and superiority of Jesus's messiahship is praised in stark contrast with the prophetic status of the Jews' patriarch, Moses. On the other hand, the second argument on Chalcedonian orthodoxy is done in a blatant antagonism against the stances of "the heretics" (*arāṭīq/'arāsīs*), namely the Jacobites, Nestorians, Monothelites and other Christian sects who do not abide by the referential orthodoxy of the church's ecumenical councils, especial-

32 See, for instance, on the Arabization of the Syro-Palestinian Christian culture Griffith, Sidney, "From Aramaic to Arabic: The Language of the Monasteries of Palestine in the Byzantine and Early Islamic Periods," in *The Beginnings of Christian Theology in Arabic: Muslim-Christian Encounters in the Early Islamic Period*, ed. by Sidney Griffith, pp. 11–31. Aldershot: Ashgate, 2002; Sahas, Daniel, "The Arab Character of the Christian Disputation with Islam: The Case of John of Damascus (ca. 655–ca. 749)," in *Religionsgespräche im Mittelalter*, ed. by Bernard Lewis and Friedrich Niewöhner, pp. 185–206. Wiesbaden: Otto Harrasowitz, 1992; Griffith, Sidney, "The Monks of Palestine and the Growth of Christian Literature in Arabic." *The Muslim World* 78 (1988): pp. 1–28; Griffith, Sidney, "The Church of Jerusalem and the 'Melkites': The Making of an 'Arab Orthodox' Christian Identity in the World of Islam (750–1050 CE)," in *Christians and Christianity in the Holy Land, from the Origins to the Latin Kingdoms*, ed. by Guy G. Stroumsa and Ora Limor, pp. 175–204. Turnhout: Brepols, 2006; Swanson, Mark, "Arabic as a Christian Language." on www.copticbook.net/books/20101102002.pdf (consulted on August, 26th, 2013); Griffith, Sydney, "What Has Constantinople to Do with Jerusalem? Palestine in the Ninth Century: Byzantine Orthodoxy in the World of Islam," in *Byzantium in the Ninth Century: Dead or Alive?*, ed. by Leslie Brubacker, pp. 181–194. Aldershot: Ashgate, 1998.
33 Abū Qurra, "Maymar fī taḥqīq nāmūs Mūsā", pp. 140–154.

ly Chalcedon and other pro-Chalcedonian ones.[34] In both sections, thus, Theodore Abū Qurra never mentions Islam or Muslims in his conversation. He rather addresses clearly assumed Jewish interlocutors, saying "So, if you said, O Jew" (*fa-in qulta yā yahūdī*)[35], "So, tell me, Jew" (*fa-akhbirnī yā yahūdī*),[36] "and you cannot say, O Jew" (*wa-lā tastaṭī' an taqūl yā yahūdī*),[37] and "this, O Jew, proves to you, if you have reason" (*hādhā yā yahūdī yadulluk in kāna laka 'aql*).[38] Abū Qurra similarly addresses the heretics and non-Melkites in his defense of Chalcedonian orthodoxy by saying, for example, "So, you O Nestorian, know that you are mistaken" (*fa-anta ayyuhā an-nusṭūrī i'lam annaka fī khaṭā'*),[39] "and, you, O Jacobite, what is wrong with you, accepting the three councils...and rejecting this fourth council" (*wa-anta yā ya'qūbī mā bāluk qabilta ath-thalāth majāmi' wa-lam taqbal hādhā al-majma' ar-rābi'*),[40] "so, what is wrong with you, O monothelite: you gladly and easily accepted the first and the second and the third councils [...] so, when you arrived to the sixth council, you seemed as if you have forgotten the teaching of the Holy Spirit" (*fa-mā bāluk yā mūnūthilitī qabilta al-majma' al-'awwal wa-th-thānī wa-th-thālith mustarsilan munbasiṭan [...] fa-lammā balaghta al-majma' as-sādis ka-annaka qad nasīta ta'līm ar-Rūḥ al-Qudus),*[41] *and "and you said, O heretic"* (*wa-anta qulta yā herṭīq*).[42] These phrases suggest that this specific *Maymar* does not actually exemplify before us any deducible apologetic methodology that is used by Christian *mutakallimūn*, like Abū Qurra, in interaction with Islam. The Muslims are not the primary counterparts at all in this *Maymar.* They are the Jews and the non-Melkite Christians.

What, more intriguingly, catches the attention in this *Maymar* is the fact that Abū Qurra seems actually not to be relying on an *arguing from scriptures* method, as one may think. My reading of this *Maymar* drives me to propose that Theodore neither argues from scriptures, nor from reason. He rather argues from both divine miracles and the Holy Spirit. In the first section on the Messiah's Gospel over-against Moses's Law alone, Theodore refers to miracles (*a'ājīb*) forty three

34 Ibid., pp. 155–180.
35 Ibid., p. 146. All the English translations of the Arabic phrases and texts from Abū Qurra's *Mayāmir* are my own translations.
36 Abū Qurra, "Maymar fī taḥqīq nāmūs Mūsā", p. 147.
37 Ibid., p. 149.
38 Ibid., p. 153.
39 Ibid., pp. 166, 172.
40 Ibid., p. 167.
41 Ibid., p. 169.
42 Ibid., pp. 171, 173–174, 176–177.

times.[43] While, in the ensuing section on the ecumenical councils, Abū Qurra uses the terms Holy Spirit (*Rūḥ al-Qudus*) and spirit (*rūḥ*) sixty times. In the first section, he uses miracles as the corroborating criterion which proves that the Messiah, Jesus, and His Gospel are the authentically true religious truth, whatever the Jews say in contrast. In the second section, this criterion becomes the Holy Spirit, as Theodore begins to argue against various Christian heretic claims that all the ecumenical councils that are adopted by the Melkites-Chalcedonians are inauthentic and ecclesially irreferential. In response to this allegation, Theodore argues that, although they were summoned by royalties, and though they come to add teaching on top of the first councils of Nicaea and Constantinople, the third, fourth, fifth and sixth councils are no less orthodox than the first two, because they were equally generated by the guidance of the same Holy Spirit, and were empowered by the same Spirit's miraculous grace, that lead the ones of Nicaea and Constantinople. This extensive speech on miracles and the Holy Spirit suggests nothing, first and foremost, but Theodore's development of two consecutive arguments: one from miracles and another from the Holy Spirit. The content of the *Maymar* verifies much less clearly any conjecturing that the apologetic method which is used here depicts an 'arguing from scriptures' rule.

3.2 In *Maymar on the Existence of the Creator and the Right Religion*

The second *Maymar* Sidney Griffith construes as an example of a Christian *kalām*'s dependence on *arguing from scriptures* apologetic methodology in debate with Muslims is Theodore's *Maymar fī wujūd al-khāliq wa-l-dīn al-qawīm (Maymar on the Existence of the Creator and the Right Religion)*. In the ensuing paragraphs, I shall be reading this text as it appears in Ignace Dick's Arabic critical

---

43 Also in his *Maymar fī wujūd al-khāliq wa-l-dīn al-qawīm (Maymar on the Existence of the Creator and the Right Religion)*, in chapter sixteen (believed to be an appendix to the original text by Ignace Dick), Abū Qurra also refers to "argument from miracles" to prove to those who even doubt the criterion of reasoning that Christianity was spread around the world and accepted by all by virtue of its miracles: Abū Qurra, *Maymar fī wujūd al-khāliq*, ch. 16, pp. 259–270. On a good analysis of this chapter and its relation to Abū Qurra's overall argument in the *Maymar*, see Swanson, Mark N., "Apology or its Evasion? Some Ninth-Century Arabic Christian Texts on Discerning the True Religion." *Currents in Theology and Mission* 37 (2010): pp. 389–399, here 391–394.

edition, the version of 1982.[44] I will examine whether or not this *Maymar* truly demonstrates Abū Qurra's representation of an *arguing from scriptures* Christian *kalām* in his discourse.[45]

In the second book of this *Maymar*, where Abū Qurra starts discussing the criterion for determining the right religion, one can realize that Theodore's methodological point of departure is far from building upon ready-made presumptions on other religions and their claims. Theodore launches, instead, an auditing process to the followers of the other religious ways, making them introduce their own beliefs without allowing himself, at this stage, to debate their saying's content.[46] At the foundation of this starter is the following main question: how do we come to determine the true religion among all the religious beliefs that we have available around us? Abū Qurra's *Maymar* states that we first depart from knowing each religion from the testimony of those who follow it, not from previous knowledge we personally structure about what these religions claim. Second, we place these religion's claims opposite to each other within a comparative, analytical framework. Abū Qurra pursues this twofold policy in the ensuing chapters of his *Maymar.* And, from this comparative analysis he eventually concludes that all religions more or less commonly speak about "deity, good and evil and reward and punishment." Theodore finds in these three elements an invitation for him to discern a noticeable, principal commonality between religions. However, this very same comparison equally suggests to him that religions beg to differ from each other in how each understands the

---

44 In my reading of the text, I follow Dick's presentation of the three parts of this *Maymar* as one textual body, or as three chapters of one and the same essay. I deem this to be more convenient to the content and systematic structure of the text, and more successful in conceiving the inner logic of Theodore's argument as one whole. In this, I do methodologically disagree with John Lamoreaux's attempt in his English translation of this *Maymar* to treat the text's three parts as three wholly independent texts, each forming independent treatise. See the introduction of

Lamoreaux, John C., *Theodore Abū Qurrah*, trans. by John C. Lamoreaux, Library of the Christian East 1. Provo: Brigham Young University Press, 2005, p. xxxiv. And see pp. 1–25, 41–47 and 165–174 for Lamoreaux's division.

45 In the following lines, I rely on my analysis of this *Maymar* in the second chapter of my monograph on Theodore Abū Qurra's theological *kalām* in debate with Muslims: *Orthodoxy in Arabic Terms: A Study of Theodore Abū Qurrah's Theology in its Islamic Context.* Boston: De Gruyter, 2015.

46 Abū Qurra, *Maymar fī wujūd al-khāliq*, ch. 7–16 (pp. 200–270), ch. 7 (pp. 200–210). Between brackets, I record the pagination of the text as it appears in Dick's edition.

above mentioned three elements and how each speaks about this understanding in the service of its own claims.[47]

The previous two stages highlighted, Abū Qurra moves yet toward a third, more crucial, stage that is related to the far more daunting challenge of deciding which religion among them all is authentically the true and accurate one. The basic logical principle that seems to be guiding Theodore's exploration here is the following: one, and only one, of these religions must be the belief that people should follow, and the thinking intellectual should find out which one it is.[48] One of the common ways, according to Abū Qurra, that people normally follow to decide which is the true religion, is to consider as reliable that religion discourse which claims that God sent to His followers a messenger (*rasūl*) and provided them with a book (*kitāb*) to guide them to the truth and drive them away from condemnation. But, Abū Qurra then comments that though this is occasionally a reliable measurement, one can easily realize that all the religions unexceptionally speak about a messenger from God and argue that the book this messenger brought to them as the message from God is alone the true book. Be that as it may, the question "which religion is the true one?" should transform now into "which religious messenger among the messengers, and which religious book among the books, are the true ones? Can we know this by the help of these religions *per se?*" Abū Qurra affirms that this is far from likely, for the religions' messengers speak against each other and each one presents an argument that aims at proving the falsehood and forgery of the others' messages.[49] What matters for Abū Qurra, in other words, is assessing the validity and plausibility of all religions' self-allegations, as well as their claims about other faiths, by means of an external, reliable criterion of investigation from without the religions *per se*. This external criterion, as Abū Qurra suggests, will enable us to see 1) which religious claims about God's attributes concur plausibly with God's reality, and which are not, and 2) which message of salvation from sinfulness expresses a

---

47 Abū Qurra, *Maymar fī wujūd al-khāliq*, ch. 8.2–3 (p. 211): "2. Now concerning their agreement, every one of them, except for one or two, claims that it has a god, and it owns right and wrong, and also reward and punishment. 3. Concerning their discrepancies, they disagree on the attributes of their deities and in their [understanding of] right and wrong and reward and punishment."

48 Abū Qurra, *Maymar fī wujūd al-khāliq*, ch. 8.6–8 (p. 212): "6. The situation [of these religions] is one of two cases: either none of them is from God, or if any is from God, it is only one. 7. There must be among them one [religion] expressive of the known providence and grace of God towards His creatures. 8. But how can one get to know [which religion is it]?"

49 Abū Qurra, *Maymar fī wujūd al-khāliq*, ch .8.20–21 (p. 214).

true and reliable interpretation of the notion of healing as we understand it, and which does not.[50]

Now, when we chase after the investigation criterion that Abū Qurra refers to, we realize that it is no else than the sheer power of reasoning, which, according to him, God bestowed upon the human and ordered her to use for knowing the divine, for knowing and doing the good and for knowing and avoiding evil.[51] One chapter before the end of his *Maymar*, Theodore Abū Qurra states explicitly that his main goal in this specific writing is to argue strictly and exclusively from reason alone, and not from religious scriptures, in defense of the authenticity of Christianity and its religious texts: "3. So we respond to this speaker: our intention in our text, this one, is to corroborate our religion from reason, and not from scriptures."[52]

What makes us know the true religion from the false one? Abū Qurra's answer is the following: it is not the scriptures of these religions, but rather human reasoning. The scriptures and their messengers' claims should both be first judged by reason:

> 37. Now, we should [...] put the scriptures aside and ask reason: how did you know the attributes of God, which the senses cannot perceive or the brains apprehend, from its image in human nature? (...)
> 38. If [reason] answered us and we knew how, we can then examine the books we have, and the books that contain the same knowledge, we would know that it is from God, we will [then] accept and confirm them, throwing others away.[53]

When in the ensuing chapters of his *Maymar* Abū Qurra applies the criterion of *arguing from reason* to verify the content of the scriptures of religions, he discovers that only the Christian Gospel contains the rational understanding of God, good and evil, and reward and punishment, to which the criterion of reason naturally leads and which it normally verifies. The Gospel speaks about God, he says, exactly as our reason analogically tells us about God from His image in our human nature.[54] The Gospel alone calls us for following the way of right-

---

50 Ibid., ch. 8.26 (p. 215).

51 "30. And the doctor, the reason he was given, and by means of it he knows God and by means of it he knows the good and works according to it and knows evil and abstain from it." See Abū Qurra, *Maymar fī wujūd al-khāliq*, ch. 8.30 (p. 30)

52 Abū Qurra, *Maymar fī wujūd al-Khāliq*, ch. 15.3 (p. 255).

53 Ibid., ch. 8.37–38 (pp. 217–218).

54 "16. And the description of the Gospel alone is from God because it brought to us what our nature in its assimilation to God has taught us, as we described above." See Abū Qurra, *Maymar fī wujūd al-khāliq*, ch. 12.4–5 (p. 241).

eousness and love as we learn about them by virtue of rationally deducing their images in our human nature.[55] The Gospel also exclusively offers the promise of being truly with God that concurs fully with what our human nature seeks rationally and aspires at from the relationship with God.[56] From studying cognitively the Christian texts and judging their content by the rules of reasoning and logic, Abū Qurra gleans the following:

> 30. Therefore, we believe in this religion, we take it upon ourselves and we abide with it, tolerating for its sake the world's tribulations, for the sake of the hope it promises, and we die with it hoping to meet the face of God on it
> 31. and we cast away anything else and reject it and consider it nothing.[57]

In the remaining of the *Maymar*, Abū Qurra moves from offering an apology for the criteriality of reason in deciding the true religion into an apologetic elaboration on the correctness and reliability of the religion of the Bible, which he concluded by now that it is the only truthful belief according to reason. So, he responds to the Marcionites' rejection of the Old Testament by means of the Gospel's witness and argues afterwards for the reliability of Christianity on the basis of miracles. One wonders here if actually Abū Qurra is contradicting his previous claim that the validity of any message should not be proved from any proof-texting or by means of the textual content of the message itself. Far from this, Abū Qurra himself states here that his purpose is to advocate Christianity from reason, not from religious texts.[58] What Abū Qurra is doing here, in fact, is relying on the Gospel's textual testimony *after*, and not before, he verified upon mere deductive reasoning that this scripture alone is congenial with the criterion of rationalism and is representative of authentic logic.

The careful reading of Abū Qurra's methodological choice in *Maymar on the Existence of the Creator and the Right Religion* clearly suggests that Theodore seems to be more secure in his reliance on reason and more certain that *arguing from reason* would undoubtedly corroborate that Christianity alone is the right religion. His rationale here seems to be the following: the written texts do not prove the truthfulness of anything. However, when reason ascertains that the content of certain texts is according to human rationality, this text is then reliable for validating any other claim. If the text is according to reason, the text can

---

55 Abū Qurra, *Maymar fī wujūd al-khāliq*, ch. 13.1 ff. (pp. 243–247).
56 Ibid., ch. 14.1–24 (pp. 238–252).
57 Ibid., ch. 14.30–31 (p. 253).
58 Ibid., ch. 15.3 (p. 255).

be used as a criterion of truth. But, one should first prove that the text is according to logical and rational verification.

3.3 Abū Qurra's *Mujādalah* in al-Ma'mūn's Court

Maymar on the Existence of the Creator and the Right Religion drives us to surmise that, to a noticeable extent, Abū Qurra seems to be less confident about the efficiency of *arguing from scriptures* to prove the validity of any religion. He seems, rather, to be more secure with the effectiveness and ability of *arguing from reason* method. It goes without saying that one cannot move such a conjecture from the level of speculation into the level of certainty, since Abū Qurra himself is no more personally available to approve or deny this possibility. This historical fact admitted, one can still, in my conviction, try to extract traces to corroborate such a total refuge in reason more than textuality from Theodore Abū Qurra's other apologies before Muslims.

There is one particular text from the extant written narrations on Theodore's theological legacy, which I have here specifically in mind, and I would like in this section to share some insights about. The text I mean here is the one which narrates a debate Abū Qurra had with Muslim interlocutors in the court of the caliph, al-Ma'mūn. In the ensuing lines of this section, I will lay out an observation from the text of *al-Mujādalah*. The critical edition of *al-Mujādalah* I use in my reading is the one that was made by Wafīq Naṣrī and published by CEDRAC in 2010.[59]

In *al-Mujādalah*'s text, there is a recording of a very intriguing incident between the Christian apologist, believed by the author of the text to be Theodore Abū Qurra, and one of the Muslims he debated with before the caliph. This event narrates a specific attitude or a position that Abū Qurra falls into, which, I believe, demonstrates that Abū Qurra is more at home with relying on logical and philosophical reasoning, and is rather ill-at-ease with using arguments from scriptures, especially when the scripture he tries to cite verses from is the Qur'ān. It is inescapably interesting that the scribe, or the copier, of the debate's moments deliberately records this moment of weakness in Abū Qurra's performance before the caliph. This stimulates, in fact, the assumption that this scribe could be an eye-witness of Abū Qurra's debate in al-Ma'mūn's court, and could not but pause at this seldom moment of hesitation in the career

59 Wafīq Naṣrī, ed., *Abū Qurrah wa-l-Ma'mūn: al-Mujādalah*. Beirut: CEDRAC, 2010.

of the Melkite *mutakallim*, who is often depicted as an impeccably and flawlessly intelligent and eloquent apologist.

At an early stage in *al-Mujādalah*'s proceedings, the scribe records for us Abū Qurra's response to an inquiry from al-Ma'mūn on the relation between the Jewish Old Testamental teaching and the Christian Gospel.[60] Abū Qurra closes up his answer to the caliph's inquiry by stating the Christian belief in Jesus Christ's divine identity in the following words:

> 47. But we conceded him [the Messiah] with well intention and we memorized his commandments with our whole hearts
> 48. and we did not doubt that He is the God of Abraham, Isaac and Jacob
> 49. and He is a God [and] Son of God, God's Word and Spirit without separation between them.[61]

As the reader can realize, Theodore here incorporates a qur'ānic terminology into his theological definition of Christ. Hence, he tries to play the game of recruiting the Muslim scriptural language in the service of his apologetic theology. What is truly worth noting here ensues directly Abū Qurra's utterance of these words. The text of *al-Mujādalah* narrates that, from the audience, a Muslim called Muḥammad b. ʿAbdullah al-Hāshimī jumps up immediately on his feet and harshly rebukes Abū Qurra's endeavor. Al-Hāshimī says:

> 51. Woe to you, Abū Qurra, Christ is God's Word and Spirit sent by Him to Mariam
> 52. and, for God, He is like Adam, created him from soil and breathed in him His Spirit.[62]

Al-Hāshimī smartly comprehends Abū Qurra's implementation of the qur'ānic language. So, he immediately takes Theodore to the inner circle of the qur'ānic attestation, developing a refutation that combines textual attestations from Q 4:171 ("God's Word and Spirit sent by Him to Maryam"); Q 3:59 ("He is like Adam, created him from soil") and the two verses of Q 15:29 and Q 38:72 ("and breathed in him His Spirit"). This was al-Hāshimī's way to warn Abū Qurra that he imposes on the qur'ānic text inappropriate connotations when he uses its language to back up the content of the Gospel. For al-Hāshimī, adhering to the qur'ānic language to speak about the Messiah requires reading this language strictly within the framework of ʿĪsā's birth from a human virgin as a

60 Abū Qurra, *al-Mujādalah*, ch. A.1–3.36–49 (pp. 106–109).
61 Ibid., ch. A.3.47–49 (p. 109).
62 Ibid., ch. B.51–52 (p. 110).

purely human child, and never reading Word and Spirit as a scriptural evidence of Jesus' super-human identity.[63]

What I want to pause at shortly in relation to this very peculiar incident is Abū Qurra's reaction to al-Hāshimī's harsh rebuke and thorough correction. The text says that Abū Qurra delves suddenly into a long and deep silence and pensive contemplation.[64] The scribe of *al-Mujādalah* tries to offer his 'way-out' to Abū Qurra's, rather, very unprecedented moment of vulnerability and wordlessness, by suggesting that Abū Qurra was politely and diplomatically waiting for the caliph's permission to comment back.[65]

Abū Qurra's attitude in the other parts of *al-Mujādalah*'s text, let alone his style in his other *Mayāmir*, does not at all depict before us a Christian *mutakallim* who busies himself with diplomatic politeness and communication protocols. He is most of the time blunt and explicit and sometimes rude and aggressive in his accent and words. Therefore, and contrary to the scribes' justification, what seems to actually have happened in his clash with al-Hāshimī is that Abū Qurra immediately realized that his previous rhetorical endeavor to rely on the Qur'ān in the service of his theology was doomed totally to failure, for al-Hāshimī demonstrated his solid knowledge of the real content of the Qur'ān. As Abū Qurra perceived this, he tried to usurp some moments to come up with another arguing strategy to compensate the failure of his earlier *arguing from scriptures* endeavor. Theodore simply concludes that he needs to resort to something other than textuality alone. He determines, thus, to press forward in his argument by bringing back into the discussion the apologetic methodology, which he is more at home with, namely *arguing from reason.*

This is exactly what the ensuing pericopies of *al-Mujādalah* inform us about, as Theodore then starts to drag his interlocutor into philosophical and logical inquiries and ideas, rather than really delving at any deeper level into scriptural hermeneutics, neither of the Gospel nor of the Qur'ān. Though the following debates in the court of the caliph involve invocation and citation of Qur'ānic and Biblical claims, by Theodore and his Muslim counterparts alike (as, for example, when Abū Qurra comments on Q 4:157 to explain the death of Christ),[66] yet these

---

63 I also derive some ideas here from chapter five of my book *Orthodoxy in Arabic Terms*. The ensuing lines are borrowed from this chapter.

64 Abū Qurra, *al-Mujādalah*, ch. B.1.53 (p. 110): "53. So Abū Qurrah remained for a long time silent and gave no answer and stared reflectively at the floor."

65 Abū Qurra, *al-Mujādalah*, ch. B.1.54–61 (pp. 110–112).

66 Abū Qurra, *al-Mujādalah*, ch. C.315, 318, 320–323 (pp. 159–160). I recently studied this attempt in the text of the debate in the caliph's court in a yet unpublished essay titled "Need the Crucifixion Happen? *An-Nisā'* 4:157–158, Theodore Abū Qurrah and His Muslim Interlocutors

scriptural invocations were conspicuously controlled by, and shaped from this moment onward after, a more dominant rational and philosophical debating framework that Abū Qurra kept pulling his opponents into. These scriptural verses would not be used in the service of a purely rational, syllogistic argument. Abū Qurra now uses not to prove the truthfulness of the Christian faith from within scriptural attestations, but to show his interlocutors the logical plausibility of this faith and the impossibility of its rational nullification. The scribe of *al-Mujādalah* even describes to us Theodore's rational strategy in his interaction with the Muslims, when, further through the debates, a Muslim interlocutor with the nickname of al-Dimashqī confesses his helplessness before Abū Qurra's rational arguments and attributes his failure to Theodore's cunning strategy of responding to his opponents' challenges with a rational explanation that is developed from within the latter's very own ideas and claims:

> 363. [ad-Dimashqī] said: because Abū Qurra shoots me with arrows from my own quiver
> 364. so, my weapon is mortal, his weapon is remaining.[67]

4 Abū Qurra, a Unique Rationalist in Early *Kalām*

My previous reading of Abū Qurra's *Maymar on Moses's Law and Chalcedonian Orthodoxy* reveals that this *Maymar* does not persuasively prove Abū Qurra's reliance on *arguing from scriptures* rule as his apologetic strategy with Islam. This *Maymar* presents to us a Melkite *mutakallim* who does not actually argue with Muslims, but the Jews and the non-Chalcedonian Christians. This *Maymar* also does not depict a Christian *mutakallim* arguing from scripture or even from reason, but one who follows, so to speak, 'arguing from miracles' and 'arguing from the Holy Spirit' methods, instead.

On the other hand, my reading of *Maymar on the Existence of the Creator and the Right Religion* shows that this text cannot as well be considered an evidence of Theodore Abū Qurra's dependence on an *arguing from scriptures* apologetic method. In this *Maymar*, as I demonstrated, Theodore personally states that he is developing a verification of Christianity "*from* reason, and not from the scriptures" (*min al-'aql wa-laysa min al-kutub*).[68] In addition, Abū Qurra subjects all the religious scriptures, the Bible included, to the criterial judgment of natu-

in al-Ma'mūn's Court," which I read in the Annual conference of the American Oriental Society, Phoenix, Arizona, March, 13–16, 2014.

67 Abū Qurra, *al-Mujādalah*, ch. D.4, 363–364 (p. 167).

68 Abū Qurra, *Maymar fī wujūd al-khāliq*, ch. 15.3 (p. 255).

ral deductive reasoning, making reason the judge that decides which religious text conveys the true faith among them all. Finally, the brief reading of Abū Qurra's clash with al-Hāshimī in the text of *al-Mujādalah* indirectly invites us to speculate about the reason behind Theodore's reliance on an *arguing from reason*, instead of an *arguing from scriptures* method in his interlocution with Muslims. The reason may presumably exceed the mere historical factors that are descriptive of Theodore's intellectual, prevalently Mu'tazilite milieu.[69] The reason may also be personal and experiential.

I am not trying here to make Theodore Abū Qurra an anti-scripture thinker or an unreserved antagonist of the use of *arguing from scriptures.* I do agree in principle with the conviction that the writings of Abū Qurra do show us in a general way the importance of using scriptures, especially the Gospel, in the Christians' engagement in dialogues with Muslims during early Islam.[70] I, for one, see this noticeably demonstrated, for instance, in Abū Qurra's *Maymar yuḥaqqiq annahu lā yalzam an-naṣārā an yaqūlū thalāthat āliha* (*Maymar that verifies that the Christians do not need to say three gods*)[71], where he starts his exposition with extensive hermeneutics of Biblical verses from the Old and the New Testaments. However, even in this *Maymar,* Abū Qurra at one point states that his Biblical argument is pursued without starting first with any rational argument on the cognitive validity of the Bible, because this later argument, as Theodore explains, has already been developed in one of his earlier *Mayāmir,* and that his Biblical argument in this *Maymar* on the Trinity comes as a continuation of

---

69 Franz Rosenthal states that the Mu'tazilites' thought was the foundational intellectual ideology of the Abbasid era. To these thinkers, he affirms, "can be ascribed the chief merit of introducing into the realm of Islamic culture the philosophical treatment of theological problems based on ideas and methods originally found in Greek philosophy." See Rosenthal, Franz, *The Classical Heritage in Islam*, trans. by Emile and Jenny Marmorstein. London: Routledge, 1994, p. 4. On the impact of Mu'tazalitism on early *kalām* and Arabic Rationalism, see for example Van Ess, Josef, *The Flowering of Muslim Theology*, trans. by Jane Marie Todd. Cambridge, MAS: Harvard University Press, 2006, pp. 153–190; Nagel, Tilman, *The History of Islamic Theology: From Muhammad to the Present*, trans. by Thomas Thornton. Princeton: Markus Wiener Publishers, 2010, pp. 82–99; Frank, Richard M., *Early Islamic Theology: The Mu'tazilites and al-Ash'ari, Texts and Studies on the Development and History of Kalam*. Aldershot: Ashgate, 2007; and Peters, *Aristotle and the Arabs*, pp. 136–157.

70 Griffith, "Arguing from Scripture", p. 47.

71 Abū Qurra, "Maymar yuḥaqqiq annahu lā yalzam an-naṣārā an yaqūlū thalāthat āliha idh yaqūlūn al-Āb ilāh wa-l-Ibn ilāh wa-r-Rūḥ al-Qudus ilāh wa-law kāna kull wāḥid minhum tāmm 'alā ḥidah" (Maymar that verifies that the Christians do not need to say three gods when they say the Father is God and the Son is God and the Holy Spirit is God, even if each one of them is perfect in its own), in *Mayāmir Thāwūdūrūs Abī Qurrah, usquf Ḥarrān, aqdam ta'līf 'arabī*, ed. by Constantine Bacha, pp. 23–47. Beirut: Al-Fawā'id, 1904.

the previous one and it builds upon it.[72] For Theodore Abū Qurra, the scriptural argument itself must first be justified and certified by an *a priori* argument from reason on the authenticity and the rational validity of the Scripture itself. Theodore Abū Qurra's methodology does not successfully work as a proof of the conviction that "the whole Christian/Muslim encounter in the early Islamic period could be characterized as a conflict over the proper understanding of the narratives in the scriptures."[73] Methodologically speaking, there is a difference between saying that an author uses scriptural attestations as one among many other means in the service of her own rational and logical argument in general, on one hand, and another author who considers arguing from scriptural textuality his basic and primary methodological argumentation strategy, on another. From the perspective of this difference between these two methodological cases, one can say that while the scriptural attestations never disappear from Theodore Abū Qurra's texts and known arguments, *arguing from scriptures* as a strategic method of argument is not, as it seems to be, his first and favorite method of argument in his debates with Muslims.

Far from centralizing textual evidentiality or exegetical accuracy, I do find in Theodore's literature various indications that the Christian-Muslim encounter in that early period was circled, instead, around a question on *epistemological foundationality:* what is the foundation of the belief in the truthfulness and plausibility of the religious scriptures? I do agree with the belief that, though Abū Qurra's emphasis on reason is common to all the other *mutakallimūn* in his era, Abū Qurra centralizes, and much more highly essentializes, the role of reason in his writings than his contemporaries do in theirs.[74] This is why I am not quite convinced that, in his apologies, Theodore Abū Qurra "*assumes* biblical, even Qur'ānic, faith in his readers, whom he *expects* to be convinced of the coincidence of meanings in the rational and scriptural languages about God which he has been investigating."[75] It is my belief, to the contrary, that, in his *Maymar on the Existence of the Creator and the Right Religion*, Abū Qurra does not really

---

72 Abū Qurra, "Maymar yuḥaqqiq annahu lā yalzam an-naṣārā", p. 27: "And the verification of what we mentioned has its clarification and summary in a *maymar* we composed on it, the one who wants self-healing can read it and be persuaded by what we said therein. And, it is not convenient to repeat what we have stated there and broadcast it in this *maymar.*"

73 Griffith, "Arguing from Scripture", p. 56.

74 See the comparison between Theodore and his other contemporary Muslims and Christian intellectuals with regard to the role of reason in Varsanyi, Orsolya, "The Role of the Intellect in Theodore Abū Qurrah's *On the True Religion* in Comparison with His Contemporaries' Use of the Term." *Parole de l'Orient* 34 (2009): pp. 51–60.

75 Griffith, "Faith and Reason", p. 27.

assume this scriptural faith. He rather *examines* it by exposing its textual sources evenly to reason, and he *aspires* at convincing his readers, rather than expects them to be convinced already. For Theodore, scriptures *per se* have no referential statues that warrant us to argue unreservedly from them. It is these scriptures' interpretation, and the guiding role of the Holy Spirit in the mind of their exegetes, that makes these texts useful as tools (not as criterion) for arguing in defense of faith.[76]

This apologetic methodology does not only bestow upon Theodore Abū Qurra, as Sidney Griffith rightly opines, the uniqueness of shifting "the theoretical grounds on which Christians themselves customarily thought of justifying [their doctrines]."[77] I do dare to go even further than Griffith in ascribing to Theodore the prerogative of exceeding the theoretical grounds on which not only the Christians, but more challengingly, the Muslims of his era customarily stand. In his reliance on *arguing from reason* to validate textual authenticity, Theodore Abū Qurra proposed to Christian and Muslim *mutakallimūn* alike the criterion of deductive reasoning alone as the sole judge of not only the Gospel's, but also the Qur'ān's, truthfulness. The Muslim *mutakallimūn* were, presumably, keen on exerting this rule on the Christian scriptures. Yet, until this very moment, many Muslim orthodox thinkers in today's Arab world, like the early Muslim *mutakallimūn* of Theodore's era, do not show readiness to expose the qur'ānic texts to the same level of rational examination.

5 Concluding Remarks

I allowed myself to detect an interesting nuance between Griffith's reading in 1994 of Theodore's methodology in *Maymar on the Existence of the Creator and the Right Religion*, and *Maymar on Moses's Law and Chalcedonian Orthodoxy*, and his appraisals of the same texts in 2005 and 2013. My reading of Abū Qurra's extant legacy invites me to stand in both feet on the side of Griffith's 1994 attention to Theodore's methodological reliance on a philosophical deductive reasoning in his attempt at arguing that Christianity is the one and only true religion. I

76 It is therefore correct to interpret Abū Qurra's argument in defense of the church's ecumenical councils in *Maymar on Moses's Law and Chalcedonian Orthodoxy* as his attempt at showing that the church councils (wherein the Christian reason produced creedal faith) are "not only the final arbiter of faith and doctrine, but also the sole hermeneutical authority on Scripture": Nassif, Bassam A., "Religious Dialogue in the Eighth Century." *Parole de l'Orient* 30 (2005): pp. 333–340, here 337.

77 Griffith, "Faith and Reason", p. 29.

am hesitant to concur with Griffith's 2005's and 2013's consideration of Theodore's argument in these mentioned *Mayāmir* as an example of an early Christian *kalām*'s *arguing from scriptures* method.

Moreover, I allowed myself in this essay to take Sidney Griffith's conclusions in his 1994's article a step further beyond merely considering Theodore just as an original and pioneering *mutakallim* among his Christian contemporaries in importing Muslim *kalām* methods or reasoning that were customary to the Christian thinkers. I argued further that Abū Qurra's originality lies in his invitation for all the *mutakallimūn* of his era, Christians and Muslims alike, to follow his initiative of judging every religious scripture (one's own, not only the opponents') by means of pure deductive reasoning. The original innovativeness of this Arabic Melkite church father may actually merit careful attention and thorough examination, far more than we have so far reflected readiness to offer.

Juan Pedro Monferrer Sala

The Lyre of Exegesis

Ibn al-Ṭayyib's Analytical Patterns of the Account of the Destruction of Sodom

1 Introduction

The saga recounted in Gen 19, which contains the story of Sodom and the daughters of Lot as part of a broader episode (Gen 18–19) narrating a whole day in the life of Abraham, in which his nephew Lot plays a major role,[1] is a textual example of what might be termed the 'shared tradition' common to the three monotheistic religions.

In the Biblical textual tradition, Gen 19 was always seen as a largely autonomous story within the Abraham-Lot narrative cycle to which it belongs. Over the centuries, the story has invariably been hailed as an archetypal account of man's depravity; yet the scope of that depravity, and the narrative details of the account, have varied not only among the three great monotheistic religions but even within a single faith (cf. Isa 1:10; 3,9; Ezek 16:49; Jer 23:14). The events narrated have not always been viewed as unnatural habits, despite the harsh claims made against the sexual practices of the inhabitants of Sodom.[2]

In structural terms, Gen 19 comprises three subthematic sections: 1. the destruction of Sodom (19:1–11); 2. the saving of Lot (19:12–29); and 3. Lot's incest with his daughters (19:30–38). This triple division also serves to highlight the major narrative elements on which commentators were later to construct their respective exegeses of the 'story': a) the judgement and destruction of Sodom as a city of sin; b) the sparing of Lot due to his kinship with Abraham; and c) the final tragic downfall of Lot, the loser figure cast into oblivion because of his incestuous relations with his daughters.[3]

1 Haag, Ernst, "Abraham und Lot in Gen 18–19," in *Mélanges bibliques et orientaux. Festschrift M. Henri Cazelles*, ed. by André Caquot and Mathias Delcor, pp. 173–179. Kevelaer: Butzon & Bercker, 1981.
2 Whybray, Norman, "Genesis," in *The Oxford Bible Commentary*, ed. by John Barton and John Muddiman, pp. 52–53. Oxford: Oxford University Press, 2007.
3 Cf. Orbe, Antonio, "Los hechos de Lot, mujer e hijas vistos por S. Ireneo (Adv. Haer. IV,31,1,15/3.71)." *Gregorianum*, 75/1 (1994): pp. 37–64.

https://doi.org/10.1515/9783110564341-008

It was this latter element that led to the virtual expunging of Lot and his adventures from the Abraham story in many Christian writings, ranging in genre from the chronographical to the apocryphal, and including Eutychius of Alexandria's *Kitāb al-Ta'rīkh* or *Annals*,[4] and Solomon of Boṣtra's *Ktābā d-Dĕbūrītā* or *The Book of the Bee*.[5]

The Nestorian Abū l-Faraj ʿAbdallāh ibn al-Ṭayyib al-ʿIrāqī (d. 1043)[6] wrote an interesting *Tafsīr* on the Book of Genesis, included in his *Kitāb firdaws al-naṣrāniyya* (*The Book of the Paradise of Christianity*), which drew heavily on a number of Syriac authors, among them Ephrem the Syrian, Moshē bar Kēphā and Īshūʿdād of Merv.[7] In its treatment of the account provided by Gen 19, Ibn al-Ṭayyib's commentary adopts a dual approach:[8] a) direct quotation of the Biblical text, accompanied in some cases by brief comments and in others by no comment at all; b) paraphrasing of the text with a view to summarising its content, accompanied by explicative notes wherever Ibn al-Ṭayyib deems it appropriate.

2 Interpretation on Biblical Quotations and Allusions

An interesting feature of Ibn al-Ṭayyib's Commentary on the double episode narrated in Gen 19 is his sparse use of direct quotations from the Bible; indeed, these are reduced to one direct quote (Gen 19:8) and three allusions (Gen 19:12, 22, 25, 29), to which he adds a twofold quote from the New Testament (Luke 17:32–33a), included – as we shall see – for exegetical purposes. The references covering the quotations from Gen 19 are as follows:

---

4 Cf. Eutychius of Alexandria, *Annals*, ed. by L. Cheikho. Beirut/Paris/Leipzig: Secrétariat du Corpus SCO, 1906, p. 22.

5 *The Book of the Bee*, ed. and trans. by E. A. Wallis Budge. Oxford: Clarendon Press, 1886, *m'-mb* (Syriac), p. 42 (English).

6 On this author, see Faultless, Julian, "Ibn al-Ṭayyib," in *Christian-Muslim Relations. A Bibliographical History. Volume 2 (900–1050)*, ed. by David Thomas et al. Leiden/Boston: Brill 2010. See also Samir, Samir Khalil, "*La place d'Ibn-at-Tayyib dans la pensée arabe.*" *Journal of Eastern Christian Studies*, 58/3–4 (2006): pp. 177–193.

7 Samir, Samir Khalil, "Christian Arabic Literature in the ʿAbbasid Period," in *Religion , Learning and Science in the ʿAbbasid Period*, ed. by M. L. Young, J. D. Latham and R. B. Serjeant, pp. 446–460, Cambridge: Cambridge University Press, 1990 p. 447; Féghali, Paul, "*Ibn At-Tayyib et son commentaire sur la Genèse.*" *Parole de l'Orient*, 16 (1990–91): pp. 149–162.

8 Ibn al-Ṭayyib, *Commentaire sur la Genèse*, ed. and trans. by J.C.J. Sanders, 2 vols. Louvain: Secrétariat du Corpus SCO, 1967, I, pp. 72–74 (Arabic), II, pp. 68–70 (French).

19:8: لانهم دخلوا ظلال قريتي “for they have come under the shadow of my roof.”
19:12: بنيه وبناته والاتين “his sons, his daughters, and the sons-in-law.”
19:22: صاغار “Zoar.”
19:25, 29: انقلاب “overthrow.”

These four references provide various exegetical insights, ranging from linguistic considerations to expansive, informative commentaries, and including both paraphrase and interpretation. In his works as exegete, therefore, Ibn al-Ṭayyib[9] combines three features characteristic of an approach to Biblical commentary that owed much to Jewish and Eastern Christian tradition.[10]

2.1 Gen 19:8

This quotation is of interest for two reasons. Firstly, because it derives from a Syriac *Vorlage*, as is evident both in the complete version of the sentence and in the two nouns used (*ẓilāl* and *qarīt*): ܡܛܠ ܕܥܠܘ ܠܛܠܠܐ ܕܩܪ̈ܝܬܝ, (lit. “for they have come under the shadow of my beam”).[11] In the phrase *zilāl qarītī*, *ẓilāl* is cognate with *ṭelal* (“shadow”)[12] while *qarīt* is a loanword from the Syriac *qarītā* (“beam”).[13] Secondly, although the author offers a loan-translation of the Syriac original, even including a Syriac term in the Arabic version, the phrase is interpreted by analogy with Song 1:17, the synecdochic value of ܩܪ̈ܝܬܐ ܒܬܝܢ (“the beams of our house”)

9 Samir, Samir Khalil, “*Nécessité de l’exégèse scientifique. Texte de ‘Abdallâh Ibn at-Tayyib.*” *Parole de l’Orient*, 5 (1974): pp. 243–279.
10 See for example Reeves, John C., “Scriptural Authority in Early Judaism,” in *Living Traditions of the Bible: Scripture in Jewish, Christian, and Muslim Practice*, ed. by J.E. Bowley, pp. 63–84. St. Louis: Chalice, 1999; Van Rompay, Lucas “Development of Biblical Interpretation in the Syrian Churches of the Middle Ages,” in *Hebrew Bible/Old Testament: The History of Its Interpretation, Volume I: From the Beginnings to the Middle Ages (Until 1300). Part 2: The Middle Ages*, ed. by Magne Sæbø, pp. 559–577. Göttingen: Vandenhoeck and Ruprecht, 2000. See also Lössl, Josef and John W. Watt, eds., *Interpreting the Bible and Aristotle in Late Antiquity: the Alexandria Commentary Tradition between Rome and Baghdad*. Farnham: Ashgate, 2011. See also Kannengiesser, Charles, *Handbook of Patristic Exegesis: The Bible in Ancient Christianity*. With special contributions by various scholars. Leiden/Boston: Brill, 2006.
11 Cf. Ginzberg, Louis, *The Legends of the Jews*, trans. from the German Manuscript by Henrietta Szold, 6 vols. Philadelphia: The Jewish Publication Society of America, 1909, I, p. 253.
12 Smith, Robert Payne, *Thesaurus syriacus: Collegerunt Stephanus M. Quatremere et al.*, 2 vols. Oxford: Clarendon Press 1879, 1901, I, p. 1469.
13 Smith, Robert Payne, *Thesaurus syriacus*, III, 3717. On the interpretation of Hebrew קורתי like “woods,” see Ibn Danān, Sĕʿadyah, *Libro de las raíces. Diccionario de hebreo bíblico*. Introduction, translation and indexes by Milagros Jiménez Sánchez. Granada: Universidad de Granada, 2004, p. 370.

being reflected in the term *bayt* in Ibn al-Ṭayyib's explanation: *yurīd bihi baytahu*, i.e., "with what his house means."[14] This latter explanation is an adaptation from Īshūʿdād of Merv's Commentary: ܡܛܠܠܗ ܕܩܘ̈ܪܝܬܐ ܠܒܝܬܝ ܗܢܘ ܕ ܠܛܠܠܐ ܕܥܠܘ „ "They entered the shadow of (my beams), in other words of my house, whose beams give shade."[15]

A similar interpretation is offered by the anonymous Commentary contained in the Ms *Diyarbakir 22*,[16] which differs from that provided by GenR, the earliest rabbinic commentary in strictly redactional terms (conventionally dated in the 5th century CE).[17] The latter gives two interpretations based on two lexical and referential equivalents: in the first, the phrase *bĕṣēl qôrātî* (בצל קורתי > ܡܛܠ، ܛܠܠܐ ܗܢܝ > ظلال قريتي) is applied to Abraham, for it is the fruit of his work (זכותו של אברהם).[18] This is an old exegesis, also to be found in Job 16:7.[19] In the second interpretation, by contrast, the phrase is applied to Lot's wife,[20] who is against Lot's practice of extending hospitality to strangers, obliging him to receive them in the men's section of the house:[21]

---

14 Ibn al-Ṭayyib, *Commentaire sur la Genèse*, ed. by J.C.J. Sanders, I, p. 72 (Arabic), II, p. 68 (French).

15 Īshūʿdād of Merv, *Commentaire d'Iṣoʿdad de Merv sur l'Ancient Testament. I. Genèse.*, ed. by J.-M. Vosté and C. van den Eynde, trans. by C. van den Eynde. Louvain: Secrétariat du Corpus SCO, 1950, 1955, I, p.163 (Syriac), II, p. 176 (French).

16 *Commentaire sur Genèse-Exode 9,32 du manuscrit (olim) Diyarbakir 22*, ed. and trans. by Lucas Van Rompay, 2 vols. Louvain: Peeters, 1986, I, p. 81 (Syriac), II, p. 104 (French).

17 Stemberger, Günter, *Einleitung in Talmud und Midrasch*. 8th edition. München: C.H. Beck, 1992, pp. 272–279.

Cf. Neusner, Jacob, *Chapters in the Formative History of Judaism. Some Current Essays on the History, Literature, and Theology of Judaism*. Lanham MD: University Press of America, 2010, pp. 61–63.

18 Cf. MidTanḥ *wayyera* 4:21: "He [Lot] had grown up in Abraham's house and had seen him receive travelers he had learned from him."

19 *The Book of Jubilees. A Critical Text*, ed. and trans. by James C. Vanderkam, 2 vols. Louvain: Peeters, 1989, I, p. 94 (Ethiopic), II, p. 95 (English).

Cf. Charles, Robert Henry, *The Ethiopic Version of the Hebrew Book of Jubilees*. Oxford: Clarendon Press, 1895, pp. 56–57.

20 On Lot's wife, see Munier, Charles, "La femme de Lot dans la literature juive et chrétienne des premiers siècles," in *Figures de l'Ancien Testament chez les Pères*, ed. by Pierre Maraval, pp. 123–142. Strasbourg: Centre d'analyse et de documentation patristiques, 1989.

21 *Midrash Bereshit Rabba*, ed. by J. Theodor and Ch. Albeck. Critical Edition with Notes and Commentary. Jerusalem: Wahrmann Books, 1965, *wayyera* 50, p. 6. For the English translation: *The Midrash Rabbah. I. Genesis*. Translated into English with notes, glossary and indices under the editorship by. H. Freedman and Maurice Simon. Oxford: Oxford University Press, 1977, p. 438.

> כי על כן באו בצל קורתי לא בזכותי אלא בזכותו של אברהם ד"א כי על כן באו בצל קורתי מלמד שהטתה את הבית עליהם א"ל אם בעית מקבלתון קבל בחלקך
>
> *Forasmuch as they are come under the shadow of my roof:* not in my merit, but in the merit of Abraham. Another interpretation is that the verse *Forasmuch as they are come under the shadow of my roof* teaches that she [Lot's wife] turned the house against them, saying to him, 'If you want to receive them, receive them in your portion'

This dual interpretation is also provided in MidTanḥ (*wayyera* 4:22), although with a different narrative development.[22] Ibn al-Ṭayyib, too, focuses on the refusal of Lot's wife to extend her own hospitality to the visitors, drawing on an anonymous exegetical tradition which he includes in his commentary (*wa-yuqāl annahā*). Unlike the rabbis, however, he uses this interpretation not to highlight Lot's wife's refusal to entertain the visitors but to account for her being turned into a pillar of salt:[23]

> ويقال انها صارت قايمة ملح لان الغربا اذا كانوا حلصوا في دار لوط جعلت بينها وبين جيرانها علامة التماس ملح لتعلمهم حصول العربا عنده
>
> It is said that she was turned into a pillar of salt because when the strangers arrived at Lot's house, she made a sign to her neighbours, asking (them) for salt, in order to tell them of the arrival of the strangers.

This is a new explanation, of an aetiological nature, intended to account for the legend according to which Lot's wife was turned into a pillar of salt; again, Ibn al-Ṭayyib takes it from Īshūʿdād of Merv's Commentary.[24] This commentary is itself the result of a long process of transmission, since it is to be found in rabbinic literature (GenR 51:5) which notes that "she sinned through salt" (שחטאה במלח),[25] and was later developed in Haggadic literature.[26] At this late stage in the exegesis, therefore, the failure to respect the rule of hospitality – the major element at the start of the commentary – is suddenly relegated to a secondary position, in order to account for the punishment that befell Lot's wife.

---

22 English translation in *Midrash Tanḥuma*. Translated into English with Introduction, Indices, and Brief Notes by John T. Townsend. Hoboken NJ: Kyav Publishing House, 1989, p. 108.

23 Ibn al-Ṭayyib, *Commentaire sur la Genèse*, ed. by J.C.J. Sanders, I, pp. 73–74 (Arabic), II, p. 69 (French).

24 Īshūʿdād of Merv, *Commentaire d'Iṣoʿdad de Merv sur l'Ancient Testament. I. Genèse*, ed. by J.-M. Vosté and C. van den Eynde, trans. by C. van den Eynde, I, p. 166 (Syriac), II, p. 179 (French).

25 *Midrash Bereshit Rabba*, ed by. J. Theodor and Ch. Albeck, *wayyera* 51, p. 5; English translation: *The Midrash Rabbah. I. Genesis*, ed. by H. Freedman and M. Simon, p. 447, cf. n. 1.

26 Ginzberg, *The Legends of the Jews*, I, p. 254. Cf. on the other hand PRE 15:5.

2.2 Gen 19:12

Although, *pace* Sanders,[27] this is not strictly speaking a translation, it is not devoid of interest, in that the succession of terms (*banīhi wa-banātihi wa-l-ātīn*, “his sons, his daughters, and his relatives”) differs from that offered by the Syriac text ܚܬܢܝܟ ܘܒܢܝܟ ܘܒܢܬܟ (“your sons-in-law, your sons, your daughters”). The third p. sg. pronominal suffix indicates that this is indeed a reference to the text, but not a quotation. The order of the three elements also differs from that of the *Peshīṭtā*, which gives an eclectic version involving TarJon (חתנך בנך ובנתך, “your son-in-law, your sons, and your daughters”) and TarNeoph (חתנין ובניך ובנותיך, “your sons-in-law, your sons, and your daughters”), whereas TarOnk offers (חַתְנָא וְבְנָךְ וְבְנָתָךְ), the version closest to the Masoretic Hebrew Text: חָתָן וּבָנֶיךָ וּבְנֹתֶיךָ, “thy son-in-law, and thy sons, and thy daughters.”[28] Interestingly, GenR has nothing to say about this issue, which was presumably of little importance for the rabbinic tradition. Ibn al-Ṭayyib’s interest in this Biblical reference derives from Īshū‘dād of Merv,[29] although his exegetical approach is different. While both Īshū‘dād and Ibn al-Ṭayyib opts to interpret the phrase *banīhi* (“his sons”), Ibn al-Ṭayyib goes further, glossing it with the term *khatanīhi* (“his sons-in-law”): *ishāra bi-l-banīn ilā khatanīhi*, i. e., “by his sons he means his sons-in-law”[30] The gloss is of interest because, although it appears to be a possible regular Arabic plural *khatanūna*, *khatanī* is a calque on the Syriac *ḥatnē* (ܚܬܢܐ, “sons-in-law”), the term used by the *Peshīṭtā* to translate its Hebrew cognate חָתָן (LXX γαμβροὶ), whose meaning is identical. The interpretation of the term enables Ibn al-Ṭayyib to explain that the sons-in-law perished along with the people of Sodom, for failing to take the visitors at their word: *wa-li-anna khatanīhi lam yuṣaddiqū bi-l-wa‘d halakū ma‘a ahl Sadūm* (“since the sons-in-law did not believe the promise, they perished with the people of Sodom”).[31]

---

27 Ibn al-Ṭayyib, *Commentaire sur la Genèse*, ed. by J.C.J. Sanders, II, p. 68 (French), cf. I, p. 73 (Arabic).

28 Ben Mobarak ben Ṣa‘īr, Šelomo, *Libro de la Facilitación. Kitāb at-Taysīr (Diccionario judeoárabe de hebreo bíblico)*, intro., ed. and trans. by José Martínez Delgado, 2 vols. Granada: Universidad de Granada, 2010, I, p. 389 (82v) s.v. On חָתָן interpreted as ‘daughter’s husband’, see Ibn Danān, *Libro de las raíces*, p. 173. (nº 667.3).

29 Cf. Īshū‘dād of Merv, *Commentaire d’Iṣo‘dad de Merv sur l’Ancient Testament. I. Genèse*, ed. by J.-M. Vosté and C. van den Eynde, trans. by C. van den Eynde, I, p. 163 (Syriac), II, p. 177 (French).

30 Ibn al-Ṭayyib, *Commentaire sur la Genèse*, ed. by J.C.J. Sanders, I, p. 73 (Arabic), II, p. 68 (French).

31 Ibid., I, p. 73 (Arabic), II, p. 68 (French).

2.3 Gen 19:22

Strictly speaking, this passage cannot be regarded as a translation either, since it is a linguistic gloss of the calque *Ṣāghār* from the Syriac *Ṣā'ār* (ܨܥܪ < Heb. צוֹעַר *Ṣô'ar*),[32] introduced by Ibn al-Ṭayyib with the formula *tafsīr: tafsīruhā al-ṣaghīr*, i.e., "its explanation is the small one."[33] Here, too, the direct source is Īshū'dād of Merv's Commentary,[34] although Ibn al-Ṭayyib simplifies Īshū'dād's explication, reducing it to a simple translation and avoiding any reference either to the Hebrew etymon or to the explanation provided in LXX.[35] This is an example of what might be termed internal exegesis – to which GenR pays no attention – drawing on Gen 19:20, where Lot pleads that "this town is near to flee to, and it is a little one" (ܡܕܝܢܬܐ ܗܕܐ ܩܪܝܒܐ ܗܝ ܠܡܥܪܩ ܠܬܡܢ ܗܝ؛ ܘܙܥܘܪܬܐ ܗܝ < הָעִיר הַזֹּאת קְרֹבָה לָנוּס שָׁמָּה-וְהִוא מִצְעָר).[36] One interesting stylistic feature is that, for explanatory purposes, Ibn al-Ṭayyib's Arabic commentary retains the play on words found in the Syriac version, which itself echoes the Hebrew text through the common exchange (ܙ < ץ):[37] צוֹעַר – מצער < ܨܥܪ – ܙܥܘܪܬܐ < صاغار – الصغير ("Zoar – small"). Despite the adaptation offered by Ibn al-Ṭayyib (صاغار) following the Syriac reading, Zughar (زغر) is the form preferred by Arabic geographers.[38]

Moreover, this gloss enables Ibn al-Ṭayyib to introduce a series of explanatory elements concerning the location of this village (ܡܕܝܢܬܐ): وبعدها من سدوم نصف فرسخ ويقال ان لوط سماها بهذا الاسم ولم يبدها الله لتكون ملجا للوط, "It lies half a parasang

32 For the correspondence between Arabic *ghayn* and Syriac *'ē*, see Costaz, Louis, *Grammaire syriaque*, 3rd edition. Beirut: Imprimerie catholique, 1994, p. 218 § 866. Al-Ḥamawī, Yāqūt, *Mu'jam al-buldān*, 5 vols. Beirut: Dār Iḥyā' al-Turāth al-'Arabī, 1399 H/1979 CE, III, 411a calls it *Ṣughar.*

33 Ibn al-Ṭayyib, *Commentaire sur la Genèse*, ed. by J.C.J. Sanders, I, p. 73 (Arabic), II, p. 68 (French).

34 Cf. another explanation in *Le Commentaire sur Genèse-Exode 9,32 du manuscrit (olim) Diyarbakir 22*, ed. and trans. by L. Van Rompay, I, p. 82 (Syriac), II, p. 105 (French), which also narrates that the explanation of Ṣa'ar is Bala' (ܒܠܥ).

35 Īshū'dād of Merv, *Commentaire d'Iṣo'dad de Merv sur l'Ancient Testament. I. Genèse*, ed. by J.-M. Vosté and C. van den Eynde, trans. by C. van den Eynde, I, pp. 163–164 (Syriac), II, pp. 175–177 (French).

36 Alter, Robert, *The Five Books of Moses*. New York/ London: W.W. Norton and Company, 2004, p. 94. On this city, one of the five "cities of the Plain," see Negev, Avraham and Shimon Gibson eds., *Archaeological Encyclopedia of the Holy Land*. Nashville: T. Nelson, 1986, p. 412

37 Wright, William, *Lectures on the Comparative Grammar of the Semitic Languages*. Cambridge: Cambridge University Press, 1890, p. 58.

38 Heidet, Loius, "Ségor," in *Dictionnaire de la Bible*, ed. by F. Vigouroux, pp. 1561–1565, V/3. Paris: Letouzé et Ané, 1912.

from Sodom. It is said that Lot called it with this name. God did not destroy it, that it might be a refuge (*malja*ʾ) for Lot."[39]

Again, this explanation draws on Īshūʿdād of Merv's Commentary,[40] although Ibn al-Ṭayyib changes the order of the three elements and the explanation of the third: "that it might be a refuge for Lot" rather than Īshūʿdād's ܠܠܘܛ ܕܬܗܘܐ ܠܬܘܪܣܝܐ ("to ensure Lot's survival").[41] The Persian loan-word *farsakh* (cf. παρασάγγης, "parasang") is used by classical authors to indicate a distance equal to thirty *stadia* or some 3 1/2 or 4 English miles, although its length varied in later times.[42] For Jewish authors, a parasang may be 3.840, 4.640 or 4.760 mt.[43] According to Arab geographers, a *farsakh* was equal to 3 miles,[44] so the distance between Zoar and Sodom would be about 3 miles, according to Ibn al-Ṭayyib. His claim that *Lūṭ sammāhā bi-hadhā l-ism* ("Lot called it with this name") is another piece of internal exegesis from Gen 19:22 (cf. ܡܛܠ ܗܢܐ ܩܪܐ ܫܡܗ̇ ܕܩܪܝܬܐ ܨܥܪ, "therefore the name of the city was called Zoar"). The same is true of *lam yubidhā Allāh li-takūna malja*ʾ *li-Lūṭ* ("God did not destroy it, that it might be a refuge for Lot"), which draws on 19:17,19,20,21, where the indirect referent of *malja*ʾ ("refuge") is the Syriac *ṭūrā* "mountain" of vv. 17,19 (cf. Aram. טורא in GenR *wayyera* 50:11, HT הרה) and the direct referent ܡܥܪܬܐ (*mĕʿartā*, "the cave" < מְעָרָה, "a cave")[45] in accordance with Gen 19:30, given the symbolic value of the cave in the OT as a refuge to which to flee (Josh 10:16; 1 Sam 13:6). The episode is echoed in both the patristic[46] and the rabbinic tradition, where David, fleeing from King Saul, seeks refuge in a cave (מערה):[47]

39 Ibn al-Ṭayyib, *Commentaire sur la Genèse*, ed. by J.C.J. Sanders, I, p. 73 (Arabic), II pp. 68–69 (French).

40 Īshūʿdād of Merv, *Commentaire d'Iṣoʿdad de Merv sur l'Ancient Testament. I. Genèse*, ed. by J.-M. Vosté and C. van den Eynde, trans. by C. van den Eynde, I, p. 164 (Syriac), II, p. 177 (French).

41 Ibid.

42 Cf. Peck, Harry Thurston, *Harpers Dictionary of Classical Antiquities*. New York: Harper and Brothers, 1898, *s.v.* 'Parasanga'.

43 Cf Bleich, J. David, *Contemporary Halakhic Problems*, 6 vols. New York: Ktav Publishing House, 1983–2012, III, p. 31, n. 3.

44 Hinz, Walther, "Farsakh," in *Encyclopédie de l'Islam*. Nouvelle édition, ed. by H.A.R. Gibb et al., 13 vol., II, p. 832. Leyde/Paris: E.J. Brill – G.-P. Maisonneuve, 1960–2009.

Cf. Gil, Moshe, *Jews in Islamic countries in the Middle Ages*. Leiden: Brill, 2004, p. 511 and Sharon, Moshe, *Corpus Inscriptionum Arabicarum Palaestinae*, 5 vols. Leiden/ Boston: Brill, 2003, III, p. 11.

45 Cf. Ibn Danān, *Libro de las raíces*, p. 252.

46 Gargano, Innocenzo, "'Lot si rifugió nella grotto con le sue due figlie' (Origene)." *Parola Spirito e Vita* 26 (1992): pp. 215–231.

47 *Midrash Bereshit Rabba*, ed. by J. Theodor and Ch. Albeck, *wayyera* 51, p. 7; English translation: *The Midrash Rabbah. I. Genesis*, ed. by H. Freedman and M. Simon, p. 447.

ויעל לוט מצוער וישב בהר הה"ד (תהלים נז) למנצח אל תשחת לדוד מכתם בברחו מפני שאול במערה אמר לפניו רבון העולמים עד שלא נכנסתי למערה עשית חסד עם אחרים בשבילי עכשיו שאני נתון במערה יהי רצון מלפניך אל תשחת

And Lot went up out Zoar, and dwelt in the mountain (19:30). Thus is written, *For the Leader, al-tashḥeth; [A Psalm] of David; Michtam; when he fled from Saul, in the cave* (Ps 57:1) He [David] prayed to Him: 'Sovereign of the Universe! Before I entered this cave Thou didst show mercy to others for my shake; now that I am in the cave, O grant that Thou destroyest not (*al tashḥeth*)!

The toponym prompts Ibn al-Ṭayyib to note that Lot gave the city its name, and that it was spared by God precisely because it was Lot's refuge, even though all its other inhabitants died, as he later stresses.[48] Although the textual source of this explanation is Gen 19:20 – 22, 30, the introductory element *wa-yuqāl anna* ("and it is said that") derives ultimately from Ephrem the Syrian.[49]

2.4 Gen 19:25, 29

Here, Ibn al-Ṭayyib departs from Īshūʿdād of Merv's Commentary,[50] in describing the destruction of Sodom he gives *maṣdar inqilāb* ("overthrow," cf. 19:21), a substantivisation of the Syriac perfective *hfaḵ* (ܗܦܰܟ < HT הָפַךְ, "to overthrow, ruin"),[51] when he could have opted for the cognate *afaka* in form VIII ("to be turned upside down"), hence the active participle *mu'tafikāt* used to refer to cities overturned by divine punishment, i.e., Sodom, Gomorrah and the neighbouring cities.[52] The reason why Ibn al-Ṭayyib chose to use *inqilāb* is simply that the Qur'ān eschews other possible options (e.g., Q 11:82; 15:74: *jaʿalnā ʿalayhā sāfilahā*, "we turned them upside down") in favour of the root *'fk* in form VIII *i'tafaka*[53]

48 Ibn al-Ṭayyib, *Commentaire sur la Genèse*, ed. by J.C.J. Sanders, I, p. 73 (Arabic), II, p. 69 (French).
49 Ibid., I, p. 73 (Arabic), II, p. 69 (French).
50 Īshūʿdād of Merv, *Commentaire d'Iṣoʿdad de Merv sur l'Ancient Testament. I. Genèse*, ed. by J.-M. Vosté and C. van den Eynde, trans. by C. van den Eynde, I, p. 166 (Syriac), II, p. 180 (French).
51 Smith, Robert Payne *Thesaurus syriacus*, I, p. 1036.
52 Freytag, Georg Wilhelm, *Lexicon arabico-latinum*, 4 vols. Halle: C.H. Schwetschke et Filium, 1830 – 37, I, p. 44a-b.
53 On the unconvincing hypothesis of the relationship of *afaka* with Gəʿez *'afākiyā*, see Zammit, Martin R., *A Comparative Lexical Study of Qur'ānic Arabic*. Leiden/Boston/Köln: Brill, 2002, p. 592. Cf. Leslau, Wolf, *Comparative Dictionary of Geʿez (Classical Ethiopic)*.Wiesbaden: Otto Harrassowitz, 1991, p. 9b.

through its active participle, in Q 9:70 (*mu'tafikāt*) and 53:53 (*mu'tafikah*).[54] Thus, Ibn al-Ṭayyib has avoided the cognate of the Syriac referent, using a nominal synonym to express the same idea. The fact that Ibn al-Ṭayyib shunned the perfective verbal form used by the *Peshīṭtā* (ܗܦܰܟ݂), does not reflect a desire to imitate the Qur'ān, since by using *inqilāb* Ibn al-Ṭayyib is directly echoing the emphatic fem. participle of Gen 19:29: ܡܗܦܟܬܐ < הֲפֵכָה ("overthrow, reverse, ruin"). Rabbinic tradition explains the destruction of the five cities of the Plain by stating that "the angel stretched out his hand and overturned them" (אחד שלח מלאך את ידו והפכן):[55]

> ויהפוך את הערים האל רבי לוי בשם רבי שמואל בר נחמן חמשת הכרכים הללו היו יושבות על צור אחד שלח מלאך את ידו והפכן הה"ד (איוב כח)
>
> *And he overthrew those cities* (19:25). Rabbi Levi said in the name of R. Samuel b. Naḥman: 'These five cities were built on one rock, so the angel stretched out his hand and overturned them, as it is written, *He putteth forth his hand upon the flinty rock, he overturneth the mountains by the roots'.* (Job 18:9)

In the *Talmud Babli* (BMeṣ 86b) we are told that Gabriel came to Abraham to inform him that he would overturn Sodom (גבריאל אזל למיהפכיה לסדום).[56] The iconographic figure of overturning the earth (*taqallaba*) is often to be found in Christian apocalyptic literature, where the task is to be undertaken by the archangel Gabriel, who will be sent by Jesus Christ as soon as the Antichrist is defeated and immediately before the Last Judgement:[57]

---

54 Dietrich, Friedrich, *Arabisch-deutsches Handwörterbuch zum Koran und Thier und Mensch vor dem König der Genien*. Leipzig: F. Hinrichs, 1894, p. 6b.

55 *Midrash Bereshit Rabba*, ed. by J. Theodor and Ch. Albeck, *wayyera* 51, p. 4; English translation: *The Midrash Rabbah. I. Genesis*, ed. by H. Freedman and M. Simon, p. 446.

56 *Babylonian Talmud; Seder Nizikin*, ed. by Isidore Epstein, trans. by E. W. Kirzner et al., Tractate Baba Mezia. London: World Federation for Mental Health, 1935. See also Cf. Ginzberg, *The Legends of the Jews*, I, p. 255.

57 On this text and its sources, see Monferrer-Sala, Juan Pedro, "'The Antichrist is coming…' The making of an apocalyptic *topos* in Arabic (Ps.-Athanasius, Vat. ar. 158 / Par. Ar. 153/32)," in *Bibel, Byzanz und christlicher Orient. Festschrift für Stephen Gerö zum 65. Geburtstag*, ed. by D. Bumazhnov et al., pp. 674–675. Louvain: Peeters, 2011 and Monferrer-Sala, J.P., "'Texto', 'subtexto' e 'hipotexto' en el 'Apocalipsis del Pseudo Atanasio' copto-árabe," in *Legendaria Medievalia en honor de Concepción Castillo Castillo*, ed. by Raif Georges Khoury, J.P. Monferrer-Sala and Mª J. Viguera Molins, pp. 427–428. Córdoba: Ediciones El Almendro – Fundación Paradigma Córdoba, 2011.

ويظهر الرب يسوع المسيح له المجد علي مدينته مع ملايكته ويضرب جبرييل الارض فتتقلب وتنشف جميع الامياه من علي وجه الارض ومن تحتها

The Lord Jesus the Messiah, glory to him, will appear upon his city with his angels. And Gabriel will beat the earth, which will turn around and the waters of the surface of the earth and under it will disappear.

In addition to these four textual references, Ibn al-Ṭayyib uses a double quotation from the Gospel of Luke, which he takes, together with the central explanatory element (Lot's wife), from Īshū'dād of Merv's Commentary,[58] although retaining only two of the three original quotations (Luke 17:32–33 and 9:62), modifying the order of the discursive elements, and adding exegetical information. It is interesting to note, in this respect, that Ibn al-Ṭayyib inserts this double quotation into the text by means of a discursive link with no exegetical value: *'inda qawlihi*, i.e., "together with his saying":[59]

Luke 17:32: اذكروا امراه لوط "Remember Lot's wife."
Luke 17:33a: من احب ان يحيي نفسه فليهلكها "Those who try to make their life secure will lose it."

It should be noted, given the importance of translation-related issues to Ibn al-Ṭayyib's exegesis, that the Arabic version derives from a Syriac original; this is evident, for example, in the use of the verb *ḥayyā–yuḥayyī* ("preserve alive"), which renders the cognate aphel *aḥī* (ܐܚܝ, "preserve alive") rather than the Greek περιποιέω ("acquire, purchase, preserve alive"), for *an yuḥayyī nafsahu* is the literal translation of the Syriac *d-naḥē nafsheh*, i.e., "to preserve his life":

Luke 17:32: ܐܬܕܟܪܘ ܠܐܢܬܬܗ ܕܠܘܛ = اذكروا امراه لوط
Luke 17:33a: ܡܢ ܕܨܒܐ ܕܢܚܐ ܢܦܫܗ ܢܘܒܕܝܗ = من احب ان يحيي نفسه فليهلكها

This double quotation functions essentially as an *argumentum ad auctoritatem*. Ibn al-Ṭayyib uses Jesus' *logion* for purely instructive purposes. The intention is that the attitude of Lot's wife should serve as an example (*mathalan*), i.e., as a warning to those who refuse to believe Christ's message, of whom "Lot's wife became an example" (*imra'at Lūṭ ṣārat mathalan*): امراة لوط صارت مثلا لمن لا يتبع سنة

58 Īshū'dād of Merv, *Commentaire d'Iṣo'dad de Merv sur l'Ancient Testament. I. Genèse*, ed. by J.-M. Vosté and C. van den Eynde, trans. by C. van den Eynde, I, p. 164 (Syriac), II, p. 177–178 (French).

59 Ibn al-Ṭayyib, *Commentaire sur la Genèse*, ed. by J.C.J. Sanders, I, p. 73 (Arabic), II, p. 69 (French).

المسيح والالهيات ويلتفت الى العالميات, "Lot's wife became an example of those who do not follow the law of the Messiah and divine grace, but cling to earthly things."[60]

The use of the *terminus rhetoricus 'mathal'* (< Aram. מתלא < Heb. משל) is particularly significant in this context, since – as elsewhere – it refers to an event that takes man unawares, thus serving to warn him of its dire consequences. It is, in short, an allegory (משל)[61] with which Ibn al-Ṭayyib explains what may happen, seeking to capture their attention by opening their eyes, shaking their souls to awaken them.[62] Interesting too, in this respect, is his use of the imperative *udhkur* ("remember"), a translation of the Syriac ethpeel imperative ܐܬܕܟܪܘ, which is in turn a rendering of the Greek μνημονεύετε,[63] whose verbal form μνημονεύω in LXX translates the Hebrew *qal* זכר.[64]

The imperative *udhkur* in these exegetical contexts acquires the force of an order, which assumes special significance in "punishment stories" like this one.[65] This narrative formula, as used in Biblical accounts of an allegorical nature, thus instils in the reader the idea of a reflection that is at once retroactive and proleptic. In other words, through this remembering (*dhikr*) of what happened to Lot's wife Ibn al-Ṭayyib predicts what will happen to those who behave as she did (cf. 19:26), for on turning back she wavered (*iltafatat shakkan*).[66] He explains her wavering and her immediate transformation into a pillar of

---

60 Ibid., I, p. 73 (Arabic), II, p. 69 (French).

61 For *mashal* with the sense of 'parable', see Stern, David, "Rhetoric and Midrash: The Case of the Mashal." *Prooftexts* 1 (1981): pp. 261–291. See also Boyarin, Daniel, "History Becomes Parable: A Reading of the Midrashic Mashal." *Bucknell Review* 33/2 (1989): pp. 54–71 (= *Mappings of the Biblical Terrain: The Bible as a Text*, ed. by Vincent L. Tollers and John Maier).

62 Evans, Craig A., *Ancient Texts for New Testament Studies. A Guide to the Background Literature.* Peabody MA: Hendrickson, 2005, p. 366. Cf. Jeremias, Joachim, *Die Gleichnisse Jesu*, 11th edition. Göttingen: Vandenhoeck and Ruprecht, 1998, p. 46 and Dalman, Gustav, *Sacred Sites and Ways. Studies in the Topography of the Gospels.* Authorised translation by Paul P. Levertoff. New York: The MacMillan Company, 1935, p. 242.

63 For the occurrences in NT, see Moulton, William F. and Alfred S. Geden, *A Concordance to the Greek Testament.* Edinburgh: T. & T. Clark, 1897, pp. 654–655. Cf. Morgenthaler, Robert, *Statistik des neutestamentlichen Wortschatzes.* Zürich/ Frankfurt am Main: Gotthelf-Verlag, 1958, p. 121 §1.

64 Muraoka, Takamitsu, *A Greek-Hebrew/Aramaic Two-way Index to the Septuagint.* Louvain: Peeters, 2010, p. 80a. Cf. Muraoka, T., *A Greek-English Lexicon of the Septuagint.* Louvain: Peeters, 2009, p. 465b.

65 On this narrative featured in the Qur'ān, see the explanation by Reynolds, Gabriel Said, *The Qur'ān and Its Biblical Subtext.* London/New York: Routledge, 2010, pp. 235–36.

66 Ibn al-Ṭayyib, *Commentaire sur la Genèse*, ed. by J.C.J. Sanders, I, p. 73 (Arabic), II, p. 69 (French).

salt (*qā'imat milḥ*)[67] by stating that she felt pity (*raḥma*) for the people of Sodom, whom she loved (*maḥabba lahum*)[68] even though they might be sinners (*khaṭa'ah*).[69] At the same time, Ibn al-Ṭayyib makes subliminal use of the turning of Lot's wife into a pillar of salt in order to introduce, in an indirect manner, an aetiological explanation of the barren setting of the five cities (*khams mudun*) which constituted the land of the Plain, noting that "it (the pillar) must have been like the salt earth, saline, where there is no fruit" (*la-takūnu ka-l-arḍ al-mā-liḥah sabikha lā thamar fīhā*).[70]

Accordingly, this moment of anagnorisis in Ibn al-Ṭayyib's exegesis would also serve to reveal the whole negative component of the account: the sins of Lot, his daughters, Sodom and its inhabitants.

3 Hermeneutical Method

Ibn al-Ṭayyib's interpretation of Gen 19 is largely analytical, and follows the verse order. It is, *prima facie*, an expository commentary with a didactic purpose. His approach is based on three strategies: a) use of a Syriac version of the Bible text which provides the basis for the translation into Arabic, and also for the interpretation of the passage concerned; b) translation of the Syriac original, which does not always feature in the commentary; c) application of a synchronic hermeneutical method, which studies the Biblical text in its final Syriac redaction, but in conjunction with exegetical contributions transmitted via the Eastern Syriac tradition.

Ibn al-Ṭayyib's Commentary aims, amongst other things, to draw attention to certain issues raised by the original text, given the lack of attention to the words of Divine Law (*lā yafham al-Sharī'ah lafẓa*) on the part of the metropolitans (*maṭārinah*) in the time of Ibn al-Ṭayyib.[71] He states as much in his introduc-

67 Ephrem and Īshū'dād used the phrase ܩܝܡܬܐ ܕܡܠܚܐ ("statue of salt"), see Ephrem Syrus, *In Genesim et in Exodum commentarii*, ed. and trans. by R.-M. Tonneau. Louvain: Secretariat du Corpus SCO, 1955, I, p. 79 (Syriac), II, p. 65; Īshū'dād of Merv, *Commentaire d'Iṣo'dad de Merv sur l'Ancient Testament. I. Genèse*, ed. by J.-M. Vosté and C. van den Eynde, trans. by C. van den Eynde, I, p. 165 (Syriac), II, p. 179 (French). Cf. Ibn Danān, *Libro de las raíces*, p. 274, who interprets נְצִיב as 'heap'; cf. Ben Mobarak ben Ṣa'īr (*Kitāb at-Taysīr*, ed. and trans. by J. Martínez Delgado, I, p. 657, n. 2) rendered it as *nuṣub* ('statue, idol').

68 Cf. Ginzberg, *The Legends of the Jews*, I, p. 255.

69 Ibn aṭ-Ṭayyib, *Commentaire sur la Genèse*, ed. by J.C.J. Sanders, I, p. 73 (Arabic), II, p. 69 (French).

70 Ibid., I, p. 73 (Arabic), II, p. 69 (French).

71 Ibid., I, p. 1 (Arabic), II, p. 1 (French).

tion to the commentary on the Book of Genesis, which is included in the *Kitāb firdaws al-naṣrāniyya*, noting that these are commentaries (*tafāsīr*) into which he has incorporated his own personal exegesis on ethical and moral questions of the Christian faith: رايت ان اجمع ساير تفاسيري الكتب الحديثة والعتيقة باختصار بالعربية ليقرا وينتبه لما اطرح, "I have opted to bring together my remaining commentaries on the old and new books (i.e., New and Old Testaments) in an epitome in Arabic, in order that it might be read and might draw attention to what I suggest [in them]."[72]

Ibn al-Ṭayyib's exegetical paraphrases may be broken down into four sections, in the following order: 19:1–9; 18:16–33; 19:24, 28; 19:31–38.

3.1 Gen 19:1–9

The summary-exegesis covering these nine verses reads as follows:[73]

> وظهور الملايكة للوط العشية لكيما يحلف عليهم للمبيت للاية التي تظهر بان افشاه الاسدومايين ما يفعلونه بالغربا من البغض لهم وعصيان الله فيهم
>
> The appearance of the angels to Lot (was) at dusk to oblige them (lit. make them swear) to spend the night (in his house) in order (to fulfil) the miracle that was made manifest, because the Sodomites revealed to him what they were going to do with the foreigners, through their hatred of them and in order to disobey God.

As indicated earlier, the Syriac text used by Ibn al-Ṭayyib plays a crucial role in his exegesis which, among other things, is based on Syriac and can only be explained by reference to it. The relationship between the Syriac text and the Arabic summary can be traced through certain correspondences, where his choice betrays his specific interests, as shown below.

19:1–2:

> وظهور الملايكة للوط العشية لكيما يحلف عليهم للمبيت = ܘܐܬܘ ܬܪܝܢ ܡܠܐܟܝܢ ܠܣܕܘܡ ܒܪܡܫܐ (...) ܘܒܝܬܘ

The arrival of the angels, whose nature will later be examined by Ibn al-Ṭayyib, takes place at dusk (*'ashiyya*), and a specific hour, i.e., sunset,[74] as indicated by

72 Ibid., I, p. 2 (Arabic), II, p. 2 (French).

73 Ibid., I, p. 72 (Arabic), II, p. 68 (French).

74 Cf. Fields, Weston W., "The Motif 'Night as Danger' associated with Three Biblical Destruction Narratives," in "*Sha'arei Talmon*": *Studies in the Bible, Qumran and the Ancient Near East*

the Hebrew word עֶרֶב (= LXX ἑσπέρα) rendered as *ramshā* (ܪܡܫܐ).[75] Ibn al-Ṭayyib opts to use the term *mabīt* ("lodging, overnight stay"), a *maṣdar* of the form *bāt-yabātu/yabītu* ("spend the night") which translates the Syriac imperative *būtū* (ܒܘܬܘ), of *būt–nebūt* ("lodge, pass the night" = HT לִינוּ; LXX καταλύσατε). Ibn al-Ṭayyib has chosen a cognate, but has shunned the imperative used in the *Peshīṭtā*, preferring instead the oath formula *ḥalafa 'alā ... li...*: *likaymā yaḥlifa 'alāyhim li-l-mabīt* ("to make them swear to spend the night").

By means of this formula, Ibn al-Ṭayyib underlines the ratification of the law of hospitality as extended by Lot to his visitors,[76] with all the social implications this law entailed in the Semite milieu;[77] this is a key testimony to Lot's righteousness in contrast to the attitude of the Sodomites.[78] Through this act of swearing, therefore, the guests are committed to spend the night under Lot's roof, thus – according to Ibn al-Ṭayyib – enabling them to work the miracle of blinding-maddening the Sodomites who were besieging Lot's house (19:11). For Ibn al-Ṭayyib this is a rational justification, since if the strangers had not spent the night there, the miracle following the siege of Lot's house could not have been brought about; Lot simply observed the law of hospitality against which the Sodomites had rebelled (19:4–5).[79]

Although this passage draws directly on the Biblical account, the remainder of the text is a conclusion drawn by Ibn al-Ṭayyib on the basis of verses 3–9. Unlike the earlier example, this is not a summary but a deduction from the reading. In it, Ibn al-Ṭayyib highlights the elements with which he is really concerned:[80]

---

presented to S. Talmon, ed. by M. Fishbane and E. Tov. Winona Lake: Eisenbrauns, 1992, pp. 17–32.

75 Cf. PRE 15:5; Ginzberg, *The Legends of the Jews*, I, p. 253.

76 Īshū'dād of Merv, *Commentaire d'Iṣo'dad de Merv sur l'Ancient Testament. I. Genèse*, ed. by J.-M. Vosté and C. van den Eynde, trans. by C. van den Eynde, I, p. 162 (Syriac), II, pp. 175–176 (French).

77 Smith, William Robertson, *The Religion of the Semites.* With a new introduction by Robert A. Segal. New Brunswick, NJ: Transaction Publishers, 2002, pp. 269–270; De Vaux, Roland, *Ancient Israel: Its Life and Institutions.* English trans. by John McHugh. Grand Rapids: W.B. Eerdmans, 1997, pp. 10–11, 74–75.

78 Alexander, T. Desmond, "Lot's Hospitality: A Clue to His Righteousness." *Journal of Biblical Literature* 104 (1985): pp. 289–291. Cf. Lasine, Stuart, "Guest and Host in Judges 19: Lot's Hospitality in an Inverted World." *Journal for the Study of the Old Testament* 29 (1984): pp. 37–59.

79 Cf. PRE 15:2–3; Ginzberg, *The Legends of the Jews*, I, p. 253.

80 Ibn al-Ṭayyib, *Commentaire sur la Genèse*, ed. by J.C.J. Sanders, I, p. 72 (Arabic), II, p. 68 (French).

> *Āya* ('sign, miracle') refers to what 'the men' do in 19:11,[81] and specifically to the final outcome, the 'twinkling, hallucination, phantom' (ܒܫܪܓܪܓܝܬܐ ܪܒ ◌ܿ),[82] a kind of visual delirium (cf. סַנְוֵרִים 'sudden blindness'[83] < Akk. *sinlurmā sinnūru*, 'day- or night-blindness')[84] with which Lot's visitors punish those laying siege to Lot's house with a view to having sexual intercourse (cf. Heb. ידע) with Lot's visitors, as a result of their mental and spiritual blindness to God's law (cf. Gr. τυφλός).

This interpretation is not far removed from the exegesis offered by the Rabbis (GenR 50, 8), who say that "they were maddened" (אלאון היך). The image of visual delirium to which Ibn al-Ṭayyib alludes with the Arabic loanword *āyah*[85] is strengthened by the verb *ẓahara-yaẓharu* ("become visible"). Again, this is an internal exegesis drawing on other passages of the Bible (Lev 18:22; 20:13; cf. Rom 1:26–27) in which this sexual practice is viewed as a capital offence.[86] Ibn al-Ṭayyib's use of the technical term *āyah* is something of a novelty, in that it adds to the literal view of his Syriac predecessors, who based their exegesis on the term *shragrāgyātā*.[87]

Al-Usdūmāyyīn, the demonym used to refer to the "men of Sodom" (19:4: ܐܢܫ̈ ܣܕܘܡ ; < אַנְשֵׁי סְדֹם; cf. 19:11: ܓܒܪ̈ܐ) is based on the Syriac forms *sdūmayē* (ܣܕܘܡܝ̈ܐ)[88] and *sdūmāyit* (ܣܕܘܡܝܐܝܬ)[89] to which a prosthetic *'aleph* has been added. The Arabic version of *Mĕʻarat Gazē* or *Spelunca Thesaurorum* (6th century CE) translates ܣܕܘܡܝܐ by *ahl Sadūm*, i.e., "the people of Sodom,"[90] as do the ver-

81 Cf. 2 Kings 6:18 referring to a nation, גוי.

82 *Shragragyathā* is glossed with Persian *abrōzišn* in *Le Commentaire sur Genèse-Exode 9,32 du manuscrit (olim) Diyarbakir 22*, ed. & trans. by L. Van Rompay, I, p. 82 (Syriac), II, p. 104 (French). Cf. Smith, R. Payne, *Thesaurus syriacus*, III, p. 4326. On Persian *abrōzišn*, see Ciancaglini, Claudia A., *Iranian Loanwords in Syriac*. Wiesbaden: Otto Harrassowitz, 2008, p. 98.

83 Cf. Ibn Danān, *Libro de las raíces*, ed. and trans. by M. Jiménez Sánchez, p. 295.

84 Stol, Marten, "Blindness and Night-Blindness in Akkadian." *Journal of Near Eastern Studies* 45 (1986): pp. 295–299.

85 Mingana, Alphonse, "Syriac Influence on the Style of the Kur'ān." *Bulletin of The John Rylands Library* 11/1 (1927): pp. 77–98, p. 86.

86 Wenham, Gordon J., "The Old Testament Attitude to Homosexuality." *Expository Times* 102 (1991): pp. 359–363.

87 Cf. Ephrem Syrus, *In Genesim*, ed. and trans. by R.-M. Tonneau, I, p. 78 (Syriac), II, p. 63 (Latin), and Īshū'dād of Merv, *Commentaire d'Iṣo'dad de Merv sur l'Ancient Testament. I. Genèse*, ed. by J.-M. Vosté and C. van den Eynde, trans. by C. van den Eynde, I, p. 163 (Syriac), II, pp. 176–177 (French).

88 *Le Commentaire sur Genèse-Exode 9,32 du manuscrit (olim) Diyarbakir 22*, ed. & trans. by L. Van Rompay, I, p. 81 (Syriac), II, p. 104 (French).

89 Smith, Robert Payne, *Thesaurus syriacus*, II, pp. 2528–29.

90 Bezold, Carl, *Die Schatzhöhle aus dem syrischen Texte dreier unedirten Handschriften*. Leipzig: J.C. Hinrich'sche Buchhandlung, 1883, pp. 166–67.

sion of the Pentateuch and *tafsīr* edited by Lagarde.[91] This is not simply an option. Ibn al-Ṭayyib has constructed his demonym on the basis of a geographical reality, identified by the Arabs as the site of ancient Sodom, *Jabal Usdūm* (< *Jabal al-Sudūm*),[92] i.e., "Mount of Sodom." He combines this geographical identification with Syriac noun and adjective forms (*sdūmayē–sdūmāyit*), which give the Arabic *usdūmāyyūn* a dual value as a demonym (*sdūmayē*), but also as a term denoting the sexual aberrations in which they engaged (*sdūmāyit*).

Ghurabā' poses something of a problem. If the reading *ghurabā'* is correct, this is another term subtly used by Ibn al-Ṭayyib with the exegetical-narrative purpose of gradually building up an interpretation of the text as a whole. He uses the plural *ghurabā'* ("foreigners") in referring to verses 1–2 on the law of hospitality. Yet verses 1–2 follow the Bible text, in which they are called angels (ملائكة = ܡܠܐܟ̈ܐ); given that these are beings sent by God, they can hardly be identified as foreigners (*ghurabā'*) to whom due hospitality is owed. This accounts for Ibn al-Ṭayyib's decision to switch from "angels" (ܡܠܐܟ̈ܐ) and "men" (ܓܒܪ̈ܐ) to "foreigners" (*ghurabā'*). Yet while *ghurabā'* works perfectly, one wonders whether غربا is not in fact the Arabic غرباء but rather the Syriac ܓܒܪ̈ܐ ("men"), which not only works in the context but also fits well with one reading of the Biblical account.[93] In that case, however, the above explanation would be meaningless, since by using a calque on the Syriac term Ibn al-Ṭayyib would simply be stressing the human, as opposed to angelic, nature of the visitors. Even so, the later reference to Lot as *gharīban* ("a foreigner")[94] would support the reading *ghurabā'*.

Bughd–'iṣyān are two negative concepts associated with those besieging Lot's house. Their hatred (*bughd*) is directed at Lot's guests, while their disobedience (*'iṣyān*) is of God's law. This is a moral appraisal of verses 7–9. Ibn

---

91 De Lagarde, Paul, *Materialien zur Kritik und Geschichte des Pentateuchs*, 2 vols. Leipzig: B.G. Teubner, 1867, pp. 122–23.

92 Driver, Samuel Rolles, *The Book of Genesis.* With Introduction and Notes, 4th edition. London: Methuen, 1905, p. 169. Cf. Avi-Yonah, Michael, "Sodom (modern Sedom) and Gomorrah," in *Encyclopaedia Judaica*, ed. by Fred Skolnik and Michael Berenbaum, 2nd edition, vol. XVIII. New York/New Haven: Holmes and Meier/ Yale University Press, 2007, p. 738. On the city of Sadūm/Sadhūm in Arab geographers, see Al-Ḥamawī, Yāqūt, *Mu'jam al-buldān*, III, pp. 200b-201a. See also the information provided by Josephus in his BJ IV, 8,4. Greek text and English translation in Josephus, Flavius, *The Jewish War, IV–VII*. Cambridge MA: Harvard University Press, 1928 (rep. 1961), pp. 140–45.

93 The reading نقريا (< ܢܘܟܪ̈ܝܐ, "foreigners") has still by no means been explained in palaeographical terms.

94 Ibn al-Ṭayyib, *Commentaire sur la Genèse*, ed. by J.C.J. Sanders, I, p. 73 (Arabic), II, p. 68 (French).

al-Ṭayyib refuses to focus on the evil of the sexual aberration (ܠܐ ܬܒܐܫܘܢ <אַל תָּרֵעוּ, "do not so wickedly") to which the Sodomites planned to subject Lot's guests.[95] Ibn al-Ṭayyib simply charges them with the dual offense of hatred for mankind and hence of rebellion against God's law.[96] Interestingly, this dual charge was expressed in almost identical terms by Josephus:[97]

> Ὑπὸ δὴ τοῦτον τὸν καιρὸν οἱ Σοδομῖται πλήθει καὶμεγέθει χρημάτων ὑπερφρονοῦντες εἴς τε ἀνθρώπους ἦσαν ὑβρισταὶκαὶ πρὸς τὸ θεῖον ἀσεβεῖς, ὡς μηκέτι μεμνῆσθαι τῶν παρ'αὐτοῦ γενομένων ὠφελειῶν, εἶναί τε μισόξενοι καὶ τὰς πρὸς ἄλλουςὁμιλίας ἐκτρέπεσθαι. χαλεπήνας οὖν ἐπὶ τούτοις ὁ θεὸς ἔγνωτιμωρήσασθαι τῆς ὑπερηφανίας αὐτοὺς καὶ τήν τε πόλιν αὐτὴνκατασκάψασθαι καὶ τὴν χώραν οὕτως ἀφανίσαι, ὡς μήτε φυτὸν ἔτιμήτε καρπὸν ἕτερον ἐξ αὐτῆς ἀναδοθῆναι
>
> Now about this time the Sodomites, overweeningly proud of their numbers and the extent of their wealth, showed themselves insolent to men and impious to the Divinity, insomuch that they no more remembered the benefits that they had received from Him, hated foreigners and declined all intercourse with others. Indignant at this conduct, God accordingly resolved to chastise them for their arrogance, and not only uproot their city, but to blast their land so completely that it should yield neither plant nor fruit whatsoever from that time forward.

But Ibn al-Ṭayyib eventually turns to a key element of that punishment, the sexual practices of the Sodomites, which he neatly describes using the term *shahwah*, i.e., sexual appetite for Lot's visitors: والتماسه يعني الاسدومايين الباب بعد حلول الشهوة بهم يدل على قساوتهم, "That the Sodomites should reach it, in other words (reach) the door after (sexual) desire was unleashed in them is what proves their cruelty."[98]

3.2 Gen 18:16–33

The foregoing has also to be seen in the light of a reference to this passage in which Ibn al-Ṭayyib offers a succinct but precise opinion on one of the major

95 Cf. *Le Commentaire sur Genèse-Exode 9,32 du manuscrit (olim) Diyarbakir 22*, ed. & trans. by L. Van Rompay, I, p. 81 (Syriac), II, p. 104 (French).

96 Cf. the expression *ajwar min qāḍī Sadūm*, "more unjust than a judge of Sodom"; James, John Courtenay, *The Language of Palestine and Adjacent Regions*. Edinburgh: T. & T. Clark, 1920, p. 225.

97 Ant. I,11,1. Josephus, Flavius, *Jewish Antiquities, books I-IV*, Greek text and English trans. by H. St. J. Thackeray, Cambridge MA: Harvard University Press, 1967, pp. 95–97.

98 Ibn al-Ṭayyib, *Commentaire sur la Genèse*, ed. by J.C.J. Sanders, I, p. 72 (Arabic), II, p. 68 (French).

themes of the Abraham cycle, also shared by the account in Gen 19[99] that of the three visitors who appeared to Abraham in Mamre, two of whom visited Lot in Sodom. Ibn al-Ṭayyib's approach is brief but explicit; he states that the man who appeared to Abraham, whom he believed to be a holy man, was in fact God, while those who appeared to Lot were apparently (*yushbihu*) two angels: علم انه اللـه والا فكان يظنه رجلا صالحا اطلعه اللـه على الخفايا ويشبه ان يكون الملايكة ظهروا على شكل ناس ابرار وكهنة,

> He told him (Abraham) that he was God, but (Abraham) thought he was a holy man sent by God to (reveal to him) the secrets. Apparently, the angels appeared to him in the form of righteous men and priests.[100]

3.3 Gen 19:24, 28

Ibn al-Ṭayyib uses these two verses to introduce a kind of allegorical exegesis on the value of the city of Sodom, which he treats as being tantamount to Gehenna. Here, as elsewhere, Ibn al-Ṭayyib draws on the explanation provided by Īshū'-dād of Merv, though in an abridged form.[101] This symbolic value is achieved through the iconography found in the narrative account in Gen 19;24,28 where "the lord rained upon Sodom ... brimstone and fire ... out of heaven" (24) and "the smoke of the land went up as the smoke of a furnace" (28). Ibn al-Ṭayyib's text reads as follows: وسدوم على مثال جهنم موضوعة في استفال من الارض وارض الوعد كان كالفردوس موضوعة في علو, "Sodom, the faithful representation of Gehenna, lies in the deepest part of the earth, whereas the land of promise, like paradise, lies in the highest part."

The Arabic *Jahannam* (Syr. ܓܗܢܐ / ܓܗܝܢܐ / Aram. גיהנם < Heb. גיהנם)[102] is a word ordinarily used for Hell in rabbinic (e.g., 'Erub 19a) and Christian texts

---

99 On the identity of Abraham's visitors in Rabbinic literature and the Church Fathers, see Emmanouela Grypeou and Helen Spurling, "Abraham's Angels: Jewish and Christian Exegesis of Genesis 18–19," in *The Exegetical Encounter between Jews and Christians in Late Antiquity*, ed. E. Grypeou and H. Spurling (Leiden–Boston: Brill, 2009), 181–203.

100 Ibn al-Ṭayyib, *Commentaire sur la Genèse*, ed. by J.C.J. Sanders, I, p. 72 (Arabic), II, p. 68 (French).

101 Īshū'dād of Merv, *Commentaire d'Iṣo'dad de Merv sur l'Ancient Testament. I. Genèse*, ed. by J.-M. Vosté and C. van den Eynde, trans. by C. van den Eynde, I, p. 164 (Syriac), II, p. 177 (French).

102 On the etymon of this loanword, see Brown, Francis, Samuel R. Driver, Charles A. Briggs, *Hebrew and English Lexicon of the Old Testament*. Boston/ New York: Houghton Mifflin Company, 1906, pp. 244b-245a. Cf. Kerr, Robert M, "Von der aramäischen Lesekultur zur arabischen Schreibkultur II. Der aramäische Wortschatz des Koran," in *Die Entstehung einer Weltreligion II.*

(SibOr 1:103; 2:291–292; 4:185),[103] although it is also the name of a specific region of Hell.[104] The image of Gehenna as a smoking furnace is commonplace in Jewish and Christian texts, where it agrees with the iconographic representation of the idea that both in Judaism and in Christianity (and also in Islam) the punishment inflicted on the wicked is principally associated with fire.[105] The idea of the "land of promise" (*arḍ al-wa'd*), i.e., Israel, as distinct from Paradise, appears to derive from an internal exegesis of the concept of the Garden of Eden (e.g., Amos 9:13–15; Ezek 36:11), to which Ibn al-Ṭayyib refers using the loanword *firdaws*. The term *firdaws* comes from the Syriac tradition, in which גַּן-בְּעֵדֶן (Gen 2:8) is translated as ܦܪܕܝܣܐ ܒܥܕܢ with the identical meaning of "a garden in Eden." Obviously, use of the Syriac *pardaysā*[106] denotes a dual topographical conceptualisation: the earthly Garden of Eden as against the heavenly Paradise. It is this dual topography that enables Ibn al-Ṭayyib to refer to the "promised land" at both earthly and celestial level, thus giving rise to an antithetical pairing (Garden of Eden–Paradise *vs.* Sodom–Gehenna), the promised land being located, in both cases, in "the highest part."[107] Ibn al-Ṭayyib clearly endows "the promised land" with an eschatological value, and his internal reading gives rise to an eschatological allegory in which "Sodom-Hell" is a consequence of man's evil, as opposed to "Garden of Eden-Paradise" (i.e., "promised land") which is viewed as the reward of the righteous.[108]

Von der koranischen Bewegung zum Frühislam, ed. by M. Groß and K.-H. Ohling, pp. 553–614. Berlin: Schiler, 2012, pp. 576–77.

103 See Milikowsky, Chaim, "Which Gehenna? Retribution and Eschatology in the Synoptic Gospels and in Early Jewish Texts." *New Testament Studies* 34 (1988): pp. 238–49.

104 See Bailey, Lloyd R., "Gehenna: The Topography of Hell." *Biblical Archaeologist* 49 (1986): pp. 187–91.

105 Grypeou, Emmanouela and J.P. Monferrer-Sala, ""A tour of the other world". A contribution to the textual and literary criticism of the "Six Books Apocalypse of the Virgin"." *Collectanea Christiana Orientalia* 6 (2009): pp. 115–65, pp. 157–58.

106 On this loanword from Old Persian, see Ciancaglini, C.A., *Iranian Loanwords in Syriac*, p. 237.

107 Monferrer-Sala, J.P., "Sacred readings, lexicographic soundings: cosmology, men, asses and gods in the Semitic Orient," in *Sacred Text: Explorations in Lexicography*, ed. by J.P. Monferrer-Sala and A. Urbán, pp. 201–18. Frankfurt am Main: Peter Lang, 2009, pp. 201–7.

108 Cf. Īshū'dād of Merv, *Commentaire d'Iṣo'dad de Merv sur l'Ancient Testament. I. Genèse*, ed. by J.-M. Vosté and C. van den Eynde, trans. by C. van den Eynde, I, p. 164 (Syriac), II, p. 177 (French).

3.4 Gen 19:31–38

Unlike Ephrem the Syrian,[109] who goes into the thorny issue of Lot's incestuous relationship with his daughters, and in contrast to the denotative commentary provided by Īshūʿdād of Merv,[110] whom he generally follows, Ibn al-Ṭayyib offers a somewhat cursory account of the episode, shunning all detail; he introduces it in a restrictive manner by means of the formula *wa-innamā akhbara al-Kitāb*, i.e., "the (holy) Book only states."[111] By this, Ibn al-Ṭayyib means that the Holy Scriptures refer only to the descendents of that relationship: the Ammonites (*al-ʿamanūnāyyīn* < ܐ̈ܡܘܢܝܐ) and the Moabites (*al-muwābāyyīn* < ܡܘܐܒܝ̈ܐ),[112] as well as many other people not mentioned in the Book (*wa-kathirīn mā yadhkurūna fī l-Kitāb*).[113] It is interesting to note that, following Ishoʿdad of Merv,[114] Ibn al-Ṭayyib indicates that the episode is a criticism (*tawbīkh*) of the *Majūs*, who wanted to marry their daughters.[115] These *Majūs* (μάγοι), an adaptation of the Syriac *Magūshē* (ܡܓܘ̈ܫܐ)[116] must be the magic-working members of the Persian priestly caste mentioned in the *Mĕʿarat Gazē*[117] and also by Muslim writers.[118]

---

109 Ephrem Syrus, *In Genesim*, ed. and trans. by R.-M. Tonneau, I, pp. 79–81 (Syriac), II, pp. 65–66 (Latin).

110 Īshūʿdād of Merv, *Commentaire d'Iṣoʿdad de Merv sur l'Ancient Testament. I. Genèse*, ed. by J.-M. Vosté and C. van den Eynde, trans. by C. van den Eynde, I, p. 167 (Syriac), II, pp. 180–181 (French).

111 Cf. Ginzberg, *The Legends of the Jews*, I, p. 257, where nothing is said about this episode.

112 Cf. Īshūʿdād of Merv, *Commentaire d'Iṣoʿdad de Merv sur l'Ancient Testament. I. Genèse*, ed. by J.-M. Vosté and C. van den Eynde, trans. by C. van den Eynde, I, pp. 166–167 (Syriac), II, p. 180 (French).

113 Ibn al-Ṭayyib, *Commentaire sur la Genèse*, ed. by J.C.J. Sanders, I, p. 74 (Arabic), II, p. 69 (French).

114 Īshūʿdād of Merv, *Commentaire d'Iṣoʿdad de Merv sur l'Ancient Testament. I. Genèse*, ed. by J.-M. Vosté and C. van den Eynde, trans. by C. van den Eynde, I, p. 167 (Syriac), II, p. 180 (French).

115 Ibn al-Ṭayyib, *Commentaire sur la Genèse*, ed. by J.C.J. Sanders, I, p. 74 (Arabic), II, pp. 69–70 (French).

116 On the term *majūs*, Schilling, Alexander Markus, *Die Anbetung der Magier und die Taufe der Sāsāniden. Zur Geistesgeschichte des iranischen Christentums in der Spätantike*. Louvain: Peeters, 2008, pp. 159–85. Cf. Boyce, Mary, *A History of Zoroastrianism*, 3 vols. Leiden/Köln: E.J. Brill, 1975, 1982, 1991, II, pp. 19–20.

117 Bezold, *Die Schatzhöhle*, p. 140 (Syriac), p. 141 (Arabic). Cf. Boyce, *A History of Zoroastrianism*, III, p. 357.

118 Van Gelder, Geert, *Close Relationships: Incest and Inbreeding in Classical Arabic Literature*. London/New York: Routledge, 2005, *passim*.

This first attempt to distract the reader's attention from the Biblical text is followed by the commentary proper, in the form of an argument intended to exculpate both Lot and the daughter wh ose idea it was to make him drink wine. In his interpretation, Ibn al-Ṭayyib therefore follows the Biblical account, avoiding the reading offered by the *Peshīṭtā* ܘܕܡܟܬ (< וַתִּשְׁכַּב, "and she lay," i.e., she had intercourse)[119] and simply judging the incident; he approves of the incest committed, even though the OT punishes incest with death (Lev 20:12). It would appear that, following the Biblical narrator and in accordance with Syriac tradition,[120] Ibn al-Ṭayyib condones the sexual act between daughters and father as being justified by the circumstances[121] even though the daughters were aware of what they were doing when they brought wine to their father in order to get him drunk, making his own behaviour unconscious. The explanation is similar to that offered by some Rabbis (GenR 51:8): "he did not indeed know of her lying down, but he did know of her arising" (שבשכבה לא ידע בקומה ידע).[122]

Ibn al-Ṭayyib, following Īshū'dād of Merv,[123] explains it in these terms:

> وها الكتاب يقول ان لوط لم يعلم ما فعله ابنتاه ومعنى هذا انه لو علم لكن ان ذاك منكرا عنده ولاجل الشراب ظن ان الذي كان منه كما يكون في الحلم ولهذا لم يقع اللوم على ابنته لان غرضها كان النسل لا الشهوة ولوط لما افاق من سكره لم يدن منهما
>
> And behold the Book says that Lot did not know what his two daughters did with him; this means that if he had known that, he would have rejected it. As a result of the drink he believed that what he did was as though in dreams. For that reason, he did not censure his daughter, for her purpose was to have descendants, rather than (sexual) desire. Lot, on recovering from his drunkenness, did not (ever) approach the two of them.[124]

Ibn al-Ṭayyib thus partly justifies the incest by regarding it as an unconscious act on Lot's part, taking place as though in drunken dreams,[125] even though Īshū'-

119 Ben Mobarak ben Ṣa'īr, *Kitāb at-Taysīr*, ed. and trans. by J. Martínez Delgado, II, p. 411 s.v.
120 Cf. Bezold, *Die Schatzhöhle*, p. 170 (Syriac), p. 171 (Arabic).
121 Curtis, Edward Lewis, "The Tribes of Israel," in *Biblical and Semitic Studies. Critical and Historical Essays by the Members of the Semitic and Biblical Faculty of Yale University.* New York/London: C. Scribner's sons, 1901, p. 12.
122 *Midrash Bereshit Rabba*, ed. by J. Theodor and Ch. Albeck, *wayyera* 51, p. 8; English trans.: *The Midrash Rabbah. I. Genesis*, ed. by H. Freedman and M. Simon, p. 448. Cf. GenR *wayyera* 51, p. 9–11.
123 Īshū'dād of Merv, *Commentaire d'Iṣo'dad de Merv sur l'Ancient Testament. I. Genèse*, ed. by J.-M. Vosté and C. van den Eynde, trans. by C. van den Eynde, I, p. 167 (Syriac), II, p. 180 (French).
124 Ibn al-Ṭayyib, *Commentaire sur la Genèse*, ed. by J.C.J. Sanders, I, p. 74 (Arabic), II, p. 70 (French).
125 The same explanation is found in Ephrem Syrus, *In Genesim*, ed. and trans. by R.-M. Tonneau, I, p. 79 (Syriac), II, p. 65 (Latin). Cf. *Le Commentaire sur Genèse-Exode 9,32 du manuscrit (olim) Diyarbakir 22*, ed. and trans. by L. Van Rompay, I, p. 84 (Syriac), II, p. 107 (French).

dād of Merv, in his more denotative commentary, explains that it was a "nocturnal pollution" (ܦܘܪܫܐ ܕܠܠܝܐ) thus giving rise to a state of impurity,[126] an explanation shunned by Ibn al-Ṭayyib. Interestingly, moreover, Ibn al-Ṭayyib opts not for the term ܚܡܪܐ used by Īshū'dād of Merv, but rather for the more ambiguous *sharāb*, which may denote both; a "juice" or "drink" – fermented or unfermented – and "wine." In doing so, he avoids the cognate form ܚܡܪܐ (cf. Aram. חַמְרָא < Heb. יַיִן), i.e., *khamr*, used unequivocally in Arabic to denote wine in any of its forms, either pure or diluted to a varying degree with water; this choice was probably made for moral reasons, given the excessive fondness of the Sodomites for wine.

The last piece of information provided by Ibn al-Ṭayyib with regard to this account is a chronological reference to the building and destruction of Sodom and Gomorrah. Although this appears to be of little exegetical interest for our present purposes, for the Bible commentators in Late Antiquity and the Middle Ages it was of the utmost importance, given the historical value with which such data endowed the events narrated; chronology, for the Christian authors, was also a philosophy of history.[127] Once again, Ibn al-Ṭayyib has taken the information from Īshū'dād of Merv, but this time he has gone further, providing a literal translation of the Syriac. The Syriac texts and their Arabic translation read as follows:[128]

وسدوم وغمورا بنيت في سنة تسع وخمسين	ܣܕܘܡ ܕܝܢ ܘܥܡܘܪܐ ܐܬܒܢܝ ܒܫܢܬ ܚܡܫܝܢ
لناحور ابن ساروغ واحرقت في سنة تسع	ܘܬܫܥ ܕܢܚܘܪ ܒܪ ܣܪܘܓ܂ ܒܫܢ ܕܝܢ ܒܫܢܬ
وتسعين لابرهيم يكون في الوسط مايتي وستة	ܬܫܥܝܢ ܘܬܫܥ ܕܐܒܪܗܡ ܕܗܘܝܢ ܒܡܨܥܬܐ
واربعين سنة	ܡܐܬܝܢ ܘܐܪܒܥܝܢ ܘܫܬ ܫܢܝܢ

"Sodom and Gomorrah were built in the fifty-ninth year of Nahor, son of Serug, and were burnt down in the ninety-ninth year of Abraham, a span of two hundred and forty-six years."

126 Cf. Eisenman, Robert H. and Michael Wise, *Dead Sea Scrolls Uncovered. The First Complete Translation and Interpretation of 50 Key Documents Withheld for over 35 Years*. New York: Penguin Books, 1993, p. 195, 272.

127 Momigliano, Arnaldo, "Pagan and Christian Historiography in the Fourth Century A.D," in *Conflict between Paganism and Christianity in the Fourth Century*, ed. by A. Momigliano, pp. 77–99, 83–4. Oxford: Clarendon Press, 1963.

128 Ibn al-Ṭayyib, *Commentaire sur la Genèse*, ed. by J.C.J. Sanders, I, p. 74 (Arabic), II, p. 70 (French). Cf. Īshū'dād of Merv, *Commentaire d'Iṣo'dad de Merv sur l'Ancient Testament. I. Genèse*, ed. by J.-M. Vosté and C. van den Eynde, trans. by C. van den Eynde, I, p. 167 (Syriac), II, pp. 180–181 (French).

The Syriac text – drawing on earlier chronicles[129] – is part of a fragment found by Īshū'dād of Merv, which provides additional chronological information not included by Ibn al-Ṭayyib since it is not relevant to the episodes narrated in Gen 19:1–38.

4 Conclusion

At no stage does Ibn al-Ṭayyib either rewrite the text or seek to imitate it; nor is he offering a kind of explicative introduction to the translation, a practice common among certain Arab authors.[130] This is not, either, a redundant exegesis that simply rehearses the evidence of the text itself. Quite the contrary, it is a reformulation of the text through which Ibn al-Ṭayyib establishes a discursive equivalence between the Biblical text and his own. In adopting this analytical approach, common among Aristotelian Eastern Christian thinkers, he seeks above all to preserve the Syriac Christian heritage in Arabic.[131]

The difference between the two texts, however, is that while Gen 19 simply sets forth the events (*narratio*) with an accompanying moral appraisal, Ibn al-Ṭayyib's analysis provides a summing-up that serves as a means of reconsidering those events. In essence, this is an exegetical metadiscourse characteristic of Late Antiquity and the Middle Ages,[132] which consists in viewing the deeds and episodes narrated with a degree of detachment; to that extent, the text continues the process of transmission generated by the Eastern Syriac tradition. This commentary, based on exegetical details, establishes a set of metadiscursive relationships through which Ibn al-Ṭayyib replaces translation by comment. This is not, as elsewhere,[133] a paraphrased translation modelled on earlier translations; nor is it a reworked copy comparable with the reworked Pentateuchs from Qum-

129 *Chronique de Michel le Syrien, Patriarche jacobite d'Antioche (1166–1199)*, ed. and trans. by Jean-Baptiste Chabot, 4 vols. Paris: Ernest Leroux, 1899, 1901, 1905, 1910, I, p. 25 says that the year of construction of Sodom and Ghomorra was year 48 of Nakhor.

130 As in the case of the Nestorian al-Ḥārith b. Sinān b. Sunbāṭ (9th-10th century CE), see Sadan, Joseph, "In the Eyes of the Christian Writer al-Ḥāriṯ ibn Sinān. Poetics and Eloquence as a Platform of Inter-Cultural Contacts and Contrasts." *Arabica* 56 (2009): pp. 1–26, p. 11.

131 Faultless "Ibn al-Ṭayyib", pp. 668–669.

132 Goldstein, Miriam, *Karaite Exegesis in Medieval Jerusalem*. Tübingen: Mohr Siebeck, 2011, pp. 50–1.

133 Ben-Shammai, Haggai, "The Tension between Literal Interpretation and Exegetical Freedom. Comparative Observations on Saadia's Method," in *With Reverence for the Word: Medieval Scriptural Exegesis in Judaism, Christianity, and Islam*, ed. by J. D. McAuliffe, B. D. Walfish, J. W. Goering, pp. 33–50. New York: Oxford University Press, 2010, p. 34.

rān, in which the 'compositional technique' applied to verses or pericopes alters the source text by rearranging or paraphrasing it.[134]

As was the case with the Rabbanite exegetes, for whom the classical *midrash* was an inherent component of the new exegetical enterprise which had to be reckoned with,[135] Ibn al-Ṭayyib, as a commentator shaped by the Eastern Syriac exegetical tradition,[136] bases his commentary on a Syriac text; but rather than translating it, he moulds it hermeneutically by adapting the narration of the Syriac Biblical narrative to an exposition-commentary into which he incorporates earlier authors whom he does not always acknowledge, presumably because their work came to him anonymously.

Ibn al-Ṭayyib's exegesis, then, does not consist in reworking the original text, adding, removing or rearranging narrative elements. Rather, it is an exegesis that reflects the basic content of the Biblical account, subtly modulated with precise comments, but without radically modifying the narrative discourse included in the exegesis.

We may therefore conclude that Ibn al-Ṭayyib provides a kind of contextualised narrative exegesis, but one which gives rise to a new text. And this para-Biblical text has a purely hermeneutic purpose, though always within the framework of a historical horizon, as part of the *Historia salutis*.

---

134 Zahn, Molly M., *Rethinking Rewritten Scripture: Composition and Exegesis in the 4QReworked Pentateuch Manuscripts*. Leiden/ Boston: Brill, 2011, pp. 12–9.

135 Polliack, Meira, "Concepts of Scripture among the Jews of the medieval Islamic World," in *Jewish Concepts of Scripture: a Comparative Introduction*, ed. by Benjamin D. Sommer, pp. 80–101. New York/ London: New York University Press, 2012, p. 92.

136 Cf. Charfi, Ayoub, *Le commentaire de Psaumes 33–60 d'Ibn aṭ-Ṭayyib, reflet de l'exégèse syriaque orientale*. Rome, 1997 (PhD Pontificia Università Gregoriana).

Alison Salvesen

"Christ has subjected us to the harsh yoke of the Arabs"

The Syriac Exegesis of Jacob of Edessa in the New World Order

1 Introduction

Jacob, bishop of Edessa, was the most prominent monophysite scholar of the early medieval period. His work covers a remarkable range, from Biblical exegesis, to the codification and transmission of philosophy and Greek scientific knowledge,[1] the development of ecclesiastical law, and historiography. All these areas of his output are interlinked in various ways. This makes it difficult for the modern scholar to isolate any particular aspect of his oeuvre.

Jacob was born near Antioch sometime in the 630s CE. Therefore his early years would have coincided with the death of the Prophet Muḥammad and the subsequent captures of Damascus (634 CE), Antioch (636) and Edessa (638). However, in his earlier works and in the scant biographical information about him there is no overt sign of what we would regard as epic changes taking place in his vicinity. Perhaps this is because, for Christians in the Middle East, disruption had become the norm, particularly since the Persian invasions in the earlier part of the 7th century and the deepening schisms within the Christian Church.

Apart from a brief period as bishop of Edessa during the 680s, a post from which he in fact resigned, Jacob spent most of his life studying and teaching in some of the great monasteries of Syria, namely Qenneshrin, Kayshum , Eusebona, and Tell 'Ada.[2] He wrote in Syriac but was thoroughly conversant in Greek.

1 For the role of Syriac Christian scholars and monasteries in transmitting Greek philosophy and science to the Arabs, see Montgomery, Scott L., *Science in Translation: Movements of Knowledge through Cultures and Time*. Chicago: University of Chicago Press, 2000, esp. p. 69; D'Ancona, Cristina, "Greek into Arabic," in *The Cambridge Companion to Arabic Philosophy*, ed. by Peter Adamson and Richard C. Taylor. Cambridge: Cambridge University Press, 2005, pp. 18–20.
2 Eusebona and Tell 'Ada were located to the south west of Beroea (modern Aleppo); Kayshum due north of Nisibis (now Nizip on the Turkish border); and archaeologists have lately identified

https://doi.org/10.1515/9783110564341-009

Many of his works still survive. They are at once conservative, preserving the best of what had preceded him from both Christian tradition and pagan Greek culture, and also cutting-edge, such as his *Chronicon*, his *Hexaemeron* (a series of treatises on the Six Days of Creation), and his own 'version' of the Old Testament.

2 Jacob and the Exegesis of Scripture

Jacob's scriptural exegesis falls into two main areas. First there is his version of the Old Testament, of which a few books survive in single manuscripts. The version is a clever amalgamation in Syriac form of the Greek *Septuagint* and Syriac *Peshitta* texts. There has been some scholarly debate over what principles he used to create this version or why he produced it.[3]

More pertinent to the theme of the present volume is Jacob's extensive work in solving various difficulties in Scripture, such as the meaning of individual words, or narrative or theological problems in the text. His solutions are preserved in various different places: collections of scholia, marginal comments, and letters to his correspondents.[4] Almost all of them are concerned with the Pentateuch and historical books of the Old Testament.[5]

The exegetical techniques that Jacob employs are mainly traditional ones. First he gives an explanation of the narrative or historical context of the passage,

---

Qenneshrin on the west bank of the Euphrates near Jarablus in northern Syria. See the map in Jullien, Florence, ed., *Le monachisme syriaque*. Paris: Geuthner, 2010, p. 10.

3 For a summary, see Salvesen, Alison, "Scholarship on the Margins: Biblical and Secular Learning in the Work of Jacob of Edessa," in *Syriac Encounters: Papers from the Sixth North American Syriac Symposium, Duke University, 26–29 June 2011*, ed. by Kyle Smith, Emilio Fiano and Maria Doerfler, pp. 327–344. Eastern Christian Studies 20. Leuven: Peeters, 2015.

4 For editions and translations of many of Jacob's scholia, see Phillips, George, *Scholia on Passages of the Old Testament by Mar Jacob Bishop of Edessa*. London: Williams and Norgate, 1864, from collections of scholia in two British Library Additional manuscripts, 17,193 and 14,483. A useful article on the problems of the scholia is that of Kruisheer, Dirk, "Reconstructing Jacob of Edessa's Scholia," in *The Book of Genesis in Jewish and Oriental Christian Interpretation: A Collection of Essays*, ed. by Judith Frishman and Lucas van Rompay, pp. 187–196. Traditio Exegetica Graeca 5. Leuven: Peeters, 1997. For some examples of Biblical exegesis in Jacob's letters see Wright, William, "Two Epistles of Mar Jacob, Bishop of Edessa." *Journal of Sacred Literature and Biblical Record* 10 (1867): pp. 430–460 (English synopsis and Syriac text), and Nau, François, "Traduction des lettres XII et XIII de Jacques d'Édesse (exégèse biblique)." *Revue de l'Orient chrétien* 10 (1905): pp. 197–208, 258–282 (French translation). Letters 12 and 13 are from British Library Add. MS 12,172.

5 There are also some marginal scholia in the manuscripts of his own version of Old Testament prophetic books (Isaiah, Ezekiel, and Daniel). These manuscripts remain largely unedited.

with explanations of apparent problems in it. He may then draw out the *theoria*, a term that means many different things for Christian authors, but for Jacob it generally refers to a deeper spiritual sense, and often a typological meaning. For instance, King Saul's two daughters may represent the Jews and the Church respectively, since the elder sister rejects David as a spouse (David being a 'type' of Christ), while the other loves and marries him.[6]

The present chapter examines whether Jacob's Biblical exegesis tried to meet the challenge of Arab political domination and the rise of Islam, or whether his approach was a fundamentally conservative one that aimed to shore up Syrian Orthodox identity without innovation. Below are three examples where Jacob's approach to Scripture could be said to reflect the impact of his historical circumstances, and where the influence of Jewish and early Islamic arguments is perceptible, in a way that would not have been seen in exegetes of the previous century.

2.1 Use of Jewish Traditions and Apocryphal Books to Solve Difficulties

William Adler has shown how Jacob used Jewish pseudepigrapha to answer questions from his correspondents on difficulties in the Book of Genesis. Despite the view of the 4th century Alexandrian patriarch Athanasius that the Book of Enoch was inauthentic, Jacob cites Enoch as an authoritative source.[7] This is because he says it was known to the apostles (in fact it is cited in the New Testament, in Jude 1:14–15). Following on from the work of Sebastian Brock, Adler also notes Jacob's use of a work that Jacob describes as "Jewish Histories." From its content this work seems to be closely linked to what we know as the *Book of Jubilees*, which like the Book of Enoch has its origins in Jewish circles of the Second Temple period, though it postdates it by a century or so.[8] However,

6 Phillips, *Scholia*, p. 29, Syriac text pp. *yodh-zayin—yodh-waw*.

7 Although the oldest portions of 1 Enoch go back to the 3rd century BCE, and there are Aramaic fragments among the Dead Sea Scrolls, and some material in Greek and Latin, today Enoch is preserved only in the ancient Ethiopian language, Ge'ez. See Nickelsburg, George W.E. and VanderKam, James C., eds., *1 Enoch: A New Translation*. Minneapolis: Fortress Press, 2004. Jacob of Edessa no doubt knew the work in its Greek form. Although a single small fragment of Syriac Enoch survives in the medieval chronicle of Michael the Syrian, the work as a whole is unlikely to have been rendered into Syriac: see Brock, Sebastian P., "A Fragment of Enoch in Syriac." *Journal of Theological Studies* 19 (1968): pp. 626–631.

8 The textual situation of *Jubilees* is similar to that of 1 Enoch in that it is preserved in full only in Ge'ez, with fragments in Hebrew and Aramaic among the Dead Sea Scrolls, some material in

the form that Jacob uses may not be identical to the *Book of Jubilees* known to us. Adler argues that Jacob elaborated and shaped the material that he found in his "Jewish Histories" in order to present a more coherent and theologically persuasive narrative.[9] In other words, Jacob acted in a similar way to the creators of Jewish haggadic Midrash, by 'filling in the gaps' of Scripture in a way that was theologically appropriate to his circle and by means of earlier traditions.

To give some examples of Jacob's use of early Jewish traditions, Jacob implicitly draws on "Jewish Histories" (Jubilees) in the account of Abraham's early life in Ur of the Chaldees.[10] Likewise, Jacob answers John the Stylite's question concerning the existence of writing before Moses with a reference to "reliable written stories transmitted by Jews,"[11] in which Moses is described as being taught Hebrew and Egyptian writing by his father Amram (= Jubilees 47:9).[12]

Other explanations by Jacob also display knowledge of traditions originating in Jewish Midrash. Two of his scholia and one of his letters refer to the identity of the widow who appealed to Elisha in 2 Kgs 4:1.

> i) From the 7th scholion, about the woman who appealed to Elisha and about the transformation of the waters to oil through the prophet's prayer... [2 Kgs 4:1][13]
>
> Firstly one should know whose wife the woman was, where she was from, and what was the reason for the debt she owed, on account of which the creditor wanted to take her children into slavery... She was the wife of the Obadiah who feared God, Ahab's steward, who loved and honoured Elijah the prophet [1 Kgs 18]. He was also numbered among the twelve holy prophets. When he saved the hundred prophets from the power of the murderess Jezebel and hid them in two caves, owing to the severity of the time and the scarcity of food, he was forced to borrow money and incur a debt on behalf of himself and his wife and children in order to feed [the prophets] during the scarcity, so that he could save them

---

Greek and Latin, and short extracts in Syriac (Tisserant, Eugène, "Fragments syriaques du livre des Jubilés." *Revue Biblique* 30 (1921): pp. 55–86, 206–232. For a translation of the Ethiopic version, see Charles, Robert H., *The Book of Jubilees.* London: Society for Promoting Christian Knowledge, 1917.

9 Adler, William, "Jewish Pseudepigrapha in Jacob of Edessa's Letters and Historical Writings," in *Jacob of Edessa and the Syriac Culture of his Day,* ed. by R. B. ter Haar Romeny, pp. 49–65. (Monographs of the Peshiṭta Institute 18. Leiden: Brill, 2008; Brock, Sebastian P., "Abraham and the Ravens: A Syriac Counterpart to Jubilees 11–12 and its implications." *Journal for the Study of Judaism* 9 (1978): pp. 135–152.

10 Phillips, *Scholia,* pp. 3–6, Syriac pp. *beth-gamal.* Similarly in Jacob's 13th letter to John the Stylite (Nau, "Traduction des lettres XII et XIII" *Revue de l'Orient chrétien* 10 (1905): pp. 203–206).

11 Wright, "Two Epistles", Syriac p. *ṭeth.*

12 Nau, "Traduction des lettres XII et XIII", pp. 207–208.

13 This scholion appears in BL Add. MS 14,483, of the 9th or 10th century, folios 27r–27v. It was not published by Phillips, probably because the end of it is damaged in the manuscript.

from famine as he had saved them from slaughter. "I saved them: let them not be neglected now and die of hunger!" This was the cause of the debt for which Obadiah's children were being led into slavery. For this reason his wife appealed to Elisha, "You know that your servant my husband ..." [a lacuna intervenes at this point]

ii) From Jacob's 16th scholion, showing how Jezebel killed the prophets, and about Obadiah, how he hid the rest.... [1 Kgs 18:3–4]

Now Obadiah, Ahab's steward and faithful, a true servant of God, true as his name,[14] and a secret God-fearer, concealed from her and the murderers one hundred of his brothers the prophets and hid them in two caves. He saved them from two deaths, one by the sword, and one by hunger that would follow. Because of the expense involved in this and the considerable debt that resulted in two of his children being led off to slavery, his wife and mother of the boys appealed to Elisha the prophet.[15]

iii) I will tell you a little about Obadiah the prophet. There exists the opinion among readers that he was the third captain of fifty who went up to Elijah and entreated that he should not perish as his predecessors had, and [Elijah] went down with him to King Ahaziah of Israel [2 Kgs 1:11–16].

To this I can add for your benefit another fact about him that is reliable and not mere supposition. This is that he was the husband of the woman who came to Elisha the prophet and said to him, "Your servant my husband has died. You know that your servant feared the Lord, and a creditor has come to take away my two sons as slaves." [Elisha] performed for her the miracle in which water was transformed into oil. For this man had borrowed money during the period of famine under Ahab, and bought food for the prophets who had fled from Jezebel.[16]

Jacob says that the woman in 2 Kgs 4 was the wife of Obadiah, Ahab's steward who feared the Lord and saved one hundred prophets by hiding them from Jezebel and feeding them throughout the famine (1 Kgs 18:3–4). By so doing Obadiah incurred a huge debt, and this is why his widow had to appeal to Elisha to prevent her sons being enslaved to pay off his creditors. Jacob also identifies the man with the prophet Obadiah.

As the original editor of the scholia George Phillips notes, a similar tradition to Jacob's explanation is found in the work of the Hellenistic Jewish writer Fla-

14 The Hebrew name Obadiah means "servant of the Lord," and this meaning would have been fairly clear in Syriac also.

15 BL Add. MS 17,193, fol. 56v-57r, edited and translated by Phillips, *Scholia*, pp. 45–47, Syriac pp. *kaph–ṭeth–lamad*. The translation above is my own. Cf. also foll. 20v and 21r of BL Add. MS 14,483 for another version of this 16th scholion.

16 Wright, "Two Epistles", p. 432, Syriac end p. *yod-heth*–p. *yod-ṭeth*, from British Library Add. MS 12,172, fol. 79–134. French translation in Nau, "Traduction des lettres XII et XIII", pp. 270–271.

vius Josephus.[17] Later Jewish sources include the *Babylonian Talmud* (bSanh 39b) and the medieval Midrashic work *Exodus Rabbah* (31.3–4). The Aramaic Targum Jonathan to 2 Kgs expands the Biblical text of 4:1, to show the widow telling Elisha,

> Your servant Obadiah my husband is dead: and you know that your servant was fearing from before the Lord, so that when Jezebel killed the prophets of the Lord, he took from them one hundred men and hid them in fifties in caves. He used to borrow money and feed them so as not to provide them with food from Ahab's wealth because it was wrongfully obtained. And now the creditor has come to take away my two sons as slaves! [additions to the biblical text italicised][18]

Despite the evident Jewish origin of the tradition, it is highly unlikely that Jacob derived his explanation directly from either the rabbinic Targum or from Josephus. An intermediary Greek Christian source is probable. Jacob no doubt approved of the identification of the two men in 1 Kgs 18 and 2 Kgs 4:1 because it was soundly based on the verbal link between the passages: Obadiah in the Biblical text of 1 Kgs is said to have "feared the Lord," the very characteristic of the dead man that his widow mentions in 2 Kgs 4. The association with the prophet Obadiah, about whom nothing much is otherwise known, is a more obvious step, and although it no doubt originates in Jewish tradition, it could also have been made independently later.

Jacob mentions a further tradition in his 13th letter to John the Stylite of Litarba. Jacob notes that opinion exists that the prophet Obadiah was the third captain of fifty men sent to arrest Elisha in 2 Kgs 1:13. This is attested in the Greek work *Lives of the Prophets* that was falsely attributed to the Christian writer

---

17 Phillips, *Scholia*, p. 47 n. 8. Flavius Josephus (d. c. 100 CE), *Antiquities of the Jews* IX.4: "For they say that the widow of Obadiah the prophet and Ahab's steward, came to [Elisha], and said that he was not ignorant how her husband had preserved the prophets that were to be slain by Jezebel, wife of Ahab; for she said that he had hidden a hundred of them, and had borrowed money for their maintenance, and that following her husband's death, she and her children were being taken away to slavery by the creditors; and she asked him to have mercy on her on account of her husband's good deed and to provide her with some assistance" [translation based on that of William Whiston (*The genuine works of Flavius Josephus, the Jewish historian*, London, 1737)].

18 Translation my own. See also Harrington, Daniel J. and Saldarini, Anthony J., *The Aramaic Bible 10. Targum Jonathan of the Former Prophets: Introduction, Translation and Notes*. Edinburgh: T & T Clark, 1987, p. 270, and Dray, Carol A., *Translation and Interpretation in Targum to the Book of Kings*. Studies in the Aramaic Interpretation of Scripture. Leiden: Brill, 2006, pp. 65–66, 114.

Epiphanius of Salamis.[19] Jacob mentions both this work and its attribution in disparaging terms earlier in Letter 13.[20] Perhaps for this reason he evinces less conviction concerning the identification of the prophet Obadiah with the captain of fifty than the tradition that he was the husband of the widow mentioned in 2 Kgs 4:1.[21]

Thus "Jewish histories," apocryphal writings ("secret books"), and other traditions are all of use to Jacob as commentary on Scripture. In view of Jacob's hostility to Jews and Judaism of his own day, expressed in his canons and other works, his use of these originally Jewish extra-canonical books to solve exegetical problems is perhaps a little surprising.

However, Jacob does not use such sources indiscriminately, but assesses whether they withstand the tests of apostolic use (for example in the case of the citation of Enoch in the Letter of Jude) or historical chronology. For instance, Jacob rejects the identification of the prophet Jonah with the boy raised by Elijah, given by Pseudo-Epiphanius, since their respective lifetimes could not have overlapped.[22] He clearly distinguishes between the unreliability of what he sometimes terms "superfluous tales" and the authority of "Jewish books."[23] In this re-

19 Hare, Douglas R. A., "The Lives of the Prophets," in *The Old Testament Pseudepigrapha II*, ed. by James Charlesworth, pp. 379–400. London: Darton and Todd, 1985.

Coincidentally the same Syriac manuscript containing the second version of Jacob's scholion above, BL Add. MS 17,683, also includes a Syriac version of the *Lives of the Prophets*. It states, "Obadiah means 'serving the Lord', according to the Hebrew language. Obadiah was from the land of Shechem, from a region called Beth Aqerem. He was Elijah's disciple, and since he bore many things on his behalf, he was saved. He was the third head of fifty, whom Elijah spared, and he went down to Ahaziah. After he left the king's service he became a prophet and when he died in peace he was buried with his fathers." (Translation my own.)

20 Wright, "Two Epistles", Syriac p. *yod-zayin*, Nau, "Traduction", p. 268.

21 The 5th century Greek Syrian writer Theodoret of Cyrrhus also identifies Obadiah with Ahab's steward and the deceased husband in 2 Kgs 4:1, and like Jacob's, his account is independent of the *Lives of the Prophets*. Fernández Marcos, Natalio and José R. Busto Saiz, *Theodoreti Cyrensis Quaestiones in Reges et Paralipomena*. Madrid: Instituto "Arias Montano", C.S.I.C, 1984, p. 202, §14.

22 Wright, "Two Epistles", Syriac p. *yod-zayin*. Jacob describes such works as "superfluous stories lacking reliability."

23 Wright, "Two Epistles", p. 430 and Syriac p. *ḥeth* using the Syriac term for *apocrypha*, *ktābē mṭašayā*, ("hidden books"). French translation Nau, "Traduction des lettres XII et XIII", pp. 206–207.

spect his approach is perhaps comparable to that of St Jerome (though it is hardly likely that Jacob would have known of the 4th century Latin writer).[24]

2.2 'Deuteronomistic' Theology: A Biblical Parallel to the Plight of the Syrian Orthodox at the End of the 7th Century

What modern scholars term the "Deuteronomistic" theological outlook on history is a pervasive concept in the Hebrew Bible. It expresses the idea that God's people enjoy his favour and his covenant conditionally: though he will never reject them entirely, he will chastise them by subjection to foreign powers and expulsion from their land if they practise idolatry and break his commandments.[25] Yet repentance and a return to right worship and living will bring about restoration of God's favour.[26]

This theology was dominant after the Babylonian Exile, and even more so after 70 CE in the works of the rabbis.[27] It is certainly far from unusual in Christian works, including Syriac texts contemporaneous with Jacob of Edessa, in the late 7th century.[28] Generally, it means that world events are interpreted according to this Biblical paradigm: oppression of the religious community or nation by an

---

24 See Salvesen, Alison, "'*Tradunt Hebraei*': The Problem of the Function and Reception of Jewish Midrash in Jerome," in *Midrash Unbound: Transformations and Innovations*, ed. by Michael Fishbane and Joanna Weinberg, pp. 57–81. Oxford/Portland, OR: Littmann Library, 2013.

25 See for example Deut 28 and 29; Ezra 9:7.

26 E.g., Deut 30:1–5.

27 See Goldenberg, Robert, "The Destruction of the Jerusalem Temple: its meaning and its consequences," in *Cambridge History of Judaism IV: The Late Roman-Rabbinic Period*, ed. by Steven T. Katz, pp. 191–205. Cambridge: Cambridge University Press, 2006, esp. pp. 196–97. The concept of divine punishment is brought out frequently in Targum, e.g., Targum Jonathan to Isa 1:3–9, esp. 1:7: "your country lies desolate, your cities are burned with fire; in your presence *the Gentiles take possession of your land* and *because of your sins it is removed from you and given to* aliens" (Chilton, Bruce D., The *Aramaic Bible 11. The Isaiah Targum: Introduction, Translation, Apparatus and Notes*. Edinburgh: T & T Clark, 1987, p. 3).

28 Compare the chronicle of disasters a few years later, covering the years 712–716 CE, in which the themes of judgment and repentance are prominent (Palmer, Andrew, *The Seventh Century in the West-Syrian Chronicles*. Translated Texts for Historians 15. Liverpool: Liverpool University Press, 1993, pp. 45–48), and a Chalcedonian writer's interpretation of recent history in Brock, Sebastian P., "North Mesopotamia in the late seventh century. Book XV of John Bar Penkayē's *Riš Mellē*." Jerusalem Studies in Arabic and Islam 9 (1987): pp. 51–75 (repr. in Brock, Sebastian P., *Studies in Syriac Christianity: History, Literature and Theology*. Variorum. Hampshire: Ashgate, 1992, ch. II).

enemy is taken to indicate God's displeasure at sin within the community. This is explicit in the famous remark in the Syriac historical work, the *Chronicle of Zuqnin*, where the writer, commenting on the *jizya* imposed by ʿAbd al-Malik in 692 CE/AG 1003, states: "From this time the Sons of Hagar began to subject the Sons of Aram to Egyptian slavery. Yet woe unto us! Because we sinned, slaves have gained authority over us!"[29] Similarly, the *Apocalypse of Pseudo-Methodius*, dated to the early 690s by Palmer and Reinink, says of "these Children of Ishmael," "it was not because God loves them that he allowed them to enter the kingdom of the Christians, but because of the wickedness and sin committed by Christians, the like of which was not done in any previous generation."[30]

In more than forty preserved scholia deriving from Jacob there is only one instance where he mentions contemporary historical circumstances. And yet, as in contemporaneous Syriac sources, Jacob uses the opportunity to make a connection between the sinful state of Syrian Orthodox Christians and their subjugation to the Arabs.[31]

> And Rehoboam son of Solomon became king over Judah. Rehoboam was forty-six years old when he became king, and he reigned seventeen years in Jerusalem, the city that the Lord chose for himself to place his name in it, out of all the tribes of Israel. His mother's name was Maʿcah the Ammonite. Rehoboam and Judah did what was evil before the Lord', etc. etc. [= 1 Kgs 14.21–22a]
>
> From these words it is clear that all the people of Israel were ready to turn away from the Lord and to go after the error and abomination of the nations who worshipped demons, even if Jeroboam himself had not made the golden calves that he did, and through which he caused Israel to sin. For the people of Judah, who were not subject to Jeroboam, acted more wickedly than Jeroboam and the people of Israel, because it was their own desire to turn aside from the Lord and to serve the gods of the nations. In addition to dishonouring God and bowing down to idols, they also dishonoured and polluted the city of Jerusalem that the Lord had chosen for himself and over which he had invoked his Name.
>
> For this reason when recounting that Rehoboam became king, the scriptural narrative wishes to demonstrate the extent of his wickedness and iniquity by describing him as the son of Solomon, the one who abandoned God and worshipped idols. It also notes that it

---

29 Syriac text: Chabot, Jean-Baptiste, ed., *Incerti Auctoris Chronicon Pseudo-Dionysianum vulgo dicto, II.* Corpus Scriptorum Christianorum Orientalium 104. Louvain: Secrӗtariat du CSCO, 1965, p. 154, lines 25–27. Translation my own.

30 Syriac text in Reinink, Gerard J., *Die syrische Apokalypse des Pseudo-Methodius.* Corpus Scriptorum Christianorum Orientalium 540, 541; 220, 221. Louvain: Peeters, 1993, texts XI.5, p. 25, lines 9–12. See Reinink's long note, n. 3*, pp. 43–44.

31 The scholion appears in Phillips, *Scholia*, pp. 39–41, Syriac pp. *kaph-heh–kaph-zayin.* Since Phillips's translation is both rather dated and occasionally misleading, I have provided a fresh version on the basis of BL Add. MS 14,483, fol. 18v–19v. I am grateful to David G.K. Taylor for assistance with a difficult section in the Syriac.

was in the city that the Lord had chosen out of all the tribes of Israel to set his name, that [Rehoboam] reigned: as one might say, he together with all the people of Israel that he ruled, dishonoured and defiled even this holy place.

Scripture also says that [Rehoboam] was the son of an Ammonite woman, who had made his father Solomon set up an idol and altar to Milcom, the abomination of the Ammonites, to bow down to it and offer sacrifices to it. [Scripture says this] in order to [show] that the paganism and error learned from his father and mother, [Rehoboam] practised more thoroughly than his father and also more than Jeroboam himself, that lawless apostate who caused Israel to sin. Thus both Rehoboam and the people of Judah who were called the Lord's portion and the house of David, and were in the Lord's holy city of Jerusalem, sinned, acted wickedly and did what was evil before the Lord, though they were called by his own Name.

Both through their name [i.e. Judah] and their small size, they portray typologically this small, confessing people, who are called and are indeed "orthodox" and acknowledge the Lord Christ, and are in the church, the city of God, Jerusalem, that the Lord chose for himself and sanctified out of all the peoples of the earth. Yet they provoke him by their deeds and lawless ways of life, "more than all the nations." This is because, despite being called by his Name, they cause everyone to stumble, even in the faith itself. As [God] would say, as well as "My own Name is blasphemed among the nations because of you," also "You have dishonoured and polluted the church, Jerusalem, the city that I chose, and my house you have made a cave of thieves."

So we are those who name ourselves true Christians and confessors of Christ and of the house of God, and "His portion, Jacob," and "Israel His inheritance," and "the people which sees God," and "the holy nation," and "the royal priesthood." Yet we have become those who sin more than anyone, and who are deprived of every virtue, and right behaviour, and of love, peace, and harmony. While these [virtues] were apparent in us, they declared us to be disciples of Christ. Their removal declares us to be opponents of Christ, as we trample upon his laws and the commandments that he taught us.

Thus, because of the wickedness of Rehoboam and of Judah, God brought on them Shushaqaym[32] king of Egypt. He took captives and scattered them, and he destroyed their cities because of their sins and their provocation, as divine Scripture recounts.

In the same way, because of our many sins and iniquities, Christ has delivered us up also. He has subjected us to the harsh yoke of the Arabs[33] who do not acknowledge him to be God and son of God—Christ, who is God and God's Son, he who bought us by his blood from the slavery of sin. He saved us through his cross from the subjection of the Accuser and of demons, and he freed us and rescued us by his death from death and corruption. He gave us true hope of resurrection from the dead, and he promised us the blessed life of the world to come, and a portion and inheritance of the kingdom of heaven. Because we did not understand all this grace and freedom that was given to us, and we were un-

32 This is closer to the form of the name found in the *Septuagint*, rather than the one in the *Peshitta*.

33 The term that Jacob uses here is *ʾArabāyē* (from the Greek Ἀραβαίοι). In other places he tends to use two other terms, *Ṭayāyē*, originally referring to the Syrian Arab tribe of the *Ṭay* but later applied more widely to Arabs as an ethnic designation; or the religious designation *Mahgrāyē*, used to denote Muslims.

grateful and deniers of grace, we have been handed over to slavery and subjection, just as ancient Judah was [handed over] to plundering and captivity.

Jacob's scholion comments on a passage that presents no textual difficulty. Thus his choice of passage is, unusually, for theological reasons, the desire to explain contemporary issues through Biblical precedent.[34] Yet the choice of 1 Kgs 14 is not an obvious one even in this respect. Though it is possible that the post-Chalcedonian split over Christology was seen as comparable to the split of the Northern Kingdom of Israel from the southern Kingdom of Judah after the reign of Solomon, Rehoboam and Shishak are not obvious ciphers for a particular Byzantine emperor and a particular caliph. Usually it is the Babylonian Exile that both Jewish and Christian authors use as the paradigm for the idea of God's chastisement. Or Jacob could have chosen a story from Judges describing Philistine oppression of the pre-monarchic tribes as divine chastisement for unfaithfulness to the Lord.

Towards the end of the scholion Jacob refers to Shushaqaym king of Egypt taking captives, scattering the people and destroying the cities because of the people's sins. However, in the Biblical account in 1 Kgs 14, Shishak comes up to Jerusalem and then merely departs with the treasures of Jerusalem. In fact Jacob's comment on the Kings passage must have been influenced by the more detailed parallel account of Rehoboam's reign found in 2 Chr 12:1–14. Unlike the version in 1 Kgs, 2 Chron describes the king of Egypt plundering and attacking cities widely in Judah, and raiding Jerusalem including both palace and Temple. Rehoboam then humbles himself before the Lord in response to a prophet who accuses him of infidelity to God. The Egyptian army therefore leaves.[35] One explanation for this is that the books of Chronicles are less prominent in Syriac churches (and are barely commented on), so that while Chronicles provided useful material for Jacob's exegesis of the passage in 1 Kgs, he may not have wanted

34 See for instance Gray, John, *I and II Kings*, 2nd rev. edition. Old Testament Library. London: SCM Press, 1970, pp. 340–346, on the Deuteronomistic editor/writer.

35 Phillips, David, "The Reception of Peshitta Chronicles: Some elements for investigation," in *The Peshitta: Its Use in Literature and Liturgy. Papers read at the 3rd Peshitta Symposium*, ed. by Robert Barend ter Haar Romeny, pp. 259–295. Monographs of the Peshitta Institute, Leiden 15. Leiden: Brill, 2006. He notes that although Chronicles was fully canonical by the 7th century and appeared in complete bibles, such as the 7th century MS 7a1, the book was in "bad company," i.e., it appears with the deuterocanonical books Judith, Ben Sira, Apocalypse of Baruch, 4 Ezra (261). But from the 8th century onwards Chronicles appears in "good company," following the books of *Kings*. Phillips comments that Jacob terms Chronicles "the Book of Missing Things [*Ḥasīrāṯā*, like the Greek name for the book, *Παραλειπόμενα*]" in a scholion in his own version of Samuel (267).

to depend on the Chronicles account explicitly. Perhaps the use of the version in Chronicles also subtly suggested the possibility of redemption through communal repentance?

The last section of Jacob's scholion has three interlocking motifs, brought out by the structure of the discourse and the vocabulary used by Jacob. First, there is the 'Deuteronomistic' paradigm of divine chastisement for sin, where he creates a parallel between the case of Judah/Jerusalem and the contemporary situation of the Syrian Orthodox Church. Secondly, there is the idea that because Christians denied God's grace in Christ, they have been handed into the power of those who deny Christ's divine status (i.e., the punishment fits the crime). Finally, there is the perception that because Christians failed to appreciate Christ's liberation of them from slavery to sin and subjugation to the devil and to demons, they are punished by being delivered into slavery and subjugation to the Arabs instead.

Another work that may be connected to Jacob's interpretation of current events is the *Testamentum Domini,* a semi-apocalyptic work prefaced to the 'Clementine' collection of church canons.[36] It purports to be an account of Jesus Christ warning his followers about the end times, and commanding his disciples to observe his commands, precisely. Then they will be preserved in the time of trial in the coming days of oppression. By "Christ's commands" the work refers to the church canons that covered many aspects of ecclesiastical life, from the structure of the church building to the character of the bishop, the form of prayers, and various other rules. Following the discourse between Jesus and the disciples the text presents the canons as spoken by Christ himself. The *Testamentum Domini* includes an apocalyptic section where Jesus describes the suffering of the end times and the coming of the Antichrist. The precedent for this is the Gospel account of Jesus warning his disciples about the eschaton before his death (Matt 24:7).

Some scholars[37] have argued that the *Testamentum Domini* was updated or even composed by Jacob, who almost certainly was responsible for translating

---

36 Vööbus, Arthur, ed., *The Synodicon in the West Syrian Tradition: vol. I,* pp. 27–32, Syriac text pp. 1–7.

37 Principally Drijvers, Han J.W., "The Testament of our Lord: Jacob of Edessa's response to Islam" in *ARAM* 6 (1994): pp. 104–114; and Drijvers, Han J.W., "The Gospel of the Twelve Apostles," in *The Byzantine and Early Islamic Near East, 1. Problems in the Literary Source Material,* ed. by Averil Cameron and Lawrence J. Conrad, pp. 189–213. Princeton, NJ: Darwin Press, 1992, esp. pp. 209–10. Michael Kohlbacher is one of the scholars who disagree (Kohlbacher, Michael, "Wessen Kirche ordnete das Testamentum Domini Nostri Jesu Christi? Anmerkungen zum historischen Kontext von CPG 1743," in *Zu Geschichte, Theologie, Liturgie und Gegenwartslage der sy-*

it from Greek to Syriac in 686/7 CE.[38] The *Testamentum Domini* is particularly concerned with canonical laxity on the part of the "evil pastors" whose vices will cause unbelief and other sins among the nations during the endtimes.[39] In the time of the Son of Perdition, Syria's fate is the first to be mentioned: "Syria will be captive and mourn for her children."[40]

The *Testamentum Domini* shares the concerns expressed in Jacob's scholion on 1 Kgs 14. It also has parallels with the account of Jacob's life which states that Jacob resigned the bishopric of Edessa because of the clergy's laxity and failure to observe church canons. He then burned a book of canons in front of the patriarch's house.[41] Han Drijvers argued that the laxity of observance that Jacob criticised was linked to the patriarch Julian's accommodation with Muslims, and that Jacob feared Christian conversion to Islam.[42] This of course may be reading too much into these sources, although as Robert Hoyland observes, apostasy to Islam is certainly of concern to Jacob in his canonical rulings.[43]

---

rischen Kirchen. Ausgewählte Vorträge des deutschen Syrologen-Symposiums vom 2.–4. Oktober 1998 in Hermannsburg, ed. by Martin Tamcke and Andreas Heinz, pp. 55–137. Studien zur Orientalischen Kirchengeschichte 9. Münster/Hamburg/London: LIT, 2000.

38 According to the subscription in some manuscripts, ascribing the work of translation to "the wretched man Jacob [*meskinā*]" and providing a date of AG 998: Rahmani, Ignatius Ephrem II, *Testamentum Domini Nostri Jesu Christi*. Mainz, 1899 [repr. Hildesheim: Olms, 1968], p. XIV. The *Testamentum* forms the first two books of the Clementine Octateuch: Vööbus, Arthur, *The Synodicon in the West Syrian Tradition*. Corpus Scriptorum Christianorum Orientalium 367 and 368, pp. 161, 162, texts 1–49, version 27–64, and Teule, Herman, "Jacob of Edessa and Canon Law," in *Jacob of Edessa and the Syriac Culture of His Day*, ed. by Robert Barend ter Haar Romeny, pp. 84–86. Monographs of the Peshiṭta Institute Leiden 18. Leiden: Brill 2008.

39 Vööbus, *Synodicon*, Syriac text, p. 3.

40 Vööbus, *Synodicon*, Syriac text, p. 5. Other nations affected by the upheavals are Cilicia, Babylon, Cappadocia, Lycia, Lycaonia, Armenia, Pontus, Bithynia, Pisidia, Judea, Phoenicia. Vööbus's translation is rather inaccurate on p. 29.

41 Jean-Baptiste Chabot, ed., *Chronique de Michel le Syrien, patriarche jacobite d'Antioche (1166–1199)*. Brussels: Culture et Civilisation, 1963 (4 vols., 1899–1910; vol. 4, pp. 445–446 [Syr.]; vol. 2, pp. 471–472 [French translation]). The shorter version of Barhebraeus can be found in Abbeloos, Jean-Baptiste and Lamy, Thomas Joseph, eds., *Gregorii Barhebraei Chronicon Ecclesiasticum*, vol. 1. Louvain: Peeters, 1872, pp. 289–294.

42 Drijvers, Han J. W., "The Testament of our Lord: Jacob of Edessa's Response to Islam," *ARAM* 1–2 (1997): pp. 104–114, and see also Drijvers, Han J. W., "The Gospel of the Twelve Apostles: A Syriac Apocalypse from the Early Islamic Period," in *The Byzantine and Early Islamic Near East: I. Problems in the Literary Source Material*, ed. by Averil Cameron and L. I. Conrad, pp. 189–213. Studies in Late Antiquity and Early Islam 1. Princeton, NJ, 1992 where he argues that the *Gospel of the Twelve Apostles* was influenced by the *Testamentum Domini* (213).

43 Hoyland, Robert G., "Jacob and Early Islamic Edessa," in *Jacob of Edessa and the Syriac Culture of his Day*, ed. by Robert Barend ter Haar Romeny, pp. 16–18. Monographs of the Peshiṭta

The *Testamentum Domini* and the scholion on 1 Kgs 14 differ a little in emphasis. In the scholion Jacob's concern is that the hardships of the late 7th century are divine chastisement for falling Christian standards of behaviour. In the *Testamentum Domini*, the writer advocates virtuous conduct and adherence to canonical norms, in order to withstand persecution and receive a divine reward. In neither text does there seem to be any hope of an imminent improvement of the earthly situation for Christians.

Given Jacob's hostility towards Jews in this period, it is ironic that both his diagnosis of the cause of his community's plight as God's punishment for their sin and his advocacy of precise observance of religious law resemble those of rabbinic Jews in Late Antiquity. But it is an attitude common to other contemporary works by Christians in this region.

3 Use of Syllogism and Chronological Arguments: The Davidic Descent of the Virgin Mary

> That the Messiah is of Davidic descent is confessed by everyone, whether Jews, *Mahgrāyē* [Muslims], or by Christians. They all confess that he was incarnate and made human by nature. So this, that the Messiah is of Davidic descent by the flesh, is the confession as was predicted by the holy prophets, and is a fundamental belief for all, whether Jews, *Mahgrāyē* or Christians. As for the Jews, I said that it was a fundamental statement even though they deny the true Messiah who has truly come. Even in the case of that false Messiah that they expect, they firmly state that he is a descendant of David and is destined to come from him. The *Mahgrāyē* too, even though they do not know or wish to say that this true Messiah (who has come and is confessed by Christians) is God and Son of God, [they admit] that he is the true Messiah who was to come and who was foretold by the prophets. All of them [i.e. Muslims] confess it firmly, and they have no dispute with us about this, but instead they disagree with Jews over it.
>
> ... I myself judge that by means of a compelling and convincing syllogism of this kind we can demonstrate to any Christian or *Mahgrāyā* who asks whether the Holy Virgin Mary, the God-bearer, is of Davidic descent, even though this cannot be demonstrated from Scripture. We should not adduce testimony about this from popular and superfluous stories cited by many people, and recorded and read, but which are not part of Holy Scripture. O lover of the truth, I know well that there are stories written by zealous people on their own initiative, without any confirmation from Scripture, that demonstrate that the Holy Virgin Mary, Mother of Christ, was daughter of Anna and the just man Joachin ...

Institute Leiden 18. Leiden: Brill 2008; and Hoyland, Robert, *Seeing Islam as Others Saw it: A Survey and Evaluation of Christian, Jewish and Zoroastrian Writings on Early Islam*. Princeton, NJ: Darwin Press, 1997, pp. 162–163.

> As I have already stated, I do not wish to prove what is required by means of superfluous proof from a popular story, but instead through this compelling and convincing syllogism.... If it should happen that the man who in conversation presses you about this, whether he is a *Mahgrāyā* or a Christian, if he is intelligent and has a rational approach, he will understand the syllogism when he hears it and testify to its truthfulness of his own accord, without disputing it ...
>
> I must now set forth for you here the words of the Prophets, which show you clearly that Christ is of Davidic descent in the flesh, and after that it will be shown to you as a refutation to the Jews that the Messiah did come at the time he was meant to, as was written about him, and that therefore [Jewish] hope is empty, since for their wickedness and blindness of heart they have been consigned to believe in falsehood and not truth....[44]

This text, from Jacob's 6th letter to his correspondent John the Stylite, reflects both Jacob's animosity towards Jews (as mentioned previously) and a perhaps surprisingly neutral attitude towards Muslims. Jacob also employs a less common method of solving a problem in Scripture, namely syllogism.

The accounts of Christ's birth in the gospels of Matthew and Luke give the genealogy of Jesus from King David by the paternal line. But for anyone who accepts the virgin birth of Jesus, Joseph was not Jesus' natural father, and so John of Litarba's question is, was Mary also of Davidic descent?

Jacob's letter is typically rather repetitious and convoluted, so I will summarise his argument. He cannot rely on the apocryphal accounts for proof of Mary's Davidic descent: although they were written by people motivated by pious intentions, they are not part of Holy Scripture. This is an interesting objection, as Jacob was only too ready to use apocryphal material for "gap-filling" and problem-solving in the case of Old Testament passages, as we saw earlier. Yet because Christians and Muslims shared a common belief in Mary's virginity and that her son Jesus was Christ/the Messiah, it is possible that Jacob was reluctant to use proofs that Muslims would certainly have rejected as being unscriptural.

i) The first premise of the syllogism is that the Messiah has already come, and at the time foretold by the Prophets. Jacob goes into this step in some chronological detail to counter Jewish arguments that the prophesied Messiah has not yet come. He demonstrates that Jewish reckonings of the number of years until the coming of the Messiah (based on Daniel) are erroneous, and that his own reckoning is correct: so Christ has already come, at the time foretold.

ii) The second premise is that the Messiah was of Davidic descent. Again, this is supported by scriptural arguments.

---

44 Nau, François, "Lettre de Jacques d'Édesse sur la généalogie de la sainte Vierge." *Revue de l'Orient Chrétien* 6 (1901): pp. 512–531, from BL Add. MS 12, 272, fols. 87v–91r. English translation my own.

iii) The conclusion of the syllogism is thus that, since the Messiah born of Mary was of Davidic descent, his mother Mary was necessarily also of Davidic descent.

Thus Jacob displays his well-known interests in chronography and philosophy, but in this instance he employs them in an exegetical context. An awareness of Muslim beliefs in relation to a central tenet of Christianity is a fairly new feature of Christian exegesis, and Jacob takes an eirenic approach: though Jacob ascribes Muslim use of the term "Spirit of God" to refer to Jesus as due to ignorance, he otherwise does not criticise them. By contrast, he attacks the obduracy of Jewish beliefs in an all-too familiar way.

4 Conclusion

Jacob depends on earlier Christian interpretative techniques such as typology and *theoria*, worked out in traditional genres such as commentary and *scholia* as well as letters. Where there are difficulties in the Biblical text he compares the Syriac *Peshitta* with the Greek *Septuagint*, and uses information from what he acknowledges are Jewish extra-canonical works but vouched for by apostolic authority.

Unsurprisingly, Jacob's interpretation of history is heavily influenced by the Deuteronomistic theology of the Old Testament that holds sway for both Christians and rabbinic Jews. So he attributes the domination of Christians by the Arabs to Christian sin, and his implied response is similar to that of the late Biblical era and the rabbis: a rigorous keeping of the "commandments"—in his case, those of Jesus Christ, rather than the precepts of the Torah.

Despite Jacob's regard for Scripture, if Drijvers is correct and Jacob was indeed responsible for the form of the *Testamentum Domini* that we have, this would suggest a lack of embarrassment about presenting a work that claimed dominical authority. The *Testamentum Domini* shows strong affinities with other apocalyptic documents of the period from the region: the *Gospel of the Twelve Apostles*[45] and the *Apocalypse of Ps-Methodius*;[46] the *Edessene Apoca-*

45 Harris, James Rendel, ed., *The Gospel of the Twelve Apostles together with the Apocalypses of Each One of them*. Cambridge: Cambridge University Press, 1900; Drijvers, "The Gospel of the Twelve Apostles", pp. 189–213.

46 See Reinink, Gerard, "Ps.-Methodius: a Concept of History in Response to the Rise of Islam," in *The Byzantine and Early Islamic Near East: I. Problems in the Literary Source Material*, ed. by Averil Cameron and L. I. Conrad, pp. 149–187. Studies in Late Antiquity and Early Islam 1. Princeton, NJ: Darwin Press, 1992.

lypse; John of Fenek's *Riš Mellē*. Though these works were written by Melkites,[47] Monophysites and Eastern Church writers, rather than by Syrian Orthodox scholars, they all appear to have been composed from the late 680s to early 690s. They respond in similar ways to the disruption of the Second Fitna, the plague and famine which followed, and the subsequent heavy taxation under ʿAbd al Malik. Most of these works, like Jacob's scholion, attribute Muslim domination to divine punishment for Christian sin.

Where Scripture failed to answer a key question, that concerning the Davidic descent of Mary the mother of Jesus, Jacob did not turn to the Church's apocryphal stories of Mary's birth to fill in this apparent gap in the Biblical account. Instead he turned to a combination of chronography and syllogism to prove his point in a way that could be countenanced by Jews, Christians and Muslims alike.

47 Middle Eastern Christians adhering to the Chalcedonian confession of the Byzantine Empire.

Haggai Ben-Shammai

From Rabbinic Homilies to Geonic Doctrinal Exegesis

The Story of the Witch of En Dor as a Test Case

1 Introduction

The geonic period (8th – 11th centuries) opens a whole new chapter in the history of Jewish learning. I have discussed elsewhere the important characteristics of the period in cultural terms, notably the emergence of Judaeo-Arabic culture, and also in terms of the communal history, namely the schism between Rabbanites and Karaites.[1]

For the present discussion, however, I wish to reiterate and elaborate two points:

1) In principle, the Karaites did not recognize the authority of the rabbinic tradition for the interpretation of the Bible. In practice, Karaite exegetes mostly followed exegetical principles and attitudes similar to those of the Rabbanites,

---

1 * Parts of this paper have been delivered as lectures at the Center for Advanced Judaic Studies of the University of Pennsylvania, Philadelphia, and at the 3rd Einstein German-Israeli Summer School, *Muslim Perceptions and Receptions of the Bible*, held at the Hebrew University, Jerusalem.

Ben-Shammai, Haggai, "Jerusalem in Early Medieval Jewish Bible Exegesis," in *Jerusalem: Its Sanctity and Centrality to Judaism, Christianity, and Islam*, ed. by Lee Israel Levine. New York: Continuum, 1999, pp. 447–48. See Vajda, Georges, "Judaeo-Arabic Literature." *Encyclopaedia of Islam*² 4, pp. 303–7; Halkin, Abraham S., "Judeo-Arabic Literature." *Encyclopaedia Judaica*² 10, pp. 410–23. Concerning halakhic contributions see Brody, Robert, *The Geonim of Babylonia and the Shaping of Medieval Jewish Culture*. New Haven: Yale University Press, 1998, esp. ch. pp. 16–17 (249–82). Regarding philosophy and Biblical exegesis, there are a number of recent publications in these areas, which have also benefited from recent findings: see my survey, "The Exegetical and Philosophical Writing of Saadia: A Leader's Endeavor." *Pe'amim* 54 (1993): pp. 63–81 (Heb. = Ben-Shammai, Haggai, *A Leader's Project: Studies in the Philosophical and Exegetical Works of Saadya Gaon*. Jerusalem: the Bialik Institute, 2015, pp. 21–36; Goodman, Lenn E., *The Book of Theodicy: Translation and Commentary on the Book of Job by Saadia ben Joseph al Fayyumi*, Yale Judaica Series 25. New Haven: Yale University Press, 1988; Sklare, David E., *Samuel ben Hofni Gaon and his Cultural World*, Études sur le judaïsme médiéval 18. Leiden: Brill, 1996; see also Frank, Daniel, *Search Scripture Well: Karaite Exegetes and the Origins of the Jewish Bible Commentary in the Islamic East*, Études sur le judaïsme médiéval 29. Leiden: Brill 2004, pp. 248–57.

https://doi.org/10.1515/9783110564341-010

and also made use of, or reacted to, similar exegetical traditions. Consequently, for the present discussion, Karaite exegesis may be considered part of the same corpus as Rabbanite exegesis.

2) In the geonic period two of the new genres that were developed by Jewish scholars were Biblical exegesis and religious philosophy. The novelty in the genre of Biblical exegesis was that unlike in Talmudic times, the study of the Biblical text now acquired a new status as a legitimate occupation in its own right. In both genres the principle that the prophets had spoken in philosophical terms and conveyed philosophical messages was applied.[2] Saʿadya applies this principle time and time again in his works, although he is not the first to do so. In fact, the first medieval Jewish author to apply this principle may have done so in a Christian context. This was David (or Dāwūd) ben Marwān al-Muqammaṣ (9th century), writing in his lengthy philosophical *Commentary on Genesis*. He lived a considerable part of his life as a Christian, and was closely connected to circles of learning and higher education of that faith.[3]

Works like this could certainly reflect the patristic approach, namely that the prophets were actually philosophers, much closer to Jewish Arabic-speaking, educated intellectuals, and therefore the entire Scripture carries a philosophical

---

2 This is comparable to the common feature in the rabbinic literature where Biblical figures are so often portrayed as Talmudic scholars.

3 Stroumsa, Sarah, "The Impact of Syriac Tradition in Early Judaeo-Arabic Exegesis." *Aram* 3 (1991): pp. 83–96, and more recently Stroumsa, Sarah, "From the Earliest Known Judaeo-Arabic Commentary on Genesis." *Jerusalem Studies in Arabic and Islam* 27 (2002): pp. 375–95. Regarding this author and his Christian connections, see Stroumsa, Sarah, *Dawūd ibn Marwān al-Muqammiṣ's Twenty Chapters*, Études sur le judaïsme médiéval 13. Leiden: Brill 1989, pp. 15–35; on his indebtedness to Christian sources and exegetical methods, see Qirqisānī's testimony in his introduction to the Pentateuch published in Hartwig Hirschfeld, *Qirqisani Studies*, Jews' College London Publication 6. London: Jews' College Publications, no. 6, 1918, pp. 40:10 ff. (Arabic); English translation in Nemoy, Leon, *Karaite Anthology*, Yale Judaica Series 7. New Haven: Yale University Press, 1952, p. 54; French translation by Vajda, Georges, "Du prologue de Qirqisani à son commentaire sur la Genèse," in *In Memoriam Paul Kahle*, Beihefte zur Zeitschrift für die alttestamentliche Wissenschaft 103, ed. by Matthew Black and Georg Fohrer. Berlin: Töpelmann, 1968, p. 224. The central position occupied by exegesis in the Syrian learning institutions is reflected in the fact that the director of such institutions in late antiquity (e.g., in Edessa and Nisibis) was the *mepashqānā*, the exegete, the interpreter. See Vööbus, Arthur, *The History of the School of Nisibis*, Corpus Scriptorum Christianorum Orientalium 266. Leuven: Peeters, 1965, pp. 10–24, pp. 57–99 and *passim*; see also summaries of previous discussions in the Ph.D. thesis of Chapin, Richard Steven, *Mesopotamian Scholasticism: A Comparison of the Jewish and Christian 'Schools'*. Cincinnati: Hebrew Union College, 1990, esp. ch. p. 3 (pp. 107–152).

message.[4] Needless to say, such a principle (or tendency) is known only vaguely, if at all, from rabbinic literature. I should mention here that between Jews and Christians in Arabic-speaking countries there was actually a 'bilingual' channel, since they could communicate in both Aramaic (Syriac) and Arabic. By applying the philosophical interpretation of the prophetic message, the Jewish scholars—whether Rabbanite Geonim or Karaite teachers—actually legitimized philosophical, or rationalistic, speculation on religious questions. This approach, which had by no means been universally accepted by all circles of the learned segment of Jewish society, evidently called for doctrinal consistency, which is completely missing in the rabbinic sources. Those sources, or more accurately, the rabbis, strive to convey an ethical message rather than a philosophical or a dogmatic one, and they do this through a rich web composed of myriad fragmentized or atomized units that lack, by nature or by definition, any doctrinal consistency. One could of course argue that the leaders and authors of each phase react to the agenda of their day. But then it would be legitimate to ask: Who set the agenda? Why? What were the sources of inspiration for this agenda? And similar questions.

In this context it is relevant to mention Alexander Altmann's study, *A Note on the Rabbinic Doctrine of Creation.*[5] There, Altmann discusses in particular the various rabbinic views or descriptions of the creation of light, and also the precursors of rabbinic views in Judaeo-Hellenistic thought. He especially concentrates[6] on a Midrash[7] in which R. Samuel b. Nahman describes (in a whisper![8]) the creation of light thus: "God wrapped Himself in a white garment and its splendour shone forth from one end of the world to the other." A reference is then made to Ps 104:2 ("wrapped in a robe of light"[9]). Altmann argues that the doctrine contained in this Midrash goes far beyond the literal meaning of Ps 104:2, and that it implies the doctrine of emanation ("effulgence"), which in the Middle Ages became a corner-stone of Neo-Platonic religious thought, in Islam as well as in Judaism. Altmann acknowledges, however, that R. Samuel

4 See Wolfson, Harry Austrin, *The Philosophy of the Church Fathers.* Cambridge: Harvard University Press, 1970, pp. 109–11.

5 Altmann, Alexander, *Studies in Religious Philosophy and Mysticism.* Ithaca: Cornell University Press, 1969, pp. 128–39 (repr. from *JJS* 7 [1956]: pp. 195–206).

6 See ibid., pp. 129–30.

7 *Genesis Rabba* ch. 3:4; Julius Theodor and Chanoch Albeck, *Bereschit Rabba.* Jerusalem: Shalem Books, 1996, p. 19.

8 Cf. Altmann's explanation of this "dramatic" element as reference to esoteric teachings.

9 Translations of Biblical verses are taken, unless otherwise indicated, from *Tanakh, A New Translation of the Holy Scriptures according to the Traditional Hebrew Text.* Philadelphia: Jewish Publication Society, 1985.

does not express himself in doctrinal terms, but in a "mythical image".[10] Further on Altmann concluded:

> We have therefore before us a somewhat veiled but unmistakable form of the doctrine of emanation. It does not suggest an emanation of the Logos-Wisdom from the divine essence but is content to allude to the emanation of the primordial light from the divine Logos, mythically described as God's garment.[11]

We see therefore that although Altmann believed that some rabbis were acquainted with philosophical doctrines, or that some Midrashic statements reflect such acquaintance, he himself pointed out more than once that the rabbis did not think or express themselves in doctrinal terms, but rather in mythical terms, unlike Philo, who "demythologized"[12] the old Oriental myths.

The purpose of the present study is to investigate the way in which different —possibly adversary—approaches towards the scriptural texts react to each other, as well as the interaction between those approaches, across communal or confessional demarcation lines. It is not impossible to define the relationship between these approaches even as one of reciprocal influence provided of course that there is no lingual barrier between the different literary expressions. Put otherwise, it may well be that specific doctrinal questions asked by members of one community with respect to certain sections of Scripture attract the interest of the members of an adjacent community, who prior to that had had no interest at all in such questions. In yet other words, it seems that against the historical background (above), and once it is recognized that a fundamental change in the approach toward the Hebrew Scripture occurred in the geonic period, it is legitimate to ask whether it is possible to point to certain similarities or parallelisms between geonic (including Karaite) and Christian exegesis on particular points, or on general questions.[13]

The story of the witch of En Dor may serve as an appropriate test case for such an investigation. It is possible to follow a very interesting line of development, and a series of turning points with regard to this story, starting with the

---

10 Altmann, *Studies*, p. 130.

11 Ibid., p. 136.

12 Ibid., p. 135.

13 By this term I mean, for example, the theme of articles of faith, meaning normative beliefs or dogmas. I discussed the principles of faith in the geonic period in "Saadya Gaon's Ten Articles of Faith." *Da'at* 37 (1996): pp. 11–26. For the Hebrew version see Ben-Shammai, *A Leaders Project*, pp. 93–109; "Kalam in Medieval Jewish Philosophy," in *History of Jewish Philosophy*, ed. by Daniel H. Frank and Oliver Leaman. London: Routledge, 1997: pp. 126–32; Hebrew version: Ben-Shammai, *A Leaders Project*, pp. 53–60.

scriptural presentation of the story and its message; then the rabbinic homiletic exposition, with its particular emphasis, which seems to originate from ancient attitudes that have their roots in the beginning of the Second Temple period, as reflected in late scriptural references; and then the homilies and interpretations of the Church Fathers, heavily impregnated with doctrinal elements and polemical background. While there is apparently no direct evidence indicating that the Christian attitudes, whose development had begun simultaneously with the rabbinic expositions, affected those latter expositions, Islamic sources seem to demonstrate that there was a Jewish response to the patristic approach towards the story. This Jewish response eventually culminated in the geonic period in a heated debate between Jewish exegetes regarding the nature of the En Dor episode.

The Jewish exegetes in the early mediaeval Islamic areas, writing mostly in Arabic, include some prominent authors who rely heavily on rabbinic sources and endeavor to establish these sources in their readers' minds, but at the same time tend to rearrange and develop classical materials from these traditions. In certain cases they also address themselves to questions which can be explained only in terms of response to, or interaction with, Christian exegesis. One such case is the story of Saul and the witch, or sorceress, or necromancer, of En Dor, and especially the appearance of the prophet Samuel in this story.

The scope of the investigation in the present paper allows only for rather generalized observations and a survey of the contents of the relevant texts. A proper and exhaustive discussion of the subject in question would also involve a presentation of the full relevant medieval texts, in the original language where this has not been published yet, and an annotated English translation of the entire corpus. Bearing in mind the volume of these texts, the size of such an undertaking would seem to be that of a book rather than an article, and as such, it will have to wait for some time. The following presentation will therefore consist of an exposition of the Biblical account, with a discussion of its literary and ethical focal points; a short discussion of the rabbinic approach, mainly on the basis of two different types of texts;[14] a survey of the main exegetical-methodological principles in the works of commentators of the geonic period, with an analysis of the focal points; and finally a survey of approaches of early eastern Christian sources to the story and the evidence of representatives of Islamic sources.

14 This discussion will concentrate on the main points only, leaving out many very interesting literary and exegetical points of the rabbinic texts.

2 The Biblical Account

The Biblical account of the story is recorded in 1 Sam 28. The context is the last phase in the reign of King Saul, the first king of Israel in Biblical times. Samuel, the prophet (*seer*)-judge, had already died, after having severed relations with the King. The fact that the prophet's death and burial is mentioned here[15] proves important later on in the story. At this point the narrator adds, in what seems to be an aside, that Saul had done away with[16] the use of divination through spirits. The Philistines, Saul's bitter enemies throughout his reign, gathered their forces for yet another encounter with the Israelites. This time they encamped in Shunem, at the eastern end of the valley of Jezreel, while Saul and his army encamped on the range of Gilboa. The site of the Philistines' encampment shows how far north they managed to extend their control, and reflects the balance of power between the two parties: the Philistines are on the ascent, while the Israelites are on the decline. Saul was afraid of the magnitude of the Philistine force and sought guidance from God through dreams, *Urim* and prophets, but with no response. He then remembered that there was still one means of communication with the Divine which he might try, namely consultation with the ghosts of the dead. He asked his servants to find him a woman who could help him in this matter, and they promptly directed him to one such woman who resided in En Dor, a nearby locality in the valley of Jezreel. Saul disguised himself, set forth and arrived there with two servants at night. He asked the woman to divine for him the ghost of the person whom he would name to her. The woman was struck with fear, because any person who resorted to the practice of divination by ghosts was liable to death according to the strict instructions (and past actions!) of King Saul. The disguised king swore to the woman that no harm would befall her as a result of her positive response to his request, and asked her to bring up (the ghost of) Samuel for him. When she saw Samuel in front of her she shrieked loudly, realizing that she had been deceived by her client, who was no other than King Saul himself. The latter pacified her and asked her what she had seen. She described "a divine being coming up from the earth" (see below, n. 35) that looked like "an old man [...] wrapped in a robe" (vv. 13–14). The verbal part of the divination is enacted in a dialogue between Saul and Samuel. The latter was angry with the King for disturbing him and bringing him up from his grave. Saul apologized and explained his difficult mili-

15 After it had already been mentioned once in 1 Sam 25:1.

16 The exact nature of Saul's action is rendered in different ways in various English translations, and a proper discussion of the matter is beyond the scope of the present study.

tary position, which had been aggravated by his lack of any means of communication which would assist him with divine guidance. To this the prophet replied with a harsh admonition:

> Why do you ask me, seeing that the Lord has turned away from you and has become your adversary? [...] The Lord has torn the kingship out of your hands and given it to your fellow, to David, because you did not obey the Lord and did not execute His wrath upon the Amalekites. That is why the Lord has done this to you today. Further, the Lord will deliver the Israelites who are with you into the hands of the Philistines. Tomorrow your sons and you will be with me; and the Lord will also deliver the Israelite forces into the hands of the Philistines. (vv. 16–19)

Having obtained the information for which he was so eager, the King was thus explicitly notified of his and his sons' imminent fatal end and the terrible defeat of his army at the hands of the Philistines. He was terrified, weak and disturbed. The woman, who had meanwhile managed to collect herself, convinced the King, with the aid of his two servants, to take some food, which he accepted; he left the same night with his servants.

The Biblical account has a special dramatic effect, which is applied in three aspects: tragedy, irony and justification. Each of the three literary aspects also carries an ethical message and lesson.

The tragedy is that of the King who is doomed, and who had actually lost his kingdom even before the defeat at Gilboa. The ethical message is obvious: the King, who was anointed by the Prophet, must obey God's commandments and especially those which apply to his position as political and military leader. Failure to do so—and the case of the battle against the Amalekites is the most conspicuous failure—must carry the severest punishment, which in this case is the loss of both kingdom and life. This is no doubt the focal point of the story.

The irony is that of the King who persecuted the witches, and ultimately is forced to seek help from a witch who survived his persecution. The mention of the persecution of the witches in 1 Sam 28:3 is probably aimed at accentuating the irony; the persecution must have taken place long before his resort to the witch at En Dor, when he had been cooperating with the priests of Nov, before he murdered them (1 Sam 22:9–23). Scripture does not give the reason for Saul's action against the witches. This is in fact the only one of Saul's actions that is not associated with political or military purposes.[17] The only point of put-

17 It is, however, associated with a number of actions; see on the historical context Elath, Moshe, "Saul at the Apex of his Success and the Beginning of His Decline (The Historiographical Significance of I Samuel 13–14)." *Tarbiz* 63 (1993–4): pp. 15.

ting it on record is apparently to show that Saul was not consistent even in this single action against idolatry.[18]

The justification is that although magic and sorcery are condemned by Mosaic Law as idolatrous practices, owing to the fact that all other means of communication with the Divine were not available to Saul, and that God probably wanted Saul to receive the divine message prior to his defeat and death, Saul's resort to the witch is at least partly justified. After all, according to the Biblical account, several years must have elapsed since Samuel last had any verbal exchange with Saul, so a final message from the prophet who anointed the King is definitely called for before the King's demise (which at this stage is evidently inevitable for all the individuals concerned and for the reader of the story as well). Saul's transgression, which when judged in itself is very grave indeed, is somewhat mitigated, in relative terms, as reflecting a genuine, positive desire to obtain the divine word.[19]

In the Biblical account the feasibility or the veracity of the occurrences as related is not questioned. It is clear that the woman was really capable of divining the ghost of Samuel, who then actually spoke to Saul. For the Biblical narrator, what should be questioned, as already mentioned, is the propriety of Saul's resort to magic, but even that is partially justified: since all alternative and permitted means of communication with the Divine were no longer available to Saul, and since his desire to obtain the word of God before the battle was in principle positive, the magic practice of the pagans was the only remaining alternative. Certainly the resurrection or revival of the person of Samuel, not to mention his body, is not an issue, nor even a possibility, in this account; his ghost was expected to appear for the specific task and duration of communicating God's message, and then return to the abode of the ghosts.

Since the focus of the present study is on the different approaches towards the appearance or the apparition of Samuel, or its exact nature, it is important to

---

18 The ironic element is retained in the rabbinic interpretation of the story, see below.

19 For a detailed, somewhat different literary analysis of the Biblical account, see Simon, Uriel, "A Balanced Story: The Stern Prophet and the Kind Witch." *Prooftexts* 8 (1988): pp. 159–71; a Hebrew version of this study in: Bar-Asher, Moshe, ed., *Rabbi Mordechai Breuer Festschrift*, vol. 1. Jerusalem: Akademon Press, 1992, pp. 113–24; additional literary and historical aspects are discussed by Garsiel, Moshe, "Torn between Prophet and Necromancer: Saul's Despair (1Sam 28:3–25)." *Beit Mikra: Journal for the Study of the Bible and Its World* 41 (1996): pp. 172–96, with a revised expanded Hebrew version in *Studies in Bible and Exegesis* Vol. *VI: Yehuda Otto Komlosh – In Memoriam*, ed. by Rimon Kasher and Moshe Zipor. Ramat-Gan: Bar-Ilan University Press, 2002 pp. 25–45.

examine this component of the Biblical account in its proper context, which is that of ancient Near Eastern civilizations.[20]

The story reflects the contemporary Mesopotamian concepts and practices of necromancy, although in that context, cases of bringing up the ghosts for rituals of consultation or divination appear to be rare. The souls of the dead exist in a world of shadows, which is situated beneath the surface of the earth (evidently on the level of graves).[21] The subterranean location occupied by the souls of the dead according to the myths that were current in the Canaanite-Israelite environment (notably Biblical *she'ol*) was quite similar. Accordingly, this is the "place" from which the witch brings up Samuel, and this is where the spirit of the prophet expects to meet Saul and his sons the following day (v. 19). In these traditions it was considered possible in principle to communicate with the dead. The story about the witch of En Dor is apparently a unique (or at least quite rare) attestation of its kind from the ancient Near East at that time (late second millennium BCE) outside the Mesopotamian civilization. The story is further unique in relation to Mesopotamia, where ghosts would mostly be retrieved in order to hurt adversaries, not as a means of communication with the Divine.[22]

---

20 I owe much of my knowledge on this matter to Dr. Tzvi Abusch.

21 See the summation of Scurlock, Joann, "Death and the Afterlife in Ancient Mesopotamian Thought," in *Civilizations of the Ancient Near East*, ed. by Jack M. Sasson, vol. 3, pp. 1883–93. New York: C. Scribner's Sons, 1995, with further bibliography. The question of the exact nature and scope of necromancy in the ancient Near East is not the issue of the present study. It has been the subject of a fair number of learned publications recently: see for example, Schmidt, Brian B., *Israel's Beneficent Dead: Ancestor Cult and Necromancy in Ancient Israelite Religion and Tradition*, Forschungen zum Alten Testament 11. Tübingen: Mohr Siebeck 1994; Nihan, Christophe L., "1Samuel 28 and the Condemnation of Necromancy in Persian Yehud," in *Magic in the Biblical World: From the Rod of Aaron to the Ring of Solomon*, ed. by Todd E. Klutz, pp. 23–54. London: T & T Clark International, 2003 (with rich documentation); the author discusses the lexical aspect of necromancy (the Biblical term *ôḇ*) as well as the textual aspect of the story, arguing that it reflects a post-Deuteronomistic, postexilic "composition" (either an oral tradition or a document) that was inserted into the Deuteronomistic history of 1 Samuel in the Persian period (there are other scholars who hold similar views: all are quoted by Nihan) as a polemic against necromancy; and see the following note. An important contribution to the study and understanding of 1 Sam 28 is Tropper, Josef, *Nekromantie: Totenbefragung im Alten Orient und im Alten Testament*, Alter Orient und Altes Testament 223. Kevelaer: Butzon & Bercker, 1989, esp. pp. 166–86 (a survey of the understanding the meaning of witchcraft in general and necromancy in particular in the ancient Near East, the ancient Greek and Aramaic Bible translations, Jewish Hellenistic as well as rabbinic perceptions); pp. 212–23 (a detailed critical analysis of the Biblical account of the En Dor episode); pp. 223–27 (analysis of the terminology used to describe the dead person's ghost and the exact nature of the witchcraft).

22 Finkel, Irving L., "Necromancy in Ancient Mesopotamia." *Archiv für Orientforschung* 29/30 (1983/4): pp. 1–17; cf. the preceding note and Farber, Walter, "Witchcraft, Magic, and Divination

Finally, one further Biblical reference to the story of the battle at Gilboa should be noted, namely 1 Chr 10:1–7. It is interesting that Saul is not mentioned there as a king.[23] Only in the following chapter (v. 2) does Scripture mention the fact that he was king; this occurs in the course of a speech made by the Israelites at the ceremony of the anointment of King David in Hebron. Especially relevant to the present discussion is 1 Chr 10:13–14. These verses list the reasons for Saul's untimely and undignified death: he "died for the trespass that he had committed against the Lord in not having fulfilled the command of the Lord; moreover he had consulted a ghost to seek advice, and did not seek the advice of God; so He had him slain and the Kingdom transferred to David son of Jesse." According to this account Saul's resort to the witch and to magical practices was an unforgivable sin, which, combined with his disobedience, in the battle against the Amalekites, justified his being slain by God. Unlike the account in 1 Sam 28, the chronicler sees Saul's resort to the witch as a transgression of the gravest order, for which there are no mitigating circumstances.[24] It seems that the reference in 1 Chr is the starting point of the attitude of the rabbinic literature.

---

in Ancient Mesopotamia," in *Civilizations:* 1895–909. On the unique status of divination by the dead in the Bible see Xella, Paolo, "Death and the Afterlife in Canaanite and Hebrew Thought," in *Civilizations*, ed. by Jack M. Sasson, pp. 2059–2072, esp. pp. 2068–69. Recently, Ritner, Robert K., "Necromancy in Ancient Egypt," in *Magic and Divination in the Ancient World*, Ancient Magic and Divination 2, ed. by Leda Ciraolo and Jonathan Seidel, pp. 89–96. Leiden: Brill 2002, compared the case of the witch of En Dor with the Egyptian traditions and argued for significant parallels. In the course of his conclusion he says: "Thus paralleled in context, methodology, practitioner, result and underlying theology, the Biblical and Egyptian practices are distinguished chiefly by legality, variety and frequency. For if Saul's consultation was illicit and unique, comparable Egyptian practices were legal, normative, multiform and omnipresent" (p. 95). It seems that this direction deserves further investigation, which is obviously beyond the scope of the present study.

23 This section is taken from 1 Sam 31, where the context makes it clear that he was king. Nor is he mentioned as such in the genealogy of the tribe of Benjamin in 1 Chr 8.

24 On this account see the analysis of Japhet, Sara, *The Ideology of the Book of Chronicles and its Place in Biblical Thought*[2], Beiträge zur Erforschung des Alten Testaments und des antiken Judentums 9. Frankfurt am Main: Peter Lang, 1997, pp. 406–411. She thinks that the chronicler recognizes Saul's kingship as the kingship of God, subsequently to be replaced and continued by David; on other views see Japhet, *Ideology*, p. 409, n. 38; see also Japhet, Sara, *I & II Chronicles: A Commentary, Old Testament Library*. London: SCM Press, 1993, pp. 229–30. On the text of 1 Chr in general, see Kalimi, Isaac, *Zur Geschichtsschreibung des Chronisten—Literarisch-historiographische Abweichungen der Chronik von ihren Paralleltexten in den Samuel-und Koenigsbuecher*, Beihefte zur Zeitschrift für die alttestamentliche Wissenschaft 226. Berlin: de Gruyter, 1994, esp. pp. 277–78 and *passim*; on our story see idem, "The Contribution of the Literary Study of Chronicles to the Solution of its Textual Problems." *Tarbiz* 62 (1992–3): pp. 471–486, esp. p. 473, and also Zalewski, Saul, "The Purpose of the Story of the Death of Saul in 1 Chronicles X."

3 The Rabbinic Approach

The main rabbinic exposition of our story seems to stem from the reference in 1 Chr 10 (see above), which is not a conventional account. The earliest source for this exposition is Lev. Rabba 26:7.[25] It is also found in some parallel sources (probably later ones),[26] and partial references or short segments of it are found in the Talmud.[27] The context is the prohibition against witchcraft, and the moral lesson is relevant to our story. Although the Midrashic works that contain this exposition are in the category of introductory homilies (*petiḥta*, as opposed to interpretive homilies, which relate to individual verses), the text at hand is actually close to a running commentary on a large part of the Biblical account (1 Sam 28:7–20), quoting the verses almost in full in the course of the discussion, and commenting mainly on key words in each verse. At the end, the concluding moral that is based on 1 Chr 10 is repeated.

The Midrashic text focuses on the religious-moral cause for Saul's failure and death, and counts no less than five sins that Saul committed and was therefore punished with death. The fact that this was the case is inferred by the homi-

Vetus Testamentum 39 (1989): pp. 449–467. According to Nihan the story in 1 Sam 28 is contemporaneous with that of 1 Chr. Consequently, one should not expect any ideological/theological differences between the two accounts. But such do exist, if the present analysis is correct. See Nihan, "1Samuel 28 and the Condemnation of Necromancy in Persian Yehud" (above n. 21).

25 *Midrash Wayyikra Rabbah*, ed. by Mordecai Margulies, 3rd printing. New York: Jewish Theological Seminary of America, 1993, pp. 598–608 (beginning of the pericope of Emor). The redaction of the work was initially completed during the 5th century. See Stemberger, Günter, *Introduction to the Talmud and Midrash*, trans. by Markus Bockmuehl. Edinburgh: T & T Clark, 1996, pp. 288–91. The text is quite problematic in places, as is evident from the editor's notes; the text of the critical edition, however, is a considerable improvement on the previous printed editions. The English translation by Slotki, Judah J., *Midrash Rabbah – Leviticus*. London: Soncino Press, 1939, pp. 330–36, was published before the critical edition and was therefore based on the traditional printed editions. A more recent English version is, however, available for the parallel text of Tanḥuma (Buber's recension; see next note), which is by and large identical, see *Midrash Tanḥuma: translated into English with introduction, indices, and brief notes (S. Buber Recension)*, trans. by John T. Townsend, vol. 2. Hoboken: Ktav, 1997, pp. 317–320.

26 *Midrash Tanḥuma* on the pericope Emor, section 2; ed. by Solomon Buber. Vilna, 1885, on the same pericope, sec. pp. 3–4, pp. 81–84. The version of both editions is very close to that of *Lev. Rabbah*, virtually identical in most of the contents, and also in the exegetical structure of the homily. It does, however, leave out some important details of the conversation between Saul and Samuel.

27 E.g., b*Ḥagiga* 4b.

list from 1 Chr 10:13 – 14.[28] The five sins are: 1) The murder of the priests of Nov ("Saul died for the trespass"); 2) His showing mercy towards Agag, the king of the Amalekites ("that he had committed against the Lord"); 3) His disobeying Samuel who ordered him to wait for him seven days in Gilgal (1 Sam 10:8) ("not having fulfilled the command of the Lord"); 4) His resort to the witch ("he had consulted a ghost to seek advice");[29] 5) He "said to the priest, 'Withdraw your hand'" (1 Sam 14:19) which, according to the homilist, corresponds to the chronicler's statement that "he did not seek advice of the Lord."

It is thus the moral-religious aspect which, on the basis of the attitude reflected in the 1Chron verses, becomes the focal point of the entire story. The homilist pays little attention to Saul's failure as a religious-political leader and does not at all consider any doctrinal questions. Furthermore, the rabbinic homily does not question the veracity of the occurrences as told in the story, namely that Samuel's ghost spoke to Saul. On the contrary, it includes a detailed record of a long conversation between Saul and Samuel, which is in itself very interesting. In this context, mention is made of the current practices of necromancy. These are mentioned in the Talmud as well,[30] and are not much different from those of the ancient Near East. There is no question of Samuel rising from the abode of the dead and appearing on the scene in person. The fact is that according to the rabbinic description (which in its turn constitutes a rather literal understanding of the relevant Biblical statements), Saul could not see Samuel but only hear him, whereas the witch could only see him but not hear him, and the others present could neither see nor hear Samuel.[31] This situation seems to fit well with the knowledge that the rabbis had of magic practices. It may be re-

---

28 As the editor correctly commented in his notes. In the following account, the segments of the verses in 1Chron that constitute the proof text for each sin are indicated in brackets. The proof text for the third sin is from 1 Sam 8:10, where Samuel orders Saul to wait for him seven days in Gilgal. It is true that this order of Samuel is not mentioned subsequently anywhere. This does not mean that Saul disobeyed the prophet. I could not find any parallel to this accusation of the homilist.

29 The discussion of this account contains two explicit expressions of irony: 1) A parable that compares Saul to a king who came to a town and ordered all roosters slaughtered. When he prepared to depart he asked: Isn't there any rooster here [to wake us up in the morning]? They told him: It was you who ordered all of them slaughtered (*Lev. Rabbah*, p. 599). 2) When Saul promises the woman, swearing by God, that she will not be punished for engaging in necromancy at his behest, the homilist remarks: "To what is Saul comparable? To a woman who was situated with her lover and swore by the life of her husband" (trans. by Townsend, *Tanḥuma* 2, p. 317).

30 b*Sanhedrin* 65b.

31 See Ginzberg, Louis, *The Legends of the Jews*, vol. 6. Philadelphia: Jewish Publication Society, 1926 (repr. 1946), p. 237, n. 77 (new edition, Philadelphia: Jewish Publication Society, 2003, 2 904b, n. 77).

marked here that Josephus in his description of the episode[32] states several times that the art of the witch was to bring up the souls (ψυχὰς) of the dead, and that she actually brought up the soul of Samuel (Σαμουῆλου ψυχὴν). Josephus does not seem to have any doubt with regard to the veracity of the event. The presence of Samuel on the scene is termed an "apparition" (ὄψιν), not a resurrected body and soul. In adjusting the location of the souls to Greek mythology, Josephus makes the witch summon Samuel from *Hades*.[33]

Nevertheless, there is one vague allusion to an eschatological theme, when the rabbis say that when Samuel heard that he was called upon to speak to Saul he was afraid that the time of the Last Judgment had arrived. He called on Moses to join him and they set out together on their way to the supposed event of the Judgment.[34] This somewhat odd and uncalled-for addition to the story helps to explain the linguistic difficulty of the plural verb in the witch's statement, "I saw gods ascending out of the earth"[35] (1 Sam 28:13). It also indicates the kind of environment in which the homilist envisaged Samuel to be located. From the fact that he takes for granted that Samuel dwelled together with Moses, it would appear that he believes this environment to be a heavenly surrounding. At the same time, this does not prevent the homilist from accepting the Biblical description that the souls of the dead rise from below—a description that is rooted in the ancient Near Eastern (and Hellenistic) belief that the persons of the dead exist in the Kingdom of Shadows that is located in subterranean surround-

---

32 Josephus, Flavius, *Josephus: in nine volumes with an English translation by Henry St. John Thackeray, Jewish Antiquities. Books V–VIII* vol. 5, trans. by Henry St. John Thackeray and Ralph Marcus. London: Heinemann, 1934, VI: 14.2–3 (= pp. 327–350) (Loeb Classical Library), pp. 330–343.

33 There are various other similarities or parallelisms between Josephus and the Midrashic sources; the discussion of those elements is beyond the scope of the present study. Brief mention may, however, be made of the laudatory comments of Josephus with respect to Saul's heroic death; these comments have interesting parallels in the rabbinic homilies referred to above. See ibid., pp. 343–350 (pp. 340–343). A striking deviation of Josephus from the Biblical account is that the witches were not killed by Saul, but expelled.

34 In addition to the sources indicated above, see b*Ḥagiga* 4b; cf. also Jer. Talmud (Venice 1520) *Ḥagiga*, 77 col. A.

35 This is the rendering of the King James Version, which faithfully retains a plural; the singular pronoun used in the *Tanakh* rendering ("I see a divine being coming up from the earth") clearly reflects the interpretation of the editors. For references to ancient translations of the term as well as suggestions of modern scholars see Tropper, *Nekromantie*, p. 163 (indicating the dead); p. 202 (consulting the dead was a ritual act); p. 211 (according to a theory that in a northern Israelite tradition the term refers to a deity who resides in the underworld of the dead [= *Hades?*]); pp. 219–21 (modern suggestions that the word refers to Samuel's name or to a deity, or to Saul's ancestors); see also p. 390, Index of Hebrew terms, *ʾlhym*.

ings (*She'ol, Hades*). As already mentioned, this belief was still current in Talmudic times.

A very interesting deviation from the rabbinic pattern is found in a late Midrashic source, *Pirke de-Rabbi Eli'ezer.* It is widely thought that this is a relatively late compilation of early Midrashic materials. Probably no earlier than the beginning of the geonic period, these materials were arranged, redacted, rephrased and often considerably augmented by contemporary materials.[36] Chapter 33 of this work deals ostensibly with charity (the proof text is Gen 26:12), but ultimately with resurrection. The reward for charity is resurrection. Here the author quotes a number of cases, such as that of the son of the woman of Zarefat whom Elijah brought back to life (1 Kgs 17:17–24),[37] and the son of the woman of Shunem, who went through a similar experience at the hands of Elisha (2 Kgs 4:8–37). The story of Saul at En Dor is then brought in. The person who deserved resurrection for charity is oddly enough King Saul who, after having eliminated the sorcerers, "turned back to love those whom he had hated." Samuel's appearance in his famous outfit is explained in accordance with the rabbinic statement that the dead are expected to rise dressed up in their garments.[38] Samuel is described as "accompanied by others who thought that the time of Resurrection arrived." Another opinion quoted there says that "many righteous came up with him [Samuel] at that time." A discussion of the story ensues, which includes the earlier Midrashic materials and a lengthy exposition of Ezek 37, of course in association with resurrection.[39] The whole chapter is thus devoted mainly to the theme of actual resurrection, not apparition, and the

---

36 See Herr, Moshe David, "Pirke de-Rabbi Eliezer," in *Encyclopaedia Judaica*² 13, pp. 182–183, who dates the work to the first half of the 8th century; Stemberger dates it to the 8th or 9th century. See Stemberger, *Introduction*, pp. 328–30.

37 Or resurrected or revived or resuscitated, see below, Additional Note. It is not at all clear whether the author is aware of the distinctions made in the Additional Note, see *Pirke de-Rabbi Eliezer.* Jerusalem 2005, pp. 274–284; the chapter is available in two translations into English: Friedlander, Gerald, *Pirḳê de-Rabbi Eliezer* (The chapters of Rabbi Eliezer the Great): according to the text of the manuscript belonging to Abraham Epstein of Vienna, 4th edition. New York: Hermon Press, 1981, pp. 239–251; Finkel, Avraham Y., *Pirkei Drebbi[!] Eliezar*, vol. 2. Scranton: Yeshivath Beth Moshe, 2009, pp. 2–11.

38 Friedlander, *Pirkê de-Rabbi Eliezer*, pp. 284–87. For the earlier rabbinic statement see b*Ketubot* 111b, with many parallels, e.g., *Gen. Rabba*, ch. 95, ed. by Theodor-Albeck, p 1186.

39 Friedlander, *Pirkê de-Rabbi Eliezer*, pp. 290–93. Regarding the place of Ezek 37 as a major Biblical proof text for the belief in resurrection, see my remarks in "The Status of Parable and Simile in the Qur'ān and Early Tafsīr: Polemic, Exegetical and Theological Aspects." *Jerusalem Studies in Arabic and Islam* 30 (2005): pp. 154–69, there pp. 167–169 (additional note). The present citation of this source is a minor correction of my dating there of the earliest reference to Ezekiel as a scriptural proof text of future resurrection.

story of Saul is a centerpiece of the discussion, with special emphasis on Samuel's resurrection. As already mentioned, it is quite evident that the plain language of the scriptural story does not warrant any such interpretation. One has a clear sense that there may well be some polemical or apologetic factor in the background of the connection between the story of Saul and the witch on the one hand and the theme of resurrection on the other.

4 Jewish Exegesis of the Geonic Period

Turning to the Jewish Judaeo-Arabic exegesis of the geonic period, one is no longer surprised at the somewhat unexpected approach of the discussion in *Pirke de-Rabbi Eliʿezer.*

It should again be emphasized that the exegesis dealt with in the present study is that which was composed by Rabbanites and Karaites alike in Islamic (Arabic-speaking) countries, roughly between the years 900–1100. Thus far no less than nine commentators are known to have dealt with the subject, either in the course of a running commentary on 1 Sam 28 or some other Biblical book, or else in a theoretical work that does not belong formally to the exegetical genre. It goes without saying that only in the writings of early authors may one find truly new positions, whereas in later writings one finds many repetitions with few new arguments or variations. In later medieval Judaeo-Arabic writings further developments of the subject may be discerned, notably in the context of the controversy between Samuel b. ʿEli (who tried to renew the Yeshiva [academy] of the Geonim in Baghdad in the second half of the 12th century) and Joseph b. Judah b. Shimʿon concerning the view of Maimonides on resurrection,[40] and in the Judaeo-Arabic commentary of Tanḥum Yerushalmi (the end of the 13th century) on Samuel.[41] It should be emphasized that works written in Hebrew in the East display much less interaction with the kind of theories discussed in the present study. Thus, for example, the homiletic-exegetical compilation entitled *Pitron Torah*, possibly written in Iraq in the 9th century, when discussing the question of witchcraft and sorcery in general and the En Dor episode in partic-

---

40 See Stroumsa, Sarah, *The Beginnings of the Maimonidean Controversy in the East: Yosef Ibn Shimʿon's Silencing Epistle Concerning the Resurrection of the Dead.* Jerusalem: Ben-Zvi Institute, 1999.

41 Tanhum ben Joseph of Jerusalem, *Commentarium arabicum ad librorum Samuelis et Regum*, ed. by Theodor Haarbruecker [with Latin tr.]. Leipzig: F. C. G. Vogel, 1844, pp. 44–49. Tanhum took a position which is similar to that of Samuel ben Ḥofni, but formulated in Maimonidean terminology.

ular, follows the old rabbinic sources very conservatively and evinces no trace of acquaintance with the theological issues discussed in Judaeo-Arabic exegesis.[42]

From the first third of the 10th century two interconnected questions occupy a central place in the Judeao-Arabic interpretation of 1 Sam 28: a) Was Samuel resurrected? b) If he was, who resurrected him?

The earliest geonic author whose view on the story has come down to us is Sa'adya Gaon (882–942, served as head of the Yeshiva of Sura, actually located in Baghdad), who discussed it in his commentary on Lev 19:31 or 20:6 or 20:27. He outlined the scope of the discussion. Sa'adya says that it may be asked whether Samuel was actually resurrected or not. It is clear from Sa'adya's wording that he does not think in terms of calling up Samuel's ghost. The Arabic term he uses is derived from the root *ḤYY* in the forth form (literally: causing to live), which clearly denotes resurrection of the dead, body and soul, from a grave. It could be argued that the question of which of the two alternatives Sa'adya had in mind—bringing up Samuel's soul or resurrection—is hypothetical, and that the first alternative was not on the agenda of the scholarly circles, of interested intellectuals. The heated debate on the subject, and additional evidence cited below, may be a clear indication that the question was not hypothetical, at least with respect to the second alternative.

Another Rabbanite authority who expressed an innovative opinion on the question was Samuel ben Ḥofni, who, like Sa'adya, headed the Yeshiva of Sura (d. 1013). No complete text of the original interpretations of these two survived.

Two 10th -century Karaites paid special attention to the En Dor episode: 1) Abū Yūsuf Ya'qūb al-Qirqisānī, a contemporary of Sa'adya who was active in Iraq. He discussed the issue in the historic-ideological sections of his voluminous halakhic compendium, most of which has survived; 2) The most prolific Bible exegete in Karaite history, Yefet ben 'Eli, who was active in Jerusalem in the second half of the 10th century. He discussed the story in his commentary on 1 Sam 28.

Four exegetes in the 11th century who wrote in Judaeo-Arabic are known to have dealt with the En Dor episode:

'Eli b. Israel Alluf, a Palestinian Rabbanite scholar of early 11th century,[43] discussed the story of the witch extensively in the course of his commentary on the

---

42 *Sefer Pitron Torah*, ed. by Ephraim E. Urbach. Jerusalem: Magnes Press, 1978, pp. 81–84. The edition is based on a unique manuscript copied in Persia at the beginning of the 14th century. On the provenance and dating of the work see also Stemberger, *Introduction*, p. 359.

43 In modern publications (such as those quoted below in this note) his name is usually given as 'Alī ben Israel. However the Biblical name 'Eli seems to be quite common in our author's

book of Samuel. So far a unique incomplete copy of this commentary has been identified.[44] The commentary on chapter 28 (which according to the author's or the scribe's division is the last chapter of the first part of Samuel) is almost complete. It comprises 13 leaves[45] (i.e., 26 rather large pages!), and provides a detailed survey of opinions of previous and contemporary commentators. It is somewhat surprising to see how little attention the author pays to rabbinic views. He is occupied almost exclusively with the views of medieval exegetes and the questions that interest them. These are classified by the author into two main groups: Sa'adya[46] and his followers on the one hand, and those who were opposed to him on the other hand, notable among them Samuel ben Ḥofni. It seems that a copy of the latter's treatise on the story of the witch of En Dor was available to 'Eli.[47]

Three 11th-century exegetes were associated with Spain. One resided there all his life; the other two moved at some stage to the eastern Mediterranean, one to Akko (Acre) in Palestine, the other to Egypt. Judah b. Samuel Ibn Bal'am, Biblical exegete and Hebrew grammarian, was active in al-Andalus around the middle

time, among both Rabbanites and Karaites (e.g., the famous Karaite exegete Yefet b. 'Eli), and there is no reason to use the Arabic form for certain people. Steinschneider, Moritz, *Die arabische Literatur der Juden*. Frankfurt am Main: Kauffmann, 1902, §68 gives little information about him. Much important information is found in Mann, Jacob, *Texts and Studies in Jewish History and Literature*, vol. 2. Cincinnati: Hebrew Union College Press, 1935; repr.: New York: Ktav, 1972, pp. 30–31, p. 58, and interesting excerpts from the commentary on Samuel appear on pp. 95–96. On 'Eli's linguistic work, see Skoss, Solomon L., ed., *The Hebrew-Arabic Dictionary of the Bible known as Kitāb Jāmiʿ al-Alfāẓ of David ben Abraham al-Fāsī* the Karaite (tenth century) vol. 1. New Haven: Yale University Press, 1936, cvi–cxx. On this matter see also the important contribution of Mann in *Texts and Studies*, 1, pp. 96–98. A first-hand examination of the photocopy of his manuscript commentary on Samuel ascertains his Palestinian origins, or at least prolonged residence there, beyond doubt. Part of 'Eli's commentary on 1 Sam 28, which contains part of his quotation from Sa'adya's commentary (fols. pp. 186–87), has been published by Shtober, Shim'on, "Ba'alei ha-'ovot be-ferusho shel rav Saadya Gaon la-torah ve-la-navi." *Sinai* 134 (2004): pp. 3–18, with Hebrew introduction and annotated translation. Shtober used for the edition the unique ms. together with other quotations of Sa'adya. However, as will be noted below, none of the quotations from Sa'adya is identical to any other.

44 Ms. in the National Russian Library in St. Petersburg (formerly Firkovich Collection II), Judaeo-Arabic, 1st series 3361.

45 Fol. 186–198.

46 The author quotes Sa'adya's view on the story directly from his commentary on Lev 19:31, where various techniques and methods of divination are prohibited.

47 See the codicological reference to that copy in the excerpt quoted by Mann, *Texts and Studies*, p. 96: it comprised three quires.

of the 11th century.[48] In his commentary on Samuel he summarized the different opinions of exegetes and thinkers on the question of the appearance of Samuel in En Dor. He does this from the conservative perspective of the belief that the Biblical account should be understood 'literally', namely that Samuel was resurrected by the witch and conveyed to Saul the divine message regarding the latter's imminent fate.[49]

Very little is known about Moses b. Joseph Ibn Kashkīl other than the quotation of his view about the occurrences in En Dor in the commentary on Samuel by Isaac b. Samuel the Spaniard al-Kanzī. The latter is known from historical documents found in the Cairo Geniza. He was born in al-Andalus. At a certain stage he immigrated to Fusṭāṭ in Egypt, where he served as a judge. He wrote a commentary on the books of Samuel, of which large sections have survived and are preserved in several libraries.[50] In this commentary he quoted and discussed the three main opinions regarding the Biblical account of the En Dor episode. He quotes Ibn Kashkīl[51] extensively, as well as Saʿadya and Samuel b. Ḥofni. His quotations of their views are presented in a polemical context mixed with their refutations. Following Ibn Kashkīl and Ibn Balʿam, he identifies with the conservative view understanding the Biblical account to include Samuel's resurrection, which, to be sure, is not the literal meaning of the Biblical text.

Saʿadya's exposition in his commentary has not survived,[52] but his comments on the subject under discussion have survived in quotations by the exe-

48 Blau, Joshua, "Ibn Balʿam, Judah ben Samuel," in *Encyclopaedia Judaica*^{2} 8, pp. 660–61. Critical editions of Ibn Balʿam's commentaries on Isaiah, Jeremiah and Ezekiel have been published since 1994.

49 His summation of all the views, with his personal preference, was published in Judaeo-Arabic with Hebrew translation in *Otzar ha-Ge'onim* 4/2: tractate *Ḥagiga*, ed. by Menashe B. Lewin. Jerusalem, 1931, pp. 2–5. A medieval Hebrew paraphrase of this summation by David Qimḥi was also printed. See Lewin, p. 5.

50 Large fragments of his commentary on the Book of Samuel have survived; see Simon, Uriel, "The Contribution of R. Isaac b. Samuel al-Kanzi to the Spanish School of Biblical Interpretation." *Journal of Jewish Studies* 34 (1983): pp. 171–178; and see Schlossberg, Eliezer, "The Commentary of R. Isaac b. Samuel al-Kanzi on the Story of the Witch of ʿEn Dor." *Studies in Bible and Exegesis* 8 (2008): pp. 193–223 (Heb.), with discussion of al-Kanzi, his view on the En Dor episode, analysis of his attitude towards the various views, and edition of his commentary on the Biblical account with an annotated Hebrew translation.

51 On Ibn Kashkīl, his commentary and Isaac b. Samuel, see Mann, Jacob, *Texts and Studies*, vol. 1, pp. 386–93 and Schlossberg, Eliezer, "The Commentary of R. Isaac b. Samuel al-Kanzi".

52 To the best of my knowledge. Fragments of his commentary on Leviticus have been identified in some collections of Geniza fragments, and a considerable number of them have been published thus far, but none of them with the text on witchcraft and necromancy.

getes discussed above, namely, ʿEli b. Israel, Ibn Balʿam, Ibn Kashkīl, Isaac b. Samuel al-Kanzī as well as a later exegete named Abraham ben Solomon, who apparently was active between the late 14th and early 15th centuries. His origins and place of residence are not clear.[53] It seems that the quotation of Saʿadya's comment on the En Dor episode closest to his original formulation is the earliest quotation found in the commentary on 1 Sam authored by ʿEli b. Israel. It survived in an apparently unique manuscript.[54] The following is a paraphrase of that quotation.[55]

Saʿadya expounded five points (or: issues, topics *maʿānī*) in this story:

1) Was it right/possible [*jāʾiz*] in Saul's mind that a woman would bring up a dead person and revive him?

2) Did Samuel really rise up alive or not?

3) If he did rise, who made him do so—the witch or God?

4) Why did the woman see Samuel whereas Saul did not?

5) Why did Saul hear his [Samuel's] speech whereas the woman sorcerer did not?

Saʿadya's answers:

1) Scripture tells us that Saul's mind had become unbalanced at times [1 Sam 16:14]. It is not impossible for a person in such a condition, in a state of imbalance, to believe that an inanimate stone could resurrect a dead person, all the more so a female sorcerer.

2) Since Scripture said twice [in this story] "Samuel said" [vv. 15–16], we must concede that Samuel actually rose up alive [from the dead; *qad qāma ḥayyan*]. It is impossible to contend that the witch was saying to Saul, "Samuel says

---

53 See Cohen, Boaz, "Quotations from Saadia's Arabic Commentary on the Bible from Two Manuscripts of Abraham ben Solomon," in *Saadia Anniversary Volume*, ed. by Boaz Cohen, pp. 75–139. New York: American Academy for Jewish Research, 1943, (Cohen thought that Abraham was active in Yemen in the second half of the 14th century), esp. pp. 96–97 (Judaeo-Arabic text), pp. 121–122 (Hebrew translation); Razaby, Yehuda, "Muvaot ḥadashot mi-perush R. Saadya la-miqra." *Sinai* 88 (2008): pp. 97–108 (Razaby thought, on the basis of a colophon, that Abraham was active between the late 14th and early 15th centuries, somewhere in the eastern Mediterranean). The quotation adduced by Razaby on p. 101 as written by Abraham belongs apparently to Isaac b. Samuel and can be found in Schlossberg, "The Commentary of R. Isaac b. Samuel al-Kanzi." Abraham's complete commentary on the Early Prophets was published by Yosef Qafiḥ in 4 volumes and Hebrew translation (Qiryat Ono: Mekhon Moshe, 1999–2006). The passage published by Cohen can be found in volume 2, (2000), pp. 302–303.

54 See above, n. 44.

55 Shtober, Shimʿon, "Baʿalei ha-ʾovot be-ferusho shel rav Saadya Gaon la-torah ve-la-navi." *Sinai* 134 (2004): pp. 3–18. The paraphrase is based on a fresh reading of the microfilm reproduction, and the English rendering is based entirely on my understanding.

to you such and such," and also to say that what is written in Scripture is according to what Saul imagined. Had we allowed this [kind of argument] our adversaries could have forced us to accept [*alzamanā*] this interpretation regarding each occurrence of "He said," "He spoke," and we would be forced to strip [such phrases] of their external [*ẓāhir*] meaning.

3) The Creator made Samuel rise [*aqāma!*] in order to demonstrate an occurrence of resurrection. Scripture quotes explicitly [the witch's] saying "Whom shall I bring up for you?" [v. 11] However, Scripture does not ascribe to her anything connected to Samuel's rising. It says only, "She shrieked loudly" [v. 12]. If she had made him rise, Scripture would have said, "The woman made Samuel rise." [Saʿadya] said further: "If somebody asks: Why did not Scripture make clear that God made Samuel rise? We will reply: It referred us to our reason, since we know that no one other than He is capable of resurrecting the dead." Then [Saʿadya] said: "If somebody still asks: [Is it possible that] God makes a prophet rise by means of a woman-witch? We will reply: She was not the means by which God made [Samuel] rise, since there is no scriptural statement to that effect, nor is there such statement that indicates that she claimed doing so."

4) There is no indication in Scripture that Saul did not see Samuel from the beginning of the affair until the end. Scripture stated that he did not see him only in relation to when Samuel first appeared and his arrival. If Saul had waited, he would not have asked her. And yet Saul hastened to ask her; consequently Scripture reported her response to him.

5) There is no scriptural statement to the effect that Saul heard [Samuel's] speech whereas the woman was present without hearing. Furthermore there is nothing at all that resembles this opinion. Rather, when Saul and Samuel met, the woman withdrew, walked away from them and left them talking. She did so out of politeness. Then Scripture says "The woman went up to Saul and, seeing how greatly disturbed he was" etc. [v. 21]

In this discussion (especially in the third reply) Saʿadya uses a somewhat 'neutral' verb, *aqāma* (made Samuel rise) rather than *aḥyā* (resurrected). It seems that Saʿadya ascribes here a rare (perhaps late) sense to the verb *aqāma* (this sense is attested in later sources, see the dictionaries of Dozy and Hava) which is still bound to remind the reader of *al-qiyāma*, a very common term for the Resurrection. Saʿadya thus provided readers of Judaeo-Arabic with a systematic theological exposition of the Biblical account. The true meaning of the sequence of events was clear even to Saul at the end. All possible questions seem to Saʿadya to have been resolved satisfactorily.

From the discussion of a Karaite contemporary of Saʿadya, Abū Yūsuf Yaʿqūb al-Qirqisānī, we learn that the question had been discussed at least one generation earlier, since he quotes the views of "our teachers" on the issue. This evidence therefore brings us towards the end of the 9th century at the latest. Al-Qirqisānī quotes two views on the question: one is identical to Saʿadya's, and the other is that the whole affair consisted of a series of psychological tricks that the witch played on Saul. Samuel was not there at all.[56] Theologically and exegetically this view poses an extremely difficult problem: if Samuel was not there, who was it that conveyed God's words to Samuel? Or were they in fact God's words? It may be that the question was dealt with by some early exegetes through the argument that the alleged Word of God was actually a clever conjecture which Saul mistook for a truthful prophecy.[57] As we shall see, this solution was taken up and developed a generation after al-Qirqisānī.

It is not clear whether al-Qirqisānī takes sides in this debate, but his discussion clearly indicates two things: a) By Saʿadya's time the question was not at all hypothetical, as might appear from the language in which Saʿadya presents it; b) By this time the two contradictory positions were well defined. Nevertheless, the conservative view, i.e., that the Biblical account should be interpreted literally, was probably still deeply rooted, albeit with the variation that the witch resurrected Samuel instead of "brought up his soul (from the underworld of the dead)."

From that time on and until the 13th or 14th century, the question of the resurrection of Samuel in En Dor dominated the Judaeo-Arabic exegesis of the story, with long repetitions and new, sophisticated arguments.

The Karaite exegete Yefet ben ʿEli polemicized against the views of Saʿadya (without mentioning his name) in his commentary on the story. It is worthwhile to expound his argument in some detail.[58]

Yefet begins by stating that he intends to comment on this story only on the basis of a sound, rationalist investigation, not a pretense (this is probably an al-

56 Al-Qirqisānī, Yaʿqūb, *Al-Anwār wa-ʾl-marāqib*, ed. by Leon Neomy, vol. 3. New York: The Alexander Kohut Memorial Foundation, 1941, pp. 582:14–583:21. See the French translation and discussion by Vajda, Georges, "Études sur Qirqisānī." *Revue des Études Juives* 107 (1946–7): pp. 52–98, esp. pp. 96–97, pp. 112–15; see also the discussion in Ben-Shammai, Haggai, "The Doctrines of Religious Thought of Abū Yūsuf Yaʿqūb al-Qirqisānī and Yefet Ben ʿEli." vol. 1. Ph.D. Thesis, The Hebrew University of Jerusalem, 1977, pp. 301–02.

57 Ginzberg, *Legends of the Jews*, vol. 6. Philadelphia: Jewish Publication Society, 1926; repr. 1946, p. 237, n. 77 (new edition, Philadelphia: Jewish Publication Society, 2003, 2 904b, n. 77) had already observed in very general terms that "the rationalistic view that necromancy, like sorcery in general, is nothing but a fraud, is first met with among authors who flourished about 900 C.E."

58 The Arabic text was edited in my "Doctrines of Religious Thought," vol. 2, pp. 201–203, and discussed with partial Hebrew translation ibid. vol. 1, pp. 308–312.

lusion to Saʿadya). Yefet lays great stress on the concept of perceptions. People are often led to accept what they perceive as true occurrences because they are so told. We must examine such alleged occurrences rationally. If we do so we may well conclude that the perceptions are wrong. A claim by a human being to have resurrected a dead person must be false, because reason dictates that resurrecting the dead is beyond the capacity of human beings. Hence God would not resurrect a dead person in association with a liar's claim, if only because this might lead a person who witnesses the resurrection to believe that it was caused by a human being. This in turn would destroy the authority of real miracles as verifiers of the truth of the prophets' messages. The only true fact of the En Dor episode was, therefore, that the woman pretended to have brought up/resurrected Samuel and to have seen him. She made the noises which Saul mistook for Samuel's voice, and the message delivered was according to her conjecture, based on common knowledge. The decision whether to describe real facts or the perception of the beholder is left, according to Yefet, to the discretion of the redactor (*mudawwin*) of the Biblical text, who is guided by prophetic inspiration (*ilhām*). The use of this exegetical device by Yefet (and some of his predecessors, such as al-Qirqisānī, as well as his followers) serves mainly to explain away theological difficulties, such as historical anachronisms, logical inconsistencies, dogmatic difficulties and the like.[59]

Samuel ben Ḥofni, the prolific Gaon of Sura/Baghdad (d. 1013) takes a similar rationalist, critical position.[60]

In the classification of ʿEli ben Israel (see above) in his commentary on 1Sam, the dividing line between the two camps runs along one doctrinal ques-

---

59 See Ben-Shammai, Haggai, "On *Mudawwin*—the Redactor of the Hebrew Bible in Judaeo-Arabic Bible Exegesis," in *From Sages to Savants: Studies Presented to Avraham Grossman*, ed. by Joseph Hacker et al., pp. 73–110 (Heb.). Jerusalem: The Zalman Shazar Center for Jewish History, 2010. For a different approach to the figure of the *mudawwin*, stressing the literary aspect, see Polliack, Meira R., "The 'voice' of the narrator and the 'voice' of the characters in the Bible commentaries of Yefet ben ʿEli," in *Birkat Shalom: Studies in the Bible, Ancient Near Eastern Literature, and Postbiblical Judaism Presented to Shalom M. Paul*, ed. by Chaim Cohen et al., vol. 2, pp. 891–915. Winona Lake: Eisenbrauns, 2008; Zawanowska, Marzena, "Was Moses the *mudawwin* of the Torah?: The Question of Authorship of the Pentateuch according to Yefet ben ʿEli," in *Studies in Judeo-Arabic Culture*, ed. by Haggai Ben-Shammai et al., pp. 7–35. Tel Aviv: Tel Aviv University, 2014.

60 See Sklare, David E., *Samuel ben Ḥofni Gaon and his Cultural World*, pp. 28–29, 42. Samuel ben Ḥofni seems to have written a separate treatise, fragments of which survived in the Cairo Geniza. See also Sklare's remarks in "Scriptural Questions: Early Texts in Judaeo-Arabic," in *A Word Fitly Spoken: Studies in Mediaeval Exegesis of the Hebrew Bible and the Qurʾān*, ed. by Meir Bar-Asher et al. Jerusalem: The Ben-Zvi Institute, 2007, p. 207 (Heb.).

tion: was Samuel really resurrected or not? ʿEli is more than willing to accept that there is no truth to the trickeries of magicians, with whose techniques he shows some acquaintance, and therefore it is impossible that the witch resurrected Samuel. However, he fully supports Saʿadya's view that God resurrected Samuel especially for the occasion so as to convey His message to Saul. ʿEli polemicizes sharply against Samuel ben Ḥofni, who rejected Saʿadya's position. At approximately the same time the most prominent Karaite scholar in Jerusalem, Yeshuʿa ben Judah (active in the middle of the 11th century) repeated, more or less, Yefet ben ʿEli's arguments against Saʿadya's position, defending the psychological, fraudulent tricks theory.[61]

All these authors thus focused their attention on one doctrinal aspect of the story (or, perhaps, should I say, such an aspect that was attached to the story or imposed on it), rather than various ethical and ritual aspects, which had previously been the focus of the story, both in its scriptural version and in the rabbinic homilies. It is noteworthy that the medieval exegetes who were responsible for this turn probably had adequate knowledge of the necromantic practices that had been rooted deeply in the Near East from antiquity. These practices continued. This is borne out, *inter alia*, by the fact that the manner in which their use continued is mentioned by the 12th-century scholar from Baghdad, Samuel b. ʿEli,[62] precisely with respect to the En Dor affair. In order to avoid the proposition that Samuel had been resurrected by the witch, he seems to have preferred the possibility that she brought up his ghost using the technique known in Arabic by the term *mandal*.[63] It is doubtful whether this turn, almost "sudden," as it were, would have been called for by internal Jewish developments.

The following table is a brief summation of the division of views between the authors of Judaeo-Arabic commentaries on the En Dor episode.

The witch resurrected Samuel, who conveyed	God actually resurrected Samuel in order to convey	Samuel was not resurrected; he was not pres-

61 Fragments of his discussion are found in Ms. St. Petersburg, RNL Evr.–Arab. 1:1989, fol. pp. 152–153, which possibly belong to his very extensive halakhic-theological commentary on the Ten Commandments (*Tafsīr ʿaseret ha-devarim*).

62 He was most prominent in the attempts to restore the office of Gaon in Baghdad in the latter part of the 12th century (died apparently 1193): see Mann, *Texts and Studies*, vol.1, pp. 214–19.

63 This is attested in the controversy between Samuel and Joseph b. Judah b. Shimʿon, the disciple of Maimonides, recorded by the latter in his *Kitāb al-iskāt* and published by Sarah Stroumsa, see above, n. 40. The discussion about *mandal* is recorded in paragraphs 80–108, pp. 34–40 in the Judeo-Arabic text, 77 in the medieval Hebrew translation, 107–13 in Stroumsa's annotated Hebrew translation, and 146–54 in Stroumsa's commentary with further references.

God's message to Saul. Samuel was there, flesh and spirit.	His message to Saul. Samuel was there, flesh and spirit, but the witch did not see him	ent there at all, neither in flesh nor in spirit. The witch deceived Saul/ played on him a psychological trick
Judah Ibn Bal'am (Spain, second half of the 11th century). Wrote a commentary on the Prior Prophets.	Sa'adya Gaon (Egypt, Israel, Baghdad, 882–942). Quoted from his commentary on Leviticus.	Abū Yūsuf Ya'qūb al-Qirqisānī (Karaite exegete, jurist and philosopher; Iraq, active 920–940?). Has a chapter in his legal code.
Moses Ibn Kashkīl (Spain and the Near East, second half of 11th century). Wrote a commentary on the Prior Prophets.	Abū Yūsuf Ya'qūb al-Qirqisānī (Karaite exegete, jurist and philosopher; Iraq, active 920–940?). Has a chapter in his legal code.	Yefet ben 'Eli (Karaite, Jerusalem, second half of 10th century). Wrote a commentary on the entire Bible.
Isaac b. Samuel (Spain and Egypt, end of 11th–Beginning of 12th century). Judge and exegete. Wrote a commentary on the Prior Prophets.	Hay Gaon (Iraq, d. 1038). Jurisprudent and communal leader. Quoted from a responsum.	Samuel b. Ḥofni Gaon (Iraq, d. 1013). Wrote a responsum/epistle/treatise on the En Dor episode.
	'Eli ben Israel Alluf (Syria? First half of 11th century). Wrote a comentary on Samuel.	

5 The Patristic Approach

The question of the nature or the definition of the episode of Samuel's alleged appearance in En Dor had been discussed by Christian commentators since the time of Origen (perhaps even earlier).[64] In a homily on 1 Sam 28, Origen dis-

64 See Smelik, Klaas A. D., "The Witch of Endor: I Samuel 28 in Rabbinic and Christian Exegesis till 800 A.D." *Vigiliae Christianae* 33:2 (1979): pp. 160–79. His classification of the positions of Christian exegetes and thinkers is based on a wide, learned reading of many sources. His analysis takes a different direction to the one suggested in the present study. His discussion of rabbinic sources, especially the early medieval ones, is quite meager. On Origen as exegete see Mart-

cusses the meaning of the Biblical account.[65] Origen begins the homily, which Nautin characterized as "didactic," by asking whether the appearance of Samuel, as depicted in the Biblical text, was real. Does it describe true reality? Did Samuel speak to Saul? He answers these questions in unequivocally affirmative terms. The Biblical text was authored by the Holy Spirit, and should therefore be accepted literally.[66]

He then turns to a discussion of the question of the "locality" from which Samuel was called to convey the divine message—or rather, sentence—to Saul. There were those who did not accept the proposition that Samuel, or rather his soul, arrived on the scene from *Hades*, the underworld of the dead. It seems that people utterly rejected this proposition due to the fact that in Origen's time, *Hades* was already identified with Hell, the abode of sinners; whereas two centuries earlier it was accepted that the witch brought Samuel up from there.[67] Origen explains that in *Hades* in Samuel's time, there was a compartment, as it were, for the righteous—the prophets actually—who were sent there by God to call the sinners to repent.[68] Those who would not accept Origen's view were left with two options: a) It was not Samuel who spoke to Saul, but rather a demon who was brought in by the witch. In other words, the whole affair was a deception. This option is refuted by Origen with the argument mentioned above, namely that the Biblical account, being authored by the Holy Spirit, must be interpreted literally;[69] b) Samuel was resurrected, or in other words, he had left *Hades* by the time of the En Dor episode. Origen refutes this option on the ground that "before the sojourn of my Lord Jesus Christ it was impossible

ens, Peter W., *Origen and Scripture: The Contours of the Exegetical Life*, Oxford Early Christian Studies. Oxford: Oxford University Press, 2012.

65 Two homilies of Origen on 1 Sam 28 survived in the original Greek text. They were published with other homilies and a French translation in the volume: Nautin, Pierre and Marie-Thérèse Nautin, eds., *Origen, Homélies sur Samuel*, Sources chrétiennes 328. Paris: Éditions du Cerf, 1986, pp. 11–30, pp. 61–89, pp. 171–230. The second homily appears to be a short follow-up on the first one. For an English translation of the first homily see Origen, *Homilies on Jeremiah; Homily on 1 Kings 28*, trans. by John Clark Smith. The Fathers of the Church 97. Washington, D.C.: Catholic University of America Press, 1998, pp. 319–33.

66 On this principle in Origen's system in general, and in relation to the Biblical account of the En Dor episode in particular, see Martens, *Origen*, pp. 94–95; pp. 58–59. The resemblance to Saʿadya's argumentation with respect to this principle is striking.

67 See above, n. 33.

68 Even Jesus was sent to *Hades* to save; see Origen, *Homélies sur Samuel*, pp. 188–91 (section 6:12–52; Origen, *Homilies on Jeremiah*, pp. 325–27 (section 6.2–4).

69 Cf. Origen's words, in Nautin's edition, p. 183 (section 4:21–23), Origen, *Homilies on Jeremiah*, pp. 322–23 (end of section 4.3): "We will say to the man who [...] has insisted that Samuel was not in Hades, *The woman saw Samuel*, the voice of the narrator said this".

for someone to come near to where the *Tree of Life* was." This is followed by a lengthy explanation that concludes the homily.[70]

The relationship between Origen and the rabbis of his time has already been studied.[71] This relationship can be characterized as a kind of dialogue. It is quite possible that the rabbis had some acquaintance with Origen's description of Samuel's appearance in En Dor and Origen's insistence on grading Samuel—and in fact all Biblical prophets—below Jesus. Those rabbis who had some knowledge of Origen's view could have responded by suggesting that Samuel had been physically resurrected, thus putting him on a par with Jesus. In rabbinic sources from the period of the Jerusalem Talmud or contemporary Midrashic sources one can find only vague allusions in this direction,[72] perhaps early attempts at responding to Origen's view. Such allusions later turned into straightforward rabbinic descriptions of resurrection.[73] In other words, Origen's view may be seen as the first step leading to the shift towards the centrality of the resurrection theme in the Jewish interpretation of the story.

The question continued to occupy the minds of Christians, both orthodox and heretics. Patristic authors were well aware (for their own theological purposes) that the story of En Dor is irrelevant to the doctrine of resurrection as developed in Judaism and Christianity since late Antiquity. Theodore Bishop of Mopsuestia, the 5th-century forerunner of Nestorianism and the exponent of literal exegesis, may have been in agreement with Origen's view in this respect.[74]

---

70 Origen, *Homélies sur Samuel*, pp. 204–09 (section 9:33–10:25), Origen, *Homilies on Jeremiah*, pp. 331–33 (9.3–end).

71 See Urbach, Efraim E., "The Homiletical Interpretations of the Sages and the Expositions of Origen on Canticles and the Jewish-Christian Disputation." *Scripta Hierosolymitana* 22 (1971): pp. 247–275; De Lange, Nicholas R.M., *Origen and the Jews: Studies in Jewish Christian Relations in Third Century Palestine*, University of Cambridge Oriental Publications 25. Cambridge: Cambridge University Press, 1976, esp. pp. 89–102, pp. 133–35; Martens, *Origen*, pp. 135–38.

72 See above, n. 34.

73 See above, n. 36 and following notes, and below in the discussion of the evidence of Islamic sources.

74 On Theodore as exegete (including his attitude to Origen), see Zaharopoulos, Dimitri Z., *Theodore of Mopsuestia on the Bible*. New York: Paulist Press, 1989, pp. 78–141. On the "translation project" of his commentaries, see Vööbus, *History*, pp. 14–24. As a "literalist" Theodore was opposed in principle to Origen's inclination towards allegorical interpretations. However, in the case under discussion here they seem to have agreed. On the central position taken by exegesis in the Syrian learning institutions, see above, n. 3. According to a quotation (for which I have apparently lost the source) based on Theodore of Mopsuestia, *Le commentaire sur les Psaumes (I–LXXX)*, ed. by Robert Devreese. Città del Vaticano: Biblioteca Apostolica Vaticana, 1939, pp. 386–87, "He [Theodore] likewise taught that the doctrine of the resurrection of the dead was an entirely unknown idea in pre-New Testament times, and it was the conception of

Notwithstanding Theodore's conformity to Origen's Christological considerations, it should be noted that his approach is also quite close to a modern one, namely that the Biblical account, and especially the Biblical presentation of the appearance of Samuel, must be interpreted according to the perceptions that had been prevalent in Biblical times.

Many writings of the great exegete *(mepashqānā)* were subsequently translated into Syriac and were extensively studied, discussed and commented on by the Christians of Syria and Mesopotamia in the geonic period.

It should also be noted that the story of the witch of En Dor attracted the attention of heretical Christians in the East in the early Middle Ages. A question on this theme is reported to have been sent to Anastasius Sinaita, patriarch of Antioch in the early 7th century.[75]

At the time of early Judaeo-Arabic Bible exegesis, the prevalent view among Christian commentators in the East regarding the En Dor episode was that of Origen, which was most probably established by Theodore of Mopsuestia. This view was based mainly on Christological considerations. If we assume that the rabbis in Origen's time had been aware of Origen's view and therefore began to develop a Jewish response, we can assume that the Jewish exegetes of the 9th or 10th century found themselves in a fairly similar position. However, they had at their disposal more sophisticated hermeneutical and philosophical means to respond to the Christian challenge.

6 The Evidence of Islamic Sources

It is well known that the Muslims did not engage in Bible exegesis. Nevertheless, Biblical stories and themes, perhaps even genres, are found in the Qur'ān. Many more Biblical heroes and stories figure extensively in extra-Qur'ānic traditions so much that these traditions began quite early on to be termed *Isrā'īliyyāt.*[76]

---

sheol—the shadowy realm of the underworld where the departed spirits of the dead remained—that prevailed in the teaching of the old dispensation." In Theodore of Mopsuestia, *Commentary on Psalms 1–81*, trans. by Robert C. Hill, Writings from the Greco-Roman World 5. Atlanta: Society of Biblical Literature, 2006, I could not find any reference to such a quotation.

75 See Rosenthal, Judah, *Hiwi al-Balkhi : A Comparative Study.* Philadelphia: Dropsie College, 1949, p. 46 and ibid., n. 197, for a short survey of the various views of the Church Fathers on the story.

76 Several attempts have been made by Muslim scholars over the centuries to minimize the number of such traditions, to purify, as it were, the collections of traditions of foreign materials. See Tottoli, Roberto, "Origin and Use of the Term Isrā'īliyyāt in Muslim Literature." Arabica 46 (1999): pp. 193–210. This interesting article still leaves some room for further research.

When reading the works of Muslim authors, one should always bear in mind that they have no commitment to the text of the Bible, whether Hebrew or other. If they consult *Isrāʾīliyyāt* or quote them (usually without characterizing them as such), they are committed to a body of traditions that loosely constitute a corpus of texts related to the Books of Moses (Torah), the Early Prophets and rabbinic (or even Christian) sources, without any differentiation between them. The relationship of alleged quotations from such sources can be close or remote paraphrases. Such quotations are sometimes ascribed to real or alleged Jewish converts, such as Kaʿb al-Aḥbār or Wahb b. Munabbih, and sometimes to transmitters whose names appear often in compilations of exegesis or tales (*Qiṣaṣ*), such as Ismāʿīl al-Suddī. These names count highly in the traditions that are quoted in the present study. The names of many other transmitters may be found in the transmission chains of exegesis or tales, including those that are quoted as historical accounts. Traditions transmitted as historical accounts that are related to Biblical figures (both of the Hebrew Bible and the New Testament) are of course important in the structure of prophetic history, which reached its climax and final stage with the appearance of Muḥammad. At the same time, such traditions may be used for edification, ethical guidance and even entertainment.

In the Qurʾān there is one relatively short passage about King Saul (Q 2:246–251), who is called there Ṭālūt. The identification of Ṭālūt with King Saul is quite simple: it stems both from the association with Jālūt, i. e., Goliath, who is named explicitly in the passage, and from explicit identifications in Islamic historiography. His name is widely interpreted as a reference to his extraordinary height (1 Sam 10:23: "When he took his place among the people he stood a head taller than all the people").[77] The entire story of Saul is exhausted in six verses. In such a succinct reference there is obviously no place for a complex story such as the En Dor episode.

Around this short Qurʾānic passage—and also as part of the world history of the prophets—one finds numerous collections of traditions and narratives. They are found in historiographic works, works of Qurʾān exegesis (*Tafsīr*), and compilations of tales of the prophets (*qiṣaṣ al-anbiyāʾ*). In later works one finds extensive collections of traditions, often contradictory ones, which are presumed to relate directly to the Qurʾānic passage.[78]

77 See Firestone, Reuven, "Ṭālūt." *Encyclopaedia of Islam*² 10, pp. 168–69.

78 Such is the case, for instance, in Fakhr ad-Dīn ar-Rāzī's *Tafsīr*. Beirut: Dār al-kutub al-ʿilmiyya 2000, part 6, pp. 144–76, or the much later *Biḥār al-Anwār*, by the important Shīʿī compiler Muḥammad Bāqir al-Majlisī. Teheran: Dār al-kutub al-islāmiyya, 1992, part 13, pp. 435–457.

In a variety of early sources (before 350AH/961CE), four main accounts of the stories of King Saul may be found.

1) The first is well represented in the world history authored by Aḥmad b. Abī Yaʿqūb b. Jaʿfar b. Wahb b. Wāḍiḥ al-Yaʿqūbī (d. 897/98), entitled *at-Taʾrīkh.*[79] This is a running historical report of Saul's kingdom (Yaʿqūbī knows that Ṭālūt's name is Shaʾūl, that Jālūt's name is Ghulyāth and that the name of the prophet who anointed Saul is Shamwīl[80]), with few Qurʾānic allusions or themes. This author is much more faithful to the Biblical account than most Islamic authors. His description contains many paraphrases of Biblical verses which are quite close to the original text (e. g., in the description of Saul's war against the Amalekites, which is almost a literal translation of the Hebrew text of 1 Sam 15:3). This fact naturally raises his possible acquaintance with Christian or Jewish Bible translations. However, there is no mention at all of the witch of En Dor and the alleged resurrection of Samuel, or of bringing him (or his soul) up from the dead. Interestingly, Saul's final war is not against the Philistines but rather against *al-ḥunafāʾ*, who are characterized as pagan worshippers of the stars (*ʿabadadatu n-nujūm*).[81] This invocation of *al-ḥunafāʾ* may indicate that Yaʿqūbī relies on a Christian version of the story, which in turn derives from a Syriac source. Saul's three sons are killed, and he himself commits suicide "so that these uncircumcised will not kill me" (*li-alla yaqtulanī haʾulāʾi l-qulf*), which is a close paraphrase of 1 Sam 31:4.

2) The second source is Muḥammad b. ʿAbdallāh al-Kisāʾī's *Qiṣaṣ al-anbiyāʾ*. The chronology of the author is not certain. The stories about King Saul/Ṭālūt are part of the chapter entitled *Ḥadīth Shamwīl wa-Ṭālūt wa-Jālūt wa-Dāwūd.*[82] The stories are told on the authority of Kaʿb al-Aḥbār. The starting points for the stories are Qurʾānic passages, mainly Q 2:212, and Q 2:246 – 251. Biblical themes and

79 Al-Yaʿqūbī, Aḥmad b. Abī Yaʿqūb b. Jaʿfar b. Wahb b. Wāḍiḥ, *al-Taʾrīkh*, ed. by Martijn T. Houtsma. Leiden: Brill, 1883; repr. 1969, pp. 50 – 53. For an evaluation of his work and his use of Jewish and Christian sources, see Zaman, Muḥammad Q., "al-Yaʿḳūbī." Encyclopaedia of Islam² 11, pp. 257– 58; Adang, Camila, *Muslim Writers on Judaism and the Hebrew Bible: From Ibn Rabban to Ibn Ḥazm*, Islamic Philosophy Theology and Science 22. Leiden: Brill, 1996, pp. 71– 5, pp. 250 – 53.

80 See Rippin, Andrew, "Shamwīl."Encyclopaedia of Islam² 9, p. 300.

81 Watt, William Montgomery, "Ḥanīf." Encylopaedia of Islam² 3, pp. 165 – 66, with reference to this particular source and to previous discussions of this use of *Ḥanīf.*

82 Al-Kisāʾī's, Muḥammad b. ʿAbdallāh, *Qiṣaṣ al-anbiyāʾ*, ed. by Isaac Eisenberg vol. 2. Leiden: Brill,1923, pp. 250 – 258. The work is available in English translation by Thackston Jr., Wheeler M., *The Tales of the Prophets of al-Kisāʾi*. Boston: Twayne Publishers, 1978, pp. 270 – 278; see also Nagel, Tilman, "al-Kisāʾī." Encyclopaedia of Islam² 5, p. 176.

phrases keep coming up, with contaminating elements related to Gideon, Samson, Saul, David, and other Biblical heroes.

There can be no doubt that written or oral Arabic paraphrastic renderings of the books of Samuel, or parts thereof, from Jewish or Christian sources were available to the author.

The end of al-Kisāʾī's description of Saul's kingdom is quite surprising.[83] Saul is deserted by the Israelites in favor of David. There is no witch there. There is a pious woman who is "known for having her prayers answered," i.e., she does not engage in any form of illegitimate craft. Saul asks her to bring Samuel to life from the dead. This she declines, saying she "has no station before her Lord" to make God resurrect the dead in answer to her prayer (*laysa manzilatī ʿinda rabbī an yuḥyiya l-mawtā bi-duʿāʾī*). She can only pray to God, who will see to it that Saul communicates with Samuel in his dream.[84] Samuel indeed appears in Saul's dream, reproaching him for his behavior. Saul loses his kingdom, which passes to David. Saul ends up in submission to David. No heroic death, no illegitimate magic, no resurrection or apparition of Samuel. All these elements, which are vital components of the Biblical story, are completely absent from this version.

3) An interesting representative of the third account is found in al-Ṭabarī's World History (*Taʾrīkh*).[85] It is noteworthy that nothing about the En Dor episode is found in his *Tafsīr* on the relevant verses.[86] Al-Ṭabarī (d. 923) relied heavily on Wahb b. Munabbih and Ismāʿīl al-Suddī.[87] Al-Ṭabarī's account has been copied or adapted by many authors. One such typical adaptation is found in the very popular work of ath-Thaʿlabī (d. ca. 1035), *Qiṣaṣ al-anbiyāʾ*, also known as

83 Al-Kisāʾī's *Qiṣaṣ*, pp. 257:19–258:14; Thackston, *The Tales of the Prophets*, pp. 277–278.

84 This may be related to that detail in the Biblical account (v. 6) that Saul could not communicate with the Divine even in his dreams.

85 Al-Ṭabarī, Abū Jaʿfar Muḥammad, *Taʾrīkh al-rusul wa-ʾl-mulūk*, ed. by Muḥammad Abū ʾl-Faḍl Ibrāhīm, vol. 1. Cairo: Dār al-maʿārif, 1960, pp. 365–75. Al-Ṭabarī's version of the En Dor episode is available in English translation in Al-Ṭabarī, *The Children of Israel, The History of al-*Tabari, vol. 3, translated and annotated by William M. Brinner. Albany: State University of New York Press, 1991, pp. 138–39.

86 Al-Ṭabarī, Abū Jaʿfar Muḥammad, *Jāmiʿ al-bayān ʿan tafsīr ʾāy al-qurʾān*, vol. 2. Cairo: Muṣṭafā al-Bābī al-Ḥalabī, 1968, pp. 595–635.

87 See on them Khoury, Raif Georges, "Wahb b. Munabbih." *Encylopaedia of Islam*² 11, pp. 34–36; Juynboll, Gautier H. A., "al-Suddī." *Encylopaedia of Islam*² 9, pp. 762; for his biography from a Shīʿite angle see Al-ʿĀmilī, Muḥsin al-Amīn, *Aʿyān al-shīʿa*, vol. 3 Beirut: Dār al-taʿāruf li-ʾl-maṭbūʿāt, 1986, pp. 378–80, including references to his Shīʿite inclination (*tashayyuʿ*).

ʿArāʾis al-Majālis.[88] The final episode described in this account is Saul's encounter with the woman who knew the Ineffable Name (*Ism Allāh al-aʿẓam*)[89] that resurrected Samuel, Samuel's instructions to King Saul and the latter's death. Ath-Thaʿlabī is mostly dependent on al-Ṭabarī.

This is a quite detailed description of the event. It is not clear whom Saul is going to fight. I would like to discuss briefly a few points:

The witches in general are referred to in the story as "wise men" (*ʿulamāʾ*).[90] The woman-necromancer in particular is referred to as a wise woman (*ʿālima*) who knows the Ineffable Name, which is theoretically a legitimate device of magic.

The parable ("the rooster's parable") that is quoted in a Midrashic source in order to emphasize the irony of Saul's action seeking a scholar, i.e., a witch, is quoted in al-Ṭabarī's account almost verbatim. It is thus clear that al-Ṭabarī's source had a Jewish (Judaeo-Arabic?) written or oral source at his disposal. That assumed source contained an Arabic close paraphrase of the rooster's parable.

Saul is accompanied in some versions of this account by a giant (*jabbār*), in others by a baker (*khabbāz*). It should be noted that in Arabic orthography the difference between the two adjectives is only in the diacritical punctuation.[91] In the Biblical account Saul is accompanied by "two men".

According to al-Ṭabarī, the woman is requested explicitly to resurrect a prophet. She takes Saul and his companion to the grave of Joshua ([!]; this is 'corrected' in ath-Thaʿlabī's version to Samuel). Since the Muslim authors were absolutely ignorant of the geography of the Biblical account, they knew nothing

88 Al-Thaʿlabī, Abū Isḥāq Aḥmad b. Muḥammad b. Ibrāhīm, *Qiṣaṣ al-anbiyāʾ*. Beirut: al-Rāʾid al-ʿArabī, 1985 [repr. of: Cairo: Muṣṭafā al-Bābī al-Ḥalabī, 1955], pp. 272–274; for an English translation see Brinner, William M., ed., *Arāʾis al-majālis fī qiṣaṣ al-anbiyāʾ, or: "Lives of the Prophets": As Recounted by Abū Isḥāq Aḥmad ibn Muḥammad ibn Ibrāhīm al-Thaʿlabi*, Studies in Arabic Literature 24. Leiden: Brill 2002, pp. 459–61; for a German translation see Busse, Heribert, *Islamische Erzählungen von Propheten und Gottesmänner: Qiṣaṣ al-anbiyāʾ oder ʿarāʾis al-maǧālis, von Abū Isḥāq Aḥmad b. Muḥammad b. Ibrāhīm aṯ-Ṯaʿlabī*, Diskurse der Arabistik 9. Wiesbaden: Harrassowitz 2005, pp. 348–50.

89 See Gimaret, Daniel, *Les noms divins en Islam: Exégèse lexicographique et théologique*. Paris: Éditions du Cerf, 1988, pp. 85–94.

90 This title can refer also to the Priests of the Tabernacle in Nov who had been killed by Saul for allegedly supporting David against him (see 1 Sam 22:6–19). Brinner suggested that the term "wise men" may be due to some early Arabic (Judaeo-Arabic?) translation of the term *yidʿōnī* in the Hebrew Bible, which is a cognate of *ōv* (necromancer). See Brinner, *Children*, p. 137, n. 714; see e.g., Lev 20:27.

91 See Brinner, *Children*, n. 735.

about the relationship between En Dor (the residence of the witch) and the battlefield of Saul and the Philistines on the one hand, and the burial places of Samuel or Joshua on the other. The Biblical account is based on forging a connection between Saul and the prophet Samuel by bringing up an apparition of the spirit of the latter, and so does not require a physical connection to his grave. Resurrection, unlike bringing up the apparition of the dead, requires physical contact with the actual grave. The episode of Saul and the witch thus turned into a proper resurrection story.

4) The diffusion of the resurrection story is attested in another Arabic-Islamic source which was much closer to philosophical circles, namely the *Rasāʾil Ikhwān aṣ-Ṣafāʾ* (*The Epistles of the Brethren of Purity*[92]).

Where the authors wish to quote something from the Jewish sources (which are often referred to in medieval Islamic sources as *tawrāh*—Torah), they say that the story is told in the "Books of chronicles of the Children of Israel, which for them have the same status as the Torah." The story as told there is found in the last epistle, the 52nd, which deals with magic, witchcraft and the like.[93] It follows the Biblical account quite faithfully. It may well be that this Arabic-Islamic version is derived from some Judaeo-Arabic rendering. There is a difference, however, in one rather crucial word. In the Hebrew Biblical version Saul asks the witch to "bring up for him" a certain person (i.e., his soul). In the Arabic Islamic version he asks her to revive/resurrect a prophet for him. The Arabic verb used here is *aḥyā*. And so, in quoting Samuel's rebuke "Why have you disturbed me and brought me up?" (1 Sam 28:15) the Arabic version has: "Why have you brought me back and revived/resurrected me (*li-mā arjaʿtanī wa-aḥyaytanī*)?" This rendering may well reflect a Jewish, i.e., Judaeo-Arabic version of the story, whose author took a clear, unequivocal position in the intra-Jewish controversy about the true nature of Samuel's appearance at En Dor. Such a Judaeo-Arabic version may have preceded the *Epistles* by several decades at least. If indeed the *Epistles* are more or less contemporaneous with Saʿadya (first half of the 10th century), this would mean that before the end of the 9th century the opposing Jewish positions had already been established. Both al-Ṭabarī's account and that

92 Or of those most worthy of being called pure; the kinship relation ("brethren") symbolizes a personification of a human/moral property. See on them Marquet, Yves, "I<u>kh</u>wān al- Ṣafāʾ." Encyclopaedia of Islam² 3, pp. 1071–76; Netton, Ian R., *Muslim Neoplatonists: An Introduction to the Thought of the Brethren of Purity (Ikhwān al-Ṣafāʾ)*. Edinburgh: Edinburgh University Press, 1991; see esp. pp. 71–7, on "Judaism and the *Rasāʾil*", and p. 75 on our story specifically.

93 *Rasāʾil Ikhwān al-Ṣafāʾ*, vol. 4. Beirut 1970, pp. 292–293. Reprint of: Ikhwān al-Ṣafāʾ, *Rasāʾil Ikhwān al-Ṣafāʾ wa khullān al-wafā*, ed. by Buṭrus Bustānī, vol. 4. Beirut: Dār Ṣādir, 1957.

of the *Ikhwān aṣ-Ṣafā* clearly ascribe the action of resurrection to the woman-witch. It seems quite probable that both accounts depend on a Judaeo-Arabic source, namely a Jewish account that clearly ascribed the bodily resurrection of Samuel to a human being.

7 Conclusion

We have seen a fairly complex development, from a focus (in the Bible) on communal or (in rabbinic sources) on personal ethics to theological questions (in Judaeo-Arabic exegesis) connected to the appearance or apparition of Samuel in En Dor, and from bringing up a soul in a séance to actual resurrection. It stands to reason that the ideological and doctrinal developments in Christian exegesis since the days of Origen could hardly have gone unnoticed by Jewish, educated intellectuals, especially in light of the bilingual channel of communication between Jews and Christians (see above in the introduction), which would have facilitated the flow of ideas across communal or confessional boundaries, and in light of the fact that both communities shared an interest in one and the same scripture. The Christian challenge alone could have set in motion some sort of Jewish response, namely, an argument that some sort of resurrection occurred there. There was also an intra-Christian response to Origen's doctrinal position, namely, Jerome's view that the entire sequence of occurrences in En Dor was a series of psychological or magical tricks played by the witch on King Saul. This view reappeared in Jewish exegesis in the 10th century. Whether those exegetes knew anything about Jerome's view I cannot say.

In addition to the Christian connection there is also the Islamic contribution to the discussion: Islamic sources show that at the time of Wahb b. Munabbih and al-Suddī, between the late 7th and mid-8th centuries CE, there already existed a Judaeo-Arabic rendering of the section in Lev. Rabba discussing the En Dor episode. This rendering also included an explicit description of the resurrection of Samuel. Whether this rendering or description had influenced or is associated in any way with the vague allusions in *Pirke de-Rabbi Eliʿezer*, or vice versa, is apparently impossible to determine. Saʿadya and his followers felt that in this situation they had to come up with a systematic rationalistic response, without rejecting the resurrection. Insisting on resurrection while rejecting that the witch had any real part in it was probably considered an appropriate response to the Christian challenge. Others preferred the rejectionist approach that denied any concrete miraculous occurrence.

If my interpretation of the development of the test case is correct, and if it is recognized that the development of Jewish approaches toward this case can hardly be explained by internal Jewish circumstances, then we have here a clue to an understanding of the origins of Jewish mediaeval Bible exegesis. This case may provide a concrete illustration of the brief and abstract reference of al-Qirqisānī to the influence of Christian (Syriac) exegetes on Jewish commentators and a primary source for the philosophical expositions in Jewish mediaeval Bible exegesis.[94] It also illustrates the extent of the revolution that Jewish thought, as represented through Biblical exegesis in our case, underwent in the geonic environment. The extent of this revolution may also be underlined through comparison to the works of exegetes (mainly in Christian Europe, such as Rashi) who lived in different environments, were content with following the old rabbinic traditions, and did not share the ideological concerns and interests of their Oriental contemporaries.

8 Additional Note: On the Terms which Denote Return to Life

It is necessary to distinguish between different terms that denote return to life one way or another.

Revival: In an intransitive sense—to come back to life; in a transitive sense—to restore a dead person to life.

Revivification: Bringing back to life a person considered dead, whose body has not been buried yet, and so has not crossed the border, as it were, between the world of the living and the world of the dead. Examples: the cases of Elijah and the son of the woman of Zarefat (1 Kgs 17:17–24), and the son of the woman of Shunem, who went through a similar experience at the hand of Elisha (2 Kgs 4:8–37).

Resuscitation: A more medical definition of the preceding, especially the cases mentioned there. The term is not used in old translations of the Bible. It relates to living organisms, human or animal bodies or plants.

Resurrection: Bringing back to life persons, body and soul, whose death has been confirmed irreversibly due to the fact that they were buried, and so passed over to the world of the dead. This was the nature of the event that occurred to Jesus. Early sections of the Hebrew Bible seem to be completely unaware of the possibility of such an event actually taking place at all; later sections may be interpreted as envisaging such an event in eschatological circumstances.

---

94 See above, n. 3. Al-Qirqisānī's remark refers to the middle of the 9th century, or perhaps even somewhat earlier, which may coincide with the chronology suggested above.

It is important to distinguish accurately between "resuscitation" and "resurrection". Smelik, p. 172, when discussing the views of Zeno, seems to confuse the two. He makes a similar mistake when, on p. 162, he uses "resuscitation" with respect to Lev. Rabba. The rabbis did not understand necromancy as an art or procedure of bringing dead and buried bodies back to life. The necromancer does not bring a revived body out of a grave, but rather brings up the soul, which is actually more like an apparition, from the underworld of the dead. Smelik, p. 163, quotes an episode from the Babylonian Talmud *Shabbat* 152b–153a:

> A certain heretic [a variant reading has "Sadducee," but the meaning is the same: a denier of the dogma of afterlife] said to Rabbi Abbahu: You maintain that the souls of the righteous are hidden under the Throne of Glory: then how did the bone[-practicing] necromancer bring up Samuel by means of her necromancy? Rabbi Abbahu replied: There it was within twelve months [of death]. For it was taught: For twelve months the body is in existence and the soul ascends and descends; after twelve months the body ceases to exist and the soul ascends but descends nevermore.

This story reflects the shift from the ancient Near Eastern perception of necromancy whereby the soul is brought *up* from the *underworld* of the dead, to an entirely different perception according to which the soul of a righteous person (such as Samuel) should *descend* by itself from heaven, not be brought up or down by a witch. Rabbi Abbahu does not seem to be disturbed by the apparent contradiction. At the next stage of the development, the question of resurrection, i.e., bringing back to life persons who have already been buried is the crux of the matter—certainly by the early Middle Ages.

Lennart Lehmhaus

“Hidden Transcripts” in Late Midrash Made Visible

Hermeneutical and Literary Processes of Borrowing in a Multi-Cultural Context.*

1 Introduction

Since the last decade of the 20th century the paradigm of ‘globalization’ dominates political and media discourses. According to this model, not only goods and information but also cultural ideas and practices travel and merge much faster in a globalized world thanks to new technologies, media, and (partly) open borders. Although the whole discourse of contemporary globalization is fraught with problems and exhibits often a good measure of historical amnesia, its ubiquity has triggered some developments in the academic world. Thus, one could observe shifting approaches and scholarly perspectives in socio-historical and cultural studies from various fields during the last 25 years. Scholars showed an increased interest in historical points of contact between traditions and in processes of cultural transfer and ‘borrowing’. These developments were not any longer conceptualized, as in earlier studies, as taking place between monolithic, often diametrically opposed entities (religions, cultures or nations), but rather as a subtle interplay between different groups or actors with a certain degree of cultural agency, often on a small-scale level.

* I would like to express my gratitude to conference organizers, editors, and the many assistants for their hospitality, efforts and their broad interest that turned this conference into inspiring, inter-disciplinary “academic crossroads”. I am grateful for the helpful questions, remarks and references I received throughout the conference and later on, from many of the other presenters.

This research has gained from my work within the Collaborative Research Center (SFB 980) “Episteme in Motion. The Transfer of Knowledge from Antiquity to Early Modern Times”, funded by the German Research Foundation (DFG). I am indebted to my colleagues at the CRC/SFB 980 and in our project A03 “The Transfer of Medical Episteme in the ‘Encyclopaedic’ Compilations of Late Antiquity”, especially to the heads of project, Philip J. van der Eijk and Markham J. Geller. I worked on the final revision of this paper as a Harry Starr Fellow in Judaica at Harvard University where I benefitted from the resources available at Widener Library and from my conversations with colleagues.

https://doi.org/10.1515/9783110564341-011

The Mediterranean in the early Muslim period (7th – 11th century), labelled recently by Fred Astren as a "cultural stew"[1], provides ample material for fruitful studies into this direction. Also 19th century scholarship, during the birth of the academic discipline called *Wissenschaft des Judentums*, witnessed a short period in which scholars have been interested in the cultural interaction between Arab-Muslim and Jewish traditions following two different approaches. Some classical studies, like Abraham Geiger's from 1833, celebrated the paradigm of linear-chronological influence, already indicated in his title *Was hat Mohammed aus dem Judenthume aufgenommen?* (*What has Muhammad taken over from Judaism?*). This kind of research question did not find many epigones in the following 150 years. Only very recently, and upon very different terms, scholars in Islamic and Semitic studies launched new research initiatives concerned with the mutually shaping and being shaped of Late Antique identities among Muslims, Jews, Christians and others.[2]

Although Ignac Goldziher as a pioneer had studied much of the pre-Islamic traditions of law and Hadith, in his comparative studies he rather looked for the adoption of Arab-Muslim elements among Jews only. Indeed, this should become the main course of research on cultural exchange, albeit focusing mainly on a much later period than the time of emerging Islam. Most studies concentrated on the classical Jewish-Muslim encounters in medieval times, often labelled as the 'Golden Age' or the 'Jewish-Arab Symbiosis'.[3] Rina Drory's study *Mod-*

1 Astren, Fred, "Islamic context of medieval Karaism," in *Karaite Judaism: A Guide to Its History and Literary Sources*, ed. by Meira Polliack, pp. 145–177. Leiden: Brill, 2003, p. 147.

2 The studies of Angelika Neuwirth as well as those of her colleagues and students address the socio-cultural and historical milieu of early Islam. For a helpful overview, see Neuwirth, Angelika, *Der Koran als Text der Spätantike: ein europäischer Zugang.* Frankfurt a. M.: Verlag der Weltreligionen, 2010; Marx, Michael and Nicolai Sinai, eds., *The Qur'an in Context: Historical and Literary Investigations Into the Qur'anic Milieu.* Leiden: Brill, 2010. An innovative and fruitful inquiry into the still early, but slightly later history of the Islamicate world is pursued in various projects designed by Sabine Schmidtke and her collaborators. Just to mention here: Schmidtke, Sabine, *The Bible in Arabic among Jews, Christians and Muslims. Intellectual History of the Islamicate World 1.* Leiden: Brill, 2013; Reynolds, Gabriel S., *The Qur'an and Its Biblical Subtext.* London/New York: Routledge, 2010; Schmidtke, Sabine and Gregor Schwab, eds., *Jewish and Christian reception(s) of Muslim theology. Intellectual History of the Islamicate World 2.* Leiden: Brill, 2014.

3 The studies into the 'Golden Age' are numerous. Some balanced accounts are given in Lewis, Bernard, *The Jews of Islam.* Princeton: Princeton University Press, 2014, esp. pp. 67–106; Stillman, Norman A., *The Jews of Arab Lands.* Philadelphia: Jewish Publication Society, 1991, esp. pp. 40–63; Frank, Daniel, *The Jews of Medieval Islam: Community, Society, and Identity.* Leiden: Brill, 1995. For a study of the rather biased scholarship into those phenomena of Muslim-Jewish-

els and Contacts, exceptional in its topic as in its quality, was the very first to transcend the disciplinary boundaries between Arabic literature and language, History of Islam and Jewish Studies. Drory tried to trace the emergence of 'fictionality' in Arabic culture and literature while comparatively examining appropriations of its cultural models in geonic Babylonia and among Jewish poets on the Iberian Peninsula.[4]

While taking up Drory's masterful and still urging research questions, I will survey the topic of cultural transfers, adaptations and negotiations here from a different angle. The following paper will focus on post-Talmudic, early medieval Hebrew texts (8th – 10th century) and their, often rather indirect, interaction with surrounding (textual) cultures, especially Arabic-Muslim patterns. Embracing the metaphor of 'Exegetical Crossroads', I am looking for those travellers, who in distance from the main crossroads and beneath the big highway junctions move along and who cross boundaries on bumpy roads and less frequented sideways. Before exploring some examples relevant to such subtle cultural transfers, I have to address two important issues in order to frame this discussion.

The first issues at stake are the nature and history of the texts under discussion from a general perspective. All of those texts, that I will discuss, are generally considered to be part of rabbinic literature – what is called the Jewish 'Oral Tradition' or 'Oral Torah (Lore)' in contrast to the 'Written Torah' referring to the *Pentateuch* or rather the Hebrew Bible. Rabbinic literature in general can be characterized by consisting of two major trends. While the so-called Talmudic literature encompasses the Torah-based religious laws and norms (*Halakha*), their elaborations and commentaries (*Mishnah*, *Tosefta*, *Talmudim*), Midrash relates more closely to the Bible with an 'exegetical' (often linear verse-by-verse) or a 'homiletical' (didactic elaboration of important biblical passages for the public reading of Torah or for the sermon in the synagogue) focus. In all Midrashim as well as in the two Talmuds one finds a varying emphasis on narrative-examplary (*Aggada*) or normative-juridical (*Halakha*) elements, which are often inextricably blended.[5] The Talmudic literature confronts every scholar who tries to

---

Christian interaction see, Lassner, Jacob, *Jews, Christians, and the Abode of Islam: Modern Scholarship, Medieval Realities*. Chicago: University of Chicago Press, 2012.

4 Drory, Rina, *Models and Contacts: Arabic Literature and Its Impact on Medieval Jewish Culture*. Leiden: Brill, 2000.

5 For an in-depth survey of the history and development of rabbinic literature, see Strack, Herrman and Günter Stemberger, *Introduction to Talmud and Midrash*. Minneapolis: Fortress Press, 1996. While the production and redaction of Talmudic literature covers several centuries spanning from around 200 (Mishnah) until the 6th or 7th centuries, Midrashic texts had their classical

establish a history of rabbinic texts with immense problems. Or, as Hayim Lapin has expressed it recently: "problems compound one another: materials of undetermined date have been reworked in larger corpora, also of undetermined date."[6] While the use for straightforward historiography seems to be very little, scholars have successfully examined the literary and discursive forms of Talmudic and rabbinic texts and emphasized structural aspects for historical studies.

According to a common assumption, the later Midrashim mark a shift in rabbinic literature from the classical 'Golden Age' of exegesis to a distinct narrative orientation. The works were described as displaying a new form but no innovation in content. Some scholars even took this as evidence for ossification, stagnation and decline of rabbinic literature.[7] Not only is the hierarchic normativity inherent to such a description problematic. Moreover, also an assumption of decline and decay conceals more than it reveals. It seems that such a pure dichotomy of classical exegesis *versus* later, simplistic narrative is rather an outdated construct that has to be thoroughly re-examined. It also disregards the actual productivity of Jewish authors from the 5th century onwards. One can rather observe a flourishing of Jewish texts in a variety of genres (*Talmud*/*Midrash*/liturgical poetry and new literary hybrids), despite the turmoil of religious persecution and violence in Byzantium, Sassanid Persia and the socio-political upheavals during the Arab expansion.[8]

---

peak in amoraic times (middle of 3rd century. to 500), but were authored also after the (Palestinian/ Babylonian) Talmudim well into medieval times.

6 Lapin, Hayim, *Rabbis as Romans: The Rabbinic Movement in Palestine, 100–400 CE*. New York: Oxford University Press, 2012, p. 40.

7 Drory, *Models and Contacts*, pp. 150/152 states: "Rabbanite literature [...] was progressively stagnating, its creative models showing signs of ossification" / ""[N]o new literary material could be admitted in a Rabbanite text unless it would be molded after the classical models." Cf. also Rubenstein, Jeffrey, "From Mythic Motifs to Sustained Myth: The Revision of Rabbinic Traditions in Medieval Midrashim." *Harvard Theological Review* 89, 2 (1996): pp. 131–159, here p. 133: "they contain few ideas not documented in classical rabbinic texts. The innovation of these Midrashim is in the use of narrative." Novelistic tendencies in line with the general decline of the classical Midrash (parable) are described by Stern, David, *Parables in Midrash. Narrative and Exegesis in Rabbinic Literature*. Cambridge, Mass.: Harvard University Press, 1991, pp. 206–224. For a rather neutral survey, see Strack/Stemberger, *Introduction*, pp. 311–350.

8 Cf. Lapin, Hayim, "Aspects of the Rabbinic Movement in Palestine 500–800 C.E," in *Shaping the Middle East. Jews, Christians, and Muslims in an Age of Transition 400–800 C.E.*, ed. by Ken G. Holum and Hayim Lapin, pp. 181–194. Bethesda: University Press of Maryland, 2011, here p. 184: "As noted, precisely this efflorescence of literary output is taken as evidence of decline: the end of the production of new traditions and the mere collection of traditions. This is an unfortunate error. It is clear in some cases that the 'collections' generate new traditions, as in the references to the Early Islamic context in PRE [...]. Moreover, the collection and redaction of rab-

Taking up the very last point, my second comment relates to the socio-historical context of changing realities in the Mediterranean world in the wake of the Arab-Muslim expansion. This context is crucial to this transitional age of Judaism in the geonic period. From the 8th to the 10th centuries the spread of Islam and the unification of a majority of Jews under its dominion brought about many political, economic, socio-cultural and religious transformations. On the one hand, these developments framed the formative period for Talmudic Judaism and the struggle of the Babylonian Geonim for cultural hegemony and maybe also political influence. One of their goals was the establishment of Babylonian religious authority based on the unique source of quasi-codified Talmudic (oral) tradition in their spheres of influence (*reshuyot*) in the Mediterranean Diaspora and beyond.[9]

On the other hand, within a new reality of a unified Mediterranean Jewry this geonic enterprise encountered several serious challenges. These included not only the traditional religious and cultural struggle with the second rabbinic centre in Palestine. Severe tensions also arose between the geonic strongholds (esp. *Babylonian Yeshivot*) and new emerging regional centers at the margins (e.g., North Africa and Iberia), which sometimes switched their loyalty and, beginning in the 9th century, developed a sense of political and religious independence.[10]

---

binic material in the fifth century and later places it in broad continuity with the codificatory impulse that generated the imperial codes, the rediscovery of philosophical diadokhai, and the formation [...] of the authority of the 'fathers'."

9 Cf. Gil, Moshe, *A History of Palestine, 634–1099*. Cambridge: Cambridge University Press, 1992, esp. pp. 527–549. This Geonic claim to power was channeled through religious and juridical expertise disseminated in Geonic responsa and a tight inter-communal network of money-collectors, messengers and representatives. See also Brody, Robert, *The Geonim of Babylonia and the Shaping of Medieval Jewish Culture*. New Haven: Yale University Press, 1998, pp. 35–53, 83–99 and 123–134. An interesting contemporary parallel is the increasing claim to leadership by Muslim religious specialists (the *Ulama*). Cf. Astren, "Islamic Contexts", pp. 156–158; Lapin, "Aspects", p. 193. Fishman, Talya, *Becoming the people of the Talmud: Oral Torah as Written Tradition in Medieval Jewish Cultures*. Philadelphia: University of Pennsylvania Press, 2011 shows that it was precisely through the endeavor of the Geonim and their medieval successors that the Babylonian Talmud became the focus of normative, rabbinic Judaism for the centuries following.

10 Ben-Sasson, Menahem, "Inter communal relations in the geonic period," in *The Jews of Medieval Islam. Community, Society and Identity*, ed. by D. Frank, pp. 17–31. Leiden: Brill, 1995 stresses the importance of local and family religious identities as well as economic networks of long distance trader dynasties who played a major role in the dissemination of Talmudic Judaism, but also challenged or influenced it by supporting marginal, sectarian traditions and movements. The trend of localization in Judaism was paralleled in the emergence of new Muslim centers (like Kairouan, and the Persian or Central Asian Jewish cities) due to economic weakness

Moreover, the self-declared, geonic 'mainstream' was confronted with a variety of local customs and cultural traditions reintroduced by masses of non-rabbinic Jews who came from socially and educationally diverse backgrounds in the peripheries, which were now parts of the new Islamicate cultural sphere.[11]

According to studies of Steven Wasserstrom and others, the plurality of early Islam, including the cultural heritage of ancient Arabian, Persian, Christian, Jewish and Gnostic cults and cultures, brought forth or enhanced syncretistic, symbiotic or hybrid lifestyles and identities of some Jews, who, thus, became important cultural intermediaries and innovators.[12] Against this backdrop, scholars started to re-think the incipient stages of the so-called 'sectarian movement' of the Karaites, which was seen as opposed to rabbinic or 'Rabbanite' Judaism, but was in itself no monolithic entity. Their formative stage (8th – 9th centuries) was contemporaneous with the production of the late Midrashic works. Scholars rather describe this branch today as a conglomerate of different sub-movements, which only gradually merged into a religious joint venture. It was formed by the alternative rabbinic party of the Ananites, coming from a learned and elitist, geonic milieu of religious and political leaders. In the 9th century they joined forces with different Jewish messianic, scripturalistic or ascetic movements (Mourners of Zion, al-Ishafani, Benjamin al-Nahwandi, Daniel al-Qumisi) and merged into the more coherent movement of Karaism that developed distinct theological and hermeneutic concepts.[13]

---

and the decline of centralistic control in the Caliphate. Cf. Astren, "Islamic Contexts", pp. 156–158.

11 Cf. Astren, "Islamic Contexts", p. 161:"In terms of law, the new "cultural stew" that resulted from the interaction of a wide variety of Jewish practice in the eigth and ninth centuries posed serious challenges for the rabbis as they sought to consolidate their own system." On the halakhic creativity, see Libson, Gideon, "*Halakha* and Reality in the Geonic Period: Taqqanah, Minhag, Tradition and Consensus: Some Observations." in *The Jews of Medieval Islam. Community, Society and Identity*, ed. by D. Frank, pp. 67–99. Leiden: Brill, 1995.

12 Cf. Wasserstrom, Steven, *Between Muslim and Jew: The Problem of Symbiosis under Early Islam*. Princeton: Princeton University Press, 1995; see also Astren, "Islamic Contexts." For a similar observation regarding the porous cultural and conceptual boundaries between Muslims and Syriac Christians during the formative period of the 7th to the 9th centuries, see Penn, Michael P., *Envisioning Islam: Syriac Christians and the Early Muslim World*. Philadelphia: University of Pennsylvania Press, 2015, esp. pp. 142–182.

13 See Gil, Moshe, "The Origins of the Karaites," in *Karaite Judaism*, ed. by M. Polliack, pp. 73–117. Leiden: Brill, 2003; Ben Shammai, Haggai, "Between Ananites and Karaites: Observations on Early Medieval Jewish Secterianism," in *Studies in Muslim-Jewish Relations Vol. 1*, ed. by R. L. Nettler, pp. 19–29. Chur: Harwood Academic Publishers, 1995; Polliack, Meira, "Medieval Karaism" in *The Oxford Handbook of Jewish Studies*, ed. by M. Goodman, 295–326. Oxford: Oxford University Press, 2005; Gil, Moshe, *History of Palestine*, pp. 777–790; Astren, Fred, *Karaite Judaism and*

Thus, the emergence of late Midrash, precisely in early Muslim and geonic times, coincides with radical historical transformations. In this formative period for the three 'people of the book', we find among Jews, Christians and Muslims different, at time dissenting schools of thought or religious branches instead of one unified, monolithic religion. Moreover, many ascetic or messianic movements as well as religiously hybrid groups stood in competition and dialogue with each other.[14] Some recent scholarship argues that rabbinic culture was but one of many facets of late antique and early medieval Jewry.[15] This rabbinic formative process as well as its struggle for influence triggered the emergence or re-awakening of various non-rabbinic trends. These groups preferred alternative modes of learning, maintained or developed different religious traditions in dialogue with their Jewish and rabbinic counterparts as well as with Christian and Islamic cultural models.[16] This setting was complicated by the internal competi-

---

Historical Understanding. Columbia: University of South Carolina Press, 2004, pp. 5–9 and 65–75; Astren, "Islamic Contexts", pp. 165–166, points to six disparate elements within the proto-Karaite milieu: 1) Annanites; 2) peripheral Jewish traditions; 3) Judeo-Muslim or other hybrid identities; 4) anti-rabbinic scripturalists (the *bene miqra* in Pirqoi ben Baboi's polemic?); 5) remnants of messianic movements (*Isawiyya*/*al-Ishafani*); 6) (semi-) ascetic trends among Palestino-centric Jews (e.g., *Mourners of Zion*). From 950 CE onwards, Muslim authors refer to thirteen, five or four different sects or schools within Judaism. For such an external heresiographical picture, see Adang, Camilla, "The Karaites as Portrayed in Medieval Islamic Sources," in *Karaite Judaism*, ed. by M. Polliack, pp. 179–197. Leiden: Brill, 2003.

14 Cf. Wasserstrom, *Between Muslim and Jew*, esp. pp. 3–91. He emphasizes the role of hybrid, so-called sectarian groups like the *Isawiyya* or *Maghriyya* and other messianic or mystical movements. See also Erder, Yoram, "The Doctrine of Abu 'Isa al- Isfahani and Its Sources." *Jerusalem Studies in Arabic and Islam* 20 (1996): pp. 162–199. Another supporting argument provides Anthony, Sean W., "Who was the Shepherd of Damascus? The Enigma of Jewish and Messianist Responses to the Islamic Conquests in Marwānid Syria and Mesopotamia," in *The Lineaments of Islam: Studies in Honor of Fred McGraw Donner*, pp. 21–59. Leiden: Brill, 2012; and Id., "Chiliastic Ideology and Nativist Rebellion in the Early ʿAbbāsid Period: Sunbādh and the Jāmāsp-Nāmah." *Journal of the American Oriental Society* 132, 4 (2012): pp. 641–655.

15 Seth Schwartz suggests a rather late process of 'rejudaization' or 'rabbinization' in late-antique Jewry (*Imperialism and Jewish Society, 200 B.C.E. to 640 C.E.* Princeton/Oxford: Princeton University Press, 2002, pp. 240–74). Others argue more restrainedly for a slow broadening of influence via the backbone-network of rabbinic households and circles in close interaction with and the support from non-rabbinic Jews. Cf. Hezser, Catherine, *The Social Structure of the Rabbinic Movement in Roman Palestine*. Tübingen: Mohr Siebeck, 1997, pp. 155–239; 307–27; Miller, Stuart S., *Sages and Commoners in Late Antique 'Erez Israel: A Philological Inquiry into Local Traditions in Talmud Yerushalmi*. Tübingen: Mohr Siebeck, 2006, pp. 446–466.

16 Ben Shammai, Haggai, "Between Ananites and Karaites: Observations on Early Medieval Jewish Secterianism" in *Studies in Muslim-Jewish Relations*, ed. by R. L. Nettler, pp. 19–29. Chur: Harwood Academic Publishers, 1995; Polliack, Meira, "Medieval Karaism," in *The Oxford Handbook of Jewish Studies*, pp. 295–326.

tion between old rabbinic strongholds of learning (the *Yeshivot*) in Palestine and Babylonia, newly arising intellectual centers (e.g., in North Africa) and a diversity of local or marginal Jewish traditions.[17]

In the remainder of this paper I will present an initial, and still preliminary, survey of the most striking literary and hermeneutic aspects in some post-Talmudic traditions (often subsumed under the umbrella-term 'Late Midrash') exhibiting a great deal of cross-fertilization between Jewish rabbinic and non-rabbinic, Arabo-Muslim and Arabo-Christian cultural models. I would like to briefly introduce and illustrate some of these features by discussing some short samples from three different traditions. Some of them exhibit a closer relation to exegesis, while others belong to the broader field of narrative and moral treatises. My special emphasis lies on *Seder Eliyahu Rabba* and *Zuta* (in the following: SER/SEZ – *The major and minor order of Elijah*), two puzzling and fascinating sibling texts of a hybrid character, most probably to be dated in the 9th or 10th century[18]. Moreover, I will amend my discussion with examples from two other traditions. First, *Pirke de-Rabbi Eliezer* (in the following: *PRE)* – a complex work that combines the re-telling and re-arrangement of biblical episodes with complex hermeneutic and exegetical strategies as well as with new 'scientific' interests. This tradition is commonly dated to the 8th century.[19] Second, the medieval *Stories of Ben Sira* (or *Alphabet of Ben Sira*; thus, in the following: ABS), dated to the 10th or 11th century, deploy the adventures of the child genius Ben Sira to enfold a negotiation of Jewish identities paired with critical allusions to Wisdom traditions or rabbinic hermeneutics and many borrowings from Arabic literary culture. Thus, the au-

17 Gil, Moshe, *A History of Palestine, 634–1099.* Cambridge: Cambridge University Press, 1992, esp. pp. 490–776. This gaonic claim to power was consolidated through religious and juridical expertise disseminated in *responsa* and a tight inter-communal network of money-collectors, messengers and representatives. See also Brody, *The Geonim*, pp. 35–53, 83–99 and 123–134. As a parallel, one can read the claim to leadership of religious specialists (the *Ulama*) in Islam. Cf. Astren, "Islamic Contexts." pp. 156–158. The interplay and struggle between center and periphery is discussed in Menahem Ben-Sasson, "Inter Communal Relations in the Gaonic Period," see above.

18 The suggestions range between 3rd-century Palestine, Babylonia in the 5th or 9th century, and11th-century Italy. The absence of attributions to rabbinic sages, confusion of two titles (*Seder Eliyahu*/*Tanna devei Eliyahu*) as well as a puzzling transmission history complicates the picture. A concise summary of the Hebrew scholarship is given in the introduction to the English translation by W. G. Braude and I. J. Kapstein, *Tanna debe Eliyyahu: The Lore of the School of Elijah.* Philadelphia: Jewish Publication Society, 1981, pp. 3–15.

19 A somewhat outdated English translation but still a pioneering introduction provides Friedlander, Gerald, *Pirḳê de Rabbi Eliezer (The chapters of Rabbi Eliezer, the Great): according to the text of the manuscript belonging to Abraham Epstein of Vienna/translated and annotated with introduction and indices by Gerald Friedlander.* New York: Bloch, 1916.

thors tried to achieve a balance between critique and distance to the new dominant culture (resp. Arab-Muslim/ Hellenistic-Pagan) on one side and great proximity and openness for literary and cultural models provided by these hegemonic systems on the other.[20]

2 Hermeneutic and Exegetical Approaches in Dialogue

2.1 Beginnings of a 'Science of Language' Between Arabic, Aramaic and Hebrew

A remarkable feature of later texts is the intensified interest in language that transcends the sphere of mere exegetical wordplays and puns. Also in SEZ one finds an increased attentiveness to linguistic features of the Hebrew language. A very instructive example can be found in SEZ, chapter 14:[21]

> Among those reading [Scripture] and reciting/repeating [Mishnah][22], who impoverishes himself (מוורש את עצמו)[23]? He who spends much time eating, drinking [and sleeping]. As

20 A critical edition of this work in Hebrew was produced by Yassif, Eli, *The Tales of Ben Sira in the Middle-Ages: A Critical Text and Literary Studies* [Hebr]. Jerusalem: Magnes Press, 1984. Based on this is Börner-Klein, Dagmar, *Das Alphabet des Ben Sira, Hebräisch-deutsche Textausgabe mit einer Interpretation.* Wiesbaden: Marix Verlag, 2007. An English translation of the first part (the so-called *Stories of Ben Sira*), which mingles the two recensions though, can be found in Stern, David and Mark Mirsky, eds., *Rabbinic fantasies.* New Haven: Yale University Press, 1998, pp. 167–202. Some of the characteristic features of ABS' discourse are discussed in Yassif, Eli, "Pseudo Ben Sira and the 'Wisdom Questions'Traditions in the Middle Ages." *Fabula* 23, 1/2 (1982): pp. 48–63; Börner-Klein, Dagmar, "Transforming rabbinic exegesis into folktale." *Trumah* 15 (2005): pp. 139–148; Lehmhaus, Lennart, "'Es ist mancher scharfsinnig, aber ein Schalk und kann die Sache drehen, wie er es haben will.' (Sirach 19,22) – Intertextuelle Kritik rabbinischer Quellenarbeit im Alphabet des Ben Sira," in *Literatur im Dialog: die Faszination von Talmud und Midrasch*, ed. by Susanne Plietzsch, pp. 127–163. Zürich: Theologischer Verlag, 2007; Id., "Ways of Wisdom – The medieval Ben Sira tradition as a Bildungsroman in its historical context," in *Proceeding of 15th World Congress of Jewish Studies 2009, Jerusalem* (online, forthcoming).

21 SEZ, Ch. 14, pp. 196–197.

22 In *Seder Elijahu* (SE) the term *mishnah* (משנה) and the related verb (שונין) seldom refer to the corpus of rabbinic law known to us as the Mishnah. In fact the reference aims at the whole tradition of rabbinic lore or Oral Torah which complements and builds upon the Written Law or Scripture (referred to in SE via the terms *Miqra*/ מקרא and reading/ קוראין).

23 To emend: מוריש. One meaning of the Hif'il is: to impoverish someone.

is said: *For the drunkard and the glutton will come to poverty (כי סבא וזולל יורש), and slumber will clothe them with rags.* (Prov 32:21) And as is said: *Love not sleep, lest you come to poverty (אל תאהב שנה פן תורש); open your eyes, and you will have plenty of bread (פקח עיניך ושבע לחם).* (Prov 20:13)

And bread does not refer to anything else than words of Torah, since it is said:

Come, eat of my bread (לכו לחמו בלחמי) [and drink of the wine I have mixed]. (Prov 9:5)

And [Scripture] says: *The Sovereign LORD has given me an instructed tongue (לשון למודים)*, to know the word that sustains the weary (לעות את יעף דבר). (Isa 50:4) Haven't you learned that words of Torah (דברי תורה) are only absorbed by one who wearies himself for them (שהוא עיף להם).

Thus the spirit of the Holy proclaimed good news to the disciples of the sages and says to them:

My children, though I have given to you a good Torah in this world, I need not to say [anything concerning the Torah] for the world-to-come. Though there is a double reward for you in this world, I need not to say [anything concerning the reward] for the world-to-come. Do not belittle the Torah and do not increase you eating and drinking, since it is said: *Return to your stronghold, Oprisoners of hope; today I declare that I will restore to you double (משנה אשיב לך).* (Zech 9:12)[24]

And [Scripture] says: *And I will repay you twice (ושלמתי לכם את השנים).* (Joel 2:25–26)[25]

And [Scripture] says: *Instead of their doubled shame (בשתכם משנה),*[26] *disgrace and dishonor in their land they shall inherit a double portion (משנה יירשו); they shall have everlasting joy* (Isa 61:7)

Haven't you learned that you will savor a *double portion in the days of the anointed (moshiah)?* Even if only a single individual of the people of Israel were dwelling at the end of the world, and a thousand rivers before him, the Holy One, blessed be he, would divide them and bring him to himself [...]

And whence [do we know/learn]? You shall know that it is like this (מנין תדע לך שכן):

For Israel will not be redeemed because of suffering, not because of the bondage, not because of [forced] migration, not because of expulsion, not because of need, not because of hunger, but because of ten men who are together; and one is reading to his fellow and his voice is not heard. Since it is said: *But on Mount Zion will be deliverance; it will be holy and the House of Jacob, they will take possession of their dispossessors (וירשו בית יעקב את מורשיהם).* (Obad 1:17) Thus, we found that our ancestors [...]

And whence [do we know/learn]? You shall know that it is like this (מנין תדע לך שכן) that Israel will be redeemed only if they increase and multiply and thus rise to a plentifulness in the world. As it is said: *For you will spread out to the right and to the left; your descendants will dispossess nations (וזרעך גוים יירש) and settle in their desolate cities.* (Isa 54:3)

Throughout this passage we find six instances in which different verb forms of the root י'ר'ש' (*ya'ra'sh*) are applied. The first appearance is in a short teaching about the moderate lifestyle of the students opens with: "Who impoverishes

24 Or: I will restore you your learning of oral lore (*mishnah*).

25 Or: I will repay you/restore you the years [of wearing yourself in study].

26 Or: For the shame [you endured because of your devotion to] *mishnah*.

himself" (*morish et atzmo*). The two answering biblical quotations provide other verb forms (Pr 23:21: "*will come to poverty [yiwaresh]*"/Pr 20,13: "*lest you come to poverty [pen tiwaresh]*") that also define the word field of *ya'ra'sh* negatively in a context of admonition. Both, question and prooftexts help to narrow down the semantic range of the verb and build a link between material and spiritual poorness. Material neediness and destitution are valued throughout SEZ and also in the following paragraph as a *midda tova* (good measure/ virtue). The next instance is devoted to God's proclamation of a double reward for those who concentrate on Torah and refrain from excessive consumption and sleeping. Here, Isa 61:7 draws on the root *ya'ra' sh* with a positive connotation. The reward of the students is described as follows: "*therefore in their land they shall inherit a double portion (משנה יירשו); they shall have everlasting joy (ḥelkam lakhen be-artzam mishne yirashu, simḥat olam tihiye lahem).*"[27] In a reappraisal this verse links the topic of abstention and austerity/deprivation back to *ya'ra'sh* with the meaning inheritance. Simultaneously, this key-word is here enmeshed in a complex wordplay, based on homonymic and homographic features, on the double reward (*mishne/shnayim/ kaful*) for the study of rabbinic lore (*mishnah*).

The next application is introduced by a teaching that highlights individual and collective study as the main precondition for divine salvation. The quoted verse describes Israel's eschatological return to Zion. The usual translation, "*and the House of Jacob, they will take possession (we-yarshu) of their inheritance (et mor**a**sheihem),*" would be just a variation of the theme of inheritance, already alluded to in the Isaiah-verse. However, a different reading, recommended in the Biblia Hebraica (BHS) and by Gesenius and other dictionaries discloses additional dimensions: "*and the House of Jacob, they will take possession of their dispossessors (we-yarshu ... et mor**i**sheihem).*" Thus, the verse adds a further meaning as "taking possession" and "dispossessing/expelling" to the semantic range of *ya'ra'sh*. Moreover, this reading bestows upon the statement a circular understanding of eschatological events. The motif of return to Zion and reward is reinforced by the reversal of present balance of power between Israel and their oppressors.

The final application of *ya'ra'sh* combines the promise of Israel growing to a powerful people with the motif of re-conquest: "*For you will spread out to the right and to the left; your descendants will dispossess nations and settle in their desolate cities.*" *(we zar'ekha goyim yirash).*

---

27 Isa 61:7: *Instead of your shame there shall be a double portion; instead of dishonor they shall rejoice in their lot; therefore in their land they shall possess a double portion; they shall have everlasting joy.* The BHS reads: תחת בשתכם משנה וכלמה ירנו חלקם לכן בארצם משנה יירשו שמחת עולם תהיה להם.

More than two third of the chapter's (SEZ 14) compositional structure depends on the key-element of *ya'ra'sh.* As connecting devices the different forms of the verb explore the semantic field of the root, while tying together ethical, intellectual and eschatological topics, using a wide range of rhetorical devices like maxims, prophecy-like speech and micro-exegesis. The literary strategy is twofold and mutual reinforcing. On the plain level, the text uses different morphological forms of a key term, merged into a strong rhetoric cluster of interconnected teachings. As such the word forms are of crucial importance for the text's construction and conveyance of meaning. At the same time, by means of application within the moral-religious context the semantic range of the root is explored and outlined to all intents and purposes.[28]

It is quite obvious that this skillful exploration of semantic fields for didactic purposes owes much to the long tradition of rabbinic exegesis in earlier Midrash. Already this classical Midrashic literature had been always preoccupied with artful 'close readings' of biblical narrative, including its language. However, the increased attentiveness to linguistic features in SE and other texts can be better understood by looking at the cultural contexts in early Muslim time. Usually Sa'adya Gaon (ca. 850 – 932) has been seen in both, Jewish tradition and modern scholarship, as the pioneer of systematic Hebrew lexicography and grammar. However, sources recently discovered by Geoffrey Khan suggest that even earlier, in the 9th century, Jewish (maybe proto-Karaite) grammarians wrote in Hebrew throughout Persia and Iraq (mainly in Isfahan, Tustar, Kufa and Basra), from where also the Arabic grammatical tradition had emerged since at least the 8th century.[29] Based on these early beginnings, the influential Jerusalem school of

---

28 Another striking example of semantic survey can be found in the opening chapter SEZ 1. In this chapter the text sets its agenda by elaborating its main literary formats (parable/aphoristic saying/ first-person-narrative etc.) and its key-concepts, among them the ambiguous term *Zedaqa* (justice/ righteousness/charity). Two intertwined lists explore the semantic range of this word by the help of cognate words or concepts like *zedeq* (justice/rigtheousness), *mishpat* (justice), or *shalom* (peace.) This minute investigation enables the text to promote righteousness and justice as preconditions for charity. And the other way round, applied charity is the concrete, material and practical expression of a righteous and just mind-set. For a more detailed discussion, see Lehmhaus, Lennart, "*Listenwissenschaft* and the encyclopedic hermeneutics of knowledge in Talmud and Midrash." in *In the Wake of the Compendia: Infrastructural Contexts and the Licensing of Empiricism in Ancient and Medieval Mesopotamia*, ed. by John C. Johnson, pp. 59 – 100. Berlin/New York: de Gruyter, 2015.

29 On the early interests in Arabic linguistics and grammar and the beginnings of a science of language in the 1st centuries of Islam, see Olszowy-Schlanger, Judith, "The Science of Language among Medieval Jews," in *Science in Medieval Jewish Culture*, ed. by Gad Freudenthal, pp. 359 – 424. Cambridge: Cambridge University Press, 2011.The scholarly accepted paradigm of a linguistic and even cultural dichotomy or rivalry between the cities Kufa and Basra has been profound-

Karaite scholars, founded mostly by migrants from the East, who engaged in linguistic and grammarian studies, flourished since the 10th century.[30] One cannot discuss these early Jewish grammarians without mentioning those Masoretic circles that engaged in the study and explanation of Scripture and created rules for vocalization, diacritical notes and rhetorical-hermeneutical exegesis in works called *Diqduqe ha-te'amim* (*Fine points of the accents*) or *Diqduqe ha-Miqra* (*Fine points of Scriptural investigation*). Although the history and origins of the Masoretes remain rather obscure – some propose a priestly, others a pharisaic-rabbinic milieu – scholars have pointed to the close ties between them and the early grammarians.[31]

Of relevance for our discussion is the distinct character of the early Jewish (mostly 'Karaite') works on grammar, as described by Geoffrey Khan and Judith Olszowy-Schlanger. These texts were still far from being well-ordered, systematic and abstract reflections on the language system. In fact, not only in those earlier, but even in later treatises on language one finds a tight nexus between linguistics, exegesis, and philosophy which served and complemented each other.[32] Thus, the early grammarian Yusuf Ibn Nuh, writing his *al-Diqduq* already in (Judeo-) Arabic with Hebrew grammatical terms, preferred an analysis of grammatical structures and morphology in relation to the biblical text through description of their literal or intentional meaning (*tafsīr* and *ma'na/murad*) and their contextual application, while relying on earlier rabbinic, Masoretic and

ly challenged by Bernard, Monique, "Medieval Muslim Scholarship and Social *Network* Analysis: A Study of the *Basra/Kufa* Dichotomy in Arabic Grammar," in *Ideas, images, and methods of portrayal: insights into classical Arabic literature and Islam*, ed. by Sebastian Günther, pp. 129–140. Leiden: Brill, 2005. In general on the early Arabic Grammarians, see Günther, Sebastian, "Pioneers of Arabic linguistic studies," in *In the Shadow of Arabic: The Centrality of Language to Arabic Culture*, ed. by Bilal Orfali, pp. 197–220. Leiden: Brill, 2011; Marogy, Amal Elesha, *The Foundations of Arabic Linguistics: Sībawayhi and Early Arabic Grammatical Theory.* Leiden: Brill, 2012 and Versteegh, Cornelis, *Arabic Grammar and Qur'ānic Exegesis in Early Islam*. Leiden: Brill, 1993, esp. pp. 1–40 and 160–190.

30 See Khan, Geoffrey, "Abu al-Faraj Harun and the early Karaite grammatical tradition." *The Journal of Jewish Studies* 48 (1997): pp. 314–334. As possible precursors to these efforts one has to mention the attempts by earlier Geonic scholars to write some dictionaries which in an encyclopaedic guise explained also ideas, concepts and terminologies from Biblical and Talmudic literature. Cf. Brisman, Shimeon, *A History and Guide to Judaic Dictionaries and Concordances*. New York: KTAV Publishing House, 2000, pp. 1–17.

31 Cf. Khan, Geoffrey, "The early eastern traditions of Hebrew grammar," in *Hebrew Scholarship and the Medieval World*, ed. by Nicholas De Lange, pp. 77–91. Cambridge: Cambridge University Press, 2001, esp. pp. 78–81.

32 Cf. Khan, Geoffrey, ed., *Exegesis and Grammar in Medieval Karaite Texts*. Oxford: Oxford University Press, 2001.

Arabic methods of linguistic inquiry.[33] This emphasis on contextualization resembles in some ways the strategy of the semantic survey in SEZ, as described above.

In fact, the early Jewish grammarians provided a grammatical commentary to the Bible focusing on difficult grammatical questions (in their Arabic terminology: *masā'il*), which seems similar to the narrative philological expositions in PRE. According to some recent studies, this Midrash transcends the exegetical or homiletic focus of its precursors while integrating different other discourses and knowledge with an increasing interest in linguistic features. Dagmar Börner-Klein and Ute Bohmeier even labelled PRE a "philological Midrash." According to their studies, PRE is particularly concerned with a clarification of biblical verses that contain a morphological, lexical, grammatical or semantic problem. Most often, a minute examination of *Hapax legomena* seems to be integrated into a narrative scaffold, based on biblical episodes.[34] As demonstrated above, also in SEZ we can detect the deployment of rare, unusual or singular words or expressions as triggers for multiple discourses. Those combine semantic inquiry of various types with moral and theological expositions.

In addition, also for Sassanian-Iranian culture and Syriac literature scholars have pointed to a "development of a multifaceted metalinguistic awareness, concerned both with formal (phonological, morphological, and grammatoogical) and semantic issues, which however did not give rise to a systematic grammatical tradition [...]." Such texts exhibit an awareness for etymological, semantical, morphological aspects as well as for the dimension of written language and the dimension of contact, evolution and competition in a multilingual culture.[35]

---

33 On Yusuf Ibn Nuh's methodology, see Khan, Geoffrey, "The Book of Hebrew Grammar by the Karaite Joseph ben Noah." *Journal of Semitic Studies* 43,2 (1998): pp. 265–286; Khan, Geoffrey, *The Early Karaite Tradition of Hebrew Grammatical Thought*. Leiden: Brill, 2000, esp. pp. 1–25 and 132–140.

34 For this approach to PRE, cf. Börner-Klein, Dagmar, *Pirke de-Rabbi Elieser. Nach der Edition Venedig 1544 unter Berücksichtigung der Edition Warschau 1852. Aufbereitet und Übersetzt*. Berlin/New York: de Gruyter, 2004, pp. XXV–XLVIII; Bohmeier, Ute, *Exegetische Methodik in Pirke de-Rabbi Elieser, Kapitel 1–24*. Frankfurt a.M. [u.a.]: Peter Lang, 2008, esp. pp. 1–34; 439–465.

35 Contini, Riccardo, "Aspects of Linguistic Thought in the Syriac Exegetical Tradition," in *Syriac Encounters. Papers from the Sixth North American Syriac Symposium, Duke University, 26–29 June 2011*, ed. by M. Doerfler et al., pp. 91–117. Leuven: Peeters, 2015, here p. 103. For the Syriac and Persian realm of a shared culture, cf. Becker, Adam H., "Beyond the Spatial and Temporal limes: Questioning the 'Parting of the Ways' outside the Roman Empire," in *The Ways that Never Parted: Jews and Christians in Late Antquity and the Early Middle Ages*, ed. by A.H. Becker and A.Y. Reed, pp. 373–392 Tübingen: Mohr Siebeck, 2003; Id., *Fear of God and the Beginning of Wisdom. The School of Nisibis and the Development of Scholastic Culture in Late Antique Mesopota-*

To sum up, one can imagine that the early Karaite-Jewish approaches to lexicography and grammar as the central aid for understanding Scripture, accompanied by similar interests in Syriac and Arabic discourses, might have appealed authors whose Midrashic discourse focused on the Bible and who displayed a keen interest in language and rhetoric. More cultural overlaps, at least of some degree, may have existed between those authors and other Jewish scripturalistic or Masoretic groups who during the 8th and 9th centuries produced Hebrew texts on grammar and hermeneutics of language.[36] I would like to suggest that those later (Midrashic) texts may indeed be regarded as a fruitful venue to study the beginnings of Hebrew language sciences, enmeshed in exegetical, ethical, and various other discursive frameworks. This proposed new line of inquiry achieves support from Syriac (and other) textual traditions, in which "fertile strands of lingusistic reflection have been developed within the context of other branches of learning, most frequently philosophy, logic, rhetoric, and the exegesis of sacred and canonical texts."[37]

2.2 Return to the Bible and Return to Hebrew

The increased linguistic interest of Jewish authors into the language of Scripture is but one of many facets of a broader cultural development evident in later Midrashic traditions. This change can be termed a 'biblical turn', or possibly 'return to Scripture'. Many later texts (like SE/PRE/Ma'asse Torah – traditions) exhibit a strong orientation towards Scripture as the sole and essential bedrock of their

mia. Philadelphia: University of Pennsylvania Press, 2006, esp. pp. 1–21 and 169–202; Elman, Ya'akov, "Middle Persian Culture and Babylonian Sages: Accommodation and Resistance in the Shaping of Rabbinic Legal Tradition," in *The Cambridge Companion to the Talmud and Rabbinic Literature*, ed. by C.E. Fonrobert and M.S. Goodman, pp. 165–197. Cambridge: Cambridge University Press, 2007.

36 An outstanding example is a text called *Diqduqe ha-Miqra* (*Fine points of Scriptural investigations*). The text has been published by Allony, Nissim, "A Karaite list of terms from the eigth century [Hebrew]," in *Writings of the Association for Biblical Research in Israel. In Memory of Dr. J.P. Korngruen*, ed. by A. Wieser and B.Z. Luria, pp. 324–63. Tel Aviv, 1964. Its contents and structure are also discussed by Khan, Geoffrey, "The Karaite Tradition of Hebrew Grammatical Thought" in *Hebrew Study from Ezra to Ben-Yehuda*, ed. by W. Horbury, pp. 186–205. Edinburgh: T&T Clark, 1999.

37 Contini, *"Aspects of linguistic thoughts"*, p. 94. For the incorporation and appropriation of Greek language science related knowledge as well as of other scientific ideas in a completely religious discourse, namely biblical translation and commentary, see Salvesen, Alison, "Scholarship on the margins: biblical and secular learning in the work of Jacob of Edessa," in *Syriac Encounters*, ed. by Doerfler et al., pp. 327–344.

discourse, while deviating from the common form of classical Midrashic compositions. In consequence, PRE and SE focus not on exegesis but rather on biblical episodes and figures, which are deployed exhaustively for illustration and legitimation of teachings.

By contrast, the classical didactic genres and rhetoric devices of rabbinic-Talmudic literature are tellingly missing. In SEZ for example, we find virtually no narratives about Talmudic sages (as exempla or proof-stories) and no precedents or case-stories (*ma'asse*). Moreover, the typical exegetical series of comments introduced by *davar acher* (lit. "another comment/opinion") and often attributed to a named rabbi or introduced as the *opinio communis* cannot be found. Thus, one should generally notice the striking absence of attribution of teachings to named rabbinic sages throughout this work. This renders SEZ' discourse quite different from the polyphonic discourse of earlier Midrash and Talmudic dialectics. I suggest that these features entail a twofold discursive strategy. While the adoption of selected, generic and structural features and terminology ensures the connection to the rabbinic chain of tradition, the eschewal of other elements serves to keep a certain distance to a socio-cultural milieu and to institutions associated too closely with the geonic academic circles. Such a noncommittal program would make the later Midrashic discourses accessible to a broader audience of interested rabbinic and non-rabbinic Jews below the level of a small scholarly elite.[38]

A second change is closely related to the 'return to Scripture' and to the linguistic attentiveness. Already Ephraim Urbach and Ya'aqov Elboim demonstrated that the pristine and flowery Hebrew of SE does not serve pseudepigraphical purposes. The choice of language seems not to aim at pretending to belong to a tradition from the biblical or early rabbinic days of old. In fact, this 'return to Hebrew' or 'Hebrew renaissance' appears to be distinctive of many later Midrashim in general. Indeed, SE's Hebrew appears to be a slightly biblicistic, literary language featuring an individual and creative style.[39] While vocabulary and terminology from biblical, or early and later rabbinic traditions is adapted, one finds neither foreign loanwords nor the mix of Hebrew and Aramaic so typical

---

38 For a more detailed discussion of these issues, see Lehmhaus, Lennart, "Making others ourselves(') – rabbinic identity formation in contexts of challenge: the tradition of Seder Eliyahu." (forthcoming); Id., "'The Birth of the Author' – literary strategies and discursive developments in later midrashic texts." (forthcoming).

39 Urbach, Ephraim, "On the question of the language and the sources of the book *Seder Eliyahu*." (Hebr.) *Leshonenu* 21 (1956): pp. 183–197, here p. 184; Elbaum, Jacob, "Zwischen Midrasch und Mussar-Literatur: Studien zu den Kapiteln 1–6 in Tanna debe Eliyahu (Hebr.)." *Jerusalem Studies in Hebrew Literature* 1 (1981): pp. 144–154, esp. pp. 144–146.

for earlier Talmudic and Midrashic traditions. This new tendency towards an exclusive use of Hebrew can be found also in several other traditions dating between the 8th to the 10th centuries. Among those texts are not only *PRE* and *ABS*, but also *Midrash Yonah*, *Midrash on the Ten Commandments*, *Midrash Tadshe*, *Mishnat R. Eliezer*, and the so-called *Ma'asse Tora*-traditions. But also other non-Midrashic, but exceptional works like the scientific texts *Baraita di-Shemuel*, *Sefer Yetzira*, *Mishnat ha-Middot* (late 8th and early 9th centuries), the traditions of *Eldad ha-Dani* (late 9th century) and even the 8th century geonic halakhic compendium called *She'iltot* (*'Questions'*) share this new and exclusive preference for Hebrew.[40]

In the specific period under discussion, Aramaic was still the standard of Talmudic-geonic culture and especially Arabic or Judeo-Arabic were to become soon the main languages of intellectual and daily life for Jews and non-Jews. Therefore, this shift to Hebrew in later Jewish texts appears to be rather programmatic. This holds true for PRE, which as a 'philological Midrash' makes extensive use of Hebrew in order to engage with many aspects of biblical language and narration. For SEZ is the dominance of Hebrew not only a literary means but also a cultural manifestation of textual focus on biblical discourse.

Both phenomena – the 'return to Scripture' and the 'Hebrew renaissance' – can be illuminated by several closely related developments in the cultural milieu of early Muslim times. Scholars have pointed time and again to the centrality of the Qur'ān and pertaining traditions of interpretation in Islamic culture.[41] This idealization was not only based religiously on its status as 'Holy Scripture' and divine revelation, but also referred to the Qur'ān as the literary ideal for ora-

---

40 *Baraita di-Shemuel* is mainly atsronomical-astrological in nature, while *Mishnat ha-Middot* discusses arithmetics and *Sefer Yetzira* provides a complex discussion of cosmology, mathematics and philosophy. On these extra-ordinary works, their preferences for Hebrew over Arabic and a possible background, see Langerman, Y. Tzvi, "On the Beginnings of Hebrew Scientific Literature and on Studying History through 'Maqbiloṯ' (Parallels)." *Aleph* 2 (2002): pp. 169–189. The obscure traditions of Eldad ha-Dani have been recently discussed by Kadari, Adiel, "'All drink from the same fountain': The Initial Acceptance of the Halakhot of Eldad Ha-Dani into the Halakhic Discourse." *Review of Rabbinic Judaism* 13, 2 (2010): pp. 211–228. Regarding the tradition of the *She'iltot*, see Brody, Robert, *The Geonim of Babylonia*, pp. 202–215.

41 An interesting comparison between early Karaite and Muslim conceptions of Holy Scripture is provided by Khan, Geoffrey, "Al-Qirqisani's opinions concerning the text of the Bible and parallel Muslim attitudes towards the text of the Qur'an." *The Jewish Quarterly Review* 81, 1–2 (1990): pp. 59–73. For a thorough survey of Jewish concepts of Scripture, see Polliack, Meira, "Concepts of Scripture among the Jews of the Medieval Islamic World," in *Jewish Concepts of Scripture. A Comparative Introduction*, ed. by Benjamin Sommer, pp. 80–101. New York: New York University Press, 2012.

tion, narration and poetry. This *topos* is accompanied by a conception of superiority of Qur'ānic Arabic (*'Arabiyya*) as the holy language of revelation. Thus, prevailed a discourse among Abbasid learned men about the "perfect speech" whose origins and confinements were fiercely discussed (like in Al-Jahiz' *Kitāb al-bayān wa-l-tabyīn*).[42] Both ideas converge into the belief in the inimitability of the Qur'ān (*i'jāz al-qur'ān*).

It seems quite likely that this new cultural formation posed a challenge to Jewish traditions and traditionalists who considered themselves as guardians of the original Holy tongue, namely Hebrew. Thus, Jews would look for discursive strategies to strengthen the position of Hebrew and their Holy Scripture vis-à-vis Arabic and the Qur'ān, while making use of the cultural models available in their immediate surroundings. Such a development would be similar to the newly increased self-consciousness and interest, which has been observed with regard to Syriac language and (biblical) literature in the same period.[43] Indeed, some scholars have compellingly argued that the new centrality of Scripture in Jewish literature and thought did not build on the model of the Qur'ān in the first place, but rather adapted ideas about the Bible and the superiority of Syriac as the eternal language in Syriac Christianity.[44]

Against this backdrop, we find multiple inner-Jewish factors pertaining to the increased engagement with Scripture. A specific 'return to Scripture' in the early Muslim period is evident in three segments of Jewish society. First, we already mentioned the long tradition of learned Masoretic circles engaged in arrangement (e. g., vocalization; conventions of recitation and writing), thorough textual study and (linguistic-exegetical) explanation of the Biblical text. Rina Drory has suggested that the Masoretic and grammarian enterprise constituted

---

42 Cf. Behzadi, Lale, "Between Theology, Philosophy, and Aesthetics. Al-Jāḥiẓ on Arabic Language," in *Center and Periphery within the Borders of Islam*, ed. by Giuseppe Contu, pp. 307–312. Leuven: Peeters 2012; and Behzadi, Lale, "Al-Jāḥiẓ and his Successors on Communication and the Levels of Language," in *Al-Jāḥiẓ: A Muslim Humanist for our Times*, ed. by Arnim Heinemann; John Lash Meloy; Tarif Khalidi, et. al., pp. 125–132. Beirut, 2009. Both papers in English give a summary of her broader study in Behzadi, Lale, *Sprache und Verstehen: al-Ǧāḥiẓ über die Vollkommenheit des Ausdrucks*. Wiesbaden: Harrassowitz, 2009.

43 Pietruschka, Ute, "Classical Heritage and New Literary Forms: Literary Activities of Christians during the Umayyad Period," in *Ideas, images, and methods of portrayal: insights into classical Arabic literature and Islam*, ed. by Sebastian Günther, pp. 17–40. Leiden: Brill, 2005. For other Syriac writers' strategies of identity-making and "othering" of Muslims in a period of mutual boundary-drawing, cf. Penn, *Envisioning Islam*, esp. pp. 53–101.

44 Cf. Stroumsa, Sarah, "Soul-searching at the Dawn of Jewish Philosophy: A Hitherto Lost Fragment of al-Muqammas's Twenty Chapters." *Ginzei Qedem* 3 (2007): pp. 137–161; and Contini, "Aspects of Linguistic Thought."

an ambivalent or neutral literary mode – accessible to and accepted by rabbinic and non-rabbinic Jews alike.[45] Second, we find scripturally oriented groups among the different branches who later merged in the Karaite movement. The Bible upon which they commented constituted the main focus of their learning. While Hebrew was their language of choice well into the 10th century (*Benjamin al-Nahwandi, Daniel al-Qumisi, Hiwi al-Balkhi*), Scripture became their authoritative source of individual halakhic inquiry and theological reasoning (*ḥippus* חיפוס lit.: "search").

Third and finally, even a Rabbanite flagship intellectual like Saʿadya Gaon, who introduced several Arabic modes of writing into geonic culture, took part in this Hebrew renaissance. Rina Drory has shown how the choice of language in the Arabic and Hebrew introduction to his poetic lexicon *Sefer ha-Egron* predetermined the type of cultural discourse he addressed. While the Arabic preface follows the common cultural pattern of philosophical-linguistic elaborations, well established in Arabic-Muslim society, the Hebrew opening is quite different. Here, Saʿadya promotes a collective return of the Jewish nation to the language of revelation and Scripture – namely Hebrew. In his eyes, Hebrew shall become the language of all learned men in the world. This agenda with its messianic overtones attests that the return to Hebrew in the later Midrashim was part of a broader trend, possibly triggered by the Arabic cultural triumph, and a forerunner of later developments among Jews across the medieval Mediterranean.[46]

In general, the Bible and its main language Hebrew constituted a common ground for most Jews – be it in the context of basic (school) education or in a liturgical framework of Torah-reading and public prayer, or in inter-communal exchange.[47] It is important to note that, in contrast to Midrashic literature, the

---

45 See Drory, *Models and contacts*, pp. 138–142. For the frequent adoption of rabbinic texts and their language into Karaite works, see Tirosh-Becker, Ofra, "The use of rabbinic sources in Karaite writings" in *Karaite Judaism: A Guide to Its History and Literary Sources*, ed. by Meira Polliack, pp. 319–338. Leiden: Brill, 2003; aed., *Rabbinic Excerpts in Medieval Karaite Literature*, 2 volumes. Jerusalem: The Bialik Institute and the Hebrew University, 2011 [Hebrew]; and aed., "Karaite sources for Rabbinic Hebrew (Tannaitic and Amoraic)," in *Encyclopedia of Hebrew Language and Linguistics. Volume 3*, ed. by G. Khan, pp. 316–319. Leiden/Boston: Brill Academic Publishers, 2013.

46 Drory, Rina, "Literary Contacts and Where to Find Them: On Arabic Literary Models in Medieval Jewish Literature." *Poetics Today* 14, 2 (1993): pp. 277–302, emphasizes also the importance of an agenda of a Hebrew revival in later works by Iberian Jews like Ibn-Ezra and al-Harizi.

47 In another paper, I have shown how prayers and blessings serve in *Seder Eliyahu Zuta* as powerful rhetorical and didactic tool for conveying the text's ethical agenda, while simultaneously creating a kind of core-liturgy for the public and domestic sphere, with which everyone regardless of education should be acquainted. The author(s) utilized texts and rituals that

Bible did not occupy such a prominent place in Talmudic traditions and in *Halakha* as they were transmitted and elaborated mainly within the geonic academies. Although the laws and norms of the Torah (the Pentateuch) constitute the core of the halakhic corpus, we only find minute exegesis or exposition of verses in Talmudic texts among many other forms of explanation and reasoning. By contrast, Scripture and its language were accessible to and accepted by a broad range of Jews from all walks of life, their collective or individual affiliations (e.g., 'Rabbanite', 'Karaite', 'messianic', 'hybrid' etc.) notwithstanding. This common Biblical heritage of 'all Israel' (i.e., the Jews) provided as well a neutral mode for cultural activities. It opened up a common 'discursive space' that could have been even shared with members of other *ahl al-kitāb*, and was less burdened with socio-political imponderabilities than the business of the *ahl al-ḥadīth* and the cognate Talmudic branch of Rabbanite Judaism.

2.3 Trends Towards Rational and Individual Exegesis and 'Philosophical' Approaches to Tradition

The aforementioned return to the Bible in several late Midrashic texts brought along also significant shifts in the applied hermeneutic approaches. In SEZ 2 one finds a dialogue between the first person narrator and one who "has knowledge of Scripture but not of rabbinic Oral Lore [Mishnah]." The questioner casts doubts on the concept of an orally transmitted, second Torah (in the sense of "body of teachings") and its divine origin that is crucial to the tradition and to the self-understanding of rabbinic Judaism by stating: "Scripture was given to us from Mount Sinai, but Mishnah was not given to us from Mount Sinai." However, the narrator holds that both were given from God as text and (insight to) exposition and he illustrates their interrelation through a parable:

> A parable: to what is the matter comparable? To a mortal king who had two servants whom he loved with utter love. To one he gave a measure of wheat and to the other he gave a measure of wheat, to one a bundle of flax and to the other a bundle of flax. What did the clever one of them do? He took the flax and wove it into a tablecloth. He took the wheat and made it into fine flour, sifting and grinding it. He kneaded [the dough] and baked it. He set [the bread] upon the table, spread the tablecloth over it and kept it

were familiar to a broader range of Jews – learned, semi-learned or unlearned alike. Cf. Lehmhaus, Lennart, "*Blessed be He* – benedictions, prayers, and narrative in the garb of late Midrashic traditions." in *"It's Better to Hear the Rebuke of the Wise Than the Song of Fools" (Qoh 7:5): Proceedings of the Midrash Section, Society of Biblical Literature, Volume 6*, ed. by Rivka Ulmer and David W. Nelson, pp. 107–151. Piscataway, NJ: Gorgias Press, 2015.

> until the king should come. And the foolish one of them did not do anything at all. After some days the king came into his house and said: My sons, bring me what I gave you. One brought the [bread of] fine flour on the table, with the tablecloth spread over it. And the [other] one brought out the wheat in a box and the bundle of flax upon it. Oh what a shame! What a disgrace! Alas for the one asking: who was [more] beloved? [Of course], he who brought the table with the [bread of] fine flour upon it.[48]

This short parable compares the twofold revelation – Written and Oral Torah – with raw products (wheat and flax) granted by the king (God) to his servants (Israel/the Jews). The contrast between the foolish and the wise servant serves to highlight the superiority of the latter's approach. The mere conservation of the basic materials is naive and subject to ridicule. Only the refinement and elaboration of all potentials of the raw material (i.e., Written Torah) generates the products (bread and tablecloth) that nourish their creators and please the King. Israel, thus, should engage in hermeneutic techniques (sifting, grounding, kneading, baking, weaving) in order to bring forth new insights (processed products). The mutual and complementary interdependence between Scripture and its exposition is illustrated by a reference to complex, hermeneutic rules (4th-7th in R. Eliezer's list) about the general and the particular (*Khelal u-frat*) aspects of understanding.

This parable introduces SEZ's hermeneutic agenda of active development, refinement and appropriation of the Written Torah in new contexts. It promotes rather an immediate and thorough intellectual engagement with Scripture. This program stands in contrast to two extreme positions in Jewish thought – usually, but too hastily, associated in scholarship, and also in early Muslim heresiological accounts about Jewish groups, with two competing factions. On the one hand, Karaites were seen as proponents of an extreme conservative and conservational Scipturalism. Rabbinic traditions or the Rabbanite Jews, on the other hand, were traditionally associated as solely relying on a chain of rabbinic Oral Traditions, almost independent from Scripture. As already mentioned, this sharp dichotomy is an anachronistic construct, which does not account for the multifacetedness of socio-historical realities and cultural diversity.[49]

Also in other dialogues (SER 14/15–15/16) where the authority and divine origin of certain traditions and customs is put into question, one can find a similar approach of the narrator to the Written Torah. While the opponent's increasing attacks clearly try to enrage him, the first-person narrator does not sharply

---

48 SEZ, ch. 2, pp. 171–172.

49 A critical survey of the scholarly consensus regarding a Rabbanite-Karaite schism provides Polliack, Meira, "Rethinking Karaism: Between Judaism and Islam." *AJS Review* 30, 1 (2006): pp. 67–93.

rebut the claims of his questioner. In fact, he rather tries patiently to emend the misunderstandings of his challenger and to instruct him ("Where did you learn this?"; "You are right concerning this ..., but if you consider that ... as well"). Without negating their first occurrence in rabbinic tradition, he shows that the teachings in question are deeply entrenched in the Written Torah, the only source of religious tradition his challenger seems to accept.[50] He deploys exactly those rabbinic exegetical techniques like analogical reasoning, inference, inner-biblical glossing and logical argumentation that were acceptable to and can be found also in later Karaite exegesis. Furthermore, the exposition by inner-biblical glossing – deriving prescriptions not mentioned in the Pentateuch from other biblical traditions (called *Qabala*) – resembles closely the predilections in later Karaite and Scripture-oriented exegesis, with which the narrator apparently tries to tie in. Simultaneously, although it would have been possible to put forth arguments from rabbinic traditions, the narrator refrains from fanciful tales, non-logic (inspired) forms of rabbinic exegesis or Talmudic literary devices (like precedent/*ma'asseh* or sage-story). [51]

The programmatic parable and the explanations of the narrator demonstrate how oral traditions can be deduced from the written by application of a rather individual and rational form of exegesis. This hermeneutic appears similar to what is known as the method of *ḥippus* among Karaite writers. According to later Karaites, this approach was invented by its founding figure Anan in the 8th century to whom they ascribed the maxim: "Search diligently in the Torah, and do not rely on my opinion."[52] However, already one of the influential texts of rabbinic Judaism displays a very similar approach to Scripture, namely

50 His approach seems to anticipate Tobia ben Moses' saying (11th century.): "Those who say that a tradition does not have scriptural support do so only because their minds are too weak to find a basis from biblical law." See Bashyachi, Elijah, *Aderet Eliyahu, Sefer ha-Mitzvot shel ha-Yehudim ha-Qara'im*. Israel: National Council of Karaite Jews, 1960, 9d, as cited in Greenspahn, Frederick E., "Sadducees and Karaites: The Rhetoric of Jewish Sectarianism." *Jewish Studies Quarterly* 18, 1 (2011): pp. 91–105.

51 The narrator derives the prohibition of cheating of others, regardless of their affiliation (Jew/Non-Jew) from Lev 19:13 (*You shall not take advantage of your neighbor*) which he proves with the prophetic verse in Ezek 18:18 (*because he heartlessly took advantage ... behold, he dies for his iniquity*).

52 This dictum, however, is first attested in Yefet ben Eli's commentary, two centuries after Anan (late 10th century). However, as a teaching attributed to the purported "founder" of Karaism it became one of the major principles of Karaite individualistic exegesis. Cf. Cohen, Martin A., "Anan ben David and Karaite Origins." *Jewish Quarterly Review* 68 (1977/78): pp. 129–45 and 224–34; see also Ben-Shammai, Haggai, "The Karaite Controversy: Scripture and Tradition in Early Karaism," in *Religionsgespräche im Mittelalter*, ed. by Bernard Lewis and Friedrich Niewöhner, pp. 11–26. Wiesbaden: Harrassowitz, 1992, esp. pp. 15–16.

in Ben BagBag's famous dictum in *Pirke Avot* (*Chapters of the Fathers*) 5, 22 about the value of Torah (-study): "Turn it, and turn it, and reflect on it for everything is in it. Reflect on it and grow old and gray with it. Don't turn from it, for nothing is better than it."

This kind of deep reflection and reasoning – bringing forth new interpretations and possibly new rules derived from Scripture – seems to have been the target of the fierce polemic of Pirqoi ben Baboi, a Babylonian scholar who lived around the year 800. With many words, he warns his co-religionists, especially his fellow-Babylonians, against individual exegesis and study that purportedly prevailed among Palestinian rabbis. Such an approach ("Everyone expounds what comes to his mind") and a focus on Written Torah seemed strange and repulsive to Baboi, who was intellectually shaped in the geonic academies and imbued with their emphasis on oral traditions transmitted within a particular context of teaching. However, in the Arab-Muslim and Syriac-Christian surroundings of this period scholars have observed a number of similar trends that emphasized and favored individual exegesis over traditional lore.[53]

Another parallel to Karaite exegetical works can be found in the exegetical approach in SE that reads Biblical verses in their wider context and reveals underlying narrative dimensions. Such an interpretation based on context and literary structures emerges already with Daniel al-Qumisi in the later 9th century, as shown by Meira Polliack and others, and can be found also in Qur'ānic exegesis.[54]

In addition, the overall style of the argumentation does not only relate to the Karaite exegesis known as *ḥippus*. In fact, the logical reasoning of the narrator resembles also approaches in the rationale theology of the *Muʿtazila* branch of Muslim thought that flourished in this period. In my dissertation I have suggested that several aspects of SEZ's theology and ethical ideals can be understood in light of the Mutazilite *Weltanschauung*. This is certainly true regarding the emphasis on human self-reliance and free will, theodicy, God's absolute justice, unlimited love and benevolence. Of special importance for SEZ's ethical discourse are Mutazilite ideas about (undeserved) suffering and divine compensation

53 For strategies of individual exegesis in early Islam, cf. Astren, "*Islamic Contexts*." pp. 162. For Baboi's critique of individual exegesis in Palestine, see Kadari, Adiel, "Talmud Torah in Seder Eliyahu – The Ideological Doctrine in its Socio-historical Context (Heb.)." *Daat* 50–52 (2003): pp. 35–59, esp. pp. 56–57.

54 For Karaite adoption and adaptation of rabbinic exegesis, see Polliack, Meira, ed., "Major Trends in Karaite Biblical Exegesis in the Tenth and Eleventh Centuries," in *Karaite Judaism*, pp. 363–413, esp. pp. 365–391; for possible inspiration from Islamic modes of reasoning, see Astren, "*Islamic Contexts*." pp. 161–169.

(*'iwaḍ*).[55] In a recent article David Sklare has argued that "Mutazilite *Kalam* had a relative early impact among the Jews," already in the 8th and 9th centuries, since "it provided a rational and systematic analytic approach to understanding the elements of religious experience."[56] He demonstrates how this rationale theology got so naturally absorbed into Jewish tradition, even into the conservative geonic milieu, that it was considered as genuine Jewish, while authors perpetuating precisely those ideas warned at the same time against what they labelled as 'Mutazilite teachings' of Jews, Muslim and Christian thinkers alike. We may detect another striking parallel in the target audience of those Mutazilite texts, some geonic treatises and many of the later Midrashim, since all of them seem to engage in an instructive discourse with a broader readership or audience of Jews who had no deeper rabbinic, or geonic theological and philosophical expertise.[57]

While the impact of Mutazilite thought and dialectics is widely attested for Syriac-Christian, Jewish geonic and proto-Karaite treatises, such adoptions and adaptations were never studied in Midrashic sources that were contemporary to these movements. However, the dialogue scenes in *Seder Eliyahu*, dis-

---

55 Cf. Goldziher, Ignac, *Introduction to Islamic Theology and Law. Translated by Andras and Ruth Hamori*. Princeton: Princeton University Press, 1991, esp. pp. 90–92; Ghaly, Mohammed, "Evil and Suffering in Islam," in *Philosophy of Religion: Selected Readings, 5th edition*, ed. by Michael Peterson, William Hasker, Bruce Reichenbach et al., pp. 383–391. Oxford: Oxford University Press, 2014; Heemskerk, Margaretha T., *Suffering in the Mu'tazilite theology : 'Abd al-Ǧabbār's teaching on pain and divine justice*, Leiden: Brill, 2000, esp. pp. 142–191.

56 Sklare, David, "The Reception of Mu'tazilism among Jews who were not Professional Theologians," in*tellectual History of the Islamicate World* 2 (2014): pp. 18–36; here: pp. 19–20. On general development of the Mutazila, see Van Ess, Josef, *Theologie und Gesellschaft im 2. und 3. Jahrhundert Hidschra: eine Geschichte des religiösen Denkens im frühen Islam, Band 4*. Berlin: de Gruyter, 1997, pp. 1–122.

57 For such a phenomenon in Geonic culture in a slightly later period, see Sklare, David E., *Samuel Ben Hofni Gaon and His Cultural World: Texts and Studies*. Leiden: Brill, 1996, pp. 90–91. Cf. also p. 95: "The new emphasis on individual scholarship – exemplified by the gaonim's beginning to compose monographs on halakhic topics, aids for beginners and theological works – represented a change not only in the conception of authorship but also in their approach to their role in meeting the religious and intellectual needs of the Jewish community. This entailed addressing sectors of the Jewish community, which had not previously come under the intellectual influence of the yeshivot. [...] In a sense, the Jews outside the yeshivot were applying pressure on the scholars of the yeshivot. They wanted authoritative and 'Jewishly' authentic responses not only to halakhic questions but also to their religious perplexities." On the interaction between Geonim, other (learned and less learned) Jews and their Arabic surroundings (in *Majalis* gatherings and other contexts), see ibid., pp. 99–141.

cussed above, might grant us a first glimpse into how a world of 'Midrashic *Kalam*' could have looked like.[58]

3 Literary Models: Contest and Contact

In the preceding sections, I have discussed how some of the hermeneutic (approaches to language) and exegetical features in late Midrash possibly have corresponded with cultural patterns in non-rabbinic (Karaite/ Masoretic) and non-Jewish milieux. However, also on a more general level one can try to discern the interplay between Persian-Arabic culture and rabbinic literature. Several texts exhibit some striking similarities regarding their innovative textual and compositional features. Since many of those elements were quite common in the emerging tradition of early, mainly Arabic literature, we might think of complex processes of borrowing or cultural negotiations that were at play. A helpful heuristic tool for my following discussion is the category of cultural 'cognates' or 'affinity'. It has been used fruitfull by Michal Bar-Asher Siegal in her study on shared discourses in monastic and Talmudic traditions. In accordance with her approach and findings, I would like to suggest also for the entanglements between the Midrashic texts under discussion and Islamicate cultural models a shared constellation of literary, structural and discursive elements and contents. This is not meant to argue for any direct genealogical relationship, but rather to propose a closer look at structural similarities and possible contact zones, while transcending the tempting but intricate idea of "influence," which has been conceptualized too often as a one-way street.[59]

---

58 For Christian *Kalam*-style literature, see Toenies Keating, Sandra, *Defending the People of Truth in the Early Islamic Period: The Christian Apologies of Abū Rāʾiṭah*. Leiden; Brill, esp. pp. 1–32 and pp. 49–55; Husseini, Sara Leila, *Early Christian-Muslim Debate on the Unity of God: Three Christian Scholars and Their Engagement with Islamic Thought (9th Century C.E.)*. Leiden: Brill, 2014, esp. pp. 187–211; Griffith, Sidney H., *The Church in the Shadow of the Mosque: Christians and Muslims in the World of Islam*. Princeton: Princeton University Pres, 2012, esp. pp. 45–128.

59 Cf. Bar-Asher Siegal, Michal, *Early Christian Monastic Literature and the Babylonian Talmud*. Ney York: Cambridge University Press, 2013, whose analysis is most instructive when she points out the shared discursive world between rabbinic and monastic texts as attested in their deployment of motifs, literary patterns and genres. Cf. also aed., "The Collection of Traditions in Monastic and Rabbinic Anthologies as a Reflection of Lived Religion." *Religion in the Roman Empire* 2 (2016): pp. 72–90.

My approach, although aiming at post-Talmudic, Midrashic texts, is shared by several recent studies of the cultural background of the Babylonian Talmud, and is pointedly summarized

3.1 Author(s) and Authority

The (figure of the) author played a central role in Arabic and Islamic literature – not so much as the ingenious and creative mind behind a certain piece of art, but all the more so as a point of reference and authority. In early Islamicate culture the chain of tradition, or rather of tradents (*isnād*) and the historical reliability of the reported accounts became the cornerstone of medieval Arabic prose writing, most obviously in the religio-historical genres like Hadith and *khabar/akhbar*. However, as can be seen in several narrative genres and certainly in *Adab* literature, the author often also occupied different and rather ambiguous roles – as a literary or rhetorical *persona* or even as a character ('actant') within the narrative or story.[60]

According to a common assumption in scholarship, the rise of authorial literature in medieval Jewish culture is deeply indebted to Arabic writing culture. Geonic responsa as well as theological-philosophical works and polemic tracts written by Karaite and Rabbanite Jews were often modelled after Arabic authorial literature. They often participated in and gained their authority from the discursive universe of Arabic culture. On the other hand, Midrash has been perceived as a 'mosaic' of teachings or exegesis produced by a collective authorship.

Also in Midrashic scholarship, however, prevailed a trend to define later texts as the work of a single author, mostly seen negatively as being an eclectic compiler rather than an ingenious author. In contrast to later, medieval treatises, one finds no formal attribution to a personal author or an individual introduction relating to the author's life or his environment. However, what can be observed, indeed, is the change from a polyphonic, multi-vocal concert of rabbinic

by Kulp, Joshua, "Review Essay. Reading the Bavli with Monks and Zoroastrians." *Prooftexts* 33, 3 (2013): pp. 381–94, here p. 382: "despite the impression the Bavli gives of having been created in splendid isolation, an internal document recording the debates that occurred in the rabbinic bet midrash, and at times even within the minds of the editors who shaped this literature, in reality Babylonian rabbis could not help but be influenced and shaped by other thinkers who shared their cultural space." For other studies focusing rather on complex or subtle entanglements, negotiations and appropriation, cf. Kalmin, Richard L. and Seth Schwartz, eds., *Jewish Culture and Society Under the Christian Roman Empire.* Leuven: Peeters, 2003, esp. pp. 1–6; see also Hasan-Rokem, Galit, *Tales of the Neighborhood: Jewish Narrative Dialogues in Late Antiquity.* London: University of California Press, 2003; Schremer, Adiel, *Brothers Estranged: Heresy, Christianity, and Jewish Identity in Late Antiquity.* Oxford: Oxford University Press, 2009.

60 Cf. the discussion of ideas about authorship and fiction in early Arabic literature in Hoyland, Robert G., "History, fiction and authorship in the first centuries of Islam," in *Writing and Representation in Medieval Islam*, ed. by Julia Bray, pp. 16–46. London: Routledge, 2006.

voices into a single, personal or authorial voice. Moreover, PRE, SE and ABS introduce characters (i.e., Eliezer ben Hyrcanos/Elijah/Ben-Sira) who function as protagonists but also as authors or narrators within the text itself. By providing some vivid, personal speech, this form lends some authenticity (personal experience) and probably greater authority to the teachings conveyed in the work. In all cases, this literary strategy entails a sophisticated play with authoritative traditions and pseudepigraphy, featuring strong allusions to Biblical, apocryphal and rabbinic 'heroes'. Thus, the perception of a great founding figure in Jewish tradition as the protagonist and/or alleged author has very likely given greater importance and authority to these texts in the course of the their transmission.[61] SE chooses for this the unique literary form of lively first-person dialogue narratives,[62] combined with attributions of ethical teachings to a certain Eliyahu, or his school (-house). This narrator or author-character oscillates between a ubiquitous wandering teacher or preacher and the prophet Elijah (*Eliyahu ha-navi*) himself. This connection is supported by the centrality of return in repentance (*teshuva*) and ethics in SEZ's discourse. The text's moral emphasis dovetail with Elijah's role as a moral teacher or a reconciling herald of the messianic age in Jewish tradition. Since the narrator interacts with his interlocutors in in-

---

61 The possibility of such an allusion or attribution was – certainly with less caution and against the backdrop of a different understanding of rabbinic literature – already mentioned by some early scholars of the *Wissenschaft des Judentums*. Theodor, Julius, "Besprechungen (zu Friedmanns Edition des Seder Eliahu)." *Monatsschrift für Geschichte und Wissenschaft des Judentums* 44, 10 (1900): pp. 550 – 561, considers these dialogues if not the whole text to be written from the perspective of the prophet (p. 554: "in der Fiction des Prophet Elia"). Also Bacher, Wilhelm, "Antikaräisches in einem jüngeren Midrasch." *Monatsschrift für Geschichte und Wissenschaft des Judentums* 23, 6 (1874): pp. 266 – 274, likens the narrator to Elijah as a wandering teacher-prophet (p. 266: "als lehrend umherziehend gedachte(n) Prophet(en) Elia"). Eliyahu was an important figure in Jewish esoteric circles and was connected to different ritual and liturgical contexts (e.g., circumcision-ceremony/brit milah in PRE 29; Pessach-Seder in ExR 18,12; or in the benediction after *haftara*-reading and in the *havdala* ceremony for Shabbat) familiar to Jews of educationally and geographically diverse backgrounds. Moreover, the prophet Elijah was also a key figure in the religious mindset of patristic, Syriac-Christian or Byzantine traditions (as a martyr, zealot, archetype of a monk) – and as such a known type and subject to *cultural borrowing*.

62 For a detailed study, see Lehmhaus, Lennart, "'Were not understanding and knowledge given to you from Heaven?' Minimal Judaism and the Unlearned 'Other' in Seder Eliyahu Zuta." *Jewish Studies Quarterly* 19, 3 (2012): pp. 230 – 258. An outstanding feature of SE is its innovation of a first-person narrator in several episodes of encounter that form SE's literary backbone. I suggest that in SEZ those narratives of instructive dialogue promote a "minimal Judaism" open to all: basic knowledge of Scripture, the most important prayers and benedictions as well as moral behavior and piety. In a liminal setting border-crossing interaction with different non-rabbinic "others," questions of Jewish and rabbinic identity are examined through the safety-screen of the narrated dialogues.

structive dialogues, the setting of the dialogues in SE might be compared to the prominent *motif* that combines wandering and instruction in the "journey in quest of knowledge" (*riḥla fī ṭalab al-ʿilm*). It constituted in early Islam not only an educational ideal, but was also a popular literary *topos*. The "journey in quest of knowledge" figures also prominently in the narrative in PRE about R. Eliezer ben Hyrcanos' beginnings as a rabbinic student. Overcoming socio-economical hardships he makes his way to Jerusalem where he studies in order to become one of the greatest scholars of his generation. Even the protagonist in ABS leaves his home for acquiring knowledge – first, with a teacher of young children and later on at the court of king Nebuchadnezzar in Babylonia.[63]

Interestingly, the patient teacher-character in SE's dialogue scenes resembles in certain regards also the *qāṣṣ (pl. quṣṣāṣ)*, a figure oscillating between public storyteller, teacher and preacher, who had important socio-cultural and literary functions in the Islamicate societies, at least from the early Islamic period onwards. Further proximity with *Seder Eliyahu* can be noticed regarding the *qāṣṣ'* agenda of instructing about righteous conduct, also by way of admonition, promoting fear of god paired with knowledge about divine grace, and teaching the basic theological and ethical concepts.[64] In his recent study, Lyell Armstrong has scrutinized the early traditions about *quṣṣāṣ*. According to his findings, the early *quṣṣāṣ* were learned men who engaged in several overlapping discourses and practices: ranging from religious instruction and counsel, transmission and development of commentary (*tafsir*) to hadith. While the *quṣṣāṣ* were neatly associated with upholding ethical values and ascetic piety, they also seem to have been involved in political agitation and some upheavals or rebellious movements in the late Umayyad and Abassid period. The author-figure in *Seder Eliyahu* and the text as a whole engages in some, if not in all of those areas mentioned. Also the deployment of narratives about Biblical prophets (Moses, Elija, Isaiah, Jeremiah, Hosea) as (counter)models of piety as well as the playful

---

63 On the popularity of the concept and its literary elaboration, see Hoyland, "History, fiction and authorship," p. 26.

64 On the *qiṣaṣ* and the figure of the *qāṣṣ* see ibid., pp. 23–24. See also Stilt, Kristen, *Islamic Law in Action: Authority, Discretion, and Everyday Experiences in Mamluk Egypt*. Oxford: Oxford University Press, 2011, here p. 81: "the storyteller (*qāṣṣ*) must only mention that which average people (*ʿāmma*) can understand and relate to, such as encouraging prayer, fasting, and paying *zakāt*." This notion of an exclusive interaction of the "storyteller" with average, unlearned people is probably late and might reflect changing (increasingly devaluing and belittling) attitudes to the figure and the function of the *qussas* (plural of *qāṣṣ*) over time. Such a gradual development has been pointed out by Armstrong, Lyall R., *The Quṣṣāṣ of Early Islam*. Leiden: Brill, 2017, who recently challenged traditional, rather negative views of the *qāṣṣ* in later Islamic traditions as well as in earlier scholarship. See following note.

allusions to Elijah the prophet as the author of moral teachings and in the role of the indulgent and patient instructor tie in well with some central features of the *quṣṣāṣ* tradition. Several accounts emphasize the didactic-religious character of *quṣṣāṣ* as a genre, traced back into and second important to the Qur'ān itself, in which exemplaric narratives and praise of pre-Islamic prophets play a central role.[65] In addition, frequently the *quṣṣāṣ* highlight the "the idea that the scholar was a living exemplar; it was his responsibility to teach the people correctly and also, perhaps more importantly, to be a model of right faith and conduct for the community."[66] The narrator-character in *Seder Eliyahu*'s dialogue scenes fits well into this self-depiction of the *quṣṣāṣ*, although this image might have been rather common and shared by several religious communities in this period.

Moreover, the association of the *quṣṣāṣ* with various religious milieux and figures, the variety of audiences and venues as well as the public, non-eliterian academic milieu of the *quṣṣāṣ* sessions, exhibits a certain degree of overlap with the thematic agenda and the different set-ups of the instructive dialogues in *Seder Eliyahu* (with a Persian priest in an urban and royal setting; with semi-learned people, commoners, and women in public and private places; with students and scholars in or close to the study house or the academy). One may surmise other points of contact between Midrashic and early Islamicate cultural models to be found in the familiarity of the *quṣṣāṣ* with Jewish and Christian traditions, and (Biblical) stories of the prophets, i.e., the *isrā'īliyyāt* and the *qiṣaṣ al-anbiyā'*, although this was not the only focus and interest in those traditions pervading the broader religious community of early Islam.[67]

65 Cf. ibid., pp. 1–13. Regarding the image of the "popular preacher to the uneducated masses," Armstrong states (p. 9): "Likewise, the *quṣṣāṣ* were not necessarily "popular" preachers, if the intended meaning of "popular" is to be understood as a teacher of frivolous learning interacting essentially with the uneducated masses." However, ibid., pp. 16–33, esp. pp. 22–26, he shows that the religious *quṣṣāṣ* were focusing on explanations of the eschaton, divine justice and mercy with strong exhortations of piety which is closely related to fear of God, righteousness, and good deeds. On the deployment of pre-Islamic prophets and the reputation of the genre in the Qur'ān itself, see ibid., pp. 33–41.

66 Ibid., p. 46.

67 Cf. ibid., pp. 90–111. On the association with other religious functions such as Qur'ān -commentators, *ḥadīth* transmitters, legal scholars and judges, formal orators (*Khuṭabā'*), pious, admonishing preachers (*wu''āẓ /mudhakkirūn*), and ascetics, cf. ibid., pp. 110–152. For the various audiences and skills, see ibid., pp. 153–189, and 280: "the early Islamic *quṣṣāṣ* held their sessions for a variety of audiences in an assortment of venues. They held sessions for the public outside of the mosque, as in the case of Abū Idrīs al-Khawlānī entertaining questions from the audience while seated on the steps of the mosque in Damascus; for smaller groups in sessions inside a mosque, apparently the most common practice of the Umayyad period; for groups

From an inner-Jewish and geonic perspective, it is noteworthy that a similar emergence of a type of fictional author-figure or literary *persona* can be found also in Sa'adya Gaon's work *Sefer ha-Egron*. This text has been usually read as an authentic, authorial treatise *par excellence*. According to Rina Drory's study, however, his two programmatic introductions in Arabic and Hebrew differ significantly. In the first version, the author enters the stage as a learned Arabic philosopher and grammarian speaking to an urban, academic audience of fellow scholars. However, in the Hebrew text, he introduces another cultural pose adopting "rather the mantle of a leader" or "[...] a Bible-type prophet who seeks to awaken the Jewish nation and to restore it to righteous ways by means of a prodigious program of reviving the Hebrew language."[68] Not only Sa'adya's creation of a fictitious authorial *persona*, but also his prophetic program of return and a Jewish-Hebrew renaissance tie in well with important features of *Seder Eliyahu*. For, this text, as has been pointed out beforehand, plays with strong allusions to the prophet Elijah who is associated in Jewish tradition with the messianic time, moral instruction and reconciliation. Moreover, *Seder Eliyahu Zuta*'s agenda focuses on the moral and spatial dimensions of *teshuva* (i.e., repentance and return), and does so in a discourse with a clear penchant for Hebrew as its language of choice.

Another famous protagonist-type is the picaresque hero in Arabic popular tales, who later became a stock-character in the *maqāma* genre. We can find some similarities to the fictional narrator-character in the ABS' medieval stories of Ben Sira (*Toldot/ Alphabeta de-Ben Sira*). Its protagonist is portrayed as a toddler who has to find his way to real wisdom only through many trials and tribulations. The opening is a 'birth of a hero'-type of narrative that recounts Ben Sira's miraculous pedigree and introduces him as an extraordinary 'Wunderkind'. The instructive dialogues with a teacher of young children reveals more facets of his superior knowledge, although due to his lack of practical experience, he seems often more witty than wise. Only in his increasingly confronta-

in homes, as did Zurāra b. Awfā and suggests that the audience was even smaller than the audience in the mosque; and, finally, even for individuals, [...]."

The author-figure in SE also exhibits in his dialogues with a broad variety of interlocutors those skills that were crucial for the *quṣṣāṣ* , as identified by Armstrong, *The quṣṣāṣ*, p. 280: "He, therefore, was not only expected to possess religious knowledge (*'ilm*), he was also supposed to be an eloquent speaker, characterized by his vocal mechanics (*lisān*) and oratorical skills (*bayān*)."

68 Drory, Rina, "Bilingualism and Cultural Images. The Hebrew and the Arabic Introductions of Saadia Gaon's Sefer ha-Egron." in *Language and Culture in the Near East*, ed. by Shlomo Izre'el and Rina Drory, pp. 11–24. Leiden: Brill,1995, here p. 20 and 22.

tional dialogue with the nemesis of the Jews, namely Nebuchadnezzar, Ben Sira succeeds to turn his coming-of-age into a 'coming of wisdom' as well. Furthermore, ABS's author directly uses and appropriates narratives and stylistic features like the 'stories within a story'-series (emboxed, Matroshka, or Russian-doll-stories) of animal tales inspired by popular traditions like *Kalilah wa-Dimna*, put into the mouth of the protagonist.[69]

Finally, I would like to touch upon an additional aspect of authorship, with which I will hopefully deal elsewhere in greater detail, soon. It pertains to the first dimension of authorship as presented above. Authenticity and a reliable and verifiable chain of tradition or transmission, as in the concept of *Isnād*, figured prominently in Arabic-Islamicate culture. In all three Midrashic texts under discussion, we can find traces of similar strategies in varying degrees. As already discussed in the last paragraph, the *Alphabeta de-Ben Sira* serves probably as the most colourful example of a text creating a fictional author-character equipped with a multi-layered authority. In the birth story, often misunderstood as a carnivalesque and vulgar satire, the protagonist Ben Sira is introduced not only as a child-prodigy equipped with Solomonic wisdom, but also as the son and the grandson of the Biblical prophet Jeremiah, all at once.[70] A second layer is added by numerous intertextual references and allusions to as well as quotation and paraphrases from Wisdom literature in general, and the tradition of the apocryphal *Sirach* (Hebrew: *Ben Sira*) in particular. Obviously, a direct connection between both traditions is established through the names of their respective authors or protagonists (*Ben Sira*). Moreover, the text ties in with numerous Tal-

---

69 A detailed analysis of ABS's transfer and adaptation of literary motifs, plots, figures and structures from Indian, Persian and Arabic traditions can be found in Yassif, *The Tales of Ben Sira*. From this we can learn that the deployment of animals in the etiological narratives in ABS does not only follow structurally the Russian-doll-principle stories in the *Khalila wa-dimna* traditions. Indeed, it also constitutes an apt elaboration of the cultural and literary conventions in its surrounding Muslim-Arabic culture as attested by the frequent use of animal narratives in Al-Jahiz', *Kitab al-Hayyawan* and similar works. Cf. Montgomery, James E., *Al-Jahiz: In praise of books*. Edinburgh: Edinburgh University Press, 2013, esp. pp. 13–14.

70 This constellation comes to happen through the impregnation of Jeremiah's daughter in the bathhouse by the semen of her father who was forced to masturbation by some men whom he rebuked for having sexual relations with each other. A possible background for this sharp and ironic depiction might be found in the Babylonian Talmud in Chagiga 15b, which presents a unique account about conception in the bathhouse via artificial insemination within a halakhic discussion on virginity and marriage. For a discussion of the birth-story see Dan, Joseph, "Ben Sira, Alphabet of," in *Encyclopaedia Judaica* 3, pp. 548–549; Stern, David, "The 'Alphabet of Ben Sira' and the early history of parody in Jewish literature," in *The Idea of Biblical Interpretation*, ed. by Hindy Najman and Judith H. Newman, pp. 423–448. Leiden: Brill, 2004.

mudic discussion, demonstrates knowledge of rabbinic hermeneutics and refers to the authority of great sages, which is critically questioned though.[71]

In contrast to this rather fanciful tradition, we cannot find a strong authorial character or narrator-figure within most of the text of *Pirke de-Rabbi Eliezer* (PRE) – except for the first two chapters about R. Eliezer's initiation into the world of rabbinic learning. Apart from this plot of "acquiring wisdom," in the remaining chapters R. Eliezer, the greatest scholar of his generation, figures most often as the reference for attributed teachings. Furthermore, in consistence with other rabbinic works R. Eliezer's name appears in many titles given to the work in the course of its transmission. Scholars have aptly argued that one should not underestimate these para-texts, since title(s) and attributions evoked the impression of authorship and had a strong impact on the reception (history) of the work.[72]

The last argument pertains to its full extent also to the *Seder Eliyahu* traditions. As explained earlier, the texts oscillate between anonymity – "Eliyahu" as some unknown (rabbinic?) preacher and teacher – and a sophisticated variant of pseudepigraphy – "Eliyahu" as the prophet Elijah. It appears that the authors played deliberately with both possibilities in order to entrench their teachings deeply in Biblical and Jewish-rabbinic authoritative traditions, while simultaneously keeping some sort of safety distance from those elements that might be deemed uniquely Talmudic-geonic or elitarian. It seems that those strategies were not uncommon in the early Islamicate culture. In one of his studies Stefan Leder has shown that even in Arabic so-called historical literature (*akhbār*) one finds the use of "unavowed authorship" as a strategy for self-legitimization. He

---

71 On the various strategies of criticizing authoritative traditions, see Börner-Klein, "Transforming Rabbinic Exegesis into Folktale"; Aed., "Narrative Kritik der rabbinischen Bibelauslegung im Alphabet des Ben Sira. " in *Literatur im Dialog: die Faszination von Talmud und Midrasch*, ed. by Susanne Plietzsch, pp. 99–125. Zürich: Theologischer Verlag, 2007; Lehmhaus, "Es ist mancher scharfsinnig, aber ein Schalk".

72 On the intertextuality between PRE and classical rabbinic texts, see Keim, Katharina, *Pirqei deRabbi Eliezer: Structure, Coherence, Intertextuality.* Leiden: Brill, 2016, esp. pp. 145–164. Keim's suggestion that the author(s)' choice of R. Eli'ezer's story as the main intertextual link that provided a suitable 'niche' for anchoring PRE deeply within rabbinic traditions supports my argument here. While she labels the reference to R. Eli'ezer who figures precisely as an outsider and 'lone voice' in many rabbinic texts, as a possible 'serendipity', I would certainly endorse her speculations that this choice might reflect deliberately a self-understanding of the author(s) and of the discursive endeavor of PRE. Sacks, Steven D.,*Midrash and Multiplicity: Pirke de-Rabbi Eliezer and the Renewal of Rabbinic Interpretive Culture*. Berlin: de Gruyter, 2009, pp. 42–59, stresses the various strategies of attribution of this tradition to R. Eli'ezer, while questioning the single-author theory taken for granted in earlier research.

describes this literary move as an "ascription to an early authority while no hints as to the author of the adaptation is given."[73]

The previously addressed examples are not to suggest that the (re-)emergence of authorial voices and author characters in later Midrash were based on a single (Arabic) model or were exclusively influenced by external cultural models. Rather, I argue that all three traditions under discussion exhibit various features shared with each other as well as with some other traditions of the early Islamicate period, which all might testify to shifts in Midrashic discourses and broader literary and cultural developments, as we shall see in the following.

3.2 Compositional Features: The Treatise Model

In *Seder Eliyahu*, as in several other later traditions, we find an artful blending of different genres, traditions and literary devices that serve superordinate thematic discourses. Such a topical arrangement differs significantly from older Midrashic paradigms that followed external structures (e.g., as a commentary or exposition on Biblical verses or homilies for the order of liturgy). This change is extremely evident in SEZ. This particular work is clearly divided into chapters and clusters discussing one particular aspect of its broader ethical program. Each unit is closely tied to a hidden threat of its main concepts through the use of recurring literary forms and key words. The impression of a systematic rhetoric structure is re-enforced when we look at the use of framing devices as compositional elements. In both works, SER and SEZ, whole chapters, thematic clusters or smaller forms serve as expositional or concluding elements. In SER this framing is temporal, linking creation and expulsion from paradise to a future, messianic return. In SEZ, the first chapter establishes all key concepts, motifs and terms of its moral discourse, and the last chapter reprises most of these elements. By the use of maxims, dialogue-narratives and the already mentioned list about *Zeddaqa*, the text explores the interplay of charity, righteousness, Torah-study, solidarity and the role of the righteous.[74]

---

73 Leder, Stefan, "The literary use of *khabasr.* A basic form of historical writing," in *The Byzantine and Early Islamic Near East*, ed. by Lawrence Conrad and Averil *Cameron, pp.* 277–315. Princeton, 1992, here p. 284.

74 One could relate this Midrashic introduction in some way to the thematical introduction of later Geonic treatises in which a theological issue of crucial importance for the following work was elaborated. Cf. Sklare, "The Reception of Mu'tazilism." p. 29; Id., *Samuel ben Hofni Gaon*, pp. 69–98.

Also other later traditions feature such framing passages. Dina Stein has shown how in PRE the first three chapters introduce not only the intellectual biography of its author character R. Eliezer ben Hyrkanos but do also familiarize the reader with its hermeneutic agenda, as SEZ does in chapter 2. [75] Likewise, the medieval *Alphabeta de-Ben Sira* opens with a miraculous birth of a hero story, a narrative on the education of its protagonist and introduces an approach to wisdom ("There is nothing new under the sun") which is 'Solomonic' in origin, but appears here as rather witty than wise.

In contrast to Ya'aqov Elboim's emphasis on the innovative character of the systematic thematic structure, Rina Drory had stressed the generic divide between rabbinic (or, in her diction: Rabbanite) Midrash as tradition-bound, loosely arranged exegetical mosaics and the Arabic-Karaite tractate-model (*ḥibbur*) with compositional and thematic unity. According to her view, the Karaite innovators resolved a cultural deadlock through adoption and appropriation of Arabic literary models. This enabled also Rabbanite, geonic scholars like Sa'adya and Shmuel ben Hofni Gaon to adopt such models that were totally alien and therefore absent from rabbinic literature.

However, the analysis of later Midrashim indicates that these developments were also reflected in the more traditional rabbinic genres. Their systematic topical arrangement and framing structures appear as the incipient stages of programmatic agendas, introductory elaborations and systematic conclusions that were to become typical of treatises in geonic and Karaite medieval Jewish literature.[76] In ways structurally similar to these monographs, our three examples from post-Talmudic, early geonic traditions deploy clear, narrative elements for rhetorical and instructive purposes. These Midrashic treatises might have been less structured, but were highly programmatic in elaborating central topics, key terms and their hermeneutic or literary agenda. Likewise, we can see a chias-

75 Cf. Stein, Dina, *Maxims, Magic, Myth: A Folkloristic Perspective on Pirkei de-Rabbi Eliezer.* Jerusalem: Magnes, 2004, pp. 115–168 (Heb.). Sacks, *Midrash and Multiplicity*, esp. pp. 42–81, questions the assumption that later Midrashim like PRE should be seen as works produced respectively by a single author. See n. 72, above.

76 Drory, *Models and Contacts*, pp. 134–138 and 153 f., stressed the great divide between all rabbinic Midrashic traditions and medieval forms of the *hibbur*-type ('composition'/'treatise'). While she described the former as a loosely connected mosaic of teachings and exegesis, the latter type is characterized by compositional and topical unity. Drory connected the *hibbur*-model with Karaite and medieval authors (i.e., Sa'adya Gaon). However, Dina Stein correctly calls attention to the structural similarities between some later Midrashim (like PRE and SE), sharing elements like introductions and conclusions, and those medieval texts of the *hibbur*-type. Cf. Stein, Dina, "Pirke de Rabbi Eliezer and Seder Eliyahu: Preliminary Notes on Poetics and Imagined landscapes (Hebr.)." *Jerusalem Studies in Hebrew Literature* 24 (2011): pp. 73–94, esp. p. 77.

tic closure of many texts in which they return to these issues and generally provide some eschatological consolation.[77]

Interestingly, this observation shares many features with those innovative, Karaite texts that were highlighted by Rina Drory as harbingers of a new era: "a thematic piece following the logic of reading, systematically divided into chapters and topics and headed by a methodological introduction, written in first person and attributed to its real author."[78] Even in Arabic literature contemporary to our Midrashim, where the personal introduction was rather the rule than the exception, one can find some striking examples of indirect introductions addressing programmatic intentions and possible readerships.[79]

3.3 *Adab* and the Broader Scope of Discourses in Later Midrash

I would like to suggest also another process of adaptation and interplay that might has been at work in the formation of later Midrashic texts. Due to the limitations of space and with regard to the scope of this contribution, I will only address some of the most striking similarities, which I intend to elaborate upon in the future.[80] In the middle of the 8th century, probably with a Persian background, emerged *Adab* as a cultural concept and literary genre in Arabic culture. According to scholarly consensus, *Adab* designated not only 'literature' but it was strongly related to the idea of urban culturality, etiquette or *belles-lettres*. Especially the ideal of general humanism and all-encompassing education (possibly coined after the Graeco-Roman ideal of *paideia*), often linked by scholars to *Adab*, exhibits many points of contact with SE's ethical discourse that forms the

---

77 For a detailed discussion of those structural features, see Lehmhaus, Lennart, *Derekh Eretz im Tora' – Seder Elijahu Zuta als universale, religiöse Ethik für rabbinische und nicht-rabbinische Juden*. Dissertation, Martin-Luther-Universität Halle-Wittenberg, 2013, pp. 329–354 and 480–505. The "chiastic closure" is mentioned by Lavee, Moshe, "Seder Eliyahu," in *Encyclopedia of Jews in the Islamic World*. Brill Online, 2015, Reference: 16 January 2015, <http://referenceworks.brillonline.com/entries/encyclopedia-of-jews-in-the-islamic-world/seder-eliyahu-SIM_0017460>.

78 Drory, *Models and contacts*, p. 154.

79 On this point, see Montgomery, James E., *Al-Jahiz: In praise of books*. Edinburgh: Edinburgh University Press, 2013, pp. 107–173 and 193–223.

80 Lehmhaus, Lennart, "Jewish *Adab*? – Midrashic traditions in early Islamicate time between narrative and ethics." (forthcoming).

backbone of the whole work.[81] In general, the description of SE and other texts (Avot/DE/ARN etc) as "rabbinic manuals of conduct" fits into the picture of *Adab* works as handbooks for general or professional etiquette.

This holds also true regarding the specialized genre of "rules of conduct for teachers and student," as manifest in some of Ibn Sahnun's or Al-Jahiz' works, which touches upon the foundational pillars of the Abassid 'learned society'. Scholars have identified three key topics of this particular *Adab*-discourse.[82] First, the texts promote a modest, indulgent and patient didactic attitude toward the student in order to enhance the process of learning. As I have shown elsewhere, exactly the same attitude prevails in SEZ whose teacher-protagonist is encouraging and forbearing, even toward those who openly provoke or mock him.[83] Second, the *Adīb* embraced a well-rounded, broad *curriculum* (*'let him seek breadth in learning'*) combining traditional education with personal experience. And the teaching would ideally aim at broader circles of students.[84] Likewise,

---

81 Probably the most famous *Adīb* was Al-Jahiz, whose *Adab*-attitude is described by scholars as strongly connected to a kind of "Arabic Humanism" (general appreciation of other human beings and their knowledge). On the "essential humanism" of *Adab* and the nature of the *gentleman-scholar,* see Kennedy, Philip F., "Preface" in *On Fiction and Adab in Medieval Arabic Literature.* Wiesbaden: Harrassowitz, 2005: i – xii; for the history of *Adab* literature, see Allen, Roger, *An Introduction to Arabic Literature.* Cambridge: Cambridge University Press, 2003, pp. 134 – 157; *Adab* as storytelling is discussed in Toorawa, Shawkat M., "Defining *Adab* by (re)defining the *Adib:* Ibn Abi Tahir Tayfu and storytelling," in *On Fiction and Adab*, ed. by Philip Kennedy, pp. 287 – 308. Wiesbaden: Harrassowitz, 2005.

82 See Günther, Sebastian, "Al-Jahiz and the Poetics of Teaching: A Ninth Century Muslim Scholar on Intellectual Education. " in *Al-Jahiz. A Humanist for Our Time*, ed. by Tarif Khalidi et al., pp. 17 – 26. Beirut, Wiesbaden: Orient-Institute of the German Oriental Society, 2009; and Günther, Sebastian. "Advice for Teachers: The 9th Century Muslim Scholars Ibn Sahnun and al-Jahiz on Pedagogy and Didactics," in *Ideas, Images and Methods of Portrayal. Insights into Classical Arabic Literature and Islam*, ed. by S. Günther, pp. 79 – 116. Leiden: Brill, 2005. Ghersetti, Antonella, "'Like the wick of the lamp, like the silkworm they are': stupid schoolteachers in classical Arabic literary sources." *Journal of Arabic and Islamic Studies* 10 (2010): pp. 75 – 100, notes Al-Jahiz' wide appreciation of the *mu'allimūn* (schoolteachers) which stood in stark contrast to the popular despise for schoolteachers as stupid fools in many popular traditions. Ghersetti's finding may also help to contextualize the motif of 'ridiculing the schoolteacher' that forms the scaffold of the first longer section in the Ben Sira narrative (Ben Sira tests the teacher of young children) in 9th and 10th centuries Abbasid society. Cf. Börner-Klein, *Das Alphabet des Ben Sira*, pp. 20 – 43; 275 – 283; Lehmhaus, "Es ist mancher scharfsinnig, aber ein Schalk".

83 Cf. Lehmhaus, "*Minimal Judaism and the Unlearned 'Other'*", pp. 253 – 258.

84 See Montgomery, *Al-Jahiz*, p. 13: "It (*adab*) appraises the third-century book as a site of cultural encounter and proposes that in our appreciation of these encounters we should not segregate philosophy from juridical thought, alchemy from poetry, belles-lettres from theology."

SEZ seeks to promote moral behavior (*Derekh Eretz*, lit. "*the way of the world,*" i.e., ethics) rooted in basic religious (*Torah*, i.e., Jewish lore) and general education as a guideline for the common Jew. Thus, the text seems to reach out to a non-rabbinic audience in order to convince them of the broad and basic principles cherished by the authors. As mentioned earlier, this agenda of instructing the masses in the basic tenets and beliefs of one's own religion appears as a parallel to the *qāṣṣ* in Muslim society and similar trends towards religious outreach in Syriac-Christian texts.[85] Third, *Adab* pedagogy values the learning, especially through the reading and writing of books, as intellectual stimulating in order to enable the student to think independently and creatively. Knowledge, reason (*ʿaql*), and personal experience (*tajriba*) should be used for critical examination (*baḥth*) of former traditions.[86] This agenda tallies well with SE's general *Kalām*-features and with its concept of individual exegesis and critical discussion of (biblical) traditions, which might resemble the Karaite exegetical hermeneutics of *ḥippus* ("search"), as emphasized above.

Moreover, important features of SE's discussion about the nexus between austerity and morality resemble the *Adab* literature and other traditional teachings that were common among early Muslim ascetic groups (*ṣūfīs*).[87] In this regard, one has to be also aware of various contacts between different ascetic movements of Christian, Jewish and Muslim background that are still awaiting further comparative studies. As already mentioned, the general agenda of ascetic *Adab* with its focus on ethical instruction and moralizing admonition would blend in well with the overarching discourse in SEZ and other Jewish ethical, but also Christian ascetic or monastic works.[88] Moreover, the similarities between

---

85 For the *qāṣṣ*, see above notes 62 to 65. For similar tendencies in Syriac Christianity, see Pietruschka, "Classical Heritage and New Literary Forms", who discusses the didactic purposes of "writings which made it possible to teach the doctrine and theological issues (*summa theologica*) in an attractive, concise and easily memorizable form [...] designed to attract a broader Christian audience with the purpose of warding off the danger of a growing number of Christian converts to Islam" (p. 27).

86 Cf. Hoyland, "History, fiction and authorship", pp. 22–23.

87 On the early Islamic ascetics, see Grämlich, Richard, *Weltverzicht: Grundlagen und Weisen islamischer Askese*. Wiesbaden: Harrassowitz, 1997; Murad, Suleiman Ali, *Early Islam Between Myth and History: Al-Ḥaṣan Al-Baṣrī (d. 110H/728CE) and the Formation of His Legacy in Classical Islamic Scholarship*. Leiden: Brill, 2006, esp. pp. 19–120.

88 Elsewhere, I have started to discuss possible links between SEZ's discourse and the ideologies of early (Syriac) Christian and early Muslim ascetic and monastic groups regarding their critical attitude towards power and wealth and their promotion of austerity, charity and righteous deeds toward the needy. See Lehmhaus, "Derekh Eretz im Torah", pp. 206–232; Aydinli, Osman, "Ascetic And Devotional Elements In The Mu'tazilite Tradition: The Sufi Mu'tazilites." *The Mus-*

SE's narrator and the preacher-storyteller in early Arabic literature would support such a comparison. The *qāṣṣ* appears often not as an elite scholar, but rather as a learned ascetic whose piety and being an outsider to the ruling class ensured his broad reach into society.[89]

4 Conclusion

The narrative proposed in Rina Drory's important study stressed the generic divide between two cultural models. On the one hand, rabbinic Midrashim lingered, in her view, as loosely arranged exegetical mosaics bound by the discursive confinements of a stagnating Talmudic tradition. On the other hand, triggered some non-rabbinic forces the development of the medieval tractate-model (*ḥibbur*) with its compositional and thematic unity, which stems from a Persian-Arabic background. Those Karaite innovators resolved a cultural deadlock through adoption and adaptation of Arabic literary models. This, in turn, facilitated the adoption of those models which were uncommon in rabbinic tradition by later geonic scholars like Sa'adya and Shmuel ben Hofni Gaon.

---

lim World 97, 2 (2007): pp. 174–189, argues for a nexus between early ascetics and the Mu'tazilite rationale school of thought. On the critical stance of Mu'tazila and other ascetics towards political power, see Yücesoy, Hayrettin, "Political Anarchism, Dissent, and Marginal Groups in the Early Ninth Century," in *The Lineaments of Islam: Studies in Honor of Fred McGraw Donner*, ed. by Paul Cobb, pp. 61–84. Leiden: Brill, 2012; Livne-Kafri, Ofner, "Early Muslim Ascetics and the World of Christian Monasticism." *Jerusalem Studies in Arabic and Islam 20* (1996): pp. 105–129, suggests a strong link between those movements; cf. here pp. 115–117 and 121: "Glorification of the poor and poverty was not merely a philosophical concept, it had at least some practical impact on the status of the poor. The fact that great ascetics would stay with the poor and diseased is a related phenomenon. Charity to the poor, the widowed and the orphaned was a very important principle among the zuhād." Cf. also Talmon-Heller, Daniella, "Charity and Repentance in Medieval Islamic Thought and Practice," in *Charity and Giving in Monotheistic Religions*, ed. by Miriam Frenkel und Yaacov Lev, pp. 265–279. Boston/Berlin: de Gruyter, 2009, here p. 278: "A discourse that idealizes the poverty and denounces wealth and social status is well known from Sufi contexts [...]. Most Sufi authors express explicit hostility towards wealth, which is held to divert man from the contemplation and worship of God, and perceive the abandonment of worldly goods as a sign of devotion towards God. Moreover, they attribute to God clear preference for the poor." Similar exchanges between Syriac monastic and rabbinic culture are discussed in Bar-Asher Siegal, Michal, "Shared Worlds: Rabbinic and Monastic literature." *Harvard Theological Review* 105, 4 (2012): pp. 423–456.

89 Cf. Hoyland, "History, fiction and authorship", pp. 23–24. For the religious scholarly background and reputation of most early quṣṣāṣ, see Armstrong, *The Quṣṣāṣ of Early Islam*.

One can hardly underestimate the role of (proto-) Karaite and other non-rabbinic groups as 'cultural brokers' in the early Islamicate world. However, Drory's rather strict division between traditional and innovative models, which builds strongly on poly-system theory, appears a little straightforward. Recent research noticed the hybridity of religious, social and cultural affiliations in early Muslim times when Islam itself was still in a stage of religio-political emergence and identity formation.[90] Such a cultural mélange allowed for much more dialogue between Arab or Arabic speaking groups and other cultural 'players' (Byzantine-Greek, Syriac-Christian, Persian, Central Asian, Jewish).

What Seth Schwartz has called "compartmentalized Jewishness"[91] with regard to Hellenized Jews, has been described by Fred Astren for the period under discussion here as follows:

> Consequently, notions of hybrid identity suggest that Jews formally could have accepted triumphant imperial Islam in various ways, while remaining participating Jews within their existing local communities. [...] From the Muslim perspective, widely embracing definitions of Muslim identity are known from this era, demonstrating that hybridity was enmeshed in the discourse of Islam itself.[92]

As the above discussion has shown, the analysis of later Midrashim indicates that these developments were already reflected in more traditional rabbinic genres, usually perceived merely as remnants of a has-been Golden Age. However, their systematic topical arrangement and framing structures can be understood as incipient stages of programmatic agendas, introductory elaborations and systematic conclusions, which became typical of Arabic, geonic and Karaite treatises in the medieval period. In ways structurally similar to these monographs, the three Midrashic examples in general, and *Seder Eliyahu Zuta* in particular, de-

90 Thus, writes Khalek, Nancy, *Damascus after the Muslim Conquest. Text and Image in Early Islam*. New York: Oxford University Press, 2011, p. 5: "The early Muslim community, always a minority in early medieval Syria, developed its initial imperial identity in a Byzantine milieu. [...] Their expressions of religious and political hegemony, whether in form of physical monuments or textual compilations, made use of and elaborated upon images and tropes that resonated with the mixed Christian and Muslim population." More boldly states Hughes, Aaron, "The formative period of Islam and the documentary approach: a prolegomenon," in *A Legacy of Learning. Essays in Honor of Jacob Neusner*, ed. by Alan J. Avery Peck et al., pp. 372–385. Leiden: Brill, 2014, here p. 377: "The result should not be the monolithic Islam of later centuries, but a host of Islams that skirmish with one another over the nature of political authority, the inheritance of this authority, and who has (and has not) a rightful claim to it." For other notions of hybridity in the early Islamicate world see above, note 13.

91 Schwartz, *Imperialism and Jewish Society*, p. 175.

92 Astren, "Islamic Contexts", p. 154.

ploy clear, narrative expositions as their introductions. These might have been less structured, but were highly programmatic in elaborating central topics, key terms and their hermeneutic, literary, or religious-ethical agenda. Moreover, these texts also feature an author or narrator character lending authenticity and authority to the work, while shifting away from traditional models of multi-vocal discussions, Talmudic dialectics and named attributions to rabbinic sages.

Moreover, these observations prompt us to reflect on Saʿadya Gaon's achievements and his role as a pioneer and a maverick who made inroads into geonic culture and transformed it from within. Talya Fishman has pointed out compellingly, that it was only his open-mindedness on his way from the cultural melting pot of Old Cairo (*Fustat*) in Egypt via Palestine replete with its Masoretic, Scripturalistic and (proto-) Karaite groups and his conversance with Arabic culture of learning that allowed him to introduce so many new features into early medieval rabbinic culture.[93] I would like to seize Fishman's suggestion for the contextualization of the Midrashic traditions in the early Islamicate period. It seems highly unlikely that Saʿadya – his designation as a revolutionary champion of tradition notwithstanding – was the first and only one who all at once and out of the blue turned Jewish geonic culture upside down. Even though the idea of an all-encompassing genius is tempting, one may assume that others before him had been traveling down this road and he could base himself on previous, rather subtle and slow cultural developments. While his precursors and contemporaries remained largely anonymous and did not gain the top of our historical consciousness, the previously discussed texts may allow us to catch a glimpse of such cultural shifts in the making.

I hope it has become clear from the previous discussion that the reference to James Scott's idea of "hidden transcripts" in my title does not serve to interpret the cultural interactions in the early Islamic period according to patterns of power relations between dominant and dominated cultures. As we have seen above, in a dense web of cultural interactions and appropriations labels like 'dominant' and 'dominated' get easily subverted when the alleged 'subaltern' not only speaks, but engages in a multi-layered and sophisticated play with iden-

93 Earlier Scholarship emphasized Sa'adya's outstanding role as a cultural and religious pioneer, who appropriated several modes of writing (treatises, Bible-commentaries, prose texts and lexical works in Hebrew, Arabic and Judeo-Arabic) and introduced new philosophical and theological ideas. Cf. Drory, *Models and Contacts*, p. 140 – 146 and 178 – 190; Brody, *The Geonim*, pp. 235 – 332. For the nuanced view, see Fishman, Talya, *Becoming the People of the Talmud Oral Torah as Written Tradition in Medieval Jewish Cultures*. Philadelphia: University of Pennsylvania Press, 2011, pp. 20 – 64, esp. p. 57 – 58.

tities, cultural models and religious ideas.[94] However, I very much appreciate Scott's argument that alongside official relations between social groups ('public transcripts'), mostly affirmative of the socio-political status quo, one finds other "discourses that takes place 'offstage'."[95] In fact, I chose "hidden transcripts" as an apt metaphor with a double frame of reference. On the one hand, I would like to indicate that the interplay between Midrash and the changing cultural context in the Islamicate '*oikumene*' have remained until recently 'hidden' from the eyes of many scholars. This was first and foremost due to misleading assumptions about a sharp dichotomy or schismatic break between a Rabbanite mainstream and a sectarian Karaite movement with their distinctive literary traditions.[96] Moreover, also bi- or multi-directional processes of intercultural and interreligious exchange and transfer have been discussed frequently based on a paradigm of 'influence' and presupposing the existence of religions or cultures (Arab-Muslim vs. Jewish vs. Syriac-Christian etc.) as monolithic entities. Such approaches ignored, by and large, the more subtle shifts and developments that took place within and between various groups that constituted this particular cultural formation of the early Islamicate world in different regions.

---

94 This word play refers to the pioneering essay of post-colonial studies by Gayatri Chakravorty Spivak. "Can the Subaltern Speak?," in *Marxism and the Interpretation of Culture*, ed. by Cary Nelson and Lawrence Grossberg, pp. 271–313. Urbana, IL: University of Illinois Press, 1988.
95 Scott, James C., *Domination and the Arts of Resistance: Hidden Transcripts*. New Haven: Yale University Press, 1990, here p. 4. Indeed, if one would like to speak about 'public and hidden transcripts' regarding power relations, this will not pertain to the relations between Jews and Muslims, but rather will discuss the hiddenness of cultural dialogues with Arabic-Muslim and Syriac-Christian culture within Judaism itself , especially in its Rabbanite or Geonic manifestations. This would correspond to those developments outlined in the chapters on decline of the yeshiva and on "Jewish culture outside the yeshivot," in *Samuel ben Hofni Gaon*, ed. by Sklare, pp. 69–141.
96 New studies show that this sharp dichotomy was by and large a scholarly and ideological construct. Probably, there was no such well-defined group as *the* Karaites. This movement and (and their precursors) were, in fact, not a strange sect on the margins of a rabbanized Jewry, but they represented a vital cultural element in a multifaceted society in which one's religious affiliation constituted but one of many aspects of identity. Cf. Marina, Rustow, *Heresy and the Politics of Community. The Jews of the Fatimid Caliphate*. New York: Cornell University Press, 2008, pp. 111–175, 239–289 and 347–355; Marina, Rustow, "Laity versus Leadership in Eleventh-Century Jerusalem: Karaites, Rabbanites, and the Affair of the Ban on the Mount of Olives," in *Rabbinic Culture and Its Critics*, ed. by Daniel Frank and Matt Goldish, pp. 195–248. Detroit: Wayne State University Press, 2008; Marina, Rustow, "Karaites Real and Imagined: Three Cases of Jewish Heresy." *Past and Present* 197 (2007): pp. 35–74; Greenspahn, Frederick E., "Sadducees and Karaites: The Rhetoric of Jewish Sectarianism." *Jewish Studies Quarterly* 18, 1 (2011): pp. 91–105; Astren, *Karaite Judaism*, pp. 5–10.

On the other hand, due to their a-historic nature and their embeddedness in a long tradition of discourse the interaction between Midrashic texts and their socio-cultural contexts are often largely hidden and not easy to grasp. The present study tried to highlight, of course in a still preliminary way, some of the areas that most probably formed a shared discursive space for people from various religious, cultural and educational backgrounds in this period.

The papers presented throughout the conference have shown how recent scholarship has substantially shifted away from the mono-lateral concepts of 'influence' in favour of more complex processes of cultural negotiation, borrowing and appropriation. With such an approach, exegetical and cultural crossroads are no longer areas of a 'parting of ways', but rather of cross-fertilisation. It seems quite likely that, in the cultural formation of the early Islamicate world, the authors of the later Midrashic traditions were receptive to and probably participating in several overlapping, and sometimes 'peripheral' discourses shaped as much in Masoretic, (proto-) Karaite or esoteric circles as in their Muslim and Christian surroundings. Particularly striking in this regard is the role that 'Persian elements' might have played in these complex interactions. This specific geographic realm apparently constituted an important contact zone of such, initially marginal, discourses that would quickly make inroads into the heart of Arabic and Jewish culture. Among those elements, we can number *Adab* and other literary or cultural models. Also the roots of the 'science of language' (Arabic and Jewish grammarians) and of Mutazilite thinking as well as the origins of early Muslim asceticism with Hasan al-Basri and Shaqiq al-Balkhi can be traced back to this region. Moreover, the ancient Persian Khorasan as well as the Khuzestan regions were not only hubs for traded goods but also fertile grounds for cultural, religious and political ideas, bringing forth some chiliastic or messianic, hybrid movements (*Abu Isa al-Ishafani, Yughdan, Maghariyya*) as well as Jewish freethinkers (*Hiwi al-Balkhi, Benjamin al-Nahwandi, Daniel al-Qumisi, Jacob Quirqisani*), who later had a large impact on nascent Karaite culture. Against this backdrop, many Arabic cultural models would not have attracted different educated Christians and Jews *per se* because of their innovative nature. In fact, it might have been their already existing acquaintance with those pre-Islamic models and cultural patterns that facilitated their appropriation in the multi-layered discourse of the early Islamicate world.[97]

---

97 On the hybrid movements in Persia, see note 13; and Choksy, Jamsheed K., *Conflict and Cooperation: Zoroastrian Subalterns and Muslim Elites in Medieval Iranian Society.* New York: Columbia Univ. Press, 1997; and Anthony, "Chiliastic Ideology and Nativist Rebellion in the Early ʿAbbāsid Period." The messianic overtones in the late Umayyad period and in early Abassid culture are discussed in Yücesoy, Hayrettin, *Messianic Beliefs and Imperial Politics in Medi-*

This humble and preliminary survey intended to make some of the "hidden transcripts of late Midrash" more visible. However, it needs further and more detailed studies to grasp the whole dimension of their transformative function as an important link in the transition from late antique to medieval Jewish culture.

---

eval Islam: The ʿAbbāsid Caliphate in the Early Ninth Century. Columbia: University of South Carolina Press, 2009, esp. pp. 18–58. For the Syrian cultural sphere and its messianic and chiliastic movements in particular, see also the instructive study by Anthony, "Who was the Shepherd of Damascus." For Jewish apocalyptic and messianic traditions from Roman Byzantine into early Islam, see Himmelfarb, Martha, *Jewish Messiahs in a Christian Empire. A History of the Book of Zerubbabel.* Cambridge, MA: Harvard University Press, 2017.

On *Shaqiq* and other Persian ascetics, see Gramlich, Richard, *Alte Vorbilder des Sufitums, Band 42, Teil 2.* Wiesbaden: Harrassowitz, 1996, esp. pp. 13–62; 249–266. On the Persian connotation of *Adab*, see Toorawa, "Defining *Adab.*"; Bosworth, Charles, "The Persian Impact on Arabic Literature," in *Arabic Literature to the End of the Umayyad Period*, ed. by Alfred F.L. Beeston et al., pp. 483–496. Cambridge: Cambridge University Press, 1983. On the familiarity with and adoption of (Sassanian) Persian cultural patterns in Eastern Christianity, see Payne, Richard, *A State of Mixture: Christians, Zoroastrians, and Iranian Political Culture in Late Antiquity.* Berkeley: University of California Press, 2015; and id. with Myles Lavan and John Weisweiler, eds., *Cosmopolitanism and Empire: Universal Rulers, Local Elites, and Cultural Integration in the Ancient Near East and Mediterranean.* Oxford: Oxford University Press, 2016. For the importance as cradles of, still overtly non-systematical, language sciences, see Contini, "Aspects of Linguistic Thought", p. 103, n. 43. Cf. also above pp. 208–211, notes 29–37.

Martin Accad

Theological Deadlocks in the Muslim-Christian Exegetical Discourse of the Medieval Orient

Identifying a Historical *Meta-Dialogue*

1 Introduction

Islam emerged in the 7th century, as the dust of the most turbulent Christian theological debates was largely settling, even though it was leaving behind it a trail of divisions and disagreements, at least as much in political as in theological realms. Arab Christianity as an identifiable theological tradition emerged as a result of Islam's encounter mainly with Syriac, Greek and Coptic Christian traditions. The resulting theological discourse in Arabic, therefore, was almost entirely apologetic in nature – a discourse born in conflict, often on the defensive. Intellectual resources immediately available to Syriac, Greek and Coptic Christians theologizing in Arabic were often either simplistic pre-Nicene language, or complex Chalcedonian arguments for which the new Arab interlocutor was not prepared. The resulting dialogical discourse is 1400 years of mostly circular and often insular arguments. Most intriguing, however, is the large amount of Biblical exegetical material that developed within the Islamic tradition. While the phenomenon itself is remarkable, an exploration of the Islamic exegetical discourse on the Bible reveals a serious epistemological challenge: the traditional understanding of the Qur'ān is the core hermeneutical key to the Muslim discourse on the Bible. Nevertheless, this exegetical discourse can no longer be ignored in the Christian academia when approaching the Bible in the world we live today. In the present paper, I will affirm the importance of recognizing the Muslim exegetical discourse as a key hermeneutical context for Christians theologizing in the setting of Islam today. But in order to do this in a new and creative way that avoids the pitfalls of history, I will propose a three-step method. First, study the *story* of the historical argument. Secondly, identify the *meta-dialogue* where the deadlock occurred. And third and finally, search for *unlocking keys* from within the Muslim discourse, recognizing at the same time its complex diversity.

https://doi.org/10.1515/9783110564341-012

2 Background and Rationale

In his book *Whose Justice? Which Rationality?* Alasdair MacIntyre defines tradition as "an argument extended through time in which certain fundamental agreements are defined and redefined"[1]. He further explains that this shaping and sharpening takes place through both internal and external debates. I would argue that, in line with MacIntyre's definition of tradition, we may refer to Arab Christianity as a theological tradition in its own right. I am aware that such identifications can be controversial, as one might ask whether the usage of the word "Arab" refers to an ethnic, linguistic, nationalist, geopolitical, or some other reality. Although I would assert that none of these categories is entirely exclusive of the others, I am using the term primarily in its linguistic sense. First of all, it is clear that there were Christians who spoke Arabic in the Arabian Peninsula before the emergence of Islam, particularly in Northern Arabia and in the Yemen region. But these groups, even though they used Arabic in every-day life, probably continued to use Syriac as their sacred language well into the Islamic period.

Evidently, Arab Christianity as a relatively independent theological tradition began to emerge when primarily Greek-, Syriac- and Coptic-speaking Christians started using the Arabic language of their 7th-century Muslim masters in their writings as well as translation of existing religious texts. If we can refer to a Greek theological tradition, as well as to Latin, Syriac, Coptic, Armenian, and other ones, then we can also rightly refer to an Arab Christian theological tradition. Though all these traditions have a common theological starting point that emerges from the text of the Bible, each tradition has developed its own theological discourse in its own language, in dialogue with other traditions that have expressed themselves with the same language in their particular context. This is a very important starting point. Every theological discourse is affected and shaped by its context, and much of it is therefore essentially apologetic in nature – a continual attempt at explaining Christian doctrines in an ever-changing world.

Inasmuch as all theological discourse is dialogical in nature, language, then, is decisive in determining the nature of that discourse. It is therefore the underlying hermeneutical questions of a particular context that largely, and often subconsciously, set the agenda of a particular theological tradition. The theological discourse of that linguistic tradition emerges as a means to address the contex-

1 MacIntyre, Alasdair, *Whose Justice? Which Rationality?* Notre Dame: University of Notre Dame Press, 1988, p. 12.

tual questions. There are, naturally, a myriad of other factors that affect the nature of any theological discourse, such as certain elements in the culture, social structures and traditions, the temperament, background and maturity of a theologian, etc. But a full analysis of the elements constituting the theological discourse of the Arab Christian tradition is well beyond the scope of the present paper. For our present purposes, it is sufficient to agree that the Arab Christian theological tradition may be studied in its own right from the perspective of intellectual history, allowing us to identify a *meta-dialogue*. What I mean by *meta-dialogue* is the overall development of the Christian-Muslim discourse through the history of encounter between the two communities. This discourse is not static; the core themes between Arab Christians and Muslims are in continuous flux across the centuries from the beginning of Islam's emergence and up to the present day. Based on this premise, we will be able to identify both the overall Islamic hermeneutical context of the *meta-dialogue*, as well as the development of the Arab Christian theological discourse within it.

3 The Muslim Exegetical Arguments

Five major themes of disagreement can be identified, which have emerged and developed through the history of encounter between Christians and Muslims. These have led to a sort of theological deadlock at this point in history. They may be summarized by the following Muslim affirmations against Christians: (1) "Your Scriptures are corrupt;" (2) "God cannot have a son;" (3) "You believe in three gods;" (4) "Jesus was not crucified;" and (5) "Muhammad was the promised Paraclete." As a result of these deadlocks, it is rare to find Christians and Muslims today who are willing to engage in theological conversations together, particularly in more official dialogue settings. Most prefer to remain within the safer confines of what has become known as a dialogue of life. Though recognizing the importance of this existential type of dialogue, I would argue that it is a mistake to stop dialoguing about core doctrines. Indeed, as we stop talking about essential areas of disagreement, we risk the multiplication of mutual stereotypes and prejudices between our two communities.

For example, I would advance that, from my study of Muslim writers' interpretation of Gospel passages between the 8th and 14th centuries, a generally far more complex and diverse theological discourse emerges than the five simple accusations stated above. The accusations as I have worded them already represent an advanced level of stereotyping and a stalling of real and dynamic dialogue. In the remaining of this paper, I will tackle only the first accusation as a case study, and will do this in three steps, which I propose as a hermeneutical method towards progress in Christian-Muslim theological dialogue. Clearly, within my space limitations, I will also only do this in broad lines, since a more thorough job requires getting into far more details. In the first step (1), I will study the *story* of the Muslim argument historically, in other words, identifying its rhetorical origins and its historical development. Secondly (2), based on this *story* I will attempt to identify instances where the deadlock has occurred within the *meta-dialogue*. And finally (3), I will propose some *keys* from within the complex and diverse Muslim discourse that may perhaps be used to unlock the theological dialogue today.

4 The Corruption of the Judeo-Christian Scriptures

4.1 State of the Argument: Some Qur'ānic Origins and Historical Developments

When we come to the familiar Muslim accusation that Christians and Jews have corrupted their own Scriptures – the accusation of *taḥrīf* – both Muslims and Christians assume that it originated in the four Qur'ānic occurrences that refer to the verb "*ḥarrafa*" or one of its derivatives. But as I have argued in an article dedicated to the history of the *taḥrīf* accusation,[2] the Qur'ānic polemic is in fact far less straightforward than we might assume. It has become well argued[3] that by the time we find it in Ibn Taymiyya's polemical work in the late 13th/early 14th century, *al-Jawāb aṣ-ṣaḥīḥ li-man baddala dīn al-Masīḥ*,[4] the accusation was recognized as having two dimensions: first, the most widely used claim that Christians and Jews had corrupted the meaning of their Scriptures (*taḥrīf al-maʿnā*), and only beginning in the 11th century, the broad belief that Christians and Jews had also corrupted the text of their Scriptures (*taḥrīf al-lafẓ*).

Though most Muslims today assume that the Judeo-Christian Scriptures have undergone textual corruption, and most Christians assume that this has always been the Muslim accusation against the Bible, the evidence indicates that even

2 Accad, Martin, "Corruption and/or Misinterpretation of the Bible: The Story of the Islamic Usage of Tahrif." *The Near East School of Theology Theological Review* 24 (2003): pp. 67–97.
3 In addition to my own work, see also Michel, Thomas, *A Muslim Theologian's Response to Christianity: Ibn Taymiyya's Al-Jawāb al-Ṣaḥīḥ*. New York: Caravan Books, 1984, and Abdullah Saeed, "The Charge of Distortion of Jewish and Christian Scriptures." *The Muslim World* 92 (2002): pp. 419–436.
4 Ibn Taymiyya, Taqī al-Dīn, *Al-Jawāb aṣ-saḥīḥ li-man baddala dīn al-Masīḥ*. Cairo: 1905.

Ibn Taymiyya, though he was aware of the existence of both arguments, nevertheless favored *taḥrīf al-ma'nā* and was not fully convinced of *taḥrīf al-lafẓ*.

As far as I have discovered, Ibn Ḥazm of Andalusia was the first to use the accusation of textual corruption extensively in the 11th century.[5] Some, such as al-Juwaynī (11th century),[6] al-Qarāfī (13th century)[7] and Ibn Qayyim al-Jawziyya (14th century)[8] followed in his footsteps. But judging from Ibn Taymiyya's position, possibly as late as the early 14th century, this hardened approach was not necessarily the only remaining option to think about the Judeo-Christian Scriptures even in the centuries that followed.

4.2 Instances of Deadlock in the *Meta-Dialogue* and their Implication for Muslims and Christians

A compilation of the *meta-dialogue* on the textual corruption of the Bible, spanning the first six centuries of Christian-Muslim engagement, seems to indicate that the accusation originated neither in the Qur'ān nor in plain logic, but rather in rhetorics, and specifically in the Muslim discourse on the proof of Muḥammad's prophethood (*tathbīt an-nubuwwa*). One particular instance of this argument's matrix can be found in the 8th-century encounter between Timothy I and the Caliph al-Mahdi.[9] At one point in this well-known dialogue, the Caliph challenges the Patriarch by asking him why he does not believe in Muḥammad, since 'Īsa had preannounced that the prophet of Islam would come after him. This is a clear reference to Q 61:6. Timothy responds that Jesus never announced the coming of Muḥammad, otherwise Christians would have believed in him. To the Caliph's invitation to consider Jesus' promise of the Paraclete in John's Gospel as having been fulfilled through the coming of Muḥammad, the Patriarch retorts that the Paraclete has nothing to do with the prophet of Islam. Herein lies the turning point in this issue. Al-Mahdī asserts that if what the Patriarch is say-

5 Ibn Ḥazm, Abū Muḥammad 'Alī b. Aḥmad, *Kitāb al-fiṣal fī l-milal wa-l-ahwā' wa-n-niḥal*, 3 parts. Cairo: al-Maṭba'a al-adabiyya, 1900.

6 Al-Juwaynī, Abū l-Ma'ālī 'Abd al-Malik b. 'Abdallah, *Shifā' al-ghalīl fī l-tabdīl; luma' fī 'awā'id ahl al-sunna. Textes apologétiques de Ğuwainî (d.1085)*, ed. and trans. by Michel Allard. Beirut: Dar al-machreq, 1968.

7 Al-Qarāfī, Shihāb al-Dīn, *al-Ajwiba al-fākhira 'an al-as'ila al-fājira*, ed. by Bakr Zakī 'Awaḍ. Cairo: Maktabat wahba, 1987.

8 Al-Jawziyya, Shams al-Dīn Muḥammad b. Abī Bakr Ibn Qayyim, *Hidāyat al-ḥayāra fī ajwibat al-yahūd wa-n-naṣāra*, ed. by Aḥmad Ḥijāzī al-Saqā. Egypt: 1980.

9 Putman, Hans, *L'église et l'Islam sous Timothée I (780–823)*. Beirut: Dar al-machreq, 1975.

ing is true, then it is either that Christians are not interpreting their Scriptures correctly (*taḥrīf al-ma'nā*), or that they must have removed the evidence that supports Muḥammad's prophethood from their Scriptures (*taḥrīf al-lafẓ*), because the Qur'ān is clear about this fore-announcement, and the Qur'ān cannot err. Timothy retreats to a defensive position and asks rhetorically: "Where is that book that is free of corruption (*taḥrīf*) from which you have learnt that we have corrupted our book? Bring it forth that we might see it and apply ourselves to it and abandon the corrupted book."[10]

This moment in the *meta-dialogue* marks a turning point, because for the next two hundred and fifty years, Muslims essentially followed the first possibility. It was thanks to Christian converts to Islam who actually knew their Scriptures well, such as 'Alī b. Rabbān al-Ṭabarī in the 9th century, that the Bible in both its testaments became a vast source of proof for the prophethood of Muḥammad. In his *Kitāb ad-dīn wa-d-dawla*, for instance, every occurrence of the expression "the Glorious One," in reference to God, is taken as a reference to the *Muḥammad*, "the most praised one." Likewise, every reference to the Paraclete in John's Gospel is taken as a promise of Muḥammad's coming.[11]

Numerous other verses are brought in as further proof. No possibility is left unexplored, to the point where Ibn Taymiyya even interprets Jesus' reference to the Devil in John 14:30 as a reference to Muḥammad: "I will not speak with you much longer, for the prince of this world is coming. He has no hold on me."[12] The absurdity of this self-reference illustrates well how insular the conversation had become.

But from the 11th century onwards, essentially beginning with the writings of Ibn Ḥazm, it is al-Mahdī's second possibility which is favored. Namely, the reason for the absence of occurrences of Muḥammad's name and announcements of his coming in the Bible is that Christians and Jews have corrupted and altered their Scriptures, removing all such prophecies from them. Though the first argument was not very sound exegetically or honest hermeneutically, at least it allowed for real text-based engagement between Christians and Muslims. Once the integrity of the text was called into question, real engagement all but stopped. There seems to be enough evidence to suggest that from the 11th century onwards, few Muslim writers actually engaged with the Biblical text. They were instead satisfied with lists of verses and parts of verses, largely taken out of

10 Putman, *L'église et l'Islam*, p. 26, stanza 125.

11 'Alī al-Ṭabarī, *Kitāb ad-dīn wa-d-dawla*, ed. by Alphonse Mingana. Manchester: Longmans, 1923.

12 Ibn Taymiyya, *Al-Jawāb aṣ-Ṣaḥīḥ*, vol. 2, part 4, p. 15: "The ruler (*arkūn*) of this world is coming and I have nothing." [Translation: M. A.]

context, which proved – at least to them – that the Judeo-Christian Scriptures had become irreparably corrupt. This virtually marked the death of any fruitful or constructive theological dialogue.

Typically, Christians have tried to counter the Muslim accusation of textual *taḥrīf* by constructing logical arguments indicating the Qur'ān's affirmation of the Judeo-Christian Scriptures in numerous places, and arguing for the logistical impossibility of the Bible's textual corruption after the advent of Islam. Alternatively, they have fallen into the trap of adopting the Islamic view of the Qur'ān's revelation as *tanzīl* and projected it on their own understanding of the doctrine of the Bible's inspiration. And on this basis, they attempt to demonstrate that the Bible is not corrupted by trying to resolve all the apparent contradictions or discrepancies between various Biblical narratives and passages.

4.3 Potential *Keys* to the Deadlock

The *meta-dialogue*, then, indicates that the Muslim accusation of textual *taḥrīf* is neither a necessary conclusion deriving from Qur'ānic exegesis, nor is it aligned with the Qur'ān's generally sweeping affirmation of the authenticity of the Biblical message. Indeed, the Qur'ān refers to the Judeo-Christian Scriptures largely as a foretaste of Muḥammad's own message that affirmed it and was a continuation and extension of it. Some keys that may help us out of this dialogical deadlock, therefore, may more readily be found within the Muslim discourse itself.

The first step is to recognize the plurality of the accusation by pointing out, as I have already done, that, until the 11th century, the hardline accusation of textual *taḥrīf* was nowhere prominent. Then within this pre-11th-century period, it is most inspiring to point out to the diversity and complexity of Muslim writers' use of the Biblical text. Al-Qāsim b. Ibrāhīm ar-Rassī, a 9th-century Zaydī Shiite from Yemen, cited extensively from the Gospels.[13] One might even argue that he developed his own adaptation of the text into the *saja'*-style of the Qur'ān in his *Kitāb ar-radd 'alā an-naṣārā*. The Shiite historian al-Ya'qūbī does the same, citing large portions from all four Gospels in his coverage of the historical period of Jesus and his immediate followers in his *Tārīkh*, which he completed in 872.[14] The writings of both these men betray no heavy agenda. The Biblical text is used as a reliable theological and historical document, with minor changes introduced, ei-

13 In Di Matteo, Ignazio, "Confutazione contro I cristiani dello zaydita Al-Qâsim b. Ibrâhîm." *Rivista degli Studi Orientali* 9 (1922): pp. 301–364.

14 Al-Ya'qūbī, Aḥmad b. Abū Ya'qūb, *Kitāb at-tārīkh*, ed. by Martijn Theodoor Houtsma, 2 vols. Leiden: Brill, 1883.

ther for stylistic reasons or to make it more palatable to their own coreligionists. Contemporaneously, during the latter half of the 9th century, the well-known and respected Qādī Ibn Qutayba goes so far as to use the Biblical text to supersede the authority of more traditional Muslim sources as he weighs in on the reliability and authenticity of questionable Islamic *aḥādīth*, particularly in his *Kitāb ta'wīl mukhtalif al-ḥadīth.*[15]

In today's engagement of Christians and Muslims in theological dialogue, inspiration can be drawn from these genuinely Islamic, yet far more creative approaches of the past. Starting out with these possibilities, Christians can begin to unpack a more authentic understanding of Biblical inspiration based on a creative and constructive theological discourse, rather than a defensive one. And if the deadlock in the *meta-dialogue* on this issue had originated in the Muslim disappointment with the Christians' unwillingness to endorse the prophethood of Muḥammad, surely today's maturing relational dialogue of life, as well as the progress of historical research into Islamic origins, opens new possibilities for Christians to affirm their respect and appreciation of Islam's prophet, and even perhaps the possibility of a more theological discourse that would consider a role for Muḥammad in a Biblical view of prophethood.

5 Conclusions

The present paper merely attempts to draw up the outline of a proposed methodology towards progress in – specifically – the theological type of dialogue between Christians and Muslims. After selecting one of the central theological themes of the Christian-Muslim disagreement – here the accusation of *taḥrīf* – we began by establishing the state of the argument by tracing the *story* of the Muslim-Christian dialogical discourse on the issue, what we have called the *meta-dialogue*. Secondly, we have attempted to identify specific instances or moments within the *meta-dialogue*, which essentially have become the matrix of the theological deadlock. And thirdly, we have proposed certain *keys* to help us come out of the theological deadlock. But these are not drawn purely from apologetics or polemics, or even from some sharp philosophical argument. Rather, they are inspired by other, more creative and constructive, instances in the same Muslim exegetical discourse on the Biblical text. These *keys* can then form the basis of a renewed discourse based on the foundations of friendship and relationship. This

15 Ibn Qutayba, Abū Muḥammad 'Abdallāh b. Muslim, *Kitāb ta'wīl mukhtalif al-ḥadīth*. Cairo: 1908.

method derives from the understanding that the deadlock in much of the Christian-Muslim dialogical discourse is not principally the result of rational inquiry, or even the conclusion of a careful exegetical exercise, but rather that it reflects a state of stagnation on an important issue, resulting from specific negative instances in the *story* of the debate on that issue through history. Due to space limitations, we have presented only one case study out of the five key issues initially identified. And this was done only in very broad lines, leaving out all specific examples and with them the actual meat of the argument.

Nicolai Sinai

Two Types of Inner-Qur'ānic Interpretation[1]

1 Introduction

The conference from which the present volume has emerged was entitled *Exegetical Crossroads*. Unlike other contributions to this book, mine will not examine intersections between post-Biblical and post-Qur'ānic scriptural interpretation in Judaism, Islam, and Christianity; rather, I shall focus on processes of interpretive engagement with Qur'ānic passages that are traceable within the Qur'ān itself. Yet this, too, will afford us the opportunity to inspect a crossroads of sorts: for one of the respects in which the Qur'ān intersects with Biblical literature is precisely insofar as it contains intriguing cases of scriptural self-interpretation. Since that phenomenon is much better researched with regard to the Bible, my main objective here is to present some of the ways in which it manifests itself in the Islamic scripture.[2] In doing so, I shall draw attention to some salient similarities and differences between the Hebrew Bible and the Qur'ān. My interest is squarely in the Qur'ān, however; I do not pretend to even remotely offer anything resembling a full account of inner-Biblical interpretation.

---

1 This book chapter was completed in January 2015 and only minor corrections were made afterwards. I would like to express my gratitude to Andrew Bannister for an extended response to a draft version of this chapter, as a result of which I have substantially refined, reconstructed, and expanded my argument in several places. I am likewise grateful to Marianna Klar for miscellaneous corrections and customarily thoughtful comments on a previous version of this chapter. To Georges Tamer I am indebted for tolerating a steady influx of supposedly final versions that were quickly superseded. – My English translations of Qur'ānic passages are based, with frequent modifications, on *The Qur'ān*, trans. by Alan Jones. Cambridge: Gibb Memorial Trust, 2007. Like Jones, I employ superscript "s" and "p" in order to mark verbs and pronouns for which singular and plural forms are undistinguishable in modern English.

2 For an overview of inner-Biblical interpretation see Fishbane, Michael, "Inner-Biblical Exegesis," in *Hebrew Bible / Old Testament: The History of Its Interpretation. From the Beginnings to the Middle Ages (Until 1300)*, ed. by Magne Sæbø, vol. 1, pp. 33–48. Göttingen: Vandenhoeck & Ruprecht, 1996, which conveniently epitomises Fishbane, Michael, *Biblical Interpretation in Ancient Israel*. Oxford: Oxford University Press, 1988; Menn, Esther, "Inner-Biblical Exegesis in the Tanak," in *A History of Biblical Interpretation*, ed. by Alan J. Hauser and Duane F. Watson, pp. 55–79. Grand Rapids: Eerdmans, 2003. For a preliminary treatment of inner-Qur'ānic interpretation see Sinai, Nicolai, *Fortschreibung und Auslegung: Studien zur frühen Koraninterpretation*, Diskurse der Arabistik 16. Wiesbaden: Harrassowitz, 2009, pp. 1–22, 59–160, with further references to previous Biblical and Qur'ānic scholarship.

https://doi.org/10.1515/9783110564341-013

Before commencing, it may be useful to point out that I take for granted the possibility of making chronological distinctions within the Qur'ānic corpus, based on the assumption that the mean verse length of Qur'ānic sūras increased over time.[3] In addition, I shall assume that the growth of the Qur'ānic corpus paralleled the prophetic career of Muḥammad. In particular, I shall be operating with the customary contrast between Meccan and Medinan sūras, or sūras proclaimed before and after the *hijra* – the latter being distinguished from the former by their consistently high verse length, the ascription to the Qur'ānic messenger of a role of political and military leadership, injunctions to religious militancy, an interest in the detailed regulation of human behaviour, explicit polemics against Judaism and Christianity, references to a prior expulsion or emigration of the Qur'ānic community from the "sacred place of prostration" (e.g., Q 2:217–218), and various terminological peculiarities related to all of the preceding thematic features.

My contribution falls into two parts addressing two different types of Qur'ānic self-interpretation. Let us assume that a given Qur'ānic passage B plays a broadly interpretive role with respect to some temporally earlier passage A, meaning that B in some sense elucidates, re-interprets, or qualifies A. If B was incorporated into the same sūra as A, we are faced with a case of interpretively motivated secondary expansion and revision, which forms the subject of Section 2. If, by contrast, B occurs in a separate and formally independent Qur'ānic sūra, I shall speak of "interpretive backreferencing," to which Section 3 is dedicated. Throughout, I employ an intentionally broad notion of interpretation.

2 Interpretation by Means of Secondary Additions

The study of how existing textual units were subsequently revised and reshaped is a staple of modern Biblical scholarship.[4] Frequently, such revision was carried

3 See Sinai, Nicolai, "Inner-Qur'anic Chronology," forthcoming in *The Oxford Handbook of Qur'anic Studies*, ed. by Muhammad Abdel Haleem and Mustafa Shah. Oxford: Oxford University Press.

4 Not being a Biblical scholar myself, I shall not attempt to intervene in the debate about how compellingly we can demonstrate specific phrases, verses, or verse groups in the Hebrew Bible – or more specifically in the Book of Amos, from which many of my subsequent Biblical examples will be drawn – to be later editorial insertions rather than integral components of the original version of the passage at hand. While such arguments always permit residual doubts, it strikes me as generally convincing to view a text like the Book of Amos as a product of editorial growth

out not by suppressing or changing an existing portion of text but rather by expanding it with new material that elucidates, updates, qualifies, complements, or amplifies the original text. In terms of their length, such additions may range from just a few words to entire chapters. At the microstructural end of the spectrum, one finds what Michael Fishbane has termed "scribal comments and corrections" consisting, for instance, in explanatory glosses of ancient toponyms.[5] Here are three Biblical examples:

> Josh 18:13: "[...] to the south side of Luz, that is [*hîʾ*], Beth El [...]"
> 1 Chr 11:4: "And David and all Israel went to Jerusalem, that is [*hîʾ*], Jebus ..."
> Gen 14:17: "The king of Sodom went out to meet him [...] at the Valley of Shaveh – that is [*hûʾ*], the King's Valley."

Such glosses are plausibly regarded as later insertions made in the process of scribal transmission in order to explain expressions that had grown obsolete.[6] At the opposite end of the spectrum we encounter macrostructural textual growth as in Isaiah 40–66, traditionally designated as Deutero-Isaiah (40–55) and Trito-Isaiah (56–66) and normally dated significantly later than the book's original nucleus.[7] At least according to one prominent scholar, the Deutero-Isaianic chapters were composed in close conversation with, and as a deliberate continuation of, the literary deposit of the original Isaiah of Jerusalem.[8]

In between the addition of brief glosses and that of entire chapters, the Hebrew Bible also exhibits what one might call mesostructural insertions consisting of individual clauses, verses, or paragraph-sized blocks of text. An example is provided by the following announcement of judgement, which occurs towards the end of the Book of Amos (9:8):

over time. For an analysis of the Book of Amos that is generally wary of invoking editorial hypotheses, see Paul, Shalom M., *Amos: A Commentary on the Book of Amos*, ed. by Frank Moore Cross. Minneapolis: Fortress Press, 1991.

5 Fishbane, *Biblical Interpretation*, pp. 23–88 (Part One).

6 The inference that glosses must be secondary interpolations is questioned in Eslinger, Lyle, "Inner-Biblical Exegesis and Inner-Biblical Allusion: The Question of Category." *Vetus Testamentum* 42 (1992): pp. 47–58. For a convincing rejoinder, see Sommer, Benjamin D., "Exegesis, Allusion and Intertextuality in the Bible: A Response." *Vetus Testamentum* 46 (1996): pp. 479–489.

7 For a brief survey of Isa 40–66 see Collins, John C., *Introduction to the Hebrew Bible*. Minneapolis: Fortress Press, 2004, pp. 379–400.

8 Williamson, Hugh G. M., *The Book Called Isaiah: Deutero-Isaiah's Role in Composition and Redaction*. Oxford: Oxford University Press, 1994, esp. the conclusion at pp. 240–244. On different accounts of the genesis of the Book of Isaiah see Jüngling, Hans-Winfried, "Das Buch Jesaja," in *Einleitung in das Alte Testament*, ed. by Erich Zenger et al., 5th edition, pp. 427–451. Stuttgart: Kohlhammer, 2004, at pp. 443–445.

> Behold, the eyes of the Lord God are upon the sinful kingdom,
> and I will destroy it from off the face of the earth;
> except [ʾ*epes kî*] that I will not utterly destroy the house of Jacob,
> says the Lord.

Many Biblical scholars would be inclined to construe the indented exceptive clause to be the product of a later editor and to have been inserted in order to mitigate the original passage's categorical announcement of judgement.[9] For a paragraph-length addition, one may point to the announcement of a future restoration of Israel concluding the Book of Amos (9:11–15).[10] Presumably, the point of the insertion was the same as in Amos 9:8, namely, to counterbalance the book's pervasive message of doom with a hope of future salvation that was appropriate to the post-exilic period of Israelite history. Another example of mesostructural expansion is provided by the oracles of judgement in Amos 1:2–2:16. The series consists of eight stanzas, some of which – especially the one directed against Judah in 2:4–5, conspicuous by its Deuteronomistic language – are often considered to be secondary.[11] Once again, the main purpose underlying this insertion would have been to adapt the original text to changing historical circumstances: an original series of oracles directed against the neighbours of the northern kingdom Israel, possibly climaxing with an announcement of the destruction of Israel itself, is updated to include an announcement of the fall of the southern kingdom Judah more than a century later.

9 Collins, *Hebrew Bible*, p. 294: "It is unthinkable that the prophet from Tekoa would have added 'except that I will not utterly destroy the house of Jacob.' To do so would have taken the sting out of the oracle of judgment. For a later editor, however, the addition was necessary. After all, Judah was also part of the house of Jacob." See, however, Paul, *Amos*, pp. 284–285, who argues that the exceptive phrase following ʾ*epes kî* is "not a subsequent addition to the text but rather an integral component part characterizing a contrasting element to the prior statement."

10 See, for example, Collins, *Introduction*, p. 295. Once more, Shalom Paul disagrees (*Amos*, p. 288, with references to previous scholarship).

11 See the summary assessment in Blenkinsopp, Joseph, *A History of Prophecy in Israel*, revised edition. Louisville: Westminster John Knox Press, 1996, p. 75. For a more detailed discussion see Schmidt, Werner H., "Die deuteronomistische Redaktion des Amosbuches." *Zeitschrift für die Alttestamentliche Wissenschaft* 77 (1965): pp. 168–193. A case in favour of the passage's genetic unity is made in Paul, *Amos*, pp. 11–27.

2.1 Interpretive Expansion in the Qur'ān: Some Examples

To what extent does the Qur'ānic corpus contain equivalent phenomena? Interestingly, glossary insertions of the sort illustrated above are not conspicuous in the canonical recension of the Qur'ān, although it is possible that future research will modify this impression.[12] Nonetheless, it is striking that not even the enigmatic reference to a place named *bakka* at Q 3:96 – which Islamic exegetes identify with Mecca – is accompanied by a gloss similar to the Biblical examples presented above. However, textual variants of a glossary character are reported to have been contained in the alternative recensions of the Qur'ān's consonantal skeleton (*rasm*) with which the Islamic tradition credits various Companions of the Prophet.[13] If Goldziher is right in viewing these variants as secondary attempts at elucidation rather than as potentially representing the original wording of the verses in question,[14] then some early transmitters of the Qur'ān would have applied techniques of glossary interpretation that are in certain respects comparable to those found in the Hebrew Bible. In contrast to the Bible, however, these glosses did not succeed in asserting themselves as part of the authoritative recension of the Islamic scripture, despite the fact that they would certainly have reduced ambiguity and facilitated comprehension. This confirms the traditional assumption that the canonical version of the Qur'ānic *rasm* stabilised much more rapidly than the text of the Hebrew Bible, a conclusion also corroborated by other considerations.[15]

---

12 For example, a recent study of mine argues that Q 5:3 contains a secondary insertion part of which can be characterised as playing a "glossary" function (although this insertion is significantly longer than the Biblical examples reviewed above and thus belongs to the category of mesostructural additions). See Sinai, Nicolai, "Processes of Literary Growth and Editorial Expansion in Two Medinan Surahs," in *Islam and its Past: Jahiliyya, Late Antiquity, and the Qur'an*, ed. by Carol Bakhos and Michael Cook. Oxford: Oxford University Press, 2017, pp. 69–119.

13 For example see Goldziher, Ignaz, *Die Richtungen der islamischen Koranauslegung*. Leiden: Brill, 1920, pp. 8–16.

14 The underlying assumption that of two textual variants the more ambiguous one is likely to be more original than the less ambiguous one is certainly plausible. Nevertheless, a fresh discussion and examination of the cases in point is now required in the light of Sadeghi, Behnam and Uwe Bergmann, "The Codex of a Companion of the Prophet and the Qurʾān of the Prophet," *Arabica* 57 (2010): pp. 343–436. Sadeghi posits that at least during the earliest stage of the Qur'ān's textual transmission scribes, whom he considers to have worked from dictation, were more likely to have inadvertently omitted brief textual segments than to have deliberately added them for interpretive purposes (ibid., pp. 387, 400–402, 403–404).

15 Sinai, Nicolai, "When Did the Consonantal Skeleton of the Quran Reach Closure?" *Bulletin of the School of Oriental and African Studies* 77 (2014): pp. 273–292 and 509–521.

The Qur'ān does however exhibit what I have called mesostructural additions. For instance, scholars have identified a number of exceptive clauses similar to Amos 9:8 as likely to constitute secondary additions.[16] One such case occurs at the end of Q 84, which belongs to the early, eschatologically dominated stratum of the Qur'ān:

> 20 What is the matter with them that they do not believe,
> 21 And that they do not bow down when the Recitation is recited to them?
> 22 No! The Unbelievers are in denial.
> 23 But God knows well what they hide.
> 24 So give them the tidings of a painful punishment.
> 25 Though that will not be so (*illā*) for those who believe and do good deeds. They will have an unfailing reward.

Read in isolation from the final verse, the perspective of the passage is grim: it may well be understood to imply that its audience consists exclusively of unbelievers destined for damnation. Verse 25 modifies this message by explicitly recognising that a positive response to the Qur'ānic preaching is possible and reassures the addressees that this will entail a corresponding reward (meaning, of course, entrance to paradise).[17] That v. 25 is in fact a later addition is indicated by a number of observations. First, the verse, coming as it does at the very end of the sūra, can obviously be removed from its context without leaving behind a gap. Secondly, the length of the verse is more than twice the average length of the sūra's remaining verses and thus stands out stylistically.[18] Thirdly, the phrase "those who believe and do good deeds" (*alladhīna āmanū wa-ʿamilū l-ṣāliḥāt*) tends to occur in sūras that are conventionally dated to the Qur'ān's late Meccan and Medinan stages. Even more strikingly, the assertion that "those who believe and do good deeds" will be given an "unfailing reward" also occurs in Q 41:8, virtually a doublet of Q 84:25.[19] The overlap suggests that both verses might be

---

16 See Neuwirth, Angelika, *Studien zur Komposition der mekkanischen Suren: Die literarische Form des Koran-ein Zeugnis seiner Historizität?*, 2nd edition. Berlin: de Gruyter, 2007, pp. 201–202.

17 The prospect of salvation is held out already by vv. 7–9, which there is no reason to view as a secondary insertion. But in what was presumably its original form, the sūra does not give any sign of acknowledging that any of its addressees belonged to those who would be saved.

18 Measuring verse length in the manner explained in Sinai, "Inner-Qur'anic Chronology" (briefly, in terms of the number of transcription letters per verse, excluding space characters and hyphens) yields a length of 57 transcription letters for v. 25 and an average length of 26 for vv. 1–24.

19 In Arabic, Q 84:25 runs *illā lladhīna āmanū wa-ʿamilū l-ṣāliḥāti lahum ajrun ghayru mamnūn*, while Q 41:8 has *inna lladhīna āmanū wa-ʿamilū l-ṣāliḥāti lahum ajrun ghayru mamnūn*. On the phrase "those who believe and do good deeds" see Nöldeke, Theodor and Friedrich Schwally,

roughly contemporary, which would date the addition to Q 84 to the later part of Muḥammad's Meccan period. The function of the presumptive addendum is obviously to tone down the exclusively minatory perspective of the original text, which may reflect the emergence of a Qur'ānic community who would naturally have expected to escape damnation. The verse thus plays a role similar to Amos 9:8, namely, to mitigate a statement that, under changed historical circumstances, would have appeared unacceptably, even unintelligibly, harsh.

Another interesting case of secondary expansion occurs in Q 37:112–113. The verses belong to the sūra's middle part, composed of a cycle of prophetic narratives whose individual episodes are separated by a refrain.[20] The longest pericope in this cycle treats Abraham's confrontation with his idolatrous compatriots (37:83–98) and the near-sacrifice of his own son (37:99–107, corresponding to Gen 22). After the episode's concluding refrain (37:108–111) but prior to the next narrative we encounter the following two verses:

> 112 And We gave him the good news of Isaac as a righteous prophet,
> 113 And We blessed him and Isaac.
> And amongst their offspring (*dhurriya*) are those who do good and those who manifestly commit wrong against themselves (*ẓālimun li-nafsihī mubīn*).

That these verses form a later insertion is already suggested by their appendix-like position following the episode's refrain,[21] on account of which they satisfy the minimum requirement that any hypothetical insertion must meet, namely, of being removable from their context without generating an untenable *non sequitur*. In addition, the couplet has a parallel in Q 2:124, which likewise insists, in the context of what appears to be a concise allusion to Abraham's readiness to sacrifice his son, that Abraham's offspring (*dhurriya*) includes wrongdoers

---

Geschichte des Qorāns, vol. 1: *Über den Ursprung des Qorāns*. Leipzig: Dieterich'sche Verlagsbuchhandlung, 1909, p. 97. Apart from Q 41:8, the formula occurs, for instance, in Q 42:22.23.26, 45:21.30, 47:12, 48:29, 65:11, and 98:7. Q 90:17, 95:6 and 103:3, which also employ the phrase but occur in early sūras, are as likely as Q 84:25 to be later insertions (in the case of Q 90, the insertion probably includes vv. 18–20 as well; see Neuwirth, *Studien*, p. 228).

20 On the structure of the sūra see Neuwirth, *Studien*, pp. 280–281.

21 See Bell, Richard, *A Commentary on the Qur'ān*, ed. by C. Edmund Bosworth and M. E. J. Richardson, vol. 2. Manchester: University of Manchester, 1991, p. 159; cf. Bell, Richard, *The Qur'ān: Translated, with a Critical Re-Arrangement of the Surahs*, vol. 2 Edinburgh: T. & T. Clark, 1937, p. 446.

(*ẓālimūn*).[22] Very likely, Q 37:112–113 are contemporary with the Abraham narrative in Q 2:124 ff., which must be considerably later than the rest of Q 37 and is best assigned to the Medinan period.

What is the purpose of the insertion? First, Q 37:112–113 make an interpretive point with respect to the preceding narrative about Abraham's near-sacrifice of his son. Since the latter is only introduced as a "prudent son" (Q 37:101), his identity remains elusive as long as the sūra is read without vv. 112–113. The latter verses, however, impose a relatively unequivocal answer: placing the annunciation and birth of Isaac after the near-sacrifice signals that the "prudent son" cannot have been Isaac, who is only introduced in v. 112, but must have been Abraham's other son, Ishmael.[23] Incidentally, this indirect identification of the "prudent son" from Q 37 with Ishmael fits well with the previously mentioned link between Q 37:112–113 and Q 2:124 ff., as Q 2:127 casts Ishmael in the important role of assisting Abraham in erecting the Meccan sanctuary. The insertion also makes a second point, though: like Q 2:124 and also Q 57:26 (likewise plausibly considered to be Medinan), it rejects the idea that the merit that Abraham acquired by obeying God's command to sacrifice his son is automatically passed on to his descendants. Given that this notion is a Rabbinic tenet and that Q 37:112–113 are likely to be contemporary with Q 2:124 ff., the expansion of Q 37 by vv. 112–113, too, is best placed in a Medinan context, where Q 37's retelling of Genesis 22 could easily have been appropriated by Jewish members of the Qur'ānic audience as supporting the idea that the physical descendants of Abraham occupy a uniquely privileged status.[24] In this sense, the insertion of Q 37:112–113 likewise responds to the need to make sense of an existing portion of text under novel historical circumstances.

22 "And [recall] when Abraham was tested by his Lord with words, and he fulfilled them. He [God] said: 'I am making you[s] a leader for the people.' He [Abraham] said: 'And of my offspring (*dhurriya*)?' He [God] said: 'My covenant does not extend to those who commit wrong.'"

23 Neuwirth, *Studien*, p. 281. Note that Reuven Firestone disagrees, insisting that "it is impossible to determine conclusively from the Qur'an which son was intended"; see Firestone, Reuven, "Abraham's Son as the Intended Sacrifice (*al-Dhabīḥ*, Qur'ān 37:99–113): Issues in Qur'ānic Exegesis." *Journal of Semitic Studies* 34 (1989): pp. 95–131, at p. 98. Based on an examination of what the post-Qur'ānic tradition has to say about the identity of the intended victim of Abraham's sacrifice, Firestone argues for a general development from reports in favour of Isaac to reports favouring Ishmael.

24 See in greater detail Sinai, *Fortschreibung und Auslegung*, pp. 139–142. In discounting the view that descent from Abraham entails any privileged status, the force of Q 37:113 is similar to the statement that Matt 3:9 ascribes to John the Baptist: "And think not to say within yourselves, We have Abraham to our father: for I say unto you, that God is able of these stones to raise up children unto Abraham."

A more extended portion of text was added at the end of Q 73. The concluding verse of the text, v. 20, stands out already by its inordinate length, which amounts to more than 50% of the entire rest of the sūra,[25] as well as by its use of terminology that is otherwise characteristic of much later texts. For example, the phrase "fighting in the way of God (*yuqātilūna fī sabīli llāh*)" and the verb *tāba* (here used of God in the sense of "to relent towards") are not otherwise found in the short and eschatologically oriented sūras to which the remainder of Q 73 would seem to belong by virtue of its verse length.[26] Coming at the end of the sūra, v. 20 can be extricated without leaving behind an unsustainable gap. Finally, it is easy to think of a motive for why the passage may have been inserted, namely, in order to mitigate the stringent liturgical requirements formulated at the beginning of the sūra, in Q 73:2–4, according to which most of the night (v. 2), or at least half of it (vv. 3–4) are to be spent in vigils. By contrast to these rather severe expectations, which are reminiscent of the devotional practices of Christian ascetics and monks, v. 20 reassures its audience that not everyone is expected to spend "two thirds of the night" in prayer and that it will be sufficient to "recite from the Qur'ān what is possible (*mā tayassara*)" and to "perform prayer and to give alms." In justification of this alleviation, the verse invokes the presence of sick persons, travellers, and fighters among the Qur'ān's addressees. Clearly, v. 20, like many Medinan passages, presupposes the existence of a broad and socially inclusive religious community and therefore adjusts the earlier practice of extended vigils accordingly.

2.2 Interpretive Expansion in the Qur'ān: General Comments

A review of all putative cases in which a Qur'ānic passage underwent secondary expansion would be far beyond the scope of this contribution. Instead, I should like to make a number of general comments. The first question that should probably be explicitly addressed is whether additions like Q 84:25, 37:112–113, and 73:20 must have been made while Muḥammad was still alive or whether they might not attest to a post-prophetic appropriation of the Qur'ānic corpus. The analysis presented above obviously leans in the former direction insofar as I have invoked a Medinan setting for Q 37:112–113 and Q 73:20, and a late Meccan

25 The total length of vv. 1–19 is 781 transcription letters, while that of v. 20 alone is 497 letters. Once again, see Sinai, "Inner-Qur'anic Chronology," for details about the measurement procedure.

26 On Q 73:20 as an insertion see Nöldeke and Schwally, *Geschichte des Qorāns*, vol. 1, p. 98; Bell, *Commentary*, vol. 2, pp. 446–447.

one for Q 84:25, based on significant similarities in content and diction between these three additions and other parts of the Qur'ān. For example, if one is content to date Q 2:124ff. to Muḥammad's Medinan stage, the insertion Q 37:112–113 is best assigned to the same period rather than seen as post-prophetic. Thus, additions of the sort discussed above may well have been made while the Qur'ānic corpus was still in a state of textual growth, rather than constituting finishing touches that a late editor put to an otherwise stable textual corpus.

In this context it may be appropriate to briefly review a recent monograph by the Biblical scholar Karl-Friedrich Pohlmann, who argues that scribal circles consisting of Jewish and Christian converts continued to shape and expand the Qur'ānic corpus after the death of Muḥammad.[27] In support of the claim that the Qur'ān underwent a significant degree of post-prophetic editing, Pohlmann cites two observations: first, he underlines the far-reaching familiarity with Jewish and Christian traditions that is displayed by many Qur'ānic passages; secondly, he points to what he sees as a pervasive presence in the Qur'ānic corpus of later additions that have been woven into their textual environment by means of literary techniques familiar from Biblical literature. For example, according to Pohlmann's analysis, many Qur'ānic passages exhibit a procedure that he calls *Wiederaufnahme*, where an existing piece of text is prefaced with a later insertion that echoes the opening of the earlier passage or its characteristic diction.

Pohlmann's general scenario of an editorial reworking of Qur'ānic material by scripturally literate converts is certainly not inherently improbable, and the pertinence of his interest in literary revision as well as the value of some of his textual observations must be unequivocally acknowledged. Nevertheless, a number of significant objections to his general argument spring to mind. First, Pohlmann's very trigger-happy approach to identifying later additions does not inspire confidence. In particular, one wonders whether many of the textual phenomena that he takes to indicate later interpolations – for example, the partial parallelism between Q 2:40 and 2:47[28] – might not equally well be accounted for without invoking later redactional interventions, for instance, as deliberate literary devices.[29] Secondly, it is far from obvious that such redactional activity must have constituted a post-prophetic phenomenon, as Pohlmann appears to assume

27 Pohlmann, Karl-Friedrich, *Die Entstehung des Korans: Neue Erkenntnisse aus Sicht der historisch-kritischen Bibelwissenschaft*. Darmstadt: Wissenschaftliche Buchgesellschaft, 2012, pp. 140–144, 184–185, 191–193.

28 Pohlmann, *Entstehung*, pp. 156–157.

29 This is not to deny that in particular the Medinan sūras exhibit numerous traits that justify a redactional analysis and that Pohlmann's work offers helpful suggestions on which such an analysis might be built.

without much argument, rather than possibly having gotten under way while Muḥammad was still alive. Thirdly, there is little reason to assume that familiarity with Jewish and Christian narrative traditions really amounts to "specialist knowledge" (*Spezialwissen*)[30] that could not have been accessible during the lifetime of Muḥammad.[31]

It is true that we ought to remain open-minded about the possibility that the Qur'ānic corpus could have undergone a certain amount of redactional work after the death of Muḥammad.[32] Yet we should also resist the temptation of simply transferring the conclusions of Biblical scholarship to the Qur'ān. Attempts to reconstruct the literary growth of one of the prophetic books of the Hebrew Bible, such as Amos, by Biblical scholars will typically involve the identification of an original core of brief mantic utterances that are assumed to have been subsequently expanded and reshaped by successive generations of scribally trained disciples. We need to be cautious about imposing this model of a linear sequence of charismatic proclamation and scribal revision on the Qur'ānic corpus, given that the latter came into being during a much shorter period and at a time when the Bible, in its Jewish and Christian incarnations, had for centuries been an object of sophisticated exegetical activity. The fact that a Hadith report depicts Muḥammad as instructing his scribes to incorporate newly revealed passages into existing sūras illustrates that early Muslims saw nothing incongruous in the notion that Muḥammad himself would cause revisions and interpolations to be made to existing revelations.[33] I am not of course suggesting that this tra-

---

30 Thus Pohlmann, *Entstehung*, p. 139.

31 True, some Medinan passages contain surprisingly specific allusions to Jewish traditions. Examples for this are a near-quotation from the Mishnah in Q 5:32, or the cross-linguistic pun in Q 2:93 that exchanges the Hebrew *šāmaʿnū ve-ʿaśīnū*, "We hear and obey" (Deut 5:24) for the phonetically similar *samiʿnā wa-ʿaṣaynā*, "We hear and disobey," and thereby polemically inverts its meaning; see Speyer, Heinrich, *Die biblischen Erzählungen im Qoran*. Hildesheim: Georg Olms Verlag, 1988, pp. 87–88 and 301. However, Islamic sources describe Medina as home to a large number of Jewish inhabitants, at least some of whom may have joined the nascent Islamic *umma* already while Muḥammad was still alive. As long as there is no compelling case for dismissing all contextual information for the Qur'ān that is derived from the Islamic tradition, there is every reason to assume that Muḥammad's followers, especially those among them who were themselves converts from Judaism, would have engaged in protracted disputes with the Medinan Jews. This tallies well with the substantial amount of anti-Jewish polemic contained in the Medinan stratum of the Qur'ān (in particular, in Q 2). As a result of such disputes, a fairly detailed knowledge of certain Rabbinic traditions could perfectly well have entered Qur'ānic discourse while Muḥammad was still active.

32 See Sinai, "Consonantal Skeleton," p. 521.

33 See al-Tirmidhī, *al-Jāmiʿ al-ṣaḥīḥ*, ed. by Aḥmad Muḥammad Shākir et al., 5 vols. Cairo: al-Ḥalabī, 1978, vol. 5, p. 272, no. 3086 (48:10: *Kitāb tafsīr al-Qurʾān*, on Q 9). Interestingly, the di-

dition may confidently be taken as historically accurate. But we certainly need to envisage the possibility that in the case of the Qur'ān the two activities of prophetic proclamation and scribal revision could at least partly have proceeded in parallel rather than having formed strictly consecutive phases. This is supported by the fact, underscored above, that insertions like Q 84:25, 37:112–113, and 73:20 are very plausibly placed during Muḥammad's lifetime. The editorial revision of existing Qur'ānic proclamations by means of secondary additions therefore began while Muḥammad was still alive, whether or not it continued after his death.

The preceding remarks imply that the three insertions discussed above, as well as similar ones, can tell us something about the status of the Qur'ānic material prior to Muḥammad's death. I would highlight three points in this regard. First, as in comparable Biblical cases, the procedure of interpretive expansion that can be observed in Q 84:37 and 73 only makes sense if the texts into which the putative additions were incorporated continued to be accessible to and recognised as authoritative by the Qur'ānic community beyond the occasion of their first recitation: texts that have dropped out of circulation or are not invested with a special authority are not generally updated and re-interpreted. It is true that all the insertions examined above are found in relatively early sūras, but there is no *prima facie* reason for suspecting that the longer and more complex sūras that were promulgated subsequently would have been treated differently, especially given that Medinan insertions have also been posited in two later Meccan sūras, Q 20 and Q 7.[34] Hence, at least those sūras displaying similar additions, but more probably all or most of Muḥammad's recitations, did not function as one-off addresses that were submerged by subsequent Qur'ānic texts.[35] Second, as in the Hebrew Bible, it appears that Qur'ānic proclamations were not generally amenable to wholesale rewriting or expurgation but had to

vision of labour emerging from this report bears a certain resemblance to the manner in which Jer 36 describes the cooperation between the eponymous prophet and his scribe Baruch: the chapter narrates the destruction of a scroll containing Jeremiah's prophecies and Baruch's subsequent production of a new scroll comprising the content of the previous one in addition to "many like words" (Jer 36:32).

34 Neuwirth, Angelika, "Meccan Texts – Medinan Additions? Politics and the Re-reading of Liturgical Communications," in *Words, Texts, and Concepts Cruising the Mediterranean Sea: Studies on the Sources, Contents, and Influences of Islamic Civilization and Arabic Philosophy and Science*, Orientalia Lovaniensia Analecta 139, ed. by Rüdiger Arnzen and Jörn Thielmann, pp. 71–93. Leuven: Peeters, 2004.

35 See Sinai, *Fortschreibung und Auslegung*, p. 81 (cf. also ibid., pp. 1–2).

be modified and re-interpreted through the insertion of additional material.[36] Once recited in public, the Qur'ānic texts must have been difficult to withdraw from communal circulation and in this sense ceased to be under the exclusive control of Muḥammad (although a number of Qur'ānic passages do acknowledge that in certain cases existing recitations were in fact suppressed and replaced[37]). Thirdly, our ability to identify specific textual segments as additions, based on stylistic and terminological discrepancies with their immediate literary context, implies that their host sūras were by and large preserved with a high degree of verbal accuracy and did not undergo incremental and subconscious adjustment in diction and content.[38]

3 Interpretive Backreferencing

As adumbrated in my introduction, I apply the term "interpretive backreferencing" to cases in which a later Qur'ānic passage B modifies the meaning of some temporally antecedent passage A or a cluster of several such passages, with A and B occurring in independent Qur'ānic proclamations or sūras.[39] Normally, B will signal the identity of A by a more or less conspicuous overlap in diction. To draw on Benjamin Sommer's helpful treatment of allusions in the Hebrew Bible, "[m]arkers (usually borrowed vocabulary) point the reader to the older text, though only if the reader is familiar with them."[40]

---

36 This is also confirmed by the case of Abraham's intercession for his idolatrous father, which is reported in Q 14:41, 19:47, and Q 26:86 and is discussed in the second part of this chapter: the pertinent passages could not be excised but had to be qualified and re-interpreted (namely, in Q 9:114 and 60:4).

37 See Q 2:106 and 16:101. Q 22:52 might also be relevant. Since such suppressed passages are by definition unavailable to us, it seems unlikely that we shall ever be in a position to ascertain the causes for suppression as opposed to interpretive expansion.

38 Admittedly, we can only speculate about how the Qur'ānic proclamations, after their initial promulgation, were preserved and reused during Muḥammad's lifetime. Even if Muḥammad's recitations were from a certain point transcribed and stored in writing, widespread communal access to written copies of Qur'ānic material appears less likely than access by means of repeated recitation and some degree of collective memorisation.

39 The phenomenon is not dissimilar to the interpretive relations obtaining between different Qur'ānic verses that al-Suyūṭī describes by the term *iqtiṣāṣ*. See Abdel Haleem, Muhammad, *Understanding the Qur'an: Themes and Styles*. London: I. B. Tauris, 1999, p. 161.

40 Cf. Sommer, Benjamin, *A Prophet Reads Scripture: Allusion in Isaiah 40–66*, Contraversions: Jews and Other Differences. Stanford: Stanford University Press, 1998. With respect to the earliest recipients of the Qur'ān, "reader" would probably need to be substituted by "hearer."

Such interpretive backreferencing also occurs in the Hebrew Bible. It has a particularly fertile breeding ground in the three major legal corpora included in the Pentateuch, normally designated as the Covenant Code (Exod 20–23), the Holiness Code (Lev 17–26), and the Deuteronomic Code (Deut 12–26). In many cases, these three collections of legal material regulate on the same issues (e.g., the release of slaves) and display conspicuous terminological and phraseological similarities yet diverge in crucial details. Assuming that the composers of the Holiness Code and of the Deuteronomic Code were not only drawing on the Covenant Code but also expected their readers to recognise this, it is plausible to see later laws as engaged in a deliberate clarification, extension, and updating of earlier ones.[41] At least broadly speaking, I find it justified to characterise this as an interpretive process.

Rather than attempting a cursory presentation of some pertinent Biblical specimens, as in the last section of this article, I shall immediately turn to the Qur'ān now and argue that interpretive backreferencing is an important feature of the Islamic scripture as well. I commence with what I hope will be seen as a relatively straightforward example and then go on to reiterate an earlier claim of mine that the phenomenon is also exhibited by the Qur'ānic Adam narratives. Throughout this section, I shall be particularly concerned to defuse what might appear to be serious objections to the general notion of targeted interpretive backreferencing.

3.1 An Initial Example: Abraham's Intercession for His Father

By way of an initial instance of interpretive backreferencing, let us consider Q 9:114: "Abraham's seeking of forgiveness (*istighfār*) for his father was only because of a promise he had made to him. When it became clear to him that his father was an enemy of God, he dissociated himself from him (*tabarra'a minhu*) – and Abraham was kind-hearted and prudent." The context in which this statement occurs is the assertion, in the immediately preceding verse, that the Prophet and the Believers may not "seek forgiveness for the Associators (*an yastaghfirū li-l-mushrikīn*), even if they were to be near kinsmen, after it has become clear to them that they are companions of the Fire." V. 114 is obviously concerned to clarify that Abraham did not violate the injunction expressed

41 See summarily Fishbane, "Inner-Biblical Exegesis," pp. 38–43. For an accessible presentation of some examples see Ska, Jean-Louis, *Introduction to Reading the Pentateuch*, trans. by Pascale Dominique. Winona Lake: Eisenbrauns, 2006, pp. 40–52.

in v. 113. However, rather than simply denying that Abraham sought forgiveness for his father at all, v. 114 emphasises that he only did so in order to honour a prior promise and that he did not hesitate to sever all ties to his father when he realised the full extent of his opposition to God. It would therefore seem, first, that v. 114 assumes its audience to be familiar with a tradition according to which Abraham pleaded with God to forgive his idolatrous father and, secondly, that the verse attempts to negate the irenic implications of this act.[42] A similar neutralisation of Abraham's intercession is contained in Q 60:4. In both cases, simple denial of Abraham's intercession does not seem to have been on the cards, although this would arguably have served the two verses' basic objective – namely, to insist on Abraham's unhesitating dissociation from his people and his father[43] – in a much simpler fashion.

Now, Abraham's act of intercession is actually described in various earlier Qur'ānic passages, namely, in Q 19:47, where Abraham announces to his father that he will "ask my Lord to forgive you" (*sa-astaghfiru laka rabbī*), and in Q 14:41 and Q 26:86, where Abraham petitions God to "forgive" (*ghafara li-*) his father and his parents. Not only is there an obvious thematic relationship between Q 9:114 and 60:4, on the one hand, and these earlier passages, on the other, but there also exists a crucial overlap in diction: all verses make prominent use of the root *gh-f-r*, with three of them (Q 9:114, 19:47, and 60:4) employing the form *istaghfara*. Furthermore, the interpretive realignment of existing sūras by means of additions, as discussed in Section 2, presupposes that many, if not most, of Muḥammad's proclamations remained in communal circulation well after their initial publication. It is therefore reasonable to posit that the addressees of Q 9:114 and 60:4 were assumed to have remained familiar with the earlier passages Q 14:41, 19:47, and 26:86. This also explains why Q 9:114 and 60:4 are, apparently, unable to simply deny that Abraham interceded on behalf of his father: earlier Qur'ānic proclamations having described that such an intercession took place, subsequent ones had to respect the basic facts of the matter. But if the audience of Q 9:114 and 60:4 were aware of the earlier passages describing Abraham's intercession for his father, they would naturally have perceived the former as comments on the latter – that is, they would have perceived them as playing a broadly interpretive role. In this sense, Q 9:114 and 60:4 qualify as instances of inner-scriptural interpretation. Interestingly, the Qur'ān itself appears to reflect this phenomenon of interpretive backreferencing, for Q 75:16–19,

42 This coheres well with the fact that Q 9 devotes significant space to chastising its audience for their unwillingness to fight the "Unbelievers" (cf. 9:38 ff.).

43 Note that Q 60:4, like 9:114, makes use of a derivative of the root *b-r-ʾ*.

a relatively early passage, emphasises that when a Qur'ānic revelation stands in need of subsequent "clarification" (*bayān*), this will be provided by a further revelation.[44]

It must be recognised that neither Q 9:114 nor 60:4 signal their likely antecedents by some sort of text-referential tag; 9:114 does not say, as one might perhaps expect it to, "Abraham's seeking of forgiveness for his father, *as mentioned in a previous sūra* [...]," or some such thing. Instead, the two earlier verses that 9:114 is latching on to are only picked out implicitly, by the occurrence of a verbal derivative of *gh-f-r* in connection with Abraham and his father. We are thus faced with a distinctly allusive kind of interpretive backreferencing. As a matter of fact, overt tagging hardly ever occurs in the Qur'ān: even Biblically derived material is almost never accompanied by explicit formulae of citation or textual reference.[45] This is plausibly taken to be symptomatic of the fact that the Qur'ānic milieu was a predominantly oral one and that within it knowledge of Biblical literature, too, was mediated orally – although the Qur'ān does of course bear witness to a clear awareness of the *existence* of earlier scriptures like the *tawrāh* and the *injīl*.[46] Situating the Qur'ān in a largely oral milieu would certainly account for its general lack of explicit formulae of citation: in a cultural environment in which the quo-

44 The passage runs as follows:

16 Do[s] not move your tongue about it to hasten it.
17 Upon Us is its putting together and its recitation.
18 When We recite it, follow[s] its recitation.
19 Upon Us is its explanation.

Not only do these verses stake out an unequivocal claim to verbal inspiration, in the sense that Muḥammad's role is envisaged as merely consisting in the faithful transmission of divine utterances, but they also assign to God the prerogative of clarifying existing revelations. Presumably, then, any interpretation of existing Qur'ān texts must occur in further Qur'ānic pronouncements. The statement is best read as an attempt to withhold interpretive control over the emerging Qur'ānic corpus from the community that was gradually forming around it. Naturally, such a model of divine self-interpretation proved to be applicable only while Muḥammad was still alive.

45 An exception is Q 21:105, which introduces a quotation of Ps 37:29 as something that "We have written in the *zabūr*." Cf. also Q 5:32 (paraphrasing Mishna Sanhedrin 4:5) and 5:45 (a variation on Exod 21:23–25 and cognate passages), both of which employ *katabnā* (although its meaning in these two cases may be closer to "We have decreed").

46 On the orality of the Qur'ānic milieu, see Griffith, Sidney, *The Bible in Arabic: The Scriptures of the "People of the Book" in the Language of Islam*. Princeton and Oxford: Princeton University Press, 2013, pp. 42–43, 55–56, and Bannister, Andrew, *An Oral-Formulaic Study of the Qur'an*. Lanham: Lexington Books, 2014, pp. 43–64, 107–128. Note, however, that part of the reason why the Qur'ān is so noticeably reticent to explicitly cite previous scriptures might be theological: given the Qur'ān's self-presentation as divine speech and its self-description as a "sending down" (*tanzīl*) of the celestial scripture (*kitāb*), it makes sense for it not to defer to previous revelations by explicitly quoting and exegeting them.

tation of fixed texts is not a widespread practice, the modern expectation that such references ought to be expressly flagged up will hardly be applicable.

It is moreover significant that in the Bible an explicit signalling of quotations seems to be most frequent in cases where a New Testament text is referring to an Old Testament / Hebrew Bible one, a good example being the so-called fulfilment citations in the Gospel of Matthew, or the citations from the Hebrew Bible in the Epistle to the Hebrews.[47] Within the Hebrew Bible, by contrast, one often finds untagged allusions.[48] It is tempting to venture the following hypothesis: where some passage B quotes or alludes to some earlier passage A, explicit signalling of the cross-reference is more likely in cases in which A is conceptualised as belonging to a closed literary corpus distinct from B, as opposed to cases in which both passages are viewed as belonging to one and the same textual corpus that is still in a state of growth. For all we can tell, the phenomenon of inner-Qur'ānic backreferencing falls into the latter rubric, which further accounts for the fact that in the Qur'ān interpretive backreferences such as Q 9:114 and Q 60:4 are generally untagged.

Assuming that the Qur'ānic proclamations are to be situated in a predominantly oral milieu, is this difficult to reconcile with the fact that they exhibit a significant amount of targeted (albeit implicit) cross-quotation and cross-allusion? I would submit that there is no need to assume that the two are incompatible. This is most easily appreciated if we take into account Angelika Neuwirth's convincing insistence that the Qur'ānic corpus documents the gradual crystallisation of a closely-knit religious community that grew increasingly distinct from its pagan environment of origin.[49] The focal point of this newly emerging collective identity would have been Muḥammad's revelatory utterances: drawing on the work of Neuwirth, we may say that the Qur'ānic community progressively withdrew from its immediate social habitat into a literary world defined by the Qur'ānic revelations. This scenario cogently explains why it makes eminent sense to view Muḥammad's followers as having possessed the intimate textual familiarity with previous Qur'ānic proclamations that my above analysis presup-

47 On Matthew's fulfilment citations see Ehrman, Bart D., *The New Testament: A Historical Introduction to the Early Christian Writings*, 5th edition. New York and Oxford: Oxford University Press, 2012, p. 118.

48 For a study of the use of allusion in Isa 40–66 to earlier portions of the Hebrew Bible, in particular to the book of Jeremiah, see Sommer, *A Prophet Reads Scripture* (who emphasises that allusion can serve many other than just exegetical functions). Many of the examples of inner-scriptural interpretation discussed in Fishbane, *Biblical Interpretation*, are also relevant.

49 See the articles collected in Neuwirth, Angelika, *Scripture, Poetry and the Making of a Community: Reading the Qur'an as a Literary Text*. Oxford: Oxford University Press, 2014.

poses, despite the fact that the cultural milieu from which they hailed would have been a largely oral one. In sum, there is no need to doubt that at least a significant part of the addressees of Q 9:114 and 60:4 would have been in a position to recognise these verses' allusive reference to earlier Qur'ānic statements about Abraham.

3.2 The Qur'ānic Adam Narratives

The obvious question arising at this point is whether the phenomenon of interpretive backreferencing exhibited by Q 9:114 and 60:4 is discernible elsewhere in the Qur'ān. In a previous publication I have argued that it is fairly widespread particularly in Qur'ānic narratives.[50] By way of an example, I shall consider the Qur'ān's retellings of the creation of Adam and his temptation by Iblīs (*diabolos*), which have previously been examined by Edmund Beck, Angelika Neuwirth, myself, Andrew Bannister, and Joseph Witztum.[51] I shall first present the relevant material, which is scattered across eight different sūras, and defend my assumption that later proclamations in this series would, in the Qur'ān's original historical context, have been understood against the background of earlier ones. The next and final section of my contribution will then consider various ways in which the passages at hand can be seen to be engaged in interpretive backreferencing.

Table 1 synoptically maps out the eight Qur'ānic version of the narrative, in a suitably summarised shape and subdivided into two parts: (1) Adam's creation and the fall of Iblīs; (2) Adam succumbs to temptation by the devil and eats from the tree of eternal life.[52] Sūras are listed, from left to right, in their probable

50 See Sinai, *Fortschreibung und Auslegung*, pp. 59–160.

51 See Sinai, *Fortschreibung und Auslegung*, pp. 86–96; Bannister, *Oral-Formulaic Study*, pp. 1–18, 243–270; Witztum, Joseph, "Variant Traditions, Relative Chronology, and the Study of Intra-Quranic Parallels," in *Islamic Cultures, Islamic Contexts: Essays in Honor of Professor Patricia Crone*, Islamic History and Civilization: Studies and Texts 114, ed. by Behnam Sadeghi, Asad Q. Ahmed, Adam Silverstein and Robert Hoyland, pp. 1–50. Leiden: Brill, 2015, pp. 14–27. For a discussion of earlier studies of the Qur'ānic Adam narratives by Edmund Beck and Angelika Neuwirth see the publications by Sinai and Witztum.

52 It is conspicuous that the first part of the story utilises the name Iblīs, whereas the second part employs *al-shayṭān*. This apparent change in name corresponds to the *Cave of Treasures*, where the devil only receives the name "Satan" after his expulsion from paradise (but prior to the temptation of Adam). See Bezold, Carl (ed. and trans.), *Die Schatzhöhle*. Leipzig: Hinrichs'-sche Buchhandlung, 1883, pp. 16f. (Arabic and Syriac) and 4 (German translation). I owe this observation to a tutorial essay by Rachel Dryden.

chronological order, based on the general assumption that the mean verse length of Qur'ānic sūras tended to increase over time.[53] (It is worthwhile to flag up that my relative dating of the texts puts me in conflict with Witztum's suggestion that the narratives about Adam and about the Israelite's worshipping of the Golden Calf in Q 20 are later than the corresponding accounts in Q 7.[54]) Note that where different sūras share a narrative element, the respective verses very often display significant literal overlap.[55] This is illustrated by Table 2, which follows the same structure and translates the openings of each version of the story.

53 Precise values for the mean verse length of the sūras in question, which underpin my chronological ordering, are provided in Sinai, "Inner-Qur'anic Chronology." In some cases, the mean verse length of adjacent sūras is so close that their chronological relationship may well have been the reverse. This is particularly true for Q 17 and Q 18. Whether the disparity in mean verse length between Q 7, on the one hand, and Q 17 and Q 18, on the other, is sufficiently large to entail that Q 7 must be the latest of the three texts also requires further consideration. However, these minor qualifications do not undermine the general developmental trajectory emerging from my table. That Q 15 is earlier than Q 38 I do not doubt, given that the 95% confidence intervals of both sūras' mean verse length (see Sinai, "Inner-Qur'anic Chronology") show only very scant overlap. On the relative dating of Q 7 and Q 20 see the following note.
54 See Witztum, "Variant Traditions." Both in the case of the story of the disobedience of Adam and that of the Golden Calf, Witztum maintains that the versions in Q 7 are in some salient regard closer to the Biblical text (especially in its Syriac version) than the versions in Q 20. For example, Witztum convincingly argues that the devil's promise in Q 7:20 is closer to Gen 3:5 than in Q 20:120 (Witztum, "Variant Traditions," pp. 25f.). He suggests that such observations are best explained by assuming a development leading away from the Bible, i.e., from the relevant passages of Q 7 to the relevant passages of Q 20. Although unable to engage in detail with Witztum's perspicacious treatment of the material, I would still insist on the possibility of secondary Biblicisation, i.e., of the Qur'ān moving closer to the Biblical text over time rather than away from it. This is far from a novel scenario; as pointed out by Witztum himself, the "classic example given for the Quran's growing awareness of the Bible is the figure of Ishmael and his relationship to Abraham, or lack thereof" (Witztum, "Variant Traditions," p. 27). Against Witztum, I am not convinced that such a model is less capable of making sense of the textual evidence than the opposite trajectory. Obviously, if Witztum's relative dating is correct, my claim that mean verse length is a reliable chronological indicator will have been refuted, given that Q 7's mean verse length is significantly higher than that of Q 20 (104.27 transcription letters as opposed to 61.04).
55 The table is an expanded and modified version of the one in Sinai, *Fortschreibung und Auslegung*, p. 96 (the latter omitting the concise reminiscences in Q 17:61–65 and 18:50, which are summary parallels to the relevant sections of Q 15 and 38). A similar breakdown that cites the relevant passages in full can be found in Bannister, *Oral-Formulaic Study*, pp. 13–18. Note that Bannister does not attempt a chronological ordering.

Table 1: A thematic synopsis of the Qur'ānic Adam-Iblīs narratives.

Q 55	Q 15	Q 38	Q 20	Q 17	Q 18	Q 7	Q 2
14–15 Humans and jinn created from clay and fire.	26–27 Introductory preview: humans and jinn created from clay and fire.						
			115 Introduction: Adam forgot to keep his covenant with God.			10 God's grace and man's ingratitude.	
			(1) The creation of Adam and the fall of Iblīs				
	28 God announces to the angels His intention to create man from clay.	71 God announces to the angels His intention to create man from clay.				11 Addressees reminded of their creation by God.	30 God announces to the angels His intention to establish a successor on earth.
							30 Protest of the angels, rejoinder by God.
							31–33 God teaches Adam "all the names" and demonstrates the angels' ignorance.
	29 The angels ordered to prostrate themselves before man.	72 The angels ordered to prostrate themselves before man.	116 The angels ordered to prostrate themselves before Adam.	61 The angels ordered to prostrate themselves before Adam.	50 The angels ordered to prostrate themselves before Adam.	11 The angels ordered to prostrate themselves before Adam.	34 The angels ordered to prostrate themselves before Adam.

Q 55	Q 15	Q 38	Q 20	Q 17	Q 18	Q 7	Q 2
	[30–31] The angels prostrate themselves, but Iblīs disobeys.	[73–74] The angels prostrate themselves, but Iblīs disobeys.	[116] The angels prostrate themselves, but Iblīs disobeys.	[61] The angels prostrate themselves, but Iblīs disobeys.	[50] The angels prostrate themselves, but Iblīs disobeys.	[11] The angels prostrate themselves, but Iblīs disobeys.	[34] The angels prostrate themselves, but Iblīs disobeys.
	[32–33] Exchange between God and Iblīs.	[75–76] Exchange between God and Iblīs.		[61] Statement by Iblīs.		[12] Exchange between God and Iblīs.	
	[34–35] Iblīs banned. [36–43] Iblīs announces his intention to lead mankind astray; God's servants are exempt.	[77–78] Iblīs banned. [79–85] Iblīs announces his intention to lead mankind astray; God's servants are exempt.		[62–65] Iblīs announces his intention to "devour" (*iḥtanaka*?) mankind; God's servants are exempt.		[13] Iblīs banned. [14–18] Iblīs announces his intention to lead mankind astray.	
(2) Adam succumbs to temptation and eats from the tree of eternal life							
			[117–119] Adam warned of the enmity of Iblīs and admonished not to forfeit paradise.			[19] Adam given usufruct of paradise but admonished not to approach "this tree"	[35] Adam given usufruct of paradise but admonished not to approach "this tree."
			[120] Adam tempted by the devil to eat from the "tree of eternal life."			[20–21] Adam and Eve tempted by the devil to eat from the tree.	[36] Adam and Eve forfeit paradise due to the devil's seduction.

Q 55	Q 15	Q 38	Q 20	Q 17	Q 18	Q 7	Q 2
			[121] Adam and Eve eat from the tree.			[22] Adam and Eve eat from the tree and are reproached by God.	
						[23] Adam and Eve incriminate themselves and ask for God's forgiveness.	
			[122] God relents towards Adam.				(cf. v. 37)
			[123] Adam and Eve driven from paradise, declaration of enmity between humans and the devil.			[24] Adam and Eve driven from paradise, declaration of enmity between humans and the devil.	[36] Adam and Eve driven from paradise, declaration of enmity between humans and the devil.
						[24–25] God decrees human mortality.	[36] God decrees human mortality.
			(cf. v. 122)				[37] God relents towards Adam. [38] Reprise: Adam and Eve driven from paradise.*
			[123–124] God exhorts humans to follow His guidance.				[38–39] God exhorts humans to follow His guidance.

* On this reprise see Witztum, "Variant Traditions", 19–21.

Table 2: Overlaps in diction between the openings of the Qur'ānic Adam-Iblīs narratives.
Note: Where a column has underlined expressions and phrases, these overlap with another column further left.[56]

Q 55	Q 15	Q 38	Q 20	Q 17	Q 18	Q 7	Q 2
14 He created man from clay (*ṣalṣāl*) like pottery [or: like a potter], 15 And He created the jinn from a mixture of fire.	26 We have created man from clay (*ṣalṣāl*), from moulded mud; 27 We created the jinn before from the fire of the scorching wind.						
						11 We created you^{p} and shaped you^{p} and then	
	28 And [recall] when your^{s} Lord said to the angels:	71 And [recall] when your Lord^{s} said to the angels:	116 And [recall] when We said to the angels:	61 And [recall] when We said to the angels:	50 And [recall] when We said to the angels:	We said to the angels:	30 And [recall] when your^{s} Lord said to the angels:
	"I am creating man from clay (*ṣalṣāl*), from moulded mud.	"I am creating man from clay (*ṭīn*).					"I am putting a successor on the earth." They said: [...]
							34 And when We said to the angels:
	29 So when I have formed him and breathed into him some of My spirit,	72 So when I have formed him and breathed into him some of My spirit,					
	then fall down in prostration to him (*fa-qa'ū lahu sājidīn*)."	then fall down in prostration to him."	"Prostrate yourselves (*usjudu*) to Adam,"	"Prostrate yourselves to Adam,"	"Prostrate yourselves to Adam,"	"Prostrate yourselves to Adam,"	"Prostrate yourselves to Adam,"

56 Note that Bannister, *Oral-Formulaic Study*, pp. 13 – 18 (who transcribes the original Arabic rather than giving a translation) makes a slightly different use of underlines.

Q 55	Q 15	Q 38	Q 20	Q 17	Q 18	Q 7	Q 2
	30 All the angels prostrated themselves, all together,	73 All the angels prostrated themselves, all together,	and they prostrated themselves,	and they prostrated themselves,	and they prostrated themselves,	and they prostrated themselves,	and they prostrated themselves,
	31 Except Iblīs;	74 Except Iblīs;	except Iblīs;	except Iblīs.	except Iblīs.	except Iblīs;	except Iblīs;
	he refused to be with those who prostrated themselves.		he refused.			he was not amongst those who prostrated themselves.	he refused
		he was haughty and one of the Unbelievers.			he was one of the jinn, and he sinned against the command of his Lord. [...]		and was haughty and one of the Unbelievers.

At the risk of slaying the slain, I would once again argue that a significant part of the addressees of each of these passages would have been aware of previous Qur'ānic proclamations touching on the same subject. This position is supported both by the phenomenon of interpretive expansion as studied in Section 2 and by my preceding analysis of Q 9:114, both of which establish that Qur'ānic proclamations continued to circulate, and to be regarded as authoritative, beyond their initial recitation and that they could become, at least on occasion, the object of later Qur'ānic clarification, comment, and qualification.

My view is also confirmed by the fact that Qur'ānic narratives often appear to assume familiarity with previous versions of the story in question, insofar as they seem content to leave narrative gaps where these can be easily filled in by someone acquainted with earlier Qur'ānic retellings. For example, Q 20 employs the brief epitome of Iblīs' act of disobedience that also occurs in Q 17, 18, 7, and 2,[57] and then goes on to narrate, for the first time, Adam's subsequent temptation by Iblīs, here referred to as "the Satan." What Q 20 omits, though, is the motif linking the first and the second part of the story, namely, Iblīs' announcement that he will henceforth strive to lead man into temptation. Plausibly, the audience of Q 20 would simply have filled this gap on the basis of their acquaintance with the earlier versions of the story contained in Q 15 and Q 38, both of which contain the motif in question. If, by contrast, later versions of a Qur'ānic narrative had been intended to replace the preceding ones rather than presupposing and complementing them, we might have expected a much more sustained attempt at incorporating essential narrative elements from earlier versions into subsequent ones. This is patently not the case, for even Q 7, which offers the most comprehensive retelling of the plot as it had so far crystallised, is far from self-sufficient: it omits God's initial creation of Adam from clay, although this is later presupposed by Iblīs' statement in v. 12 ("I am better than him; You created me from fire and him from clay"); and it reports Adam and Eve's self-incrimination and their prayer for God's forgiveness (Q 7:23), but omits God's renewed election of Adam as reported in Q 20:122.[58] We are therefore jus-

57 Q 20:116: "And [recall] when We said to the angels: 'Prostrate yourselves to Adam', and they prostrated themselves, except Iblīs; he refused."

58 The background knowledge presupposed by Qur'ānic narratives or narrative reminiscences is certainly not always itself Qur'ānic. For example, Q 73:15–18, 79:15–26, 85:17–18, 87:18–19, 89:6–14, and 91:11–15 would all appear to assume some amount of prior knowledge on the part of their audience about Noah, Abraham, and Moses, as well as about the destruction of Thamūd, ʿĀd, and Iram; and given that these passages are to be dated earlier than the more comprehensive treatment of these figures and peoples in other sūras, this background knowl-

tified in positing that a substantial part the audience of each successive version of the Adam-Iblīs narrative would have had some awareness of the preceding ones: later versions would generally have been understood in the light of earlier ones. At the same time, however, earlier ones, when re-recited, would also have been understood in the light of later ones.[59]

edge must originally have consisted in extra-Qur'ānic narrative traditions. (For an example of what these may have looked like, see Sinai, Nicolai, "Religious Poetry from the Quranic Milieu: Umayya b. Abī l-Ṣalt on the Fate of the Thamūd." *Bulletin of the School of Oriental and African Studies* 74 (2011): pp. 397–416.) One might therefore object that the gap occasioned by Q 20's omission of Iblīs' statement at the end of the first part of the narrative may well have been filled by recourse to pre-Qur'ānic traditions as documented, for instance, in the *Life of Adam and Eve* (see Speyer, *Die biblischen Erzählungen*, pp. 54–58). In other words, why assume that the gap was filled specifically by recourse to Q 15 and 38? The objection does not hold, though. Where a Qur'ānic text presupposes information not contained in an earlier sūra, we must certainly assume that Muḥammad's audience was familiar with extra-Qur'ānic traditions that would have allowed them to understand what was being said. Yet once Qur'ānic discourse had addressed a given topic – say, the disobedience of Iblīs – it is far more likely that the Qur'ānic community's background knowledge about the matter at hand would have been primarily governed by previous Qur'ānic statements, over and against any pertinent extra-Qur'ānic traditions. After all, the Qur'ānic texts claim to be divine speech. At least in the eyes of Muḥammad's adherents, Qur'ānic pronouncements would thus have been clearly superior in authority to non-Qur'ānic traditions or texts.

59 Witztum notes that my monograph inclines towards a complementary, even "harmonistic" reading of different Qur'ānic versions of the same narrative, a stance regarding which he expresses some reservations (Witztum, "Variant Traditions," pp. 12 and 43). It may be useful to underscore that I do not of course mean to deny that different Qur'ānic sūras tell stories from significantly different angles. An obvious case in point are the Calf narratives in Q 7 and Q 20, the latter being distinguished from the former by the appearance of the notoriously enigmatic figure of al-Sāmirī, who is entirely absent from Q 7 (see Witztum, "Variant Traditions," p. 35). However, even in the view of such conspicuous divergence we can still ask how the Qur'ān's recipients would have been likely to respond to it. Assuming that Q 7 is posterior to Q 20 (which Witztum denies), how would Muḥammad's hearers have perceived Q 7's omission of al-Sāmirī? I maintain that they would have naturally assumed the later sūra to *complement* rather than to *contradict* the earlier one; that is, they would have naturally tended to view Q 7 as simply presupposing the existence of al-Sāmirī rather than as implicitly denying his involvement in, and even his ultimate responsibility for, the making of the Calf. To adduce a further example, consider the different ways in which God admonishes Adam in Q 20:117 ("Let him not drive the two of you from the garden") and in Q 7:19 ("Do not approach this tree"). Here, too, I would submit that after the proclamation of Q 7, the Qur'ānic audience is likely to have assumed that the admonishment in Q 20:117 was to all intents and purposes equivalent to the more specific prohibition contained in the later verse Q 7:19. (This again assumes, against Witztum, that Q 7 contains the later account.) Such an attitude of implicit harmonisation on the part of Muḥammad's adherents would have been decisively facilitated by the fact that more often than not, later Qur'ānic narratives do not directly contradict earlier ones. For example, *pace* Witztum, Q 7:148 does not ac-

Still, the claim that the various texts making up the Adam-Iblīs complex exhibit interpretive backreferencing might seem vulnerable to an objection rooted in the highly formulaic character of many parts of the Qur'ān. This feature of the text has recently been studied by Andrew Bannister[60] and may well be reflected in the Qur'ānic self-description as a "scripture consimilar in its oft-repeated" (Q 39:23), as Arberry translates the enigmatic statement that God has sent down a *kitāban muthashābihan mathāniya*.[61] Against this background, one might wonder whether even conspicuous overlaps between an earlier and a later verse or passage are really sufficient to establish that the later text is *alluding to* or even *quoting* the earlier one.

I would respond, first of all, by insisting on the usefulness of a diachronic approach to distinguishing between instances of genuine allusion and of formu-

tually state that the Israelites themselves *made* the Calf, which would jar with the statement in Q 20:88 that al-Sāmirī "produced for them a calf, a body that made a lowing sound (*fa-akhraja lahum ʿijlan jasadan lahū khuwārun*)." Rather, what Q 7:148 says is that the Israelites "*adopted* a calf [made] from their ornaments (*wa-ttakhadha qawmu mūsā min baʿdihī min ḥuliyyihim ʿijlan jasadan lahū khuwārun*)." Strictly speaking, then, Q 7:148 does not have to be construed as directly disagreeing with Q 20:88. I do not of course have any apologetic interest in denying that the Qur'ānic corpus does contain substantial internal tensions and even unequivocal inconsistencies. But I find it significant that obvious contradictions appear to be relatively infrequent between different Qur'ānic retellings of one and the same story. This, I maintain, would have encouraged the Qur'ānic community to process such variant narratives in a complementary or harmonistic manner (which is not to say that we, as critical scholars, should overlook the different theological and anthropological messages that may be expressed by different versions of the same narrative). The hypothesis that the Qur'ānic hearers would have processed variant stories in a complementary manner is also corroborated by the fact that the Qur'ān itself stakes out an explicit claim to be devoid of inconsistency (*ikhtilāf*) in Q 4:82. This latter verse is also highlighted by Witztum, "Variant Traditions", pp. 43 f., who rightly insists that we must "distinguish between what the text tells us it is doing and what it actually does." Yet what the text tells us it is doing may permit conjectures about how it would have been viewed and interpreted by the Qur'ānic *Urgemeinde*, who was presumably committed to the basic truth of what the text was telling them.

60 Bannister, *Oral-Formulaic Study.*

61 Building on previous scholars' construal of the *mathānī* as referring to the Qur'ānic punishment stories, an earlier publication of mine links the phrase primarily to the relationship of complementarity that may be seen to obtain between different Qur'ānic versions of the same narrative; see Sinai, "Qur'ānic Self-Referentiality as a Strategy of Self-Authorization," in *Self-Referentiality in the Qur'ān*, ed. by Stefan Wild, pp. 103–134. Wiesbaden: Harrassowitz, 2006, at pp. 130–131. Yet a broader interpretation, which takes the phrase to highlight the formulaic character of Qur'ānic discourse more generally, is perhaps equally possible. The merit of such a broader reading of the phrase was brought to my attention by a presentation entitled "The Self-Similar Qur'an", which was given by Giuliano Lancioni at the 2014 Annual Meeting of the International Qur'anic Studies Association (21–24 November, San Diego).

laic diction. By way of an example, let us consider the opening of the Adam narrative as given in Table 2, "And [recall] when your^s Lord said to the angels [...] (*wa-idh qāla rabbuka li-l-malāʾikati* ...)." If my dating of the material is sound, this overture first occurred in Q 15:28. When it was subsequently repeated in Q 38:71, it may well have been perceived as a targeted allusion to Q 15:28. Yet subsequently, the phrase – like the sequence *usjudū li-Ādama fa-sajadū illā Iblīsa* ("'Prostrate yourselves to Adam', and they prostrated themselves, except Iblīs") – clearly established itself as a stereotypical component of all further versions of the story. As a result, it would have become increasingly unsuitable as a means of gesturing towards a specific earlier sūra and instead would have merely pointed, at most, towards a whole cluster of thematically related passages. Hence, the formulaic quality of Qur'ānic discourse does not as such rule out the presence of targeted backreferencing, as long as the potential term or phrase had not yet achieved formulaic status at the time of the occurrence at hand. In order to determine whether the latter was or was not the case, we must examine the occurrence of the phrase at hand, not in the entire Qur'ānic corpus, but in the subset of those Qur'ānic proclamations that are likely to predate or to be roughly contemporary with the passage at hand.

But might not already the earliest Qur'ānic occurrence of a phrase or a "formulaic system"[62] recurring in later proclamations have been grounded in a pre-Qur'ānic stock of formulaic diction? Do phrases really ever *achieve* formulaic status in the Qur'ān, or do they perhaps possess this status from the start? Once more, let us consider the earliest Qur'ānic attestations of the phrase "And [recall] when your^s Lord said to the angels [...]" in Q 15:28 and Q 38:71. Perhaps this was simply how oral retellings of the creation of Adam had commenced for decades and decades before Muḥammad recited the earliest Qur'ānic sūra? Clearly, it would be foolish to rule out that Qur'ānic discourse might be reliant, even heavily reliant, on ambient phraseology and patterns of expression. For example, hymnic locutions starting with *tabāraka* and *al-ḥamdu li-llāh*, followed by a relative clause or an epithet like *rabb al-ʿālamīn*, patently correspond to the form of standard Jewish and Christian eulogies and doxologies.[63] While all of our evidence for these latter is in Hebrew, Greek, and Aramaic, it is highly credible that such phrases had become well entrenched in Arabic already in pre-Qur'ānic

---

62 On "formulaic systems" see Bannister, *Oral-Formulaic Study*, pp. 207–242. Bannister's standard example for a formulaic system – essentially, a substitution pattern – is the sequence consisting in a derivative of *j-h-d* + a particle + a derivative of *s-b-l* + Allāh.

63 Baumstark, Anton, "Jüdischer und christlicher Gebetstypus im Koran." *Der Islam* 16 (1927): pp. 229–248.

times. Possibly, the same might apply to the opening phrase of the Qur'ānic Adam pericopes, "And [recall] when your^{s} Lord said to the angels [...]"

Still, the abstract possibility that this might have been the case does not justify the default assumption that *any* multiply occurrent Qur'ānic concatenation was routinely used in pre-Qur'ānic times. For the formulaic quality of the Qur'ān could at least in part be due to the Qur'ānic proclamations gradually building up an increasingly extensive repertoire of formulaic diction, resulting from the fact that later texts tended to echo the phraseology of earlier ones. This hypothesis is attractive because a comparable trajectory of innovation and standardisation may also be discerned with respect to the compositional structure of the Qur'ānic texts: whereas the earliest Qur'ānic sūras are brief and structurally fairly simple, later Meccan texts increasingly conform to a tripartite pattern centred around a narrative middle part. At the same time, a range of standard literary forms emerged, which included, for instance, introductory self-descriptions or superscripts containing the terms *kitāb* and/or *nazzala/anzala* (cf. Q 44:2, 15:1, or 40:2 and their multiple parallels).[64] The supposition that the Qur'ānic recitations are not simply relying on a well-established thesaurus of formulae but are themselves engaged in building up such a thesaurus is also buttressed by the fact that when the Qur'ānic corpus is viewed diachronically, many formulae and formulaic systems are divided across the corpus very unevenly: they seem to emerge at certain points in time and lose popularity at others.[65] To assume that each and every phrase that achieves formulaic status at some point during the genesis of the Qur'ān had already possessed this status in pre-Qur'ānic times incurs the risk of underestimating the importance of histor-

---

64 Since Muḥammad's revelations were very likely recited orally (whether or not they were also composed orally), the gradual crystallisation of a repertoire of stock phrases would have well suited their primary mode of display – and it may of course, over time, have facilitated a much higher degree of spontaneous composition than at the beginning of the process.

65 For example, the transition from the Meccan to the Medinan sūras appears to have entailed a palpable rupture in the Qur'ān's formulaic repertoire as it had crystallised so far. This is evident already from the formulaic systems examined in Bannister, *Oral-Formulaic Study*, pp. 220 – 236, many of which, when viewed diachronically, present cases of arrested development: they occur in Meccan sūras and then cease to be productive. This applies, for instance, to systems no. 1, 2, and 4, whose thematic habitat are the so-called *āyāt* passages (affirmations of miscellaneous manifestations of God's power and grace in nature). The reason for the phraseological rupture must be that *āyāt* passages are generally less frequent in the Medinan Qur'ān; not unlike the Meccan punishment legends studied in Marshall, David, *God, Muhammad and the Unbelievers: A Qur'anic Study*. Richmond: Curzon Press, 1999, they cease to be relevant – probably because their appellatory function was no longer a pressing need in the Medinan environment, where the notion of a unique and omnipotent creator god did not need to be argued.

ical development. By way of an objection, one might of course adduce Milman Parry's insistence that the formulaic systems found in the Homeric epics must be the work of generations of bards rather than of one individual.[66] However, Parry's strongest argument for this claim is surely that the development of a formulaic repertoire that would permit the poet to respect the metrical requirements of different positions in Homeric hexameter requires an amount of literary ingenuity and invention that is best seen as being the result of a long tradition. Given that the Qur'ān is not a metrical text, the formulaic phrases and systems populating it could certainly have emerged much more rapidly.

In any event, to insist that the Qur'ānic corpus exhibits targeted backreferencing should not be understood as requiring a denial of its formulaic character. For even if the latter is acknowledged, as I think it needs to be, it seems plausible that there would still remain cases in which a parallel between two passages, all things considered, is better accounted for by assuming a deliberate allusion than reliance on shared formulaic diction. In my view, the overlap between Q 55:14–15 and Q 15:26–27 (quoted in Table 2 and again in Section 3.3.1 below) is such a case, given that the juxtaposition of the creation of man from clay with that of the jinn from fire has no further Qur'ānic parallel. Of course, we might need to opt for a different assessment if, for instance, two or three other Qur'ānic retellings of the creation of Adam opened with similar couplets, or if the same juxtaposition could be documented, say, in pre-Islamic poetry. I would therefore propose the following rules of thumb for distinguishing between formulaic diction and genuine allusions: In the absence of any relevant pre-Qur'ānic intertexts, if a pair of Qur'ānic passages displays lexical overlap that is conspicuous enough in order not to be accidental, then we are probably faced with a targeted allusion.[67] Again barring any relevant pre-Qur'ānic intertexts, if such overlap is

---

66 See Bannister, *Oral-Formulaic Study*, pp. 69–71.

67 I should highlight that this is deliberately incompatible with the low repeat threshold that Bannister tends to demand for considering a phrase to be a formula (cf. Bannister, *Oral-Formulaic Study*, pp. 140, where a formula is preliminarily defined as "a sequence of three Arabic bases [= words] that occurs at least twice in the qur'anic text"). Notwithstanding the arguments put forward in Bannister, *Oral-Formulaic Study*, pp. 146–149, I have grave doubts whether two occurrences of a particular phrase within the Qur'ān suffice to endow it the status of a genuine "formula." It is true, of course, that the Qur'ānic corpus is not large (Bannister, *Oral-Formulaic Study*, pp. 148–149), but I am not convinced by Bannister's proposal that we should compensate for the ensuing difficulty of demonstrating the formulaic character of a given phrase by in effect building an oral-formulaic prejudice into our analysis. For doing so would effectively foreclose any possibility that the Qur'ānic corpus might be found to contain not only formulaic language, but also targeted allusions and deliberate self-quotations. Especially in the light of the relatively

displayed by three or four passages, then we are either faced with a manifestation of pre-Qur'ānic formulaic diction or a case of a gradual crystallisation of formulaic diction resulting from later Qur'ānic passages echoing or alluding to earlier ones. Against Bannister's preference for the former possibility, I confess to a certain prejudice in favour of the latter (although appropriate pre-Qur'ānic evidence would be sufficient to prove me wrong). This inclines me to view the Adam narrative in Q 38 as intentionally reproducing part of the earlier one in Q 15 while also including interpretively motivated variations. Yet I would certainly agree that by the time we arrive at the Adam narrative in Q 2, we have definitely crossed the line from targeted allusion and quotation to the redeployment of stock diction.

3.3 The Interpretive Dimension of the Qur'ānic Adam Narratives

Thus far, I have argued, with special reference to the Qur'ānic Adam-Iblīs complex, that we may assume at least a substantial part of the addressees of a given text in the series to have been familiar with the texts preceding it, and that conspicuous overlaps between a later text and an earlier one could in principle constitute a case of deliberate allusion. I shall now attempt to highlight some specific aspects in which the Qur'ānic Adam narratives can be seen to be, in part, engaged in interpretive backreferencing. It is convenient to distinguish three main stages in the Qur'ānic development of the Adam-Iblīs complex, leaving out the brief allusion in Q 18:

(1) Q 15, 38, and 17 recount the first part of the story consisting in the creation of Adam and the fall of Iblīs.
(2) In a second stage of development, which partly overlaps with (1), Q 7 and Q 20 append a sequel to the story, consisting in Satan's successful temptation of Adam and Eve.
(3) Finally, Q 2 expands the first part by narrating the angels' opposition to the creation of Adam and their subsequent humiliation by God.

---

unambiguous cases of Qur'ānic self-interpretation presented above, such a methodological decision strikes me as highly problematic.

3.3.1 Q 15, Q 38, and Q 17

The Adam pericope in Q 15 presents itself relatively unequivocally as latching on to and unpacking Q 55:14–15's concise reference to the creation of man from clay (overlaps underlined):

Q 55:14–15	Q 15:26–28
14 He created man *(al-insān)* from clay like pottery [or: like a potter],	26 We have created man *(al-insān)* from clay, from moulded mud;
15 And He created the jinn from a mixture of fire.	27 We created the jinn before from the fire of the scorching wind.
	28 And [recall] when your[s] Lord said to the angels: I am creating a human *(basharan)* from clay, from moulded mud.
	[...]
	33 He [Iblīs] said: 'I am not one to prostrate myself to a human *(li-basharin)* whom You have created from clay, from moulded mud.'

In light of what was said above, we can assume that at least part of Muḥammad's audience would have recognised this reprise, thus endowing the putatively earliest Qur'ānic Adam narrative with the status of a midrash-like amplification of an earlier couplet about the creation of man and the jinn. It is noteworthy that man's creation from clay is not only echoed by Q 15:26 and then reiterated in Q 15:28 but also invoked by Iblīs in justification of his refusal to prostrate himself to Adam (Q 15:33), thus tying the Adam episode in Q 15 even more closely to Q 55:14–15. An additional link is created in Q 38:76, where Iblīs explicitly contrasts Adam's creation from clay with his own creation from fire, thus drawing Q 55:15, too, into the narrative. Furthermore, while the Adam episodes in Q 15, 38, and 17 display far-reaching overlap and may thus largely appear as variants of each other, some of the divergences between Q 15's version of the narrative and the two subsequent accounts can also be read as having an interpretive significance.[68]

68 To adduce but two respects in which the Adam episode in Q 38, a large part of which simply reproduces that in Q 15, might be seen as engaged in interpretively motivated modification: (1) While Q 15:31 simply reports that Iblīs "refused to be among those who prostrated themselves" to Adam, Q 38:74 makes explicit the motivation underlying this refusal and its consequence: "He was haughty (*istakbara*) and one of the unbelievers." The charge of haughtiness also recurs in God's statement in Q 38:75. Note that in terms of rhyme, it would have been unproblematic to simply reproduce the wording of Q 15:31, making it probable that the point of the variant is interpretive. (2) Q 38 replaces the phrase *min ṣalṣālin min ḥama'in masnūn* ("from clay,

To avoid misunderstanding, I should like to concede that the question of how far we can press minute differences in wording to yield an interpretive point is apposite; the line between close reading and over-interpretation is, as usual, a fine one. It would also be exaggerated to suggest that the Adam sections in Q 15, 38, and 17 are doing *nothing but* interpreting earlier Qur'ānic passages. They also employ the Adam narrative as an appropriate medium for making a general theological point that is not interpretive in nature, namely, in order to draw a rigid boundary between God's "servants" (*ʿibād*), who are presented as immune against temptation by Iblīs, and the rest of humanity, who will be consigned to hell.[69]

3.3.2 Q 20 and Q 7

The Adam pericopes in Q 20 and 7 signal their link to the earlier Adam narratives in Q 15 and 38 by opening with a tightly abbreviated reprise of Q 15:28–31 / Q 38:71–74, which is likely to have marked them as a follow-up to these earlier texts. Q 20 and 7 then focus on the second part of the story, Adam's succumbing to the temptation to eat from the "tree of eternal life" (*shajarat al-khuld*, Q 20:120). This provides the story of Adam's creation with a sequel reporting how Iblīs' climactic announcement that he would from now on strive to lead man astray (*aghwā*), as narrated in Q 15:36–43 and 38:79–85, plays out (note that Q 20:121 uses the first-form verb *ghawā*). Of particular interest is the introductory verse of the Adam pericope in Q 20, which places the story of Adam under the heading of man's inherent forgetfulness and thus imposes on the narrative a highly explicit interpretation of its basic message (Q 20:115): "Formerly We made a covenant with Adam, but he forgot and We found in him no constancy." As a matter of fact, the verb *nasiya*, "to forget," reverberates throughout

from moulded mud") as it appears in Q 15 by *min ṭīn* (cf. 38:71.76 with 15:28.33). This might bespeak an attempt at lexical simplification, given that the phrase *min ṣalṣālin min ḥamaʾin masnūn* does not recur subsequently to Q 15; the word *ṭīn* prevails. Again, this substitution cannot have been due to requirements of rhyme.

69 See Q 15:40.42, Q 38:83, and Q 17:65; cf. Sinai, *Fortschreibung und Auslegung*, pp. 86–87. In all three sūras, the term *ʿibād* also occurs outside the Adam pericopes: see Q 15:49, Q 38:45, and Q 17:5.17.30.53.96 (although some of the occurrences in Q 17 would appear to refer indiscriminately to all humans). See the general remarks in Neuwirth, Angelika "Referentiality and Textuality in *Sūrat al-Ḥijr* (Q. 15): Observations on the Qur'ānic 'Canonical Process' and the Emergence of a Community," in Angelika Neuwirth, *Scripture, Poetry and the Making of a Community: Reading the Qur'an as a Literary Text*, pp. 184–215. Oxford: Oxford University Press, 2014, at pp. 206–207.

Q 20; in this sense, the basic anthropological message of Q 20's account of Adam's temptation is well integrated into the sūra as a whole.[70]

In contrast to Q 15 and 38, which present God's faithful "servants" as virtually immune to temptation, the Adam pericopes of Q 7 and 20 therefore recognise that all humans are prone to temporary lapses but may still receive God's merciful forgiveness, as Adam does in Q 20:122.[71] Q 7, which expands the narrative's second part by slotting in a quotation of Adam and Eve's dramatic self-incrimation (Q 7:23),[72] makes the further point that divine forgiveness and mercy need to be earned by an act of genuine repentance, thus forestalling the impression that it will simply be granted as a matter of course. Membership in the community of God's faithful "servants" is thus not predicated on total immunity to temptation but merely on the willingness to repent in a timely manner. Just as the verb *nasiya* occurs at several junctures in Q 20, so the topic of human repentance and divine forgiveness pervades the entire Q 7, meaning that, again, the particular angle from which the story of Adam is treated in Q 7 is integrated into the text as a whole.[73]

3.3.3 Q 2

The latest version of the Adam story in Q 2 amplifies the opening element of the narrative, God's announcement to the angels to create man. In a noticeable de-

---

70 In addition to v. 115, see also vv. 52, 88, and 126.

71 Q 20:122: "Then his Lord chose him [Adam] and He [God] relented towards him [Adam] and guided him." Jones' translation ("... and he [Adam] turned to Him [God] in repentance ...") is decidedly less likely, for *tāba ʿalā* as used in the verse at hand appears to be reserved for God as subject, whereas human repentance is described by *tāba ilā;* see Ambros, Arne A. and Stephan Procházka, *A Concise Dictionary of Koranic Arabic.* Wiesbaden: Reichert, 2004, p. 51.

72 Q 7:23: "They said: 'Our Lord, we have wronged ourselves. If You do not forgive us and have mercy on us, we shall be among the losers.'"

73 V. 23 is partially identical with v. 149, the Israelites' plea for forgiveness after worshipping the Golden Calf; cf. also Aaron's plea for forgiveness in v. 151. The underlying message of Q 7's retelling of Adam's temptation and the story of the Golden Calf is explicitly stated in v. 153: "Those who do evil deeds and subsequently repent and believe – your Lord is subsequently forgiving and merciful." Note that according to Neuwirth, "Meccan Texts – Medinan Additions?," Q 7's treatment of the Golden Calf underwent a significant amount of secondary expansion, although the precise delineation of what is original and what is secondary may perhaps not yet be finally settled (Neuwirth herself views v. 153 as secondary, but not v. 149 and v. 151). In any case, divine forgiveness is clearly a major topic of the sūra, as documented by the multiple occurrences of the root *gh-f-r* in it (see vv. 23, 149, 151, 153, 155, 161, 167, and 169). See also Sinai, *Fortschreibung und Auslegung*, pp. 88–90.

parture from the opening clause of the Qur'ān Adam narratives in Q 15 and 38, Q 2:30 modifies God's announcement from "I am creating a human from clay" to "I am establishing a successor on earth" – which is to be understood in the sense that man substitutes the angels as the highest-ranking creature.[74] This is not of course to be read as denying the earlier statement but as motivating it: God's creation of a being from clay is explained as serving the purpose of establishing a successor. From the outset, the Adam episode in Q 2 endows man with a markedly more elevated status than its antecedents, which anticipates the statement made in v. 31 that Adam participates in God's knowledge ("And He taught Adam all the names"). This statement occurs in a kind of prelude to God's command of prostration as reported in earlier recitations: the angels object to God's plan of creating man, whereupon God teaches them a lesson about the extent of their ignorance: God orders the angels to name the things created by Him, but only Adam is able to do so because God had previously "taught Adam all the names" (Q 2:31). The interposition of this episode between God's announcement of Adam's creation and His command to the angels to prostrate to Adam clarifies the ultimate meaning of the angel's prostration: it is not an act of worship directed at a creature, which would flagrantly clash with the Qur'ān's repeated injunctions not to "associate" any creature with God, but rather an act of submission to God's unfathomable knowledge – a message underlined by the multiple occurrences of *ʿ-l-m* that permeate Q 2:30–33.[75] The episode concludes with a brief retelling of the story's second part.

4 Conclusion

Virtually all component elements of the Qur'ānic Adam-Iblīs narratives have parallels in earlier Christian and Rabbinic traditions.[76] In this sense, it is certainly

---

74 On the meaning of the Arabic term *khalīfa* see Paret, Rudi, "Signification coranique de *ḫalīfa* et d'autre dérivés de la racine *ḫalafa*." *Studia Islamica* 31 (1970): pp. 211–217. Paret convincingly argues that the word's meaning within the Qur'ān is more accurately rendered by translating it as "successor" rather than "vicegerent."

75 As argued in Sinai, *Fortschreibung und Auslegung*, pp. 90–91, the Adam narrative in Q 2 may also be read as the vehicle of an anthropological reflection on human freedom: man's special status seems to be intimately bound up with his susceptibility to temptation, as illustrated in Q 7 and 20, and to "cause corruption" on the earth and to "shed blood" (Q 2:30).

76 As always, a convenient conspectus is provided in Speyer, *Die biblischen Erzählungen*, pp. 41–83. For a discussion of the relationship between Q 2:30–33 and its Rabbinic precedents see Sinai, Nicolai, *The Qur'an: A Historical-Critical Introduction*, New Edinburgh Islamic Surveys. Edinburgh: Edinburgh University Press, 2017, pp. 148–150.

the case that the passages examined above draw on pre-existing traditions, which are likely to have reached the Qur'ānic milieu by oral transmission, probably in Arabic and perhaps in a shape that already included some of the lexical features that surface in the Qur'ān. Nevertheless, the way in which the Qur'ānic proclamations successively appropriate this stock of narrative material partly serves the function of interpreting earlier Qur'ānic material – both by clarifying and elucidating it and by teasing new theological and anthropological aspects out of a story that had already been narrated in earlier recitations. The Adam-Iblīs complex thus illustrates that in studying inner-Qur'ānic interpretation we must not limit our perspective to insertions of the sort discussed in Section 2.

To reiterate the basic claim implicit in my chapter title, Qur'ānic self-interpretation comes in two modalities, interpretive expansion and interpretive back-referencing. Both of these modalities are not unique to the Qur'ān: as previous scholarship has amply demonstrated, they are found in the Hebrew Bible, and there is good reason to expect that they can also be traced in other literary corpora that are distinguished by the two following characteristics: first, having emerged over an extended period of time (whether that period spanned decades or centuries); and second, having emerged in close interaction with a community of hearers or readers who were committed to the truth and exceptional religious, moral, or philosophical significance of the textual corpus at hand, and who related to it accordingly.

The present study thus raises two main avenues for further research. On the one hand, we clearly do not yet possess an understanding of Qur'ānic self-interpretation that is on a par with that of Biblical self-interpretation; we require a close study of many more cases of the two types of Qur'ānic self-interpretation distinguished above, which will also afford us opportunities to refine the analytic categories I have proposed. On the other hand, there is reason to expect that it would be fruitful to study such phenomena against a broad comparative horizon that is not limited to Biblical literature but also includes texts from other periods and cultures. Of course, the objective of adopting such a comparative perspective should not consist merely in the identification of relevant commonalities – for instance, in attempting to show that the Bible and the Qur'ān share certain typological features. Rather, a comparative backdrop will ultimately help to throw into much sharper relief the peculiarities of each individual corpus.[77]

77 An example for this would be the above observation that scribal comments and glosses are much more infrequent in the Qur'ānic corpus than in the Hebrew Bible.

Gabriel Said Reynolds

Moses, Son of Pharaoh

A Study of Qurʾān 26 and Its Exegesis

1 Introduction

In the 26th of the Qurʾān's 114 sūras, entitled *al-Shuʿarāʾ* ("The Poets"), God calls on Moses and his brother Aaron to preach to the people of Pharaoh (Q 26:15–17). A conversation then transpires between Moses and Pharaoh:

> 18He [Pharaoh] said, 'Did we not rear you as a child among us, and did you not stay with us for years of your life? 19Then you committed that deed of yours, and you are an ingrate.' 20He said, 'I did that when I was astray. 21So I fled from you, as I was afraid of you. Then my Lord gave me sound judgement and made me one of the apostles. 22That you have enslaved the Children of Israel—is that the favour with which you reproach me?'[1]

The questions which Pharaoh asks of Moses ("Did we not rear you as a child among us, and did you not stay with us for years of your life?") suggest that he is Moses's adoptive father. Pharaoh seems to be upbraiding Moses as a father would a disobedient child. The notion that Pharaoh and Moses are father and son in the Qurʾān is confirmed in Q 28:7–9, a passage which tells the story of Moses' adoption:

> We revealed to Moses' mother, [saying], 'Nurse him; then, when you fear for him, cast him into the river, and do not fear or grieve, for We will restore him to you and make him one of the apostles.' Then Pharaoh's kinsmen picked him up that he might be to them an enemy and a cause of grief. Indeed Pharaoh and Hāmān and their hosts were iniquitous. Pharaoh's wife said [to Pharaoh], '[This infant will be] a [source of] comfort to me and to you. Do not kill him. Maybe he will benefit us, or we will adopt him as a son.' And they were not aware (Q 28:7–9; cf. Q 20:40).[2]

---

1 All Qurʾān translations are from Quli Qaraʾi unless otherwise noted: *The Qur'an with Phrase-by-Phrase English Translation*, ed. and trans. by Ali Quli Qara'i. New York: Tahrike Taarsile Qurʾān, 2007.

2 The character of Haman appears as the vizier of the Persian king Ahasuerus in the Biblical *Book of Esther*. In the Qurʾān he appears instead as the vizier of Pharaoh. The reasons for this shift, which are connected with an originally Ancient Near Eastern legend centered on the wisdom figure of Aḥīqar, have been meticulously examined by: Silverstein, A., "Haman's Transition from the Jahiliyya to Islam," *JSAI* 34 (2008): pp. 285–308.

https://doi.org/10.1515/9783110564341-014

To those who are familiar with the account of Moses from the Biblical Book of Exodus this passage might seem surprising. In *Exodus* it is not Pharaoh, or Pharaoh's wife (as in Q 28), who adopts Moses. According to Exodus it is Pharaoh's *daughter* who finds Moses and adopts him. She discovers Moses in a basket (Exod 2:5) and confides the boy with his own mother to be nursed.[3] Moses' mother later takes him back to Pharaoh's daughter who adopts him as a son: "And the child grew, and she brought him to Pharaoh's daughter, and he became her son; and she named him Moses, for she said, 'Because I drew him out of the water.'" (Exod 2:10; cf. Acts 7:21).[4]

Moreover, in Exodus the Pharaoh whom Moses confronts, the Pharaoh who refuses to let the Israelites go (until Egypt is struck by ten plagues) is *not* the Pharaoh of Moses' childhood. In Exodus Moses flees to Midian after killing an Egyptian and burying him in the sand because he is afraid of the vengeance of Pharaoh (Exod 2:12–15). He returns to Egypt only when he learns that this Pharaoh has died and a new one has taken his place: "And the LORD said to Moses in Midian, 'Go back to Egypt; for all the men who were seeking your life are dead'" (Exod 4:19).[5]

I am hardly the first to notice the contrast between the Bible and the Qur'ān as concerns the relationship between Pharaoh (or the Pharaohs) and Moses. Already Abraham Geiger, in his 1833 work *Was hat Mohammed aus dem Judenthume aufgenommen*,[6] notes the discrepancy. According to Geiger, however, the way in which the Qur'ān has the same Pharaoh on the throne is best understood as Muhammad's confused reception of a Midrashic tradition.[7] In the present brief study I would like to suggest that the Qur'ān's author has not unwittingly, but rather intentionally, diverged from the Exodus account. In my opinion the

---

3 A tradition in the Babylonian Talmud (b. *Sotah* 12b) explains the detail in the Exodus account which has Pharaoh's daughter seek out a wet-nurse among the Hebrew women by insisting that Moses refused the milk of Egyptian women. This has an echo in Q 28:12.

4 Cf. Jub 47:9, which has God recount to Moses: "And after this when you had grown they brought you to the daughter of Pharaoh and you became her son." Trans. by O.S. Wintermute in *The Old Testament Pseudepigrapha*, ed. by J.H. Charlesworth, vol. 2. Garden City, NY: Doubleday, 1985.

5 All Bible translations are taken from the Revised Standard Version, unless otherwise noted.

6 Geiger, A., *Was hat Mohammed aus dem Judenthume aufgenommen*, 2nd edition. Leipzig: Kaufmann, 1902, pp. 155–56. See the English translation of Geiger's work by Young, F.M., *Judaism and Islam*, Madras: M.D.C.S.P.C.K. Press, 1898, pp. 123–25.

7 Geiger explains (*Judaism and Islam*, p. 124) this as a point of confusion resulting from the influence on Muhammad of a tradition in *Exodus Rabbah* which makes the "death" of the first Pharaoh only an allusion to appearance of leprosy in his body. However, *Exodus Rabbah* is best dated to the 11th or 12th century and should not be taken as the source of Qur'ānic passages.

Qur'ānic author's very purpose for this scene is to have a confrontation between father and son, that is, to have Moses choose his obligation to God above his obligation to his father. This follows from the Qur'ān's larger concern to emphasize the primacy of faith over family.

Before turning to an examination of the Qur'ān itself I will begin by examining several Qur'ānic commentaries for their perspective on Moses' confrontation with Pharaoh in Q 26.

2 Islamic Exegesis on the Confrontation between Moses and Pharaoh

On Q 26:18, which has Pharaoh allude to his raising of Moses, the early *tafsīrs* attributed to Ibn ʿAbbās (d. 68/687) and Muqātil b. Sulaymān (d. 150/767) mention only that Moses stayed with Pharaoh for 30 years.[8] Fakhr al-Dīn al-Rāzī (d. 606/1210) makes the same remark (which is meant to explain why Pharaoh specifically mentions to Moses in this verse "Did you not stay with us for *years* of your life"). To this Rāzī adds an anecdote which describes the scene of the encounter between Moses and Pharaoh:

> The doorman said, 'Here is a man who says that he is a messenger of the Lord of the worlds.' [Pharaoh] said, 'Let him come in that we might have fun with him.' So they [Moses and Aaron] delivered the message. He recognized Moses and first recounted his favors done to him and then secondly recounted the wrong which Moses had done to him.[9]

The mention of a "wrong" at the end of this passage here is connected to Q 26:19 (cf. 20:40; 28:15, 33), in which Pharaoh alludes to Moses' killing of an Egyptian: "Then you committed that deed of yours, and you are an ingrate" [or "unbeliever"]."

8 Ibn Sulaymān, Muqātil, *Tafsīr*, ed. by ʿAbdallāh Muḥammad al-Shiḥāta. Beirut: Dār al-Turāth al-ʿArabī, 2002; reprint of: Cairo: Muʾassasat al-Ḥalabī, n.d., 3:260, ad Q 26:18. *Tanwīr al-Miqbās min Tafsīr Ibn ʿAbbās*, trans. by M. Guezzou. Amman: Royal Aal al-Bayt Institute, 2007, p. 407, ad Q 26:18. On the authorship and dating of this work, which is perhaps best attributed to ʿAbdallāh b. al-Mubārak al-Dināwārī (d. 308/920), see Pregill, M., "Methodologies for the Dating of Exegetical Works and Traditions: Can the Lost Tafsir of al-Kalbi be Recovered from Tafsir Ibn Abbas (also known as *al-Wāḍiḥ*)?" in *Aims, Methods and Contexts of Qur'anic Exegesis (2nd/8th-9th/15th century.)*, ed. by Karen Bauer. Oxford: Oxford University Press, 2013, pp. 393–453.

9 Rāzī, *Mafātīḥ al-ghayb*, ed. by Muḥammad Bayḍūn. Beirut: Dār al-Kutub al-ʿIlmiyya, 1421/2000, 24:108–9 ad Q 26:18.

To Ibn Kathīr (d. 774/1373), the tradition-minded Shāfiʿī commentator, the main tension between Pharaoh and Moses in this scene is not the demand which Moses makes of Pharaoh to let the Israelites leave Egypt but rather Pharaoh's memory of this killing. Ibn Kathīr paraphrases Q 26:18–19 by having Pharaoh declare to him: "Are you not the one we raised among us and in our midst and on our bed, the one whom we showed favor to for a long time, and then after you repaid this goodness from us with that deed by killing one of our men? You renounced our favor to you."[10]

The account of Moses' killing of an Egyptian, something only alluded to by Pharaoh in Q 26,[11] is found in Q 28:

> [15] [Moses] entered the city at a time when its people were not likely to take notice. He found there two men fighting, this one from among his followers and that one from his enemies. The one who was from his followers sought his help against him who was from his enemies. So Moses hit him with his fist, whereupon he expired. He said, 'This is of Satan's doing. He is indeed clearly a misleading enemy.' [16]He said, 'My Lord! I have wronged myself. Forgive me!' So He forgave him. Indeed, He is the All-forgiving, the All-merciful. [17]He said, 'My Lord! As You have blessed me, I will never be a supporter of the guilty.' (Q 28:15–17).[12]

Unlike Exodus, the Qurʾān has Moses blame the deed on Satan and seek forgiveness for it. In addition, Q 28:15–17 seems to make it clear that Moses' killing of the Egyptian was wrong. This is a departure from the Bible. *Exodus* has Moses hide the Egyptian's body in the sand (2:12), and it relates that Moses was afraid when he discovered that the deed had become known (Exod 2:14) but it does not reprove Moses for the killing. According to an opinion in the late Midrashic text *Exodus Rabbah*, Moses act was fully justified. He had studied the conduct of the Egyptian and "found that he deserved death."[13]

---

10 Ibn Kathīr, *Tafsīr*, ed. by Muḥammad Bayḍūn. Beirut: Dār al-Kutub al-ʿIlmiyya, 1424/2004, 3:312 ad Q 26:18.

11 This case might be seen as an example of Qurʾānic intra-textuality. One could imagine that the logic of Q 26 demands Q 28 to have preceded it, so that the audience would understand the allusion to Moses' "deed" (although Nöldeke makes Q 28 a "late" Meccan sūra and Q 26 a "middle" Meccan sūra). On the other hand it is certainly possible that the Qurʾān's audience would have known the story of the killing independently of Q 28.

12 Later in Q 28 the Qurʾān has Moses express fear that this killing will lead the Egyptians to seek vengeance against him. He declares to God: "My Lord! I have killed one of their men, so I fear they will kill me" (Q 28:33). This might be seen as a reflection of the "fright" which Moses feels in Exodus 2:14 when he learns that his killing of an Egyptian has come to light.

13 *Exodus Rabbah* (para. 5). See Geiger, A., *Judaism and Islam*, p. 123.

The Qur'ānic notion that Moses' killing of an Egyptian was a murder adds tension to his confrontation Pharaoh in Q 26.[14] Moses' position does not depend on his moral superiority. Nor does it depend on any obligation of Pharaoh towards him. Moses' position depends entirely on the legitimacy of his claim to be a prophet sent by God.

Like Ibn Kathīr, the Andalusian Mālikī al-Qurṭubī (d. 671/1273) is interested in the way that Pharaoh seeks to take the moral high ground in his debate with Moses. He concludes that Pharaoh mentions his adoption of Moses in Q 26:18 in order to remind Moses that he saved his life when other Israelite children were being massacred.[15] The same point is made by al-Biqāʿī (d. 885/1480). He paraphrases Pharaoh: "'We have over you in this a certain right which should prohibit you from addressing us in this way' ... because the threat of massacre of children which threatened him passed by him."[16]

Connected to all of this is the way in which Pharaoh, at the end of Q 26:19, accuses of Moses of being *min al-kāfirīn*. The Ḥanbalī Ibn al-Jawzī (d. 597/1200), in his work *Zād al-masīr*, records a debate which took place among earlier interpreters regarding this phrase. At issue was what exactly Pharaoh meant by accusing Moses of *kufr*. Inasmuch as *kufr* in the Qur'ān can function (for example in Q 14:34; 16:83; 22:66, passim) as a the opposite of *shukr* ("gratitude"), certain interpreters (Ibn al-Jawzī names Ibn ʿAbbās, Ibn Jubayr, ʿAṭāʾ, al-Ḍaḥḥāk, and Ibn Zayd) believed that Pharaoh was accusing Moses of being ungrateful "of his favor" (*niʿma*).

However, *kufr* also regularly functions as the opposite of *īmān* ("belief") in the Qur'ān. On this basis a second interpretation (supported by Ḥasan al-Baṣrī and al-Suddī) was offered, by which Pharaoh meant to accuse Moses of being an unbeliever (*kāfir*) in God. As Ibn al-Jawzī puts it, according to this interpretation what Pharaoh was really saying to Moses was, "You were a disbeliever in

14 Pharaoh reasonably refers both to his care for Moses and to Moses' offense. Pharaoh has a case to press. Moses, notably, does not seek to defend his action. He simply explains in response, "I did that when I was astray" (Q 26:20). Ibn al-Jawzī includes three different opinions of what Moses might mean when he said "I was astray" (*wa-anā min al-ḍāllīn*): first, that he was "ignorant" (*min al-jāhilīn*), not yet having received prophetic revelation; second, that he was "sinful" or "a wrongdoer" (*min al-khāṭiʾīn*), having wrongly killed a person; and third, that he was forgetful (*min al-nāsīn*). Ibn al-Jawzī, *Zād al-masīr fī ʿilm al-tafsīr*. Beirut: al-Maktab al-Islamī, 1404/1984, 6:119.

15 Abū ʿAbdallāh Muḥammad al-Qurṭubī, *Al-Jāmiʿ li-aḥkām al-Qur'ān*, ed. by ʿAbd al-Razzāq al-Mahdī. Beirut: Dār al-Kitāb al-ʿArabī 1433/2012, 13:91.

16 Ibrāhīm al-Biqāʿī, *Naẓm al-durar fī tanāsub al-āyāt wa-l-suwar*, ed. by ʿAbd al-Razzāq Ghālib al-Mahdī. Beirut: Dār al-Kutub al-ʿIlmiyya, 1432/2011, 5:353.

your god because you were among us in our religion which you now consider erroneous."[17]

It seems to me that the first interpretation is right, although it is interesting to note that by accusing Moses of being an ingrate Pharaoh is acting like God. In several passages the Qurʾān first recounts the bounty which God has given to man in nature (e.g., Q 14:32–33; 22:65) and then accuses man of being an ingrate (Q 14:34; 22:66).[18] In Q 26 Pharaoh imitates the divine character by reminding Moses of the favor he has given him and then accusing him of ingratitude. This imitation of Allah is found again later in that same passage where Pharaoh demands to be recognized as the only true god. Turning to his assistants Pharaoh declares, "If you take up any god other than me, I will surely make you a prisoner!" (Q 26:29).

As for modern Qurʾān commentators, a number of them show a distinct interest in understanding the psychology of Pharaoh in this passage. The 20th century Pakistani scholar Muḥammad Shafīʿ (d. 1976) argues that Pharaoh raises the issue of Moses' ingratitude (by calling him *min al-kāfirīn*) because he is not prepared to debate the merits of Moses' claims to prophethood. He comments:

> When a sharp opponent is not properly equipped with the correct arguments, he normally tries to switch the conversation towards the person of the addressee in order to find faults with him. This tactic is employed to embarrass the opponent and to make him look small before the audience. Hence, the Pharaoh also came out with two such points. First, 'We have brought you up in our household and have done so many favours to you. So, how can you have the face to speak before us.' Second, 'You have killed an Egyptian for no fault of his. This is not only cruelty but also ingratitude toward those among whom you are raised to your manhood.'[19]

Like Muḥammad Shafīʿ, the Egyptian rigorist Sayyid Quṭb (d. 1966) is interested in the rhetorical strategies employed by both Pharaoh and Moses. On Pharaoh's address he comments:

> Pharaoh is sarcastic, asking in an affected air of surprise: 'Did we not bring you up when you were an infant? And did you not stay with us many years of your life? Yet you have done that deed of yours while being an unbeliever. Is this how you repay our kindness as we looked after you when you were a young child? Is it fair that you come today profess-

---

17 Ibn al-Jawzī, 6:119.

18 In Q 14:34 the Qurʾān accuses man of being *kaffār* and in 22:66 of being *kafūr*. These variations of *kāfir* can be attributed to the rhyme patterns of those passages and need not be imagined to have some particular nuance of meaning.

19 Shafi, Muhammad, *Maariful Qur'an*, trans. by M. Shameen and M.W. Razi. Karachi: Maktaba-e-Darul-Uloom, 1996–2003, 6:528–29.

ing a religion other than ours, rebelling against the authority of the king who brought you up in his palace, and calling on people to abandon his worship?'[20]

Notably Quṭb finds sarcasm also in the way Moses responds to Pharaoh ("That you have enslaved the Children of Israel—is that the favour with which you reproach me?" Q 26:22). Quṭb comments:

Moses then uses a touch of sarcasm (*tahakkum*) in reply to Pharaoh's own sarcastic remarks, but he only states the truth: What sort of favour is this you are taunting me with: was it not because you had enslaved the Children of Israel? The fact that I was reared in your palace came about only as a result of your enslavement of the Children of Israel, and your killing of their children. This was the reason why my mother put me in a basket to float along the Nile. When your people found me, I was brought up in your palace, not in my parents' home. What favour is this that you press against me?[21]

That Quṭb concedes attributes sarcasm (*tahakkum*, which might also mean "scorn" or "derision") in his reply to Pharaoh is noteworthy. As a rule classical commentators do not attribute *tahakkum* to Moses (or – to my knowledge – to the prophets generally), although this is a rhetorical mode of address which God seems to employ in the Qur'ān.[22]

The Iranian Shi'ite commentator Ṭabāṭabā'ī (d. 1982) focuses on a different point. To Ṭabāṭabā'ī Pharaoh's intention is not to remind Moses of the favor done for him, or to remind Moses of his killing of an Egyptian. Pharaoh's point is simply that he knew Moses before the latter claimed to be a prophet and therefore he finds these claims of prophethood unbelievable:

The point and goal of this is, first in regard to his claim of prophethood, he says: "You are the one whom we raised when you were the little boy (*walīd*) and you remained with us from many years of your life. We know you by your name your character. We did not forget

---

20 Quṭb, Sayyid, *Fī ẓilāl al-Qur'ān*, 17th edition. Cairo: Dār al-Shurūq, 1412/1992, 5:2591. The English translation is taken from *In the Shade of the Qur'ān*, ed. and trans. by A. Salahi. Leicester, UK Foundation, 2003, 13:22–24.

21 Ibid.

22 This much is granted by the Qur'ān scholar Badr al-Dīn al-Zarkashī (d. 794/1392) in his *al-Burhān fī 'ulūm al-Qur'ān*. Zarkashī includes *tahakkum* among the various types of qur'ānic discourse. The examples of divine sarcasm he cites have God address the residents of hell (Q 56:42–43, 52–56, 93–94), or those to be condemned to hell (Q 9:34; 44:49). For example, Q 9:34 has God declare, "Give them the 'good news' (*fa-bashshirhum*) of a painful punishment." On this see Gwynn, R.W., "Patterns of Address." in *Blackwell Companion to the Qur'ān*, ed. by A. Rippin, pp. 73–87. Malden, MA: Blackwell, 2006, p. 78. I am grateful to Liran Yadgar for this reference.

> anything of your affairs. So how could you have this message when you are the one whom we know and we are not ignorant of your origins?"[23]

Ṭabāṭabā'ī's perspective reflects the influence of other passages in the Qur'ān, where unbelieving peoples find it impossible that someone whom they know from among them could be a prophet sent by God (see e.g., Q 7:63; 69; 38:4; 50:2).

Thus we can detect among our 20th -century commentators a certain interest in the psychological strategies of the Qur'ān's protagonists. Whereas classical commentators are principally interested in the meaning of Qur'ānic vocabulary, or an explanation of the Qur'ānic allusions, modern commentators have something like a psychological interest in decoding the rhetoric of the protagonists in this scene. None of them, however, think about the rhetorical strategies, or the psychology, of the Qur'ān's author. That is, they imagine this conversation to be precise transcript of a historical conversation. They see no possibility that the Qur'ān's author has shaped this scene for his own purposes. In a sense, the author of the Qur'ān disappears altogether in their analysis.

It is also worth noting that none of the commentators, modern or medieval, show any awareness or interest in the differences between the Bible and the Qur'ān in regard to the adoption of Moses. And indeed it is an appreciation of those differences which is the key to understanding the Qur'ānic passage before us.

3 Pharaoh in the Qur'ān

Before offering an interpretation of this passage it will be important to clarify some basic points about Pharaoh in the Qur'ān, most of which have been elucidated by Adam Silverstein in an excellent article on the subject.[24] The first point that Silverstein makes is that the very reference to the Egyptian ruler as "Pharaoh" is Biblical. In ancient Egyptian the term *par'o* (later rendered into Greek as φαραώ and Hebrew as *par'ōh*) means "Great House" and originally referred only to the residence of Pharaoh. Only around the year 1200 BCE (which according to a Biblical chronology would be several centuries after the lifetime of Moses) was the term applied to the Egyptian ruler.

23 Al- Ṭabaṭabā'ī, Muḥammad Ḥ., *Al-Mizān fī tafsīr al-Qur'ān*. Beirut: Mu'assasat al-'Ālamī li-l-Maṭbū'āt, 1418/1997, 15:259.

24 Silverstein, A.,"The Qur'ānic Pharaoh." in *New Perspectives on the Qur'ān: The Qur'ān in Its Historical Context 2*, ed. by G.S. Reynolds, pp. 467–77. London: Routledge, 2011.

In the Bible Pharaoh is used as a title for the Egyptian ruler (whose name never appears) in the stories of both Joseph (in Genesis) and Moses (in Exodus). In the Qurʾān, however, Pharaoh (*Firʿawn*) appears only in the Moses material. In the Qurʾānic sūra on Joseph the Egyptian ruler is referred to simply as the king (*malik*).[25] The reason for this is simple, although it is generally overlooked. The Qurʾānic author does not understand Pharaoh to be a title at all. In the Qurʾān Firʿawn is simply the *name* of the ruler of Egypt in the time of Moses (accordingly the ruler in the time of Joseph must be referred to otherwise).[26] As Silverstein puts it: "The Bible understands 'Pharaoh' to be a regnal title while the Qurʾān takes Firʿawn to be a more sharply defined historical character."[27] In order to keep this point clear, I will refer in what follows to the Qurʾānic character not as "Pharaoh" but rather as Firʿawn.[28]

4 The Qurʾānic Context

It is important for the Qurʾān's author that Firʿawn be the ruler of Egypt both in the time of Moses' childhood and his adulthood. The Qurʾān thereby makes the encounter between Firʿawn and Moses a family affair and advances one of its central arguments, namely that faithfulness to God should come before faithfulness to one's family. It is also the Qurʾān's concern with this argument which explains why it has Firʿawn's wife, and not his daughter, adopt Moses. This detail (Q 28:9) contradicts both the account of Exod 2:10, as already mentioned, and the retelling of the Exodus story in Stephen's speech in Acts 7:20 – 23. Yet in order to portray the later encounter between Firʿawn and Moses as an encounter between father and son it is necessary for the Qurʾān's author to have Firʿawn and his wife adopt Moses. Pharaoh's daughter must give way to Firʿawn's wife.

The Qurʾān's concern with the principle that faith comes before family is seen in the way it often portrays the unbelievers as a people who do things

25 The Bible refers to the Egyptian ruler also as "the king" (*ha-melek*) both in the Joseph story (Gen 40:1, 5) and in the Moses story (Exod 1:15, 18). That the Qurʾān refrains from calling the Egyptian ruler in the time of Joseph is occasionally the subject of Islamic apologetics (the point being that the Qurʾān is more historically accurate than the Bible) but in fact both the Bible and the Qurʾān are anachronistic in their usage of Pharaoh for the Egyptian ruler in the time of Moses.

26 This shift may be paralleled by the way in which the Biblical Potiphar, who is described in Gen 39:1 as the "captain of the guard," becomes in the Qurʾān *al-ʿazīz* ("the mighty one" 12:30, 51), a term which (despite the definite article) seems to be used as a nickname, and not a title.

27 Silverstein, p. 468.

28 In so doing I am following the precedent of Silverstein.

the other way around, who stubbornly cling to the false religion of their forefathers:

> When they are told, "Follow what Allah has sent down," they say, "We will rather follow what we have found our fathers following." What, even if their fathers neither applied any reason nor were guided?! (Q 2:170).[29]
>
> And when they are told, "Come to what Allah has sent down and [come] to the Apostle," they say, "Sufficient for us is what we have found our fathers following." What, even if their fathers did not know anything and were not guided?! (Q 5:104)

Similarly, in Q 23:24 Noah's opponents reject him with the declaration: "We have never heard of such a thing among our forefathers."[30] Not only does the Qurʾān warn its audience not to follow the false religion of family members, it commands its audience to separate themselves entirely from unbelieving members of their family: "O you who have faith! Do not befriend your fathers and brothers if they prefer faithlessness to faith. Those of you who befriend them—it is they who are the wrongdoers" (Q 9:23).

For our purposes it is interesting to note that the Qurʾān seems to shape its telling of Biblical narratives according to this topos. In its passages on Abraham, for example, the Qurʾān has an extraordinary interest in the conflict between Abraham and his unbelieving father, a story from Abraham's childhood which does not appear in the Bible but is well known from Jewish and Christian literature.[31] In a number of different passages the Qurʾān has Abraham preach to his father, and his father's people, demanding that they abandon idol worship. Thus Q 21:51–54:

> Certainly We had given Abraham his rectitude before, and We knew him when he said to his father and his people, 'What are these images to which you keep on clinging?' They said, 'We found our fathers worshipping them.' He said, "Certainly you and your fathers have been in manifest error."

---

29 Cf. Q 6:148; 7:28, 70–71, 173; 10:78 passim.

30 On the other hand when the Qurʾān has the prophet Elijah demand that his people – the Israelites – worship Allah he exclaims, "Do you invoke Baal and abandon the best of creators, Allah, your Lord *and Lord of your forefathers?*" (Q 37:125–26).

31 It is found, for example, in the 2nd century BCE Jewish text *Jubilees* 12 (2–3) and in the 2nd century CE *Apocalypse of Abraham* 7:11–12. For the English translations see *Jubilees*, trans. by Wintermute, O.S., in *The Old Testament Pseudepigrapha: Apocalyptic Literature and Testaments*, ed. by J.H. Charlesworth. Garden City, NY: Doubleday, 1985, 2:35–142; and, for the *Apocalypse of Abraham:* Kulik, A., *Retroverting Slavonic Pseudepigrapha: Toward the Original of the Apocalypse of Abraham*, ed. by J.R. Adair. Leiden: Brill, 2004, pp. 9–35.

Elsewhere the Qurʾān has Abraham promise to pray for his father: "He said, 'Peace be to you! I shall plead with my Lord to forgive you. Indeed He is gracious to me. I dissociate myself from you and whatever you invoke besides Allah. I will supplicate my Lord. Hopefully, I will not be disappointed in supplicating my Lord'" (Q 19:47–48; cf. Q 14:41; 60:4). The Qurʾān, however, tells its audience that believers should *not* pray for their unbelieving family members:

> The Prophet and the faithful may not plead for the forgiveness of the polytheists, even if they should be [their] relatives, after it has become clear to them that they will be the inmates of hell. Abraham's pleading forgiveness for his father was only to fulfill a promise he had made him. So when it became manifest to him that he was an enemy of God, he repudiated him. Indeed Abraham was most plaintive and forbearing (Q 9:113–14).

Similar to the Qurʾān's portrayal of Abraham's division with his father is its portrayal of Noah's discussion with a son who is ultimately lost in the flood. This son (the only son of Noah who appears in the Qurʾān) is unknown to the Genesis account of Noah, which speaks of three sons of Noah, all of whom enter into the ark.[32] In Q 11 the Qurʾān has this son refuse to board the ark. When he is drowned with the other unbelievers the Qurʾān has Noah intercede for his son. For this he is reprimanded:

> 45 Noah called out to his Lord, and said, "My Lord! My son is indeed from my family. Your promise is indeed true, and You are the fairest of all judges."
>
> 46 Said He, "O Noah! Indeed He is not of your family. Indeed this is a wrongful deed. So do not ask Me [something] of which you have no knowledge. I advise you lest you should be among the ignorant."
>
> 47 He said, "My Lord! I seek Your protection lest I should ask You something of which I have no knowledge. If You do not forgive me and have mercy upon me I shall be among the losers" (Q 11:42–47).[33]

Finally it should be noted, as Joseph Witztum of Hebrew University has pointed out, that the portrayal of Moses in the Qurʾān (and in particular in Q 28) is parallel to the portrayal of Joseph in Q 12.[34] Of particular interest for our purposes is

---

32 On the Qurʾānic account of Noah's lost son and its Biblical origins see now Reynolds, G.S., "Noah's Lost Son in the Qurʾān," *Arabica* 64 (2017): pp. 1–20.

33 By rendering "this is a wrongful deed" in v. 46 (for *innahu ʿamalun ghayru ṣāliḥin*) I have departed from Quli Qaraʾi's translation of this phrase. He renders this phrase, "He is indeed [a personification of] unrighteous conduct" in order to suggest that it is a reprimand not of Noah but of Noah's son.

This passage might be compared to Q 46:15–18.

34 See Witztum, J., *The Syriac Milieu of the Qurʾan*, Ph.D. Dissertation. Princeton University, 2011, pp. 281–92.

the parallel between the way Potiphar speaks to his wife in Q 12:21 of adopting Joseph and the way Firʿawn's wife speaks to her husband in Q 28:9 of adopting Moses.

> The man from Egypt who had bought him said to his wife, "Give him an honourable place [in the household]. Maybe he will be useful to us (*ʿasā an yanfaʿanā*), or we may adopt him as a son" (Q 12:21).
>
> Pharaoh's wife said [to Firʿawn], "[This infant will be] a [source of] comfort to me and to you. Do not kill him. Maybe he will benefit us (*ʿasā an yanfaʿanā*), or we will adopt him as a son." And they were not aware (Q 28:9).

In both cases the prophet ultimately separates himself from his adopted parents. After his conflict with Potiphar's wife (who hardly acts in a maternal manner) Joseph is sent to prison (Q 12:35) according to his own wish (v. 33), even though Potiphar recognizes his innocence (vv. 28–29). After his conflict with an Egyptian (Q 28:15), Moses flees to Midian (v. 22).[35]

Thus the Qurʾān means to present the story of Moses and Firʿawn according to a certain topos – seen also with Noah, Abraham, and Joseph – according to which prophets choose God over family. The point of this – and this is always the point for the Qurʾān – is to deliver a message to its own audience: follow the example of the prophets who preached to, confronted, and even abandoned their family members in order to dedicate their lives to God.[36]

---

35 And just as Potiphar is a righteous figure so too is Firʿawn's wife, who elsewhere in the Qurʾān declares her faith in God and asks for deliverance from Firʿawn (Q 66:11). On this cf. Speyer, H., *Die biblischen Erzählungen im Qoran*, Gräfenhainichen: Schulze, 1931; reprint: Hildesheim: Olms, 1961, pp. 272–73.

36 The Qurʾānic theme of choosing God over family is not unfamiliar to the New Testament. Matthew has Jesus predict that his message will divide family members against each other and he insists that only those who love him more than father or mother are worthy of him: "He who loves father or mother more than me is not worthy of me; and he who loves son or daughter more than me is not worthy of me" (Matt 10:37). In another Gospel tradition Jesus declares: "Truly, I say to you, there is no man who has left house or wife or brothers or parents or children, for the sake of the kingdom of God, who will not receive manifold more in this time, and in the age to come eternal life." (Luke 18:29–30). Or one might reflect on how Jesus responds to the man who requests that he bury his own father before he follows Jesus: "Leave the dead to bury their own dead; but as for you, go and proclaim the kingdom of God" (Luke 9:57–60; cf. Matt 8:22).

5 Conclusion

This brief study suggests that there is no direct textual relationship between the Qur'ān and the Bible. It is true that the case of Moses' confrontation of Firʿawn shows that the Qur'ān depends on its audience's knowledge of the Bible. One might note, for example, the abrupt transition in Q 26 between verses 16–17 and 18 and what follows. Verses 16–17 are the end of a conversation between God and Moses (and Aaron) in Midian. Suddenly in verse 18 Moses is no longer in Midian but – apparently (the Qur'ān does not say so explicitly) – in Firʿawn's court. The Qur'ān presumably feels free to skip the action in between because the audience knows the general plot. One might note as well how the Qur'ān in verse 22 has Moses refer to Firʿawn's enslaving the Israelites, although nowhere in the Qur'ān is there any description of when, how, or why Pharaoh has done so. This sort of allusion implies that the Biblical subtext, the story of a Pharaoh who did not know Joseph, was familiar to the Qur'ān's audience.

However, we have also seen that the Qur'ān departs from the details of the Biblical account in its portrayal of the relationship of Pharaoh and Moses. According to the Qur'ān there is only one ruler in Egypt and his name is Firʿawn. This Firʿawn is the ruler in Moses' childhood, and he is still the ruler in Moses' adulthood. It has to be this way, for the drama of Q 26 account consists in making the confrontation between Firʿawn and Moses a family reunion, in illustrating how Moses chose God over his own father.

These observations suggest that in the Qur'ān's original milieu Biblical narratives were transmitted orally and that – as is typical with oral accounts – those narratives were shaped by each new storyteller. The author of the Qur'ān was not restricted by the written text of the Bible as he composed his own story of Moses and Pharaoh. He was accordingly free to turn Pharaoh into Firʿawn, the father of Moses.

Stefan Wild

Unity and Coherence in the Qur'ān

1 A Lack of Coherence and a New Branch of Exegesis

Muslim and non-Muslim views of the aesthetic value of language and style of the Qurʾān and its recitation are not as divided as they sometimes seem. But a majority of non-Muslim scholars for a long time did not see much literary value in the Qurʾān. Claude Gilliot and Pierre Larcher tell us:

> "The literary structure and arrangement or construction [...] of the Qurʾān is far from being self-evident" and: "Even its weaknesses are viewed as wonderful, if not miraculous [...]. Read with eyes other than those of faith, the qurʾānic style is generally not assessed as being particularly clear [...]."[1]

The beauty and the majesty of Qurʾānic recitation are, of course, in the ear of the believing listener. Both may be slightly less impressive in the eyes of the reader. But "that inimitable symphony, the very sounds of which move men to tears and ecstasy"[2] cannot be explained away. Navid Kermani in his book *Gott ist schön* has alerted us to the central aesthetic values of the *fascinans* and the *tremendum* comprised in the miracle of Qurʾānic recitation: "[...] throughout its history of reception, the Qurʾān has been reported to have an aesthetic effect uncontested by any other text in world literature."[3] Thomas Carlyle on the other hand had written: "Nothing but a sense of duty could carry any European through the Qurʾān, it is [...] a wearysome, confused jumble, crude, incondite, endless iterations, long-windedness, entanglement—insupportable stupidity, in short."[4] Such

1 Gilliot, Claude and Pierre Larcher, "Language and style of the Qur'ān," in *Encyclopedia of the Qur'ān* 3, ed. by Jane D. McAuliffe, pp. 108a and 126a. Leiden: Brill, 2003.

2 Pickthall, Marmaduke, *The Meaning of the Glorious Qurʾan*. New York: Dorset Press, 1930, p. vii.

3 Kermani, Navid "The Aesthetic Reception of the Qurʾān as Reflected in Early Muslim History," in *Literary Structures of Religious Meaning in the Qur'ān*, ed. by Issa J. Boullata, pp. 255–276. Richmond Surrey: Curzon Press, 2000, p. 255.

4 Carlyle, Thomas, *On Heroes, Hero Worship, and the Heroic in History*. London: Chapman & Hall, 1872, p. 40; Wild, Stefan, "'Die schauerliche Öde des heiligen Buches'. Westliche Wertungen des koranischen Stils" in: *Gott ist schön und Er liebt die Schönheit. God is beautiful and He loves beauty. Festschrift in honour of Annemarie Schimmel presented by students, friends and colleagues*

https://doi.org/10.1515/9783110564341-015

a judgment was obviously misinformed, crudely generalizing and polemical. Some more measured non-Muslim voices found fault with the Qurʾānic style, but used a slightly less alarming tone. However, R. A. Nicholson maintained that the Qurʾān "is obscure, tiresome, uninteresting, a farrago of long-winded narratives and prosaic exhortations."[5] John Wansbrough says this about "the fragmentary character of Muslim scripture": "Exhibiting a limited number of themes, the exempla achieve a kind of stylistic uniformity by resort to a scarcely varied stock of rhetorical conventions [...]" and calls their effect "mechanically linked" and "tedious in the extreme."[6]

All of this was scarcely a major problem for Muslim scholars, who knew that the Qurʾānic revelations had not come down in a single pronouncement, but in installments, spread out over twenty-three years, piecemeal as it were (Q 25:32; *munajjaman* as the commentators later said). They also knew that Meccan passages could appear anywhere in Medinan sūras and vice versa. And they were not surprised that non-Muslims often were immune or even hostile toward what many Muslims felt to be the matchless beauty and the irresistible power of the Qurʾānic voice. It was only when Western non-Muslims in the 19th and 20th centuries complained about or ridiculed what they considered a lack of "unity" or "coherence" in the sūras, that Muslims were made aware of what had not been a problem before. While the deep impression of a Qurʾānic recitation on the Muslim listener was and is a fact, a theological problem for the modern Muslim scholar arose. An overriding horizon of aesthetic judgments regarding the Qurʾān, its sūras, its verses, their structure and composition, and the arrangement and order of the sūras was and is for Muslim scholars the dogma of the "inimitability" of the Qurʾān (*iʿjāz al-qurʾān*). This dogma had its roots in self-referential verses of the Qurʾān insisting on its own matchlessness. Muslim scholars developed this dogma further into a firewall around the Qurʾān that forbade finding any shortcomings in and around the text, its recitation or its codification. The Qurʾānic language was and still is claimed to be matchless and as such seen as the model of the purest, most eloquent Arabic speech. Arabic poetry and prose, however, never followed all the peculiarities of the Qurʾānic text.

---

on April 7, 1992, ed. by Alma Giese und Johann C. Bürgel, pp. 429–447. Bern: Peter Lang, 1994, pp. 433–436.

5 Boullata, Issa J., "Literary Structures of the Qurʾān," in *Encyclopedia of the Qurʾān* 3, ed. by Jane D. McAuliffe, pp. 192–204. Leiden: Brill, 2003, pp. 195–196.

6 Wansbrough, John, *Quranic Studies. Sources and methods of scriptural interpretation*. Oxford: University Press, 1977, p. 18; see also Gilliot, Claude and Pierre Larcher, "Language and style of the Qur'ān," in *Encyclopedia of the Qur'ān* 3, ed. by Jane D. McAuliffe, pp. 110a-111b. Leiden: Brill, 2003.

But how to react to non-Muslim scholarship that seemed to undermine the dogma of inimitability by calling the Qur'ānic text "disjointed," "disconnected," or "incoherent"? Even some Muslim scholars in the 19th and 20th century had accepted that the Qur'ān was "lacking in coherent composition." Richard Bell's (d. 1952) was one of the first voices to describe the Qur'anic "style" as "disjointed":

> A real characteristic of Qur'anic style (is) that it is disjointed. Only seldom do we find in it evidence of sustained unified composition at any great length [...] Some of the narratives [...] especially accounts of Moses and of Abraham run to considerable length; but they tend to fall into separate incidents instead of being recounted straightforwardly [...] The distinctness of the separate pieces, however, is more obvious than their unity.[7]

A conservative Muslim point of view explains some of the stylistic peculiarities in the Qur'ān in the following way:

> The Qur'ān may present, in the same *sūra*, material about the unity and grace of God, regulations and laws, stories of earlier prophets and nations [...]. This stylistic feature serves to reinforce the message to persuade and to dissuade. This technique may appear to bring repetition of the same themes or stories in different *sūras* but, as the Qur'an is above all a book of guidance, each *sūra* adds to the fuller picture and the effectiveness of the guidance [...]. This technique compresses many aspects of the Qur'ānic message into any one *sūra*, each forming self-contained lessons.[8]

All this notwithstanding, numerous Muslim scholars felt stung by statements on the "disjointed style" of some sūras and read this criticism as a stab at Qur'ānic inimitability. Most of the more sophisticated ensuing attempts to prove the "inner coherence" of the sūras and of the whole of the Qur'ān were probably caused by these non-Muslim pronouncements that "criticized" the Qur'ān in the 19th and early 20th centuries.[9] The quest to show the "unity" or the "coherence" of the Qur'ān was, therefore, primarily the wish to refute what had been called its "disjointedness." It was at the same time an act of preserving the dogma of the Qur'ān's inimitability and an attempt to modernize exegesis. It has meanwhile become a new branch of Qur'ānic exegesis.

---

7 Watt, William M. ed., *Bell's Introduction to the Qur'ān completely revised and enlarged.* Edinburgh: University Press, 1970, pp. 73–74.

8 Abdel Haleem, Muhammad A., *The Qur'an. A new translation.* Oxford: Oxford University Press, 2004, p. xix.

9 El-Tahry, Nevin Reda, *Textual Integrity and Coherence in the Qur'an. Repetition and Narrative Structure in Surat al-Baqara.* Toronto: PhD. Thesis, 2012, p. 41.

Problems of the composition and stylistic organization (*naẓm*) of the Qurʾān and the correlations of its sūras and verses (*munāsabāt*) had already been discussed in the classical period. A good summary of these pre-modern discussions can be found in Cuypers' most recent book.[10] But in the 20th century, a new urgency appeared. The first Muslim voices trying to refute any stylistic critique of the Qurʾānic voice and attempting to find coherence or unity in all the sūras came from South Asian Islam. Abdullah Yusuf Ali (b. 1872 in British India, d. 1953) may have been the first to react. His best-known work is his widely acclaimed book *The Holy Qur'an: Text, Translation, and Commentary* (Lahore 1938). Here, he wrote in a comment on Q 4:82 ("Do they not ponder on the Qurʾān? Had it been from other than Allah, they would surely have found therein much discrepancy"):

The unity of the Qurʾān is admittedly greater than that of any other sacred book. And yet, how can we account for it except through the unity of Allah's purpose and design? From a mere human point of view, we should have expected much discrepancy, because 1) the Messenger who promulgated it was not a learned man or philosopher, 2) it was promulgated at various times and in various circumstances, and 3) it is addressed to all grades of mankind. Yet, when properly understood, its various pieces fit together well even when arranged without any regard to chronological order. There was just the One Inspirer and the One Inspired.[11]

This statement seems to encapsulate the problem. The first author to systematically address what he saw as a "lack of coherence" in the Qurʾān was Mustansir Mir. He criticized the "atomistic" approach of many non-Muslim and some Muslim exegetes and offered the following diagnosis: "The dominant view about the Qurʾān has been that it is lacking in coherent composition and that whatever composition it may have it is not very significant. This view is shared by Muslim and Orientalist scholars."[12] It was not immediately clear whether this "lack" affected only some sūras or indeed the whole Qurʾānic text (including the order of the sūras?) or what this diagnosis meant for the dogma of the Qurʾānic inimitability. In a later statement, Mir, partly disavowing his earlier inclusion of Muslim scholars in the group of people who found the Qurʾān "lacking in composition," explained:

---

10 Cuypers, Michel, *The Banquet. A Reading of the Fifth Sura of the Qur'an.* Preface by Muhammad Ali Amir-Moezzi. Miami: Convivium Press, 2009, pp. 493–502.

11 Ali, Abdullah Y., *The Holy Qur'ān. English Translation of the Meanings and Commentary.* Medina: King Fahd Complex, 1990, pp. 237–8.

12 Mir, Mustansir, *Coherence in the Qur'an: A study of Iṣāḥī's concept of naẓm in Tadabbur-i Qur'an.* Indianapolis: American Trust Publications, 1986, pp. 2–3.

> Western scholars have often spoken of the "disconnectedness" of the Qur'ān [...] most Muslim scholars have not raised the issue at all [...] A distinction must be made between connection and unity, the former may be defined as a link – strong or weak, integral or tangential that is seen to exist between the components of a text [...] whereas unity arises from a perception of a given text's coherence and integration and from its being subject to a centralizing perspective [...].[13]

This was the beginning of a new branch of Qur'ānic exegetical scholarship setting out to prove that there was "coherence" or "unity" or "connection" throughout the Qur'ānic text. Mir had claimed—possibly prematurely—that all sūras are "units" or "unities." But he was certainly right in insisting that the quest for Qur'ānic unity had become a sustained attempt of some Muslim and later also non-Muslim scholars, who rejected the idea that the Qur'ān lacked coherence or unity. Mir was also right in stating that such an approach was a "definite break with the traditional style of exegesis [...] from the early Islamic centuries to the end of the 19th century. One manifestation of this break is the view that the qur'ānic *sūra*s are unities."[14] While many Meccan sūras incontrovertibly show structure and composition, a certain "lack of coherence" seems evident in the case of the long Medinan sūras. In the meantime, numerous serious and widely divergent attempts to discover a compositional structure in these long Medinan sūras have been made. Some of these attempts, highlighted in this article, evidently have an apologetic touch.

2 The Sūra as a "coherent unit"

What makes a sūra a sūra? Formally speaking, each liturgically recited and/or textually read passage in the Qur'ān that has a marked beginning and a marked end is a sūra. The marked beginning is the *basmala* and its end is marked by the *basmala* of the following sūra. Apparently, there were proto-types of the Qur'ānic text that did not have the *basmala* as a separating element between the sūra.[15] The existence of the *basmala* as a sūra-divisor is not apparent in the self-refer-

13 Mir, Mustansir, "Unity of the Text of the Qur'an," in *Encyclopedia of the Qur'ān* 5, ed. by Jane D. McAuliffe, pp. 405–406. Leiden: Brill, 2006.

14 Mir, Mustansir, "The sūra as unity. A Twentieth century development in Qur'ān-exegesis," in *Approaches to the Qur'ān*, ed. by Gérard R. Hawting and Abdul-Kader A. Shareef, pp. 211–224. London: Routledge, 1993, p. 211; Boullata, Isa, "Literary Structures of the Qur'an," in *Encyclopedia of the Qur'ān* 3, ed. by Jane D. McAuliffe, pp. 192–204. Leiden: Brill, 2003.

15 Graham, William A., "Basmala," in *Encyclopedia of the Qur'ān* 1, ed. by Jane D. McAuliffe, pp. 207–212. Leiden: Brill, 2001, p. 210a.

ential Qur'ānic use of the word sūra. In the Qur'ān, the word sūra evidently could not encompass all the sūras in our text.[16] Angelika Neuwirth says: "It is highly questionable if the term *sūra* was used during the Prophet's lifetime to denote the chapters of the Qur'ān in general."[17] As there are no sūras except the 114 "chapters" that we know, the literary genre "sūra" does not go beyond these 114 *specimina*. I disregard here the pious or sacrilegious pseudo-"sūra" that mimic Qur'ānic sūras.[18] I also ignore late Imāmī Shī'ite falsifications such as the "Sūra of the Two Lights" (*sūrat an-nūrayn*)."[19]

But is there a common literary or structural denominator of the genre "sūra"? In my opinion, there is no such common denominator. All sūras are sūras, because they were given this label under the theological umbrella of the early collectors and codifiers of the Qur'ān. One of their aims seems to have been to preserve precious divine utterances that had not found a place anywhere else. Angelika Neuwirth argued "most of the long *sūras* cease to be neatly structured compositions and appear to be the result of a process of collection that we cannot yet reconstruct."[20] More or less unstructured sūras, of course, easily coexist with perfectly structured ones. Mir quotes a number of modern Muslim exegetes who aimed at showing that all sūras in some way are "coherent wholes." He named the following proponents as herald for this quest: Ashraf 'Ali Thanavi (d. 1943) in his commentary *Bayān al-Qur'ān* (first published in twelve volumes in Karachi and Lahore in 1908), says that every verse in the Qur'ān is connected with the preceding and the following verses. Others followed: Sayyid Quṭb (d. 1966), in his revolutionary and widely read *Fī ẓilāl al-qur'ān* (20 volumes, Cairo 1954), claims that each sūra has an "axis" (*miḥwar);* i.e., each sūra revolves around a principal idea and is to be understood with reference to it, and that the sūras are sub-divided into "sections"; Hamid al-din al-Farrāḥī (d. 1930) and his student Amin Ahsan Islahi (d. 1997) agree in *Tadabbur-i Qur'an* (8 volumes, Lahore 1967–80) that each sūra has a central theme or a "pillar" ('*amūd*) around which all verses revolve:[21]

---

16 Mir, "The sūra as unity", particularly pp. 211 f., 217 ff.

17 Neuwirth, Angelika, "Sūra(s)," in *Encyclopedia of the Qur'ān* 5, ed. by Jane D. McAuliffe, pp. 166–177. Leiden: Brill, p. 167b.

18 Wild, Stefan, "The Koran as Subtext in modern Arabic Poetry," in *Representations of the Divine in Arabic Poetry*, ed. by Gert Borg and Ed de Moor, pp. 139–160. Amsterdam: Rodopi, 2001, 157 ff.

19 Amir-Moezzi, Mohammed A. and David Streight, trans., *The Divine Guide in Early Shiism. The Sources of Esoterism in Islam*. Albany: New York University Press, 1994, p. 80.

20 Neuwirth, "Sūra(s)", p. 173.

21 Farahi, Hamid al-Din and Mahmood Hashmi, trans., *Exordium to Coherence in the Qur'an. An English Translation of Fatihah Nizam al-Qur'an*. Lahore: Al-Mawrid, 2008.

> Each chapter of the Qur'ān is a well-structured unit. It is only lack of consideration and analysis on our part that they seem disjointed and incoherent [...]. Each chapter imparts a specific message as its central theme. The completion of this theme marks the end of the chapter. If there were no such specific conclusion intended to be dealt with in each chapter there would be no need to divide the Qur'an in chapters. Rather the whole Qur'an would be a single chapter [...].We see that a set of verses has been placed together and named "*sūra*" the way a city is built with a wall erected round it. A single wall must contain a single city in it. What is the use of a wall encompassing different cities?[22]

Islahi contributed the idea that the sūras exist in pairs, the two sūras of any pair being complementary to each other and together constituting a unit. Furthermore, according to him the 114 sūras fall into seven groups, each group containing one or more Meccan sūras with one or more Medinan sūras. Muḥammad ʿIzzat Darwaza (d. 1984) in *At-Tafsīr al-ḥadīth* (12 vols., Cairo 1962–64) opines that most of the verses and passages in the sūras are interconnected (*mutarābiṭ wa-munsajim*); the quest for unity in Darwaza's many exegetical works was marginal;[23] Muḥammad Ḥusayn al-Ṭabāṭabāʾī (d. 1981), *al-Mīzān fī tafsīr al-Qurʾān*, (20 vols. Beirut 1973–74) says that each sūra has an aim (*gharaḍ*) with an opening (*badʾ*), a discussion (*siyāq jārin*), and a conclusion (*khitām).*[24] Often, scholars arrived at opting for this or that arrangement, but the factor of arbitrariness could not be overlooked. This is clearly expressed in El-Awa's repeated critical statements. According to her, Sayyid Quṭb and Islahi

> leave their readers with a puzzling methodological question: that is, at the stage of deciding the [...] underlying theme of a *sūra*, what is the tool used for discovering that theme? The only suggested method for determining the "axis" of a *sūra* is 'reading thoroughly', 'deep reflection' and 'several thoughtful readings' [...] This leaves us with a countless number of possible suggestions [...].[25]

El-Awa divides the 73 verses of Q 33 into ten "passages," which are then developed into seven "concepts." At the end of El-Awa's reading of Q 33, she resumes the classification of these "concepts": "Each of these concepts has been developed simultaneously and not separately in a tight texture of relevance relations between several passages where the contexts were introduced and then gradual-

22 Ibid., p. 56

23 Poonawala, Ismail K., "Muḥammad ʿIzzat Darwaza's principles of modern exegesis: a contribution toward quranic hermeneutics," in *Approaches to the Qurʾān*, ed. by Gérard R. Hawting and Abdul-Kader A. Shareef, pp. 225–246. London: Routledge, 1993, p. 238.

24 Mir, "The sūra as unity", pp. 212–214.

25 El-Awa, Salwa M., *Textual Relations in the Qurʾān. Relevance, coherence and structure.* Abingdon: Routledge, 2006, p. 21.

ly expanded and enhanced which makes it fairly difficult to see the distinction and the borderlines between them."[26] In her case and elsewhere, there is often a blend of touching piety with advanced linguistics. Farid Esack resumes the problem of "sections," "passages," "units" – think fairly —: "These divisions [...] come across as arbitrary and depend rather unduly on what the reader chooses to see. It is also somewhat difficult to imagine the Prophet and the Companions working their way through an elaborate system of textual divisions as presented above."[27] And Issa J. Boullata is right when he warns of unfocused and pre-mature claims: "The study of the qur'ānic *sūra* as a unit with coherent unity is still in need of focused, philological elaboration in modern scholarship."[28] What seems to be clear is the impossibility that all of the various proposals suggested – and there are many more – can be equally right.

3 *Sūrat al-Baqara* as an Example

It is generally acknowledged that the bulk of the Medinan sūras belong to the more complex and relatively understudied part of the Qur'ān. Many Meccan sūras are hymnal, often short and apocalyptic unities, bound by a common topic, by short verses, rhyme-like assonances and introductory formulas. Frequently they have a fairly rigorous compositional structure, as is best shown in the work of Angelika Neuwirth. In the case of the Medinan sūras, the situation is different: assonances are still there, but the reader or listener is often left with almost prosaic, frequently long-winded codifications of loosely composed and often conceptually unframed passages. Montgomery Watt had already warned: "The longer *sūras* contain short pieces which are complete in themselves and could be removed without serious derangement of the context."[29] In contradistinction to most Meccan sūras, the concept of sūra becomes extremely vague when applied to the Medinan sūras. The "Medinan regulations do not display any structured composition nor do they form part of neatly composed units; they suggest, rather, later insertions into loosely connected contexts [...]"[30] Prophetic utterances that at some time were embedded in a distinct historical con-

26 Ibid., p. 99.
27 Esack, Farid, *The Qur'an. A User's Guide*. Oxford: Oneworld Publications, 2005, p. 66.
28 Boullata, "Literary Structures of the Qur'ān." *Encyclopaedia of the Qur'ān*, vol. 3, pp. 196a; Mir, "Unity of the Text of the Qur'an", p. 405.
29 Watt, *Bell's Introduction*, p. 74.
30 Neuwirth, Angelika, "Form and structure of the Qur'ān," in *Encyclopedia of the Qur'ān* 2, ed. by Jane D. McAuliffe, pp. 245–266. Leiden: Brill, p.262b.

text lost this embedding. Contexts are switched, allusions, often fragmentary and vague, abound, so that many Medinan passages become virtually de-contextualized. With the Qur'ān's final codification, these Qur'ānic passages even seem "de-historicized."[31] While we may assume that the implied *audience* of the Prophet had had at some point no serious problem to understand a recited passage in its context, the implied *readership* after codification and canonization—decades later—may have often lost its bearing.

Neal Robinson, Mathias A. H. Zahniser and Michel Cuypers attempt in different ways to prove that the long Medinan sūras do show "coherent structures," "principal sections," "blocks of verses" etc. Nobody denies that there are coherent passages in these sūras. But there is nothing that ultimately keeps these passages together as a sūra but the *basmala* at its beginning and the *basmala* at the beginning of the following sūra. The case of Q 8 (*al-Anfāl*) and 9 (*at-Tawba*) is instructive. Apparently, there was a dissent between two factions among those who were responsible for the final compilation, redaction, codification, and publication of the Qur'ān. Did the passages of *al-Anfāl* and *at-Tawba* form one sūra or two sūras? The committee responsible for the canonization of the qur'ānic text apparently neither would combine the two texts into one sūra, nor dared completely separate the two texts. The compromise was: the lack of the *basmala* in Q 9 makes its intimate connection to Q 8 obvious, but Q 9 is still counted as a separate sūra.

As Q 2 (*al-baqara*) is the longest in the Qur'ān, the problems of its "unity" are especially interesting. The confusion that a un-initiated reader might suffer when starting to read this sūra, the "giant of the *sūra*s," is graphically described by Michael Sells: "For those familiar with the Bible, it would be as if the second page opened with a combination of the legal discussions in Leviticus, the historical polemic in the book of Judges, and apocalyptic allusions from Revelation, with the various topics mixed together and beginning in mid-topic."[32] Sayyid Qutb, in his famous rebellious and huge commentary *Fī ẓilāl al-qur'ān* believes

> [...] that Q 2 has a double-lined theme whose two lines are strongly bound together. The first thematic line revolves around the hostile attitude of the Jews [...] to Islam in Medina [...] and their friendly relations with the Arabian polytheists and hypocrites [...] The second thematic line revolves around the corresponding attitude of the Muslims in Medina and their growth as a believing community prepared to carry the responsibility of God's call after Jewish rejection. Both lines are complementary and tightly bound together throughout the *sūra*, which eventually ends as it began: by exhorting [...] human beings to belief in God [...]

31 Ibid., p. 247a.

32 Sells, Michael, *Approaching the Qur'ān. The Early Revelations.* Oregon: White Cloud Press, 1999, p. 3.

> his prophets, his scriptures [...] and the metaphysical unseen world [...]. From beginning to end, the several topics of the *sura* are related to this double-lined theme.[33]

This looks like an example of a pseudo-discovery: that a sūra exhorts mankind to belief in God, the Prophets, and the scriptures is hardly a specific "line" that could characterize a sūra.

When Robinson tries to show that Q 2 has a "coherent structure," his immediate approach is completely and interestingly different from the one of Sayyid Quṭb.[34] Adopting an idea of Jacques Berque (d. 1995), he refers to the fact that an important verse (Q 2:143) stating that the Muslim community was a community of "the middle" (*umma wasaṭ*) is placed in the middle of Q 2 (with 286 verses).[35] Berque had further speculated that the "middle" of the whole Qur'ān could be Q 57, its "phonological middle" (Q 74:18), and its "lexical middle" (Q 35:26). He was convinced that focusing on the "middle" of Qur'ānic passages would lead to important results.[36] I do not agree. Most pre-modern Muslim scholars of the Qur'ān would never refer to a verse by its number. They did not need to, because most of them knew the holy text by heart anyway and did not depend on the mechanics of numbering.[37] In any case, the idea, that a verse in the "middle" of a sūra or a passage would give it a hidden structural meaning seems almost cabbalistic and not far away from Rashad Khalifa's (d. 1990) famous decoding the number 19 in the Arabic letters of the Qur'ānic text as proof of its miraculous character.

In a second and different approach, Robinson counts five "principal sections" in Q 2. Following the Urdu commentary *Tadabbur-i Qur'an* by Amin Ahsan Islahi, he structures Q 2 as follows: 1. Prologue (verse 1–39), 2. Address to the Israelites (verse 40–121), 3. The Abrahamic Legacy (verse 122–152), 4. The *Sharia* or Law (verse 153–242), 5. Liberation of the Ka'ba (verse 243–283) and a short epilogue (verse 284–286). Robinson does admit: "The *sūra* contains a number of parentheses in which subsidiary topics are developed at

---

33 Boullata, "Literary Structures", p. 195.

34 Robinson, Neal, *Discovering the Qur'an. A Contemporary Approach to a Veiled Text.* Washington D.C.: Georgetown University Press, 2003, p. 201.

35 Berque, Jacques, *Relire le Coran.* Paris: Editions Albin Michel, 1998, p. 38.

36 For some other attempts of Berque's to deal with the Qur'ānic text, see Gilliot, Claude and Pierre Larcher, "Language and style of the Qur'ān," in *Encyclopedia of the Qur'ān* 3, ed. by Jane D. McAuliffe, pp. 109–135. Leiden: Brill, p. 110b. See also Cuypers, *The Banquet*, pp. 508–512.

37 Spitaler, Anton, *Die Verszählung des Koran nach islamischer Überlieferung.* München: Verlag der Bayerischen Akademie der Wissenschaften, 1935, pp. 12–3.

length, or that on occasion's successive *āyahs* appear to contain legislation dating from different periods."[38]

Bertram Schmitz reads Q 2 as consistently and uniformly related to the Torah. In his view, what he calls its historical passages (Q 2:30 – 73 and 2:124–141) mirror parts of the Pentateuch. He counts five "law-complexes" in Q 2:142–274 that parallel verses in Deuteronomy (12; 14; 16; 21 f, 24 f; 26). According to him, Q 2:163–283 is so close to Deuteronomy that one could almost call these verses a "Tritonomyum." He translates or rather interprets *umma wasaṭ* as "a community in the center," i.e., Medina in the geographical center between Mecca with the Kaʿba in the south and Jerusalem in the North. Schmitz also refers shortly to the problem of "unity." He insists that the second sūra has to be treated as a closed unit ("eine geschlossene Einheit"), even if must historically be seen as a compilation. His main argument is that the believer considers it a closed unit, in other words a sūra—which, of course, nobody doubts.[39]

Abdullah Yusuf Ali (d. 1953), in the 1930s an early proponent of the theory of the "unity of the Qur'an," divided in his commentary *The Holy Qur'an. English Translation of the Meanings and Commentary,* all sūras from Q 2 to Q 80 into "sections." He was possibly the first exegete to introduce the idea of "sections" into Muslim exegesis. According to him, Q 2 has 40 "sections." Ali does not explain how he defines them. The vagueness of the term is evident. But equally vague are other terms such as "connection," "integral and tangential links," "integration," "centralizing-perspective," "double-lined theme," etc. Even "coherence" and "unity" can be pressed until they lose their meaning. More and more complex theories were developed: Islahi argued that on one level "each *sūra* is an integrated whole," while on another level "all *sūras* exist in the form of pairs composed of two closely matched *sūras* and distinct from the other pairs."[40] This is a far-reaching and risky theory. It groups together seven large thematic groups of sūras, which Islahi bravely identifies with *sabʿan mina l-mathānī* (Q 15:87), a difficult conjunction that is usually reserved for the seven verses of the *Fātiḥa*.

Mathias A. H. Zahniser states clearly that all his findings depend on the *assumption* of the unity and coherence of Q 2. His conclusion is:

> Although this chapter has been written on the basis of the assumption that the long *sūras* in their canonical form possess a coherent unified order, I do not intend to deny their ob-

---

38 Robinson, *Discovering*, p. 202.

39 Schmitz, Bertram, *Der Koran: Sure 2 "Die Kuh". Ein religionshistorischer Kommentar.* Stuttgart: Kohlhammer, 2009, p. 9.

40 Mir, *Coherence*, p. 87.

> vious loose structure, their parenthetical verses and verse groups, their laconic compactness or their frequent digressions, or even that they may be a collection of independently composed verses and verse groups, and that the "distinctness of the separate pieces [...] is more obvious than their unity.[41]

Most Muslim scholars would agree that *sūrat al-Baqara* contains material revealed over several years. Angelika Neuwirth put it less reverently: the long sūras often serve as "collection baskets for isolated verse groups."[42]

4 The Thimble of Pedestrian Analysis

The most famous non-Muslim voice to argue for the universal unity of the sūras and the Qur'ān was perhaps unexpectedly Arthur J. Arberry in his introduction to his rightly famous translation *The Koran Interpreted:*

> The reader of the Koran, particularly if he has to depend upon a version, however accurate linguistically, is certain to be puzzled and dismayed by the apparently random nature of many of the Suras. This famous inconsequence has often been attributed to clumsy patchwork on the parts of the first editors. I believe it to be rather the very nature of the Book itself. In many passages it is stated that the Koran had been sent down "confirming what was before it", by which was meant the Torah and the Gospel [...] All truth was thus present simultaneously within the Prophet's enraptured soul; all truth, however fragmented, revealed itself in his inspired utterance. The reader of the Muslim scriptures must strive to attain the same all-embracing apprehension. The sudden fluctuations of theme and mood will then no longer present such difficulties as have bewildered critics ambitious to measure the ocean of prophetic eloquence with the thimble of pedestrian analysis. Each Sura will now be seen as a unity within itself, and the whole Qur'an will be recognized as a single revelation, self-consistent to the highest degree. Though half a mortal life-time was needed for the message to be received and communicated, the message itself, being of the eternal, is one message in eternity, however heterogeneous its temporal expression may appear to be [...] So the pattern of each Sura can be methodically analyzed into its component parts, seen as *motives* common to the whole Koran, treated in each context individually and with an astonishing wealth and variety of rhetoric and rhythm.[43]

---

41 Zahniser, A. H. Mathias, "Major transitions and Thematic Borders in Two Long Sūras: al-Baqara and al-Nisā'," in *Literary Structures of Religious Meaning in the Qur'an*, ed. by Issa J. Boullata, pp. 26–55. Abingdon: Routledge, 2000, p. 45.
42 Neuwirth, Angelika, "Vom Rezitationstext über die Liturgie zum Kanon: Zur Entstehung und Wiederauflosung der Surenkomposition im Verlauf der Entwicklung eines Islamischen Kultus," in *The Qur'an as text.* Islamic Philosophy, Theology and Science: Texts and Studies 27, ed. by Stefan Wild, pp. 69–105. Leiden: Brill, 1996, p. 98.
43 Arberry, Arthur J.: *The Koran Interpreted. Translated with an Introduction.* Oxford: University Press, 1964. Pp. xi–xii.

This is wonderfully convincing prose. But for the moment, I am afraid, we have to stick to "the thimble of pedestrian analysis," if we want to come to a better understanding of "the heterogeneous temporal expression" of the Qur'ānic revelations.

The boldest and most consistent attempt yet to prove that the long sūras are coherent units may be the work of Michel Cuypers, who claims to have solved not only "the riddle of *al-Baqara*'s internal organization" but also to have achieved an all-embracing approach that works for (nearly) all sūras.[44] Referring to Q 2 he says:

> The apparent disorder is not the result of a lack of composition, but on the contrary the result of a very sophisticated composition, according to a rhetoric widespread in the antique world of the Middle East, but later forgotten, even by the Arabs, most probably under the influence of Hellenistic culture.[45]

Cuypers sees in most of the Mekkan sūras and in all of the long Medinan sūras examples of what he calls "Semitic rhetoric." Cuypers finds this rhetoric first in Biblical Hebrew, but also in Ugaritic and Akkadian. His latest example is Qur'ānic Arabic. As far as the Old Testament is concerned, *parallelismus membrorum* and *chiasmus* are its best-known and uncontested rhetorical devices. But Roland Meynet (b. 1923), professor of New Testament Exegesis at the Gregorian University Rome, and in his footsteps Michel Cuypers, pushed the theory and application of "rhetorical analysis" (sometimes also called "structural analysis") much further. "Applying rhetorical analysis to the short and medium Meccan *sūra*s of the Qur'ān immediately demonstrated that it was the perfect tool for decoding their composition,"[46] Cuypers says. Furthermore, the "pertinence of rhetorical analysis for the long Medinan *sūra*s, clearly more complex and more disordered than the brief Meccan *sūra*s, remained to be demonstrated."[47] The author stands by his thesis, that the whole Qur'ān was composed in the same way, following the same "rhetoric." Cuypers' approach to Semitic Rhetorical Analysis is based on what he calls "the three Semitic symmetries": "parallelism, concentrism, and mirror constructions." [48] These are further fanned out to total and partial symmetries, outer terms, initial terms, central terms, final terms, me-

---

44 Cuypers, *Banquet*, p. 517 and id., *The Composition of the Qur'an. Rhetorical Analysis*. Bloomsbury Academic 2010.
45 Cuypers, Michel, "Semitic Rhetoric as a Key to the Question of the *naẓm* of the Qur'ān," in *Journal of Qur'anic Studies* 13 (2011): pp. 1–24, p. 5.
46 Cuypers, *Banquet*, p. 29.
47 Ibid.
48 Ibid., p. 35.

dian terms, and macro- and micro-units. The relationship between these terms can be identity, synonymy, anti-thesis, paronymy, or homography. These indicators and symmetries must be distinguished from each other starting from the lower to a higher level: a *member* (usually corresponding to a *syntagma*); the *segment* that consists of one to three levels; a *piece* consisting of one to three segments; the *part* consists of one to three pieces. The "superior levels" are: passage, sequence, section and book, with additional intermediary levels: subpart, sub-sequence and sub-section. This rigorously formalized grid is laid over the English text of the Qur'ān in a gush of abbreviations, bold letters, capital and small capital letters, lower case, Roman letters, italics, indents, etc. While admiring the rigorous methodical refractions of Cuypers' approach, I cannot conceal that the high level of his extremely complicated analyses seem to me to ignore the aurality as well as the orality of much of the Qurʾān. And I wonder if there is not a circular argument here. Cuypers asks: "Are the different fragments, which make it (the qurʾānic text)[49] up, arranged according to a certain internal logic which brings coherence and unity, that is, a greater intelligibility, to the text?"[50] Cuypers' unqualified answer is: yes. The internal logic of his "Semitic rhetoric" goes much further than the parallelisms and chiasms of the Old Testament, which have been known for ages. However, the echo to Meynet's and Cuypers' notion of an all-embracing "Semitic rhetoric" (also called "Rhetorical Analysis") in Biblical scholarship is, as far as I can see, faint. Its applicability to the Qurʾān is still highly doubtful. Remembering that the qurʾānic revelations were first recited and not written I doubt that Cuypers' complicated and writing-centered hypotheses can help us much.

Cuypers concedes that the pertinence of "rhetorical analysis" for the long Medinan sūras is more complex and more disordered than for the brief Meccan sūras. Nevertheless, he addresses

> the riddle of *al-Baqara*'s internal organization, utilizing new insights from literary theory and Biblical Studies [...] The overall structure of the *sūra* emerges as chiastic [...] A new reading framework for the *sura* is developed, utilizing in part some of the theories of the Russian literary theorist and philosopher Michael Bakhtin [...] The added value in approaching the *sūras* as a whole, as a *totality* is in seeing how each theme is progressively developed and elaborated by everyone of the *sūras* various panels and how these themes hold the *sūra* together as a unit.

49 Own addition.
50 Cuypers, *Banquet*, p. 25.

Cuypers also claims that in applying "Semitic rhetoric" to Q 5 (*al-Māʾida* / "The Table," or—in Cuypers' translation—"The Banquet") as well as to numerous other sūras, he thinks he has proven that "the whole Qurʾān was composed in the same way, following the same 'rhetoric'."[51]

Cuypers' book *The Banquet* is his most elaborate and far-reaching vision of Semitic rhetoric, here applied to Q 5. The book is preceded by a short preface by Muhammad Ali Amir-Moezzi. It is rare for the author of such a preface to combine the greatest admiration for the book's author with a distinct skepticism toward the author's method and conclusions. Amir-Moezzi underlines that he, unlike Cuypers, is a follower of the historical-philological method. And he warns:

> [...] works devoted to the Qur'an's structures of composition, at least as Cuypers sheds light on them, can be supported by very little classical Muslim work [...] among the hundreds and hundreds of commentators on the Qur'an. [...] the number of Muslim scholars who have studied the stylistic features of the Qur'an whom Cuypers quotes can virtually be counted on the fingers of one hand. What is more, on his own admission, from Abū Bakr al-Nīsābūrī, al-Zarkashī and al-Biqāʿī in the Middle Ages, to Amīn Aḥsan Iṣlāḥī and Saʿīd Ḥawwā [...] none of these unusual and largely unknown authors have managed to come up with objectively convincing results" [...]. "How is it that, for almost a millennium and a half, no Muslim scholar turned to the examination of Semitic rhetoric in general and Arabic rhetoric in particular, to explain the Qur'an's 'incoherence' which always struck literary scholars?"[52] Cuypers' hypothesis to explain this vast lacuna is "that at the time when Muslim scholars began to be interested in the Qur'an's stylistic organization, Semitic rhetoric had already been completely forgotten, covered over by the influence of late-Hellenistic rhetoric [...]. From then on, Arabic rhetoric was powerless to resolve the questions that scholars were asking about the text's organization. I must confess that, for me, the question remains open."[53]

It looks like a volte-face when Amir-Moezzi then praises Cuypers' "rigorous methodology, systematic reasoning, and implacable logic" and claims that Cuypers has demonstrated "the Qur'an has a literary unity and coherence which make sense." Amir-Moezzi ends with a judgment of Solomon saying that, although "convinced of the validity [...] of the historical critical method" he has to admit, "rhetorical analysis can be just as reliable a tool for understanding the qurʾānic text as others. I do not yet know exactly how [...]"[54]

Amir Moezzi's points are well taken. Pre-Islamic Arabic poetry of the *jāhilīya* period, the language of the Arab soothsayers, the language of early contracts at

51 Ibid., p. 29.
52 Ibid., p. 16–17.
53 Ibid., p. 17.
54 Ibid., p. 18.

the time of the Prophet, and the *ayyām al-ʿarab* seem much closer to the Qurʾān than a hypothetical "Semitic rhetoric." Amir-Moezzi's argument is, however, not fully conclusive. New methods of treating texts can produce and have produced new and unexpected results. This is also true for the Qurʾān. Angelika Neuwirth's "literary analyses" of the Meccan sūras are in no way founded on earlier Muslim exegesis of the Qurʾān. To unveil the compositional character and structure of these sūras, she adopts approaches of modern literary theory and of the literary interpretation of the Qurʾān (*tafsīr adabī*). She embeds her findings in an interscriptural context with Jewish religion and against the backdrop of Late Antiquity. This is certainly a radically new approach. It does bring us "greater intelligibility" to the Qurʾānic text. Most claims regarding the structural "unity" of all sūras in the Qurʾān, however, in my view cannot yet be admitted as acceptable.

Farid Esack reminds us that "disjunctions" in the Qurʾān may be functional:

> Because the Qur'ān is the recited word in addition of being the written word, this seeming disjuncture is of little consequence to most Muslims. Repetitions are seen as God's repeated reminders, legal texts in the middle of a narrative as God drawing our attention to what has to be learnt from the text, breaks in a narrative reflect God's freedom from human literary patterns [...]. Most Muslims see the seeming absence of structure or classification in its *suras* itself as signaling a demarcation value and as reflective of the Qur'an's role in the universe.[55]

Two young Iranian scholars, Amer Gheituri and Arsalan Golfam, go even further. Their unexpected solution to the problem of unity and coherence in the Qurʾān looks like a radicalization of Esack's points. They argue:

> To unveil the non-linearity (*of the Qur'an*)[56] we should consider revelations in relation to God who reveals Himself in language not as an ordinary speaker whose speech is limited to a certain context with a beginning and an end. We should not expect God to speak like a man. The Qurʾān, thus, is seen here as a non-temporal, non-linear text that reflects its divine origin by systematically destructing the spatio-temporal context and the linear order of language.[57]

We seem to have come full circle. The "disjointedness" of the Qurʾānic text has become the seal of its divine origin. The last word on the problem of Qurʾānic "coherence" and "unity" evidently has not yet been said.

---

55 Esack, *The Qur'an*, p. 66.
56 Own addition.
57 Gheyturi, Amer and Arsalan Golfam, "The Qur'ān as a Non-Linear Text: Rethinking Coherence," in *The International Journal of Humanities of the Islamic Republic of Iran*, 15 (2008): pp. 119–133, here p. 121.

Berenike Metzler

Qur'ānic Exegesis as an Exclusive Art – Diving for the Starting Point of Ṣūfī Tafsīr

> I plunged in the above-named sciences [that means *tafsīr*, *ḥadīṯ*, *fiqh*, *naḥw*, *ma'ānī* etc., BM] so deeply, that one can't see its bottom.[1]

This citation originates from the autobiography of the infamous Egyptian scholar Jalāl al-Dīn al-Suyūṭī who perceived himself as *mujtahid* and even *mujaddid* of his era. His self-image, namely to be well versed in diverse branches of knowledge, is of course not limited to the domain of *tafsīr*. What led me to pick this citation as the beginning of my paper is the observation, that he uses the metaphor of plunging in the sea of knowledge, which will accompany us during my explanations concerning the emergence of *Ṣūfī tafsīr*, and which, in an excellent way, suits our meeting place: the 'Water Hall' of the Orangery of Erlangen. Furthermore, Suyūṭī's self-confident statement seems to contradict what we know about the early cautiousness – or, resentment in some cases – towards the human capacity to understand God's word in a proper way. So let's plunge in the matter and shed some light on the question how a mainly skeptical attitude towards *tafsīr* in early Islam could at least partly evolve into the *Ṣūfī's* own view of exegetical exclusivism.

For the question of attitudes towards *tafsīr* in early Islam I would like to shortly address the quite old and probably to all well-known debate between Harris Birkeland and Nabia Abbott about the potential opposition towards *tafsīr* in early Islam.[2] The details of this controversy are of no matter for our specific question, but what is important is that Birkeland denies an opposition towards *tafsīr* during the 1st century while claiming a strong opposition for the 2nd century. Abbott on the other hand assumes already from the beginning of Islam an opposition towards *tafsīr*, indeed not against *tafsīr* all in all, but against interpreting the *mutashābihāt*, the ambiguous verses. What both positions have in common is an interpretation of a story, which seems to be a central reference point for the whole discussion: A certain Ibn Ṣabīgh asked 'Umar I. about the interpretation of some Qur'ānic verses. The caliph's answer included whippings

---

1 Al-Suyūṭī, Jalāl al-Dīn, *Kitāb at-Taḥadduth bi-ni'mat Allāh*, ed. by Elizabeth M. Sartain. Cambridge: Cambridge University Press, 1975, p. 203.

2 See Abbott, Nabia, "The Early Development of Tafsīr." and Birkeland, Harris, "Old Muslim Opposition against Interpretation of the Koran," in *The Qur'ān: Formative Interpretation*, ed. by Andrew Rippin, pp. 29–80. Aldershot: Ashgate, 1999.

https://doi.org/10.1515/9783110564341-016

and finally banishment.[3] Although Birkeland and Abbott differ in the question, if this story is of historical reference and in the dimension and the chronological localization of opposition towards *tafsīr* in early Islam, they agree in the point, that the mentioned Ibn Ṣabīgh wasn't punished because he asked about Qur'ānic interpretation in general, but because he was asking concerning the *mutashābihāt*.[4] What speaks for this argument is that 'Umar I. approved Ibn 'Abbās as the "ocean of *tafsīr*."[5]

The term *mutashābihāt* is for our specific question of great importance. Leah Kinberg, who wrote an article in 1988 about the perception of the conceptual pair *muḥkamāt* and *mutashābihāt* by several Qur'ānic exegetes, unknowingly mediated between Birkeland and Abbott, by summarizing Birkeland but without taking note of Abbott's position. Kinberg lists Birkeland's reasons for the opposition to the interpretation of the Qur'ān as follows:

> 1. The reluctance to use independent reasoning (*ra'y*) to understand qur'ānic verses.
> 2. The humbleness the believer should feel toward his Lord.[6]

As a result of her own analysis of the conceptual pair of the *muḥkamāt* and *mutashābihāt*, Kinberg filters out quite similar reasons for the opposition to the interpretation of the *mutashābihāt:*

> 1. Interpretation based on tendentious usage of these verses might create dissent and lead their followers astray.
> 2. The believer should be humble and recognize that these verses are the secret of the Lord concealed in the Koran.[7]

We can summarize that skepticism concerning the human reasoning, humbleness and the fear of dissent are the linking issues between the early opposition of Qur'ānic interpretation as a whole and the later exegetes' cautiousness towards interpreting the *mutashābihāt*. Central for this discussion is Q 3:7:

> It is He who sent down upon thee the Book, wherein are verses clear that are the Essence of the Book, and others ambiguous. As for those in whose hearts is swerving, they follow the ambiguous part, desiring dissension, and desiring its interpretation; and none knows its

3 See Goldziher, Ignaz, *Die Richtungen der islamischen Koranauslegung*. Brill: Leiden, 1920, pp. 55–56.
4 See Abbott, *Early Development*, p. 32–33; Birkeland, *Old Muslim Opposition*, p. 51.
5 See Abbott, *Early Development*, p. 30; Birkeland, *Old Muslim Opposition*, p. 63.
6 Kinberg, Leah, "*Muḥkamāt* and *Mutashābihāt* (Koran 3/7): Implication of a Koranic Pair of Terms in Medieval Exegesis." *Arabica* 35 (1988): pp. 143–172, p. 165.
7 See ibid., p. 165.

interpretation, save only God. And those firmly rooted in knowledge say, 'We believe in it; all is from our Lord'; yet none remembers, but men possessed of minds.[8]

Along this verse the exegetes dealt with the question, if only God is entitled to know the Qur'ān's real interpretation (*ta'wīl*) as it is translated here by Arberry. The second possibility concerning this qur'ānic verse is that "those firmly rooted in knowledge" are included in the precedent sentence and so together with God are knowing the Qur'ān's interpretation; that means, even the interpretation of the *mutashābihāt*, which are according to the cited sūra not the Essence of the Book like the *muḥkamāt* but ambiguous and could cause dissension. Most exegetes were quite cautious to engross the ability to interpret the *mutashābihāt* for themselves while others limited this ability to a very small group of experts. I will handle this question through an examination of the early Islamic ascetic and theologian al-Ḥārith b. Asad al-Muḥāsibī, especially his until now not thoroughly analyzed work *Kitāb Fahm al-Qur'ān*, which I discussed in my dissertation thesis. In this work traditional skepticism towards the human capacity to understand God's word collides with the author's own practice as well as with the emerging idea of Qur'ānic exegesis as an exclusive art.

Al-Muḥāsibī, the author of *Kitāb Fahm al-Qur'ān*, was an early Islamic scholar of the 9th century from Baghdad. During his lifetime he witnessed the rise and decline of several caliphs as well as the rise and decline of several theological positions like the rationalistic theology of the Mu'tazilites or the traditionalist point of view of Ibn Ḥanbal. The struggles between the different theological parties led to the *miḥna*, known as "Islamic Inquisition" about the question, whether God's word is created or eternal and non-created. In spite of the fact that Muḥāsibī conducted a pious life and advocated the traditionalist point of view regarding the eternity of God's word, he was urged by Ibn Ḥanbal to leave Baghdad and to stop teaching, as he probably used the rationalist methodology of *kalām* too intensively.[9] In his 200 pages book entitled *Kitāb Fahm al-Qur'ān*, Muḥāsibī dwells on nearly all aspects of the question how to deal with God's word. What is interesting is the fact that the author does not begin his book with the notion of the Qur'ān itself, but with the role the human intellect plays in God's communication with man. Certainly the possibilities of the intellect (*'aql*) are limited and have to be accompanied by pious life conduct. Recitation (*tilāwa*), listening (*samā'*), remembering (*dhikr*) and thinking (*fikr*) belong to

8 Q 3:7, trans. by Arthur J. Arberry, *The Koran Interpreted*. London: Oxford University Press, 1964.
9 See al-Khaṭīb al-Baghdādī, *Ta'rīkh Baghdād aw Madīnat as-Salām*. 8 vols. Beirut: Dār al-kitāb al-'arabī, 1966–1980, pp. 215–216; Picken, Gavin, "Ibn Ḥanbal and al-Muhasibi: A Study of Early Conflicting Scholarly Methodologies." *Arabica* 55 (2008): pp. 337–361, p. 352.

the proper understanding of the Qur'ān[10], aside the presence of the heart (*ḥuḍūr al-qalb*)[11]. Following this, Muḥāsibī stresses the importance of the Qur'ān for man's life in this world as well as for the afterworld. In accordance with the controversial issues of his time, he defends the doctrine of the eternalness of God's Word against his theological opponents. While doing so, he deals with the quality of God's attributes and concentrates on the hermeneutical principle of abrogation which was an important argument in these debates. The juristic part of the book about different variants of abrogation deals with the practical application of Muḥāsibī's abrogation theory. Finally the author deals with some linguistic characteristics of the Qur'ān.

Chapter 4 of the *Kitāb Fahm al-Qur'ān* is of significant importance for our discussion: Here Muḥāsibī enlarges on several methods that one should know for the proper interpretation of the Qur'ān. Included in these methods that all belong to the field of *uṣūl al-fiqh* is the abrogation of verses by others, the distinction of *muḥkamāt* and *mutashābihāt* (definite and ambiguous verses), the Qur'ānic word order, the differentiation between the generic and the specific meaning of verses (*'āmm wa-khāṣṣ*) and the analysis of odd words with regard to language.[12] Furthermore, one has to consult the prophet's Sunna and the consensus (*ijmā'*) of the scholars to open up the true meaning.[13] But, as Muḥāsibī stresses, man can only know the *muḥkamāt* (definite verses) and under very restricted conditions apply these hermeneutic principles to the *mutashābihāt* (ambiguous verses). However, the very true extensive and deeper meaning knows only God; the absolute sovereignty of interpretation is His privilege. Muḥāsibī cites and exemplifies here a Hadith of Ibn Mas'ūd concerning the fourfold exegesis of the Qur'ān:

> 'Abdallāh b. Mas'ūd said: Each verse of the Qur'ān has a back (*ẓahr*) and a belly (*baṭn*), a border (*ḥadd*) and a lookout point (*maṭla'*).[14]
>
> Muḥāsibī said: "Concerning *ẓahr*, it is the recitation, concerning *baṭn*, it is its interpretation, concerning *ḥadd*, it is the termination of understanding. [...] But concerning the lookout point, *maṭla'*, it is transgression of the border by exaggeration, deep intrusion, immorality and disobedience.[15]

---

10 Al-Muḥāsibī, Al-Ḥārith b. Asad, *Kitāb Fahm al-Qur'ān*, Ms 22 Sel 951. Edirne: Selimiye Kütüphanesi, fol. 94v.

11 Ibid., fol. 93r.

12 Ibid., fol. 96r-96v.

13 Ibid., fol. 96v.

14 See Sands, Kristin Z., *Ṣūfī Commentaries on the Qur'ān in Classical Islam*. London: Routledge, 2006, p. 8.

15 Muḥāsibī, *Kitāb Fahm al-Qur'ān*, fol. 96v.

This fourfold exegesis has of course Christian predecessors, but Muḥāsibī, according to van Ess, is the first theologian who incorporates this idea in Islamic thinking.[16] What is interesting is the fact that Muḥāsibī, who is often seen as one of the first Ṣūfīs or at least their predecessor, militates against *maṭla'*, a term that will become quite important for the Ṣūfī dimension of Qur'ānic interpretation. Knowing this, it isn't surprising that he in Q 3:7 doesn't include "those firmly rooted in knowledge" in the interpretation of the *mutashābihāt* next to God. So he follows the slogan of the mainstream arguing that the *mutashābihāt* belong to that what cannot be known: "*mā lā sabīla ilā ma'rifatihī.*"[17] In terms of terminology it is interesting here that Muḥāsibī ascribes *tafsīr* to human explanation and, as one can see in his other works,[18] to the outer meaning of a verse, while in the Qur'ānic verse the interpretation of the inner meaning is already ascribed to God.

Contrary to these statements there are other signs, which indicate that Muḥāsibī in fact required the human ability to interpret the *mutashābihāt* – at least for himself. On this side of argumentation we find his own exegetical practice. In the middle part of *Kitāb Fahm al-Qur'ān* he interprets God's attributes in the manner of *kalām*. With regard to God's abode, Muḥāsibī lists verses about God's exceptional highness, as it is mentioned in the Qur'ān that He is sitting on His heavenly throne, next to verses that mention an exceptional closeness of God to man, as Q 50:16 shows: "We are nearer to him than the jugular vein."[19] In this regard, Muḥāsibī holds on God's transcendentality (*tanzīh*): God is on His heavenly throne not in a physical sense but in the sense that He is sublime above everything. Just as God doesn't lower himself to everything but is in the same time nearer with His wisdom to all things than they are to themselves. God is by (*'inda*) or with (*ma'a*) everything but never in it (*fī*).[20]

The strongest evidence for a beginning exclusivism with regard to Qur'ānic exegesis can be found in chapter 1 of *Kitāb Fahm al-Qur'ān*. Although Muḥāsibī emphasizes, in the introduction of this book, God's transcendence and the inability of human reason to comprehend His essence, he appreciates the human reason to a great extent as an instrument for understanding God's

---

16 See van Ess, Josef, *Die Gedankenwelt des Ḥāriṯ al-Muḥāsibī*. Bonn: Orientalisches Seminar der Universität Bonn, 1961, pp. 210 – 211 and van Ess, Josef, *Theologie und Gesellschaft im 2. und 3. Jahrhundert Hidschra*. 6 vols. Vol. 4. Berlin: de Gruyter, 1997, p. 648.

17 See Kinberg, "Muḥkamāt", p. 155.

18 See Muḥāsibī, *Kitāb ar-Ri'āya li-ḥuqūq Allāh*, ed. by Margaret Smith. London: Luzac & Co, 1940, p. 222.

19 Q 50:16, trans. by Arthur J. Arberry, *The Koran Interpreted*.

20 See Muḥāsibī, *Kitāb Fahm al-Qur'ān*, fol., 101r.

word. As the notion of reason (*ʿaql*) plays hardly any role in Muḥāsibī's era, not even in the works of the rationalistic school of the Muʿtazilites, al-Juwaynī commented on him as follows: "None of our scholars dealt with reason except al-Ḥārith al-Muḥāsibī."[21] It is possible that his notion of reason in *Kitāb Fahm al-Qurʾān* has been a cause for the above mentioned conflict with Ibn Ḥanbal. With regard to the creation Muḥāsibī doesn't consider that the creator is in need of his creation but that the creator wants to provide his creation with his mercy. In order to follow God's instructions that save human beings from punishment in hell and assure them admission to paradise he endows them with the disposition (*fiṭra*) of the intellect (*ʿaql*):

> So He chose Adam and his descendants and received a convention from them by endowing them with an accountable intellect, mental power and comprehension so that they could ponder on the evident examples of the world order and the provisions of the ruling.[22]

Thus the intellect is the canal by which God is communicating with man. But God's word is not the first thing that human intellect perceives: By looking at the world, human being witnesses the mastery of its creator. After that God sends messengers and prophets to all people to confirm His will. Muḥāsibī proves the demand of using one's intellect with several verses of the Qurʾān, in which it is written that man should think about God's commands.[23] But for Muḥāsibī using one's intellect is not a purpose in itself but it is bound to certain religious duties and pious and humble behavior. He characterizes different groups of persons in their usage of their rational abilities in terms of a climax. I already mentioned Muḥāsibī's assumption that all human beings are endowed with an intellect. In his view, God chose a first group out of the mass of all human beings, who not only understand God's Word but also believe in it.[24] These are the common believers. Out of this group God chose another selection of people who not only understand God's word and believe in it but also act upon His commands.[25] These are, in my opinion, truly pious ascetics who set themselves apart the mass of the ordinary believers but also apart sanctimonious people of their own craft. As a matter of fact, this group of experts developed

---

21 As-Subkī, Taqī al-Dīn ʿAlī b. ʿAbd al-Kāfī, *Ṭabaqāt ash-shāfiʿiyya al-kubrā*, ed. by ʿAbd al-Fattāḥ Muḥammad al-Ḥilū, 10 vols. Vol. 2. Cairo: Dār Iḥyāʾ al-kutub al-ʿarabiyya, 1970, p. 283 and van Ess, *Gedankenwelt*, p. 74.

22 Muḥāsibī, *Kitāb Fahm al-Qurʾān*, fol. 82v.

23 See ibid., fol. 83r with regard to Q 13:19; 2:164; 10:24 etc.

24 See ibid., fol. 83r.

25 See ibid., fol. 83r-83v.

later on to a group which was explicitly denoted with the term *Ṣūfī*, in *Kitāb al-Luma'* of Abū Naṣr as-Sarrāj, who lists three groups of men according to their knowledge and concedes the first rank to those who have knowledge of God (*'aqala 'an Allāh*). He finally equalizes them with "those who are firmly rooted in knowledge" (*ar-rāsikhūna fī l-'ilm*).[26] In this context Sarrāj cites Abū Bakr al-Wāsiṭī, who said:

> Those firmly rooted in knowledge are these, who are with their spirits rooted in the most hidden and most secret. [...] Through their understanding they plunge to the sea of knowledge to seek for more.[27]

Although Muḥāsibī laid the cornerstone for exclusivistic exegesis, there is a major difference between Muḥāsibī and his Ṣūfī successors: The fundament for understanding God's word in a proper way, which is limited to a small group of pious experts, is still the human intellect (*'aql*). Though Muḥāsibī differentiates between several functions of *'aql* – he calls the highest form, i.e., *'aqala 'an allāh, baṣīra*,[28] – he still holds on to the human intellect as the only instrument to decode God's commands. He, thus, rejects that an adept would achieve certainty through a vision of God. Only one generation later, al-Tustarī and al-Junayd, probably one of Muḥāsibī's students, already take one step further and detach themselves from his strong emphasis on *'aql*. Tustarī's argumentation runs as follows: As there is no end for God's uncreated word, the human intellect cannot come to an end in understanding it.[29] Like Muḥāsibī, Tustarī does not include "those firmly rooted in knowledge" in the interpretation of the *mutashābihāt*. He mentions several times tripartitions of the knowledge of God and cites the statement about the fourfold exegesis. But while Muḥāsibī was still very cautious concerning the *maṭla'* or "lookout point," Tustarī opens the door for a new dimension of understanding and not just accepts this as a "point of transcendency of the "lectio"[30] as Böwering calls it, but even perceives the hidden or inner sense of the script as the domain of the mystic man (*khāṣṣ*), namely to bring to light

---

26 See as-Sarrāj, Abū Naṣr 'Abdallāh b. 'Alī, *The Kitāb al-Luma' fī t-Taṣawwuf of Abū Naṣr 'Abdallāh b. 'Alī as-Sarrāj aṭ-Ṭūsī*, ed. by Reynold Alleyne Nicholson. London: Luzac, 1963, pp. 77–79.
27 Ibid., p. 113.
28 See van Ess, *Gedankenwelt*, p. 75 and with regard to Muḥāsibī's conception of *'aql* in general Crussol, Yolande de, *La role de la raison dans la réflexion éthique d'Al-Muḥāsibī*. Paris: Consep, 2002.
29 See Sarrāj, *The Kitāb al-Luma'*, p. 107.
30 Böwering, Gerhard, *The Mystical Vision of Existence in Classical Islam. The Qur'ānic Hermeneutics of the Ṣūfī Sahl at-Tustarī (d. 283/896)*. Berlin: de Gruyter, 1980, p. 140.

what God really intended. Junayd, who was the first who brought the term *fanā'* into the scene, knows three stages of knowledge as well: firstly the *vita activa*, secondly the cutting off from all the worldly pleasures and thirdly the vision of God, i.e., *al-fanā' wa-l-baqā'*.[31] While on the first stage the intellect is still relevant for the knowledge of God, on the higher stages the intellect has no place.[32] According to Junayd the mystic's goal is neither rational knowledge (*'ilm*) nor intuitive knowledge (*ma'rifa*) but an experience: the unification with God (*tauḥīd*).[33]

Al-Ghazālī is commonly known as the one who gave Ṣūfism its place in orthodox Islam and who systematized Ṣūfī thinking. With regard to our question of the human ability to understand God's word in a proper way, he is even the strictest:

> Those who are permitted to interpret the difficult passages of the Qur'ān are those who devote themselves exclusively to learning to swim in the seas of religious gnosis (*ma'rifa*); who restrict their lives to Him alone; who turn their faces from this world and the appetites; who turn their backs on money and fame, mankind, and all other pleasures; who devote themselves to God in the different types of knowledge and actions; who act in accordance with all the ordinances of the religious law and its courtesies (*ādāb*) in performing obedience and avoiding the objectionable; who have emptied out their hearts from everything except God; who despise the world and even the Hereafter and the Highest Paradise next to love of God. They are the divers in the sea of gnosis.[34]

It is surely no coincidence that, at the end of our dive cruise, we arrive at Ghazālī: He himself mentions in his autobiography *Al-Munqidh min aḍ-Ḍalāl* Muḥāsibī as one of his Ṣūfī ideals.[35] It is quite probable that Ghazālī had in mind the beginning of Muḥāsibī's autobiography *al-Waṣāya*, when the former composed his own autobiography: In dependence on the prophetic Hadith "My nation will divide into seventy-three sects, and only one of them will be saved,"[36] they both

---

31 See al-Qādir, 'Alī Ḥasan 'Abd, *The Life, Personality and Writings of al-Junayd. A Study of a 3rd/9th Century Mystic with an Edition and Translation of his Writings.* London: Luzac, 1962, p. 81–82.

32 See ibid., p. 101.

33 See ibid., p. 102.

34 Al-Ghazālī, Abū Ḥāmid M., *Iljām al-'awāmm 'an 'ilm al-kalām*. Beirut: Dār al-kitāb al-'arabī, 1985, pp. 67–68; Sands, *Ṣūfī Commentaries*, pp. 22–23.

35 See al-Ghazālī, Abū Ḥamīd M., *Deliverance from Error and Mystical Union with the Almighty*, ed. by Muḥammad Abūlaylah and Nurshīf Abdul-Rahīm Rif'at. Washington: The Council of Research in Values and Philosophy, 2001, p. 241.

36 Abū Dawūd, *Sunna 1*; Tirmiḏī, *Īmān 18*; Ibn Māğah, *Fitan 18***; Ibn Ḥanbal, *2,332 and 3,145* in: Wensinck, Arent Jan, *Concordance et indices de la tradition musulmane: les six livres, le Musnad*

describe the complexity of different religions and beliefs in their times as a deep sea, in which Muḥāsibī fears to drown while searching for the right-guided group (*al-firqa an-nājiya*).[37] Ghazālī, on the other hand, parallel to the over the times tightened concept of his exegetical exclusivism, assures his readers:

> I have not ceased to delve into the depths of the deep ocean (of the various beliefs of humankind), to plunge into its depths boldly, not as a cautious coward; to bury myself in obscure questions, eagerly seizing upon difficulties and leaping bravely into difficult and obscure issues; and to scrutinize the beliefs of each sect, examining from the doctrinal point of view the hidden aspects of every religious group.[38]

---

d'al-Dārimī, le Muwaṭṭa' de Mālik, le Musnad de Aḥmad Ibn Ḥanbal, 2. edition, 6 Vols. Leiden, Brill: 1992.

37 See al-Muḥāsibī, al-Ḥārith b. Asad, *al-Waṣāya (al-Qaṣd wa-r-rujū' ilā llāh – Bad' man anāba ilā llāh – Fahm aṣ-Ṣalāh – at-Tawahhum)*, ed. by 'Abd al-Qādir A. 'Aṭā. Beirut: Dār al-kutub al-'ilmiyya, 1986, pp, 61–62.

38 Ghazālī, *Deliverance*, p. 62.

Reza Pourjavady

Ibn Kammūna's Knowledge of, and Attitude toward, the Qur'ān*

In her remarkable study, *Intertwined Worlds: Medieval World and Bible Criticism* (published in 1993), Hava Lazarus-Yafeh describes what Muslims knew about the Hebrew Bible and how they may have acquired this knowledge.[1] At the end of this book and as an appendix, Lazarus-Yafeh describes briefly how much Jews knew of the Qur'ān. There she rightly points to the fact that among the medieval Jewish thinkers, Ibn Kammūna's (d. 1284) knowledge of the Qur'ān is "especially wide." To my knowledge, this aspect of Lazarus-Yafeh's study has not been pursued much after this work. At least, this is certainly true of her observation on Ibn Kammūna. In what follows, I am going to present various aspects of Ibn Kammūna's knowledge of the Qur'ān and his attitude towards it which I agree with Lazarus-Yafeh in calling it special.

Born to a Jewish family in Baghdad, Ibn Kammūna seems to have received a thorough education in both Jewish and Islamic letters.[2] We know about his knowledge of Jewish literature through his treatise on the differences of the Rabbanites and the Karaites. This treatise is replete with Talmudic and rabbinic references, while his main source was Judah ha-Levi's (d. 1141) *al-Kitāb al-Khazarī*.[3] Although Ibn Kammūna was familiar with the Jewish literature, his scholarly attention was mainly devoted to Muslim literature. In his writings, he boasts that he is among the few scholars well versed in both Jewish and Islamic literature. In the Islamic domain, his focus was mainly on philosophy and speculative theology, on which he produced several works. For the field of philosophy he states

* I would like to thank Sarah Stroumsa and Sabine Schmidtke for their comments on the draft of this article.

1 Lazarus-Yafeh, Hava, *Intertwined worlds: Medieval Islam and Bible criticism*. Princeton: Princeton University Press, 1992; reprint Princeton: Princeton University Press, 2014, p. 147.

2 On Ibn Kammūna's life and works, see Pourjavady, Reza and Sabine Schmidtke, *A Jewish Philosopher of Baghdad: 'Izz al-Dawla Ibn Kammūna (d. 683/1284) and His Writings*, Islamic Philosophy, Theology, and Science 65. Leiden: Brill, 2006.

3 For the detailed references to Ibn Kammūna's sources in this work, see "Ibn Kammunah's Treatise on the Difference between the Rabbanits and the Karaites," *Proceeding of the American Academy for Jewish Research* 36 (1968), pp. 107–65 (160–65). Cf. Pourjavady and Schmidtke, *A Jewish Philosopher of Baghdad*, pp. 8–9.

https://doi.org/10.1515/9783110564341-017

that he was self-taught.[4] One can argue that three Muslim scholars had crucial role in the formation of his intellectual thought: Shihāb al-Dīn al-Suhrawardī (d. 1191), Fakhr al-Dīn al-Rāzī (d. 1209) and Naṣīr al-Dīn al-Ṭūsī (d. 1274). Ibn Kammūna's main contribution is his commentary on Suhrawardī's *Kitāb al-Talwīḥāt* which is the first commentary ever written on Suhrawardī's philosophical works. This commentary became the basis for later studies on Suhrawardī's philosophy.

Although Ibn Kammūna's philosophical works were received later on by some Jewish scholars, it is evident that he wrote all his philosophical works mainly for Muslim audience.[5] Late in his life and as one of his last contributions, Ibn Kammūna wrote a book on the three monotheistic religions: Judaism, Christianity, and Islam. This work, titled *Examination of the Three Faiths (Tanqīḥ al-abḥāth li-l-milal al-thalāth)* is completed in 1280, four years before Ibn Kammūna's death in 1284. In the introduction to this work, Ibn Kammmūna explains his intention of writing this work as follows:

> Recent discussions have induced me to compose this tract as a critical inquiry into the three faiths, that is, Judaism, Christianity, and Islam. I have prefaced it with a general survey of prophethood, followed by a discussion of these religions in chronological order. Thus I began with the oldest, that is, Judaism, proceeded to the intermediate, Christianity, and concluded with the youngest, Islam. For each of these I have cited the fundamentals of its creed, without going into the particulars, as it would have been impossible to treat them all. I have followed this with an exposition of the arguments of the adherents of each faith for supporting the true prophethhod of the respective founder of each. In addition, I have adduced the objections commonly raised and their rebuttals, and have drawn attention to the main issues, distinguishing the valid points from the invalid. I have not been swayed by mere personal inclination, nor have I ventured to show preference for one faith over the other.[6]

The first chapter of the work, on the nature of the prophethood, is a cosmological explanation of prophethood based on Avicenna's (d. 1037) theory and its later development by Abū Ḥāmid al-Ghazālī (d. 1111), Maimonides (d. 1204) and

4 Ibn Kammūna, "Risāla fī azaliyyat al-nafs wa-baqāʾihā," in *Azaliyyat al-nafs wa-baqāʾuhā*, ed. by Insiyya Barkhwāh. Tehran: Kitābkhāna, Mūzia, va Markaz-i Asnād-i Majlis-i Shūrā-yi Islāmī, 2006, p. 138.

5 On the reception of Ibn Kammūna's works by later Jewish and Muslim scholars, see Pourjavady and Schmidtke, *A Jewish Philosopher of Baghdad*, pp. 28 – 57.

6 Ibn Kammūna, *Ibn Kammūna's Examination of the Three Faiths: A Thirteen-Century Essay in the Comparative Study of Religion,* trans. by Moshe Perlemann. Berkley: University of California Press, 1971, p. 11.

Fakhr al-Dīn al-Rāzī (d. 1209).[7] Following this chapter Ibn Kammūna addresses Judaism, Christianity and Islam in three separate chapters. As mentioned in the above quotation, each chapter starts with the arguments of the adherent of that faith and it follows with the commonly raised objections. For instance, in the Chapter Two on Judaism, the author used Judah ha-Levi's *Kitāb al-Khuzarī* for the arguments for Judaism and used *Ifḥām al-yahūd* by Jewish convert Samaw'al al-Maghribī's (d. 1175) for the arguments against it.[8]

Moshe Perlmann, who edited and translated it in 1971, rightly observed that this book was also written for Muslim circles. Hence, the chapter of this treatise dealing with Islam is arguably the 'decisive' chapter. The arguments of this chapter were written in response to the arguments of Fakhr al-Dīn al-Rāzī in the latter's theological works, such as *Kitāb al-Muḥaṣṣal, Kitāb al-Maʿālim* and *Nihāyat al-ʿuqūl.* Ibn Kammūna challenged in this chapter Muslim orthodox beliefs on several matters. Many of these arguments are related to the Qur'ān in one way or another and since they are our main source for understanding Ibn Kammūna's knowledge of the Qur'ān, I will provide here a summary of them.

One of the questions that Ibn Kammūna raises in this chapter regards the authenticity of some Qur'ānic stories of antiquity. He argues that unlike what Muslim theologians claim, some of these stories are not fully in accordance with what has been transmitted by Jews and Christians. The two stories that Ibn Kammūna explicitly mentions are (1) the story of Solomon and (2) the story of Jesus.

1) According to Ibn Kammūna many aspects of Solomon's account in the Qur'ān are opposed to what has been transmitted by Jews, for example, the way Solomon subdued the wind and the *jinn*, how he knew the language of the birds, conversed with the hoopoe, and sent it to the queen of Sheba and how the people learned that the *jinn* had not known of Solomon's death until he collapsed.[9]
2) Concerning the story of Jesus, Ibn Kammūna explains that neither the Jews nor the Christians doubt that Jesus, the son of Mary, was crucified, and they transmit the story of his crucifixion with the same certainty as that of his existence. Moreover, the name of the father of Jesus's mother Mary was according to the Christians, Joachim (Joakhin) and Mary had no brother.[10]

In addition to these two stories, Ibn Kammūna indicates the Qurʿānic claim that the Jews declared ʿUzayr the son of God, whereas there is no tradition in Judaism

7 Ibid., p. 4.
8 Ibid., p. 5.
9 Ibid., p. 131.
10 Ibid., p. 132.

according to which ʿUzayr is declared the son of God.[11] He also undermines the Qurʾānic statement that the Jews consider God's hand to be fettered (*maghlūla*).[12]

It was because of these discrepancies on the transmitted matters, Ibn Kammūna argues, that the Jews as well as the Christians of Arab descent at the time of Muḥammad did not contest the Qurʾān in so far as it narrates what is purportedly their own traditions.[13]

With all these inaccuracies, Ibn Kammūna suggests that Muḥammad might have learned the modicum of these narratives during his stay in the Levant. He argues that the duration of Muḥammad's trip to the Levant was not too brief for him to do so. Indeed it was more time than he needed.[14] He further argues that according to the Qurʾān Muḥammad was suspected by the unbelievers of having done so as the Qurʾān narrates their word that "these are nothing but tales of the ancients which he has written for himself! They are recited to him morning and evening" (Q 25:6).[15]

Ibn Kammūna also contests the Qurʾānic claim that there exist tidings about the advent of Muḥammad in the Old and New Testaments.[16] Refuting this Muslim claim is a common practice in Jewish polemical works against Islam. But Ibn Kammūna specifically refutes the arguments put forward by Fakhr al-Dīn al-Rāzī and Samawʾal al-Maghribī. He also criticizes the Muslim-orthodox belief that the Qurʾān is a miracle of the Prophet. He challenges this view by saying that this dogma is not in agreement with the text of the Qurʾān itself. There are indications in the Qurʾān, Ibn Kammūna states, which suggest that Muḥammad was perceived to have worked no miracle. Ibn Kammūna cites several quotations from the Qurʾān supporting his argument.[17] One of these verses is the following: "Why are not signs from his Lord send down to him? Say: signs are with God only and I am a clear warner, hath it not sufficed them that we have sent down to thee the Book" (Q 29:49–50)?[18]

---

11 On ʿUzayr and his hypothetical identifications in Jewish tradition see Lazarus-Yafeh, "ʿUzayr." *Encyclopaedia of Islam² 10*, p. 960.

12 Ibn Kammūna, *Ibn Kammūna's Examination of the Three Faiths*, p. 132.

13 Ibid., p. 133.

14 Ibn Kammūna, *Ibn Kammūna's Examination of the Three Faiths*, p. 153.

15 Ibid., p. 130.

16 Ibid., p. 137–43.

17 Ibid., p. 135–37.

18 Other verses that he cites are the followings: (1) "Nothing has prevented Us sending the signs, but that the people of long ago counted them false" (Q 17:61); (2) "Those who have disbelieved say: Why has not a sign been sent down to him from his Lord? Thou art only a warner, and for every people there is a guide" (Q 13:8); (3) "They said: 'O God, if this be the truth from Thee, rain upon us stones from the heaven, or come to us with a painful punishment; But God was not

Ibn Kammūna argues that if indeed the Qurʾān was perceived at the time of the Prophet as his miracle, the latter could argue that he had brought them a sign. But the Qurʾān reveals that it was not considered as a sign.

Another argument for the Qurʾān being a miracle is drawn from the predications of the future events in the Qurʾān. Ibn Kammūna argues against this that prediction could be only considered a miracle if it is exceptional rather than habitual. Indeed the Sufis, people given to devotional exercises (*aṣḥāb al-riyāḍāt*) and the "Barāhima" and many others are reported to have uttered the type of prediction that one can find in the Qurʾān. Moreover, some of the Qurʾānic prediction are those of optimism and are intended to strengthen the heart of followers. Of this type is the verse: "The Romans have been defeated in the nearer part of the land" (Q 30:1).

The main arguments of Muslim theologians, though, for considering the Qurʾān as a miracle are the verses of challenge (*taḥaddī*) in the Qurʾān and the fact that no one among the Arabs, who were known for eloquence, could produce a similar text. There are all together four verses in which this challenge has been pronounced and Ibn Kammūna cites them one by one.[19] In refuting this argument Ibn Kammūna draws the reader's attention to the history of the Qurʾān's canonization. He argues that although there is no doubt that the Qurʾān has been transmitted, it is not unlikely that some particular verses might have been inaccurately transmitted or even added later to the text. To him, the verses concerning the challenge (*taḥaddī*) might be among these. Supporting this argument, he refers to the report that in the Prophet's lifetime the Qurʾān was memorized in its entirety by only six or seven persons. The fact that these six or seven memorizers were the only ones who could have identified a change and tampered-with passages shows the vulnerability of the text, particularly when it is recalled that these people were not in agreement with each other. He reminds the reader of the report that Ibn Masʿūd, who was one of the early authorities on the

---

one to punish them whilst thou wert amongst them" (Q 8:32–33); (4) "He is but a crazy poet, or let him bring a miracle as those of olden time." (Q 21:5).

19 The verses of challenge that he cites are the followings:

1) "Say: verily if men and jinn agree to produce the like of this Qurʾān, they will not produce the like of it though one to the other were backer" (Q 17:90).
2) "Or do they say: 'he has invented it' say [thou]: 'Then bring ten *sūra*s like it which have been invented'" (Q 10:39).
3) "Or do they say: 'He has invented it?' Say: 'Then produce a *sūra* like it, and call upon whomsoever you are able [to call] apart from God'" (Q 2:21).
4) "If you are in doubt about what we have sent down to our servant, bring forward a *sura* like it, and call your witnesses apart from God, if you speak the truth [...] if you do not do so- nor will you do it" (Q 2:21–22).

Qur'ān, denied that the first sūra and the last two sūras belonged to the Qur'ān. Moreover, Ibn Mas'ūd put "in the name of the merciful and compassionate God" at the head of the ninth sūra. But two other Qur'ān authorities of the time, Ubayy b. Ka'b and Zayd b. Thābit did not. Ubayy included in his copy five sūras which Ibn Mas'ūd excluded, and of which Zayd excluded two. They disagreed on whether "in the name of the merciful and compassionate God" is or is not a verse at the opening of the sūras. Finally, each declared the other's copy a forgery. Ibn Kammūna goes on at great length over his dispute about the canonization of the Qur'ān. He reports that there were dissensions regarding the words of the Qur'ān, their order, and textual addition and deletion as well as the meaning. When at the time of the third Caliph, 'Uthmān (r. 644–656) these dissensions increased among the people, he burned all the versions except one.[20]

Another Qur'ānic subject of discussion is the eloquence of the Qur'ān. Ibn Kammūna does not contest the eloquence of the Qur'ān alltogether. However, he expresses his doubts regarding the Muslim argument that the difference between the eloquence of the Qur'ān and any other eloquent speech is unbridgeable.[21]

Ibn Kammūna also deals with the exegetical method and practice. He believes that some Qur'ānic verses need an interpretation which goes beyond the literal meaning. In other words, there are expressions referring to God, particularly those of anthropomorphism and corporeality, which should not be taken literally. However, not all Muslims, especially among the early authorities on tradition (*al-salaf min aṣḥāb al-ḥadīth*), seem to accept the fact that there are theological difficulties raised by a literal understanding of these verses.[22] When referring to the early authorities on tradition, Ibn Kammūna means Aḥmad b. Ḥanbal (d. 855) and his early followers. He blames them for their anthropomorphic perception of God, for their literal understanding of expressions like God's hand and head in the Qur'ān and in the traditions. According to Ibn Kammūna, only a metaphorical understanding of these expressions would prevent the readers from an unsound understanding of their creed. Ibn Kammūna adds that, due to the impact of philosophy on some of these scholars, their tone gradually became more rational. This might be another indication that he supports some rational and metaphorical understanding of these verses.

In one particular phrase in his *Examination of the Three Faiths*, Ibn Kammūna reveals his own idea about the origination of the Qur'ān as follows:

---

20 Ibn Kammūna, *Ibn Kammūna's Examination of the Three Faiths*, p. 110.

21 Ibid., p. 122.

22 Ibid., p. 145–146.

> It is possible that Muḥammad had read or heard the books of earlier prophets and had selected and compiled what was best in them; or that, attentive to the words of men, he studied them, chose and collected the more remarkable expressions (*kalimāt rā'iqa*) and fine points (*nukat fā'iqa*) and thus produced the Qur'ān.[23]

Ibn Kammūna narrates a tradition which supports his argument, according to which once when the Prophet was dictating to his scribe, ʿAbd Allāh b. Saʿd b. Abī Sarḥ the following divine words: "then we created the drop a clot" (Q 23:14), Saʿd said: "blessed be God, the best of creators" and the Prophet said "write that down, so it was revealed."

This statement, regardless of its suggestion that the Prophet collected these expressions from here and there, reveals Ibn Kammūna's belief that the Qur'ān does contain some "remarkable expressions" and "fine words." Indeed his occasional quotations of Qur'ānic expressions in his other writings can only be explained as a positive approach from his side. Expressions such as:

> He is the First and the Last, the Evident and the Immanent (Q 57:3).[24]
> And my success cannot be attained except from God, in Him I trust and unto Him I repent (Q 11:88).[25]

This was more or less what we can acquire from Ibn Kammūna's knowledge of the Qur'ān on the basis of his *Examination of the Three Faiths*. However, we should bear in mind that Ibn Kammūna in this work dealt with the polemical literature. Therefore, in order to gain a balanced understanding of Ibn Kammūna's attitude towards the Qur'ān, it would be better to take other works of his into account. Until recently, very few works of Ibn Kammūna were at our disposal, and the general impression, based on his *Examination of the Three Faiths*, was that Ibn Kammūna had no specific interest in the Qur'ān. Now, we know that this judgment needs some revision. In his most extensive work, his commentary on Suhrawardī's *al-Talwīḥāt*, Ibn Kammūna would seem to have been genuinely attracted by some exegetical literature. For instance, in interpreting Suhrawardī's expression *al-fāriqīn* (those who separate). Ibn Kammūna explains that Suhrawardī is either referring to souls who became detached from this world after their death or even before that or he is referring to the angels who separate the right from wrong. He explains that *al-fāriqāt* (those who separate) in the

---

23 Ibid., p. 105.
24 In his *Kalimat al-wajīza mushtamila ʿalā nukat laṭīfa fī l-ʿilm wa-l-ʿamal*, edited in Reza Pourjavady and Sabine Schmidtke, *A Jewish philosopher of Baghdad*, p. 146.
25 See his *Kalimat al-wajīza mushtamila ʿalā nukat laṭīfa fī l-ʿilm wa-l-ʿamal*, edited in ibid., p. 141.

verse, "and those who separate who bring criterion (*fa-l-fāriqāt farqan*) (Q 77:4)" was interpreted in the latter way by some exegetes. One of the Qur'ān exegetes who interpreted this verse accordingly is Fakhr al-Dīn al-Rāzī and it is not unlikely that Ibn Kammūna, who studied Rāzī's theological and philosophical works closely, was also familiar with the latter's work of *Tafsīr*.[26]

On another occasion, Ibn Kammūna quotes the following verse of the Qur'ān:

> Like darknesses in a fathomless ocean covered by a wave above which is a wave above which are clouds, darkness piled one upon the other (Q 24:40).

He then uses Abū Ḥāmid al-Ghazālī's *The Niche of Lights* (*Mishkāt al-anwār*) to interpret this verse. According to the latter the "fathomless ocean" is this world, because within it are destructive dangers, harmful occupations, and blinding murkiness. The first "wave" is the wave of the appetites which answers to man's bestial attributes, his occupation with sensory pleasures, and his striving after mundane achievements. The second "wave" is the attributes of predatoriness, which sends forth anger and many other despised dispositions. This is the higher wave, because, more often than not, anger takes control away from the appetites. The "clouds" are the loathsome beliefs, lying opinions and corrupt imaginings that have veiled man from the knowledge of the Truth, in the same way that the clouds veil the radiance of sunlight. Then Ibn Kammūna feels free to add to Ghazālī's interpretation the following: "These are all the 'darknesses', likewise are the appetite and anger. The appetite is darkness because an attachment to things makes you blind and deaf. The anger likewise is darkness because it destroys the intellect."[27] The question is how we can explain that Ibn Kammūna quoted the above mentioned passage from Ghazālī's *Mishkāt al-anwār* without considering that he had some agreements with Ghazālī on this matter and that he found his interpretation appealing. I would like to go one step further and argue that Ibn Kammūna was not only attracted to this interpretation, but also at least to some extent to the verse itself. Moreover, he did not consider this interpretation to be inappropriate for the verse. My argument for this is based on another location in the commentary, where Ibn Kammūna shows himself to be sensitive to interpreting Qur'ānic verses in an appropriate way. There

---

26 See Fakhr al-Dīn al-Rāzī, *Tafsīr al-kabīr*, Beirut: Dar al-Fikr, 1981, vol. 30, pp. 264–268.

27 Ibn Kammūna, *Sharḥ al-Talwīḥāt al-lawḥiyya wa-l-ʿarshiyya*, ed. by Najaf-Qulī Ḥabībī, vol. 3. Tehran: Mīrāth-i Maktūb, 2008, p. 524; Abū Ḥāmid al-Ghazālī, *The Niche of Lights: A Parallel English-Arabic Text*, trans. by David Buchman. Provo: Brigham Young University Press, 1998, p. 42.

Ibn Kammūna objects to Suhrawardī's interpretation of a Qur'ānic verse, not because of any doctrinal disagreement with him, but because he thinks that his interpretation overloads the verse. The verse under discussion is the following: "Everything perishes except for His Face" (Q 28:88). Suhrawardī explains the meaning of "His face" to be "by means of necessity through God (*bi-jihat al-wujūb bihī*)." Ibn Kammūna first explains what this interpretation would entail. It would entail that nothing exists by its essence except by means of the necessity through God. Because it is by means of His necessity that everything exists as otherwise a thing based on its own essence does not merit existence. Ibn Kammūna opposes this interpretation saying that it is too free, and if it is not confirmed by a supporting tradition or any other supporting indication, it has been over-interpreted (*taḥakkum*) and hence inappropriate.[28]

In conclusion, it is clear that the Qur'ān was not a holy book for Ibn Kammūna. Nevertheless, he thinks that it is an eloquent text and that it includes numerous remarkable expressions and fine points. For Ibn Kammūna, who was deeply attracted to Suhrawardī's metaphorical philosophy, some Qur'ānic verses have the potential to be interpreted in a metaphorical way. Some of these verses along with their metaphorical interpretations done by some Qur'ān exegetes seem to have been particularly appealing to him.

28 Ibn Kammūna, *Sharḥ al-Talwīḥāt*, vol. 3, p. 224. It is noteworthy that Shams al-Dīn al-Shahrazūrī (d. after 1288), who was familiar with this criticism of Ibn Kammūna, tried to justify Suhrawardī's interpretation. See his *Rasā'il al-Shajarat al-ilāhiyya fī 'ulūm al-ḥaqā'iq al-rabbāniyya*, ed. by Najaf-Qulī Ḥabībī, vol. 3. Tehran: Mu'assasa-i Pazhūhishī-i Ḥikmat u Falsafa-yi Īrān, 2006, p. 309.

Bibliography

Al-ʿĀmilī, Muḥsin al-Amīn, *Aʿyān al-shīʿa*, vol. 3 Beirut: Dār al-Taʿāruf li-l-Maṭbūʿāt, 1986.

Abbott, Nabia, "The Early Development of Tafsīr." in *The Qur'ān: Formative Interpretation*, ed. by Andrew Rippin, pp. 29–40. The formation of the classical Islamic world. Aldershot: Ashgate, 1999.

ʿAbd al-Masīḥ, Yassā, Burmester, Oswald Hugh Edward, Atiya, Aziz Suryal et al., *History of the Patriarchs of the Egyptian Church, Known as the History of the Holy Church, by Sawīrus ibn al-Muḳaffaʿ, Bishop of al-Ašmūnīn*, vol. 2–4. Cairo: Société d'archéologie copte, 1943–1974.

Abdel Haleem, Muhammad, *Understanding the Qur'an: Themes and Styles*. London: I. B. Tauris, 1999.

—, *The Qur'an. A New Translation*. Oxford: Oxford University Press, 2004.

Abdel-Kader, Ali Hassan, *The Life, Personality and Writings of al-Junayd. A Study of a 3rd/9th Century Mystic with an Edition and Translation of his Writings*. London: Luzac, 1962.

Abū l-Baqāʾ Ṣāliḥ ibn al-Husayn al-Jaʿfarī, *ar-Radd ʿalā n-naṣārā*, ed. by M. Muḥammad Ḥasanayn. Cairo: Maktabat Wahbah, 1409/1988.

Abū Qurra, Theodore, "Maymar fī taḥqīq nāmūs Mūsā l-muqaddas wa-l-anbiyāʾ alladhīna tanabbaʾū ʿalā l-Masīḥ wa-l-Injīl al-ṭāhir alladhī naqalahu ilā l-umam talāmīdh al-Masīḥ al-mawlūd min Maryam al-ʿadhrāʾ wa-taḥqīq al-urthūdhuksiyya allatī yansubuhā l-nās ilā l-Khalkīdūniyya wa-ibṭāl kull milla tattakhidh al-Naṣrāniyya siwā hādhī l-milla" (Maymar on the Verification of the Holy Law of Moses and the Prophets who prophesized about the Messiah, and on the Gospel, was conveyed to the nations by the disciples of the Messiah who is born from the Virgin Maryam, and on verifying the orthodoxy, which people attribute to Chalcedonianism and nullifying every religious group that arrogates Christianity other than this one), in *Mayāmir Thāwūdūrūs Abī Qurra, usquf Ḥarrān, aqdam taʾlīf ʿarabī*, ed. by Constantine Bacha, pp. 140–179. Beirut: Al-Fawāʾid, 1904.

—, "Maymar fī wujūd al-khāliq wa-l-dīn al-qawīm" (Maymar on the Existence of the Creator and the Right Religion), ed. by Ignace Dick. Jounieh: Libraire St. Paul, 1982.

—, "Maymar yuḥaqqiq annahu lā yalzam an-naṣārā an yaqūlū thalāthat āliha idh yaqūlūn al-Āb ilāh wa-l-Ibn ilāh wa-l-Rūḥ al-Qudus ilāh wa-law kāna kull wāḥid minhum tāmm ʿalā ḥidah" (Maymar that verifies that the Christians do not need to say three gods when they say the Father is God and the Son is God and the Holy Spirit is God, even if each one of them is perfect in its own), in *Mayāmir Thāwūdūrūs Abī Qurrah, usquf Ḥarrān, aqdam taʾlīf ʿarabī*, ed. by Constantine Bacha, pp. 23–47. Beirut: Al-Fawāʾid, 1904.

—, "Opusculum 20." in *Johannes Damaskenos, helgon*, ed. by Reinhold Glei und Adel-Théodore Khoury, p. 15–17. Corpus Islamo-Christianum, Series Latina 3. Würzburg: ECHTER; Altenberge: Oros, 1995.

Accad, Martin, "Corruption and/or Misinterpretation of the Bible: The Story of the Islamic Usage of Tahrif." *The Near East School of Theology Theological Review* 24 (2003): pp. 67–97.

—, "The Gospels in the Muslim Discourse of the Ninth to the Fourteenth Centuries: An Exegetical Inventorial table (Part I)." *Islam and Christian-Muslim Relations* 14 (2003): pp. 67–91.

https://doi.org/10.1515/9783110564341-018

—, “The Ultimate Proof-Text: the Interpretation of John 20.17 in Muslim-Christian Dialogue (Second/Eighth-Eighth/Fourtneeth Centuries).” in *Christians at the Heart of Islamic Rule: Church Life and Scholarship in ‘Abbasid Iraq*, ed. by David Thomas, pp. 199–214, The history of Christian-Muslim relations 1. Leiden: Brill, 2003.

Adang, Camilla, *Muslim Writers on Judaism and the Hebrew Bible: From Ibn Rabban to Ibn Ḥazm*, Islamic Philosophy, Theology, and Science 22. Leiden: Brill, 1996.

—, “The Karaites as Portrayed in Medieval Islamic Sources.” in *Karaite Judaism: A Guide to its History and Literary Sources*, ed. by Meira Polliack, pp. 179–197, Handbook of Oriental Studies, Section 1—The Near and the Middle East 73. Leiden and Boston: Brill, 2003.

Adler, William, “Jewish Pseudepigrapha in Jacob of Edessa’s Letters and Historical Writings.” in *Jacob of Edessa and the Syriac Culture of his Day*, ed. by Robert Bas ter Haar Romeny, pp. 49–65, Monographs of the Peshiṭta Institute 18. Leiden: Brill, 2008.

Albayrak, Ismail, “Reading the Bible in the Light of Muslim Sources: From *Isrā’īliyyāt* to *Islāmiyyāt*.” *Islam and Christian-Muslim Relations* 23 (2012): pp. 113–128.

Alexander, T. Desmond, “Lot’s Hospitality: A Clue to His Righteousness.” *Journal of Biblical Literature* 104 (1985): pp. 289–291.

Ali, Abdullah Y., *The Holy Qur’ān. English Translation of the Meanings and Commentary.* Medina: King Fahd Complex, 1990.

Allen, Roger, *An Introduction to Arabic Literature.* Cambridge: Cambridge University Press, 2003.

Allony, Nissim, “A Karaite List of Terms from the Eigth Century [Hebrew].” in *Writings of the Association for Biblical Research in Israel. In Memory of Dr. J. P. Korngruen*, ed. by A. Wieser and B. Z. Luria, pp. 324–63. Tel Aviv: 1964.

Alter, Robert, *The Five Books of Moses.* New York and London: W.W. Norton and Company, 2004.

Altmann, Alexander, *Studies in Religious Philosophy and Mysticism.* Ithaca: Cornell University Press, 1969.

Ambros, Arne A. and Stephan Procházka, *A Concise Dictionary of Koranic Arabic.* Wiesbaden: Reichert, 2004.

Ambrose of Milan, *On Abraham*, trans. by Theodosia Tomkinson. Etna: Center for Traditionalist Orthodox Studies, 2000.

Amir-Moezzi, Mohammed A. and David Streight, trans., *The Divine Guide in Early Shiism. The Sources of Esoterism in Islam.* Albany: New York University Press, 1994.

Anthony, Sean W., “Chiliastic Ideology and Nativist Rebellion in the Early ‘Abbāsid Period: Sunbādh and the Jāmāsp-Nāmah.” *Journal of the American Oriental Society* 132/4 (2012): pp. 641–655.

—, “Who was the Shepherd of Damascus? The Enigma of Jewish and Messianist Responses to the Islamic Conquests in Marwānid Syria and Mesopotamia.” in *The Lineaments of Islam: Studies in Honor of Fred McGraw Donner*, ed. by Paul M. Cobb, pp. 21–59, Islamic history and civilization 95. Leiden: Brill, 2012.

Arberry, Arthur J., *The Koran Interpreted: Translated with an Introduction.* London: Oxford University Press, 1964.

Armstrong, Lyall R., *The Quṣṣāṣ of Early Islam*, Islamic history and civilization 139. Leiden: Brill, 2017.

Astren, Fred, "Islamic Context of Medieval Karaism." in *Karaite Judaism: A Guide to Its History and Literary Sources*, ed. by Meira Polliack, pp. 145–177, Handbook of Oriental Studies, Section 1—The near and the Middle East 73. Leiden and Boston: Brill, 2003.

—, *Karaite Judaism and Historical Understanding*, Studies in Comparative Religion. Columbia: University of South Carolina Press, 2004.

Atiya, Aziz Suryal, "Sāwīrus ibn al-Muqaffaʿ." in *Coptic Encyclopedia*, vol. 7, ed. by Aziz Suryal Atiya, pp. 2100–2102. New York: Macmillan, 1991.

Augustine, "On Christian Doctrine." in *A Select Library of the Nicene and Post-Nicene Fathers of the Christian Church*, ed. by Philip Schaff, vol. 2. Grand Rapids: Eerdmans, 1977–1986.

—, *The City of God*, abridged and trans. by John W. C. Wand, vol. 1. London and New York: Oxford University Press, 1963.

Avi-Yonah, Michael, "Sodom (modern Sedom) and Gomorrah." in *Encyclopaedia Judaica*, ed. by Fred Skolnik and Michael Berenbaum, 2nd edition, vol. XVIII. New York/ New Haven: Holmes and Meier/Yale University Press, 2007.

El-Awa, Salwa M., *Textual Relations in the Qurʾān: Relevance, coherence and structure*, Routledge studies in the Quran. Abingdon: Routledge, 2006.

Awad, Najib George, "Need the Crucifixion Happen? *An-Nisā'* 4:157–158, Theodore Abū Qurrah and His Muslim Interlocutors in al-Ma'mūn's Court." (unpublished)

—, *Orthodoxy in Arabic Terms: A Study of Theodore Abū Qurrah's Theology in its Islamic Context*, Judaism, Christianity, and Islam—Tension, Transmission, Transformation 3. Boston: de Gruyter, 2015.

Awad, Wadi, "Al-Ṣafī ibn al-ʿAssāl." in *Christian-Muslim Relations: A Bibliographical History*, vol. 4, ed. by David Thomas, Barbara Roggema and Juan Pedro Monferrer-Sala, pp. 538–551. Leiden: Brill, 2012.

Aydinli, Osman, "Ascetic and Devotional Elements in the Muʿtazilite Tradition: The Sufi Muʿtazilites." *The Muslim World* 97/2 (2007): pp. 174–189.

Bacha, Constantine, *Les Œevres arabes de Théodore Abou-Kurra évêque d'Haran*. Beirut: Al-Fawā'id, 1904.

Bacher, Wilhelm, "Antikaräisches in einem jüngeren Midrasch." *Monatsschrift für Geschichte und Wissenschaft des Judentums* 23, 6 (1874): pp. 266–274.

Al-Baghdādī, ʿAlī al-Khaṭīb, *Taʾrīkh Baghdād aw Madīnat as-Salām*, 8 vols. Beirut: Dār al-Kitāb al-ʿarabī, 1966–1980.

Bailey, Lloyd R., "Gehenna: The Topography of Hell." *Biblical Archaeologist* 49 (1986): pp. 187–191.

Bannister, Andrew, *An Oral-Formulaic Study of the Qurʾan*. Lanham: Lexington Books, 2014.

Bar-Asher, Moshe, ed., *Rabbi Mordechai Breuer Festschrift*, vol. 1. Jerusalem: Akademon Press, 1992.

Bar-Asher Siegal, Michal, *Early Christian Monastic Literature and the Babylonian Talmud*. Ney York: Cambridge University Press, 2013.

—, "Shared Worlds: Rabbinic and Monastic literature." *Harvard Theological Review* 105/4 (2012): pp. 423–456.

—, "The Collection of Traditions in Monastic and Rabbinic Anthologies as a Reflection of Lived Religion." *Religion in the Roman Empire* 2 (2016): pp. 72–90.

Bar-Hebraeus, Gregorius, *Gregorii Barhebraei Chronicon Ecclesiasticum*, vol. 1, ed. by Jean-Baptiste Abbeloos and Thomas Joseph Lamy. Louvain: Peeters, 1872.

Bashyachi, Elijah, *Aderet Eliyahu, Sefer ha-Mitzvot shel ha-Yehudim ha-Qara'im*. Israel: National Council of Karaite Jews, 1960.

Basset, René, *Le synaxaire arabe jacobite (rédaction copte), V: Les mois de Baouneh, Abib, Mesoré et jours complémentaires*, Patrologia Orientalis 17. Paris, 1923.

Bauer, Johannes B., "Wunder Jesu in den Apokryphen." in *Heilungen und Wunder: Theologische, historische und medizinische Zugänge*, ed. by Josef Pichler and Christoph Heil, in Zusammenarbeit mit Thomas Klampfl, pp. 203–214. Darmstadt: Wissenschaftliche Buchgesellschaft, 2007.

Baumstark, Anton, "Jüdischer und christlicher Gebetstypus im Koran." *Der Islam* 16 (1927): pp. 229–248.

Beaumont, Mark Ivor, "Early Christian Interpretation of the Qur'ān." *Transformation* 22 (2005): pp. 195–203.

—, "Muslim Readings of John's Gospel in the 'Abbasid Period." *Islam and Christian-Muslim Relations* 19 (2008): pp. 179–197.

Becker, Adam H., "Beyond the Spatial and Temporal Limes: Questioning the 'Parting of the Ways' outside the Roman Empire." in *The Ways that Never Parted: Jews and Christians in Late Antquity and the Early Middle Ages*, ed. by Adam H. Becker and AnnetteYoshiko Reed, pp. 373–392, Texte und Studien zum antiken Judentum 95. Tübingen: Mohr Siebeck, 2003.

—, *Fear of God and the Beginning of Wisdom. The School of Nisibis and the Development of Scholastic Culture in Late Antique* Mesopotamia, Divinations. Philadelphia: University of Pennsylvania Press, 2006

Behr, John, *The Case against Diodore and Theodore: Texts and their Contexts*. Oxford Early Christian Texts. Oxford: Oxford University Press, 2011.

Behzadi, Lale, "Al-Jāḥiẓ and his Successors on Communication and the Levels of Language." in *Al-Jāḥiẓ: A Muslim Humanist for our Times*, ed. by Arnim Heinemann; John Lash Meloy; Tarif Khalidi, et. al., pp. 125–132, Beiruter Texte und Studien 119. Beirut, 2009.

—, "Between Theology, Philosophy, and Aesthetics. Al-Jāḥiẓ on Arabic Language." in *Center and Periphery within the Borders of Islam*, ed. by Giuseppe Contu, pp. 307–312, Orientalia Lovaniensia Analecta 207. Leuven: Peeters 2012.

—, *Sprache und Verstehen: al-Ǧāḥiẓ über die Vollkommenheit des Ausdrucks*, Diskurse der Arabistik 14. Wiesbaden: Harrassowitz, 2009.

Bell, Richard, *A Commentary on the Qur'ān*, ed. by C. Edmund Bosworth and M. E. J. Richardson, vol. 2, Journal of Semitic studies/Monograph 14. Manchester: University of Manchester, 1991.

—, *The Qur'ān: Translated, with a Critical Re-Arrangement of the Surahs*, vol. 2 Edinburgh: T. & T. Clark, 1937.

Ben Mobarak ben Ṣa'īr, Šelomo, *Libro de la Facilitación: Kitāb at-Taysīr (Diccionario judeoárabe de hebreo bíblico)*, intro., ed. and trans. by José Martínez Delgado, 2 vols. Granada: Universidad De Granada, 2010.

Ben-Sasson, Menahem, "Inter Communal Relations in the Geonic Period." in *The Jews of Medieval Islam. Community, Society and Identity*, ed. by Daniel Frank, pp. 17–31, Etudes sur le judaïsme médiéval 16. Leiden and New York: Brill, 1995.

Ben-Shammai, Haggai, *A Leader's Project: Studies in the Philosophical and Exegetical Works of Saadya Gaon*. Jerusalem: The Bialik Institute, 2015 (in Hebrew).

—, "Between Ananites and Karaites: Observations on Early Medieval Jewish Secterianism." in *Studies in Muslim-Jewish Relations Vol. 1*, ed. by Ronald L. Nettler, pp. 19–29. Chur: Harwood Academic Publishers, 1995.
—, "Jerusalem in Early Medieval Jewish Bible Exegesis." in *Jerusalem: Its Sanctity and Centrality to Judaism, Christianity, and Islam*, ed. by Lee Israel Levine, pp. 447–448. New York: Continuum, 1999.
—, "Kalam in Medieval Jewish Philosophy." in *History of Jewish Philosophy*, ed. by Daniel H. Frank and Oliver Leaman, pp. 126–32, Routledge history of world philosophies 2. London: Routledge, 1997.
—, "On *Mudawwin* – the Redactor of the Hebrew Bible in Judaeo-Arabic Bible Exegesis." in *From Sages to Savants: Studies Presented to Avraham Grossman*, ed. by Avraham Grossman, Joseph Hacker, Yosef Kaplan et al., pp. 73–110. Jerusalem: The Zalman Shazar Center for Jewish History, 2010 (in Hebrew).
—, "Saadya Gaon's Ten Articles of Faith." *Da'at* 37 (1996): pp. 11–26.
—, "The Doctrines of Religious Thought of Abū Yūsuf Ya'qūb al-Qirqisānī and Yefet Ben 'Eli." 2 vols., Ph.D. Thesis The Hebrew University of Jerusalem, 1977.
—, "The Exegetical and Philosophical Writing of Saadia: A Leader's Endeavor." *Pe'amim* 54 (1993): pp. 63–81.
—, "The Karaite Controversy: Scripture and Tradition in Early Karaism." in *Religionsgespräche im Mittelalter*, ed. by Bernard Lewis and Friedrich Niewöhner, pp. 11–26, Wolfenbütteler Mittelalter-Studien 4. Wiesbaden: Harrassowitz, 1992.
—, "The Status of Parable and Simile in the Qur'ān and Early Tafsīr: Polemic, Exegetical and Theological Aspects." *Jerusalem Studies in Arabic and Islam* 30 (2005): pp. 154–169.
—, "The Tension between Literal Interpretation and Exegetical Freedom. Comparative Observations on Saadia's Method." in *With Reverence for the Word: Medieval Scriptural Exegesis in Judaism, Christianity, and Islam*, ed. by Jane Dammen McAuliffe, Barry Walfish and Joseph Ward Goering, pp. 33–50. New York: Oxford University Press, 2010.
Bernard, Monique, "Medieval Muslim Scholarship and Social *Network* Analysis: A Study of the *Basra/Kufa* Dichotomy in Arabic Grammar." in *Ideas, Images, and Methods of Portrayal: Insights into Classical Arabic Literature and Islam*, ed. by Sebastian Günther, pp. 129–140, Islamic History and Civilization 58. Leiden: Brill, 2005.
Berque, Jacques, *Relire le Coran*, Paris: Bibliotheque Albin Michel, 1998.
Bertaina, David, "The Development of Testimony Collections in Early Christian Apologetics with Islam." in *The Bible in Arab Christianity*, ed. by David Thomas, pp. 151–173, The History of Christian-Muslim Relations 6. Leiden: Brill, 2007.
Berti, Vittorio, *L'au-delà de l'âme et l'en-deça du corps: Approches d'anthropologie chrétienne de la mort dans l'eglise syro-orientale*, Paradosis 57. Fribourg: Academic Press, 2015.
—, *Vita e studi di Timoteo I (†823) patriarca cristiano di Baghdad: Ricerche sull'epistolario e sulle fonti contigue*, Studia Iranica 41. Paris: Association pour l'avancement des études iraniennes, 2009.
Bettiolo, Paolo, Alda Giambelluca Kossova, Claudio Leonardi et al., eds., *Ascensio Isaiae: Textus*, Corpus Christianorum, Series Apocryphorum 7. Turnhout: Brepols, 1995.
Bezold, Carl, *Die Schatzhöhle aus dem syrischen Texte dreier unedirten Handschriften*. Leipzig: J. C. Hinrich'sche Buchhandlung, 1883.

Al-Biqāʿī, Ibrāhīm, *Naẓm al-durar fī tanāsub al-āyāt wa-l-suwar*, ed. by ʿAbd al-Razzāq Ghālib al-Mahdī. Beirut: Dār al-Kutub al-ʿIlmiyya, 1432/2011.

Birkeland, Harris, "Old Muslim Opposition Against Interpretation of the Koran." in *The Qurʾān: Formative Interpretation*, ed. by Andrew Rippin, pp. 41–80, The formation of the classical Islamic world. Aldershot: Ashgate, 1999.

Blau, Joshua, "Ibn Balʿam, Judah ben Samuel." in *Encyclopaedia Judaica* 8, pp. 660–661. Jerusalem: Keter, 1971.

Bleich, J. David, *Contemporary Halakhic Problems*, 6 vols. New York: KTAV Publishing House, 1983–2012.

Blenkinsopp, Joseph, *A History of Prophecy in Israel*, revised edition. Louisville: Westminster John Knox Press, 1996.

Börner-Klein, Dagmar, *Das Alphabet des Ben Sira, Hebräisch-deutsche Textausgabe mit einer Interpretation*. Wiesbaden: Marix Verlag, 2007.

—, "Narrative Kritik der rabbinischen Bibelauslegung im Alphabet des Ben Sira." in *Literatur im Dialog: die Faszination von Talmud und Midrasch*, ed. by Susanne Plietzsch, pp. 99–125. Zürich: Theologischer Verlag, 2007.

—, *Pirke de-Rabbi Elieser. Nach der Edition Venedig 1544 unter Berücksichtigung der Edition Warschau 1852. Aufbereitet und übersetzt*, Studia Judaica 26. Berlin and New York: de Gruyter, 2004.

—, "Transforming Rabbinic Exegesis into Folktale." *Trumah* 15 (2005): pp. 139–148.

Böwering, Gerhard, *The Mystical Vision of Existence in Classical Islam. The Qurʾānic Hermeneutics of the Ṣūfī Sahl at-Tustarī (d. 283/896)*, Studien zur Sprache, Geschichte und Kultur des islamischen Orients 9. Berlin: de Gruyter, 1980.

Bohmeier, Ute, *Exegetische Methodik in Pirke de-Rabbi Elieser, Kapitel 1–24*, Judentum und Umwelt 79. Frankfurt a. M.: Peter Lang, 2008.

Bosworth, Charles, "The Persian Impact on Arabic Literature." in *Arabic Literature to the End of the Umayyad Period*, ed. by Alfred F.L. Beeston et al., pp. 483–496, The Cambridge History of Arabic Literature. Cambridge: Cambridge University Press, 1983.

Boullata, Issa J., "Literary Structures of the Qurʾān." in *Encyclopedia of the Qurʾān* 3, ed. by Jane D. McAuliffe, pp. 192–204. Leiden: Brill, 2003.

Bovon, François, Pierre Geoltrain, Sever Voicu, et al., eds., *Les écrits apocryphes chrétiens*, Bibliothèque de la Pléiade 442 and 516. Paris: Gallimard, 1997.

Boyarin, Daniel, "History Becomes Parable: A Reading of the Midrashic Mashal." in *Mappings of the Biblical Terrain: The Bible as a Text*, ed. by Vincent L. Tollers and John Maier, pp. 51–71, Bucknell review 33/2. Lewisburg: Bucknell University Press 1989.

Boyce, Mary, *A History of Zoroastrianism*, 3 vols. Leiden/ Köln: Brill, 1975, 1982, 1991.

Braun, Oskar, ed., *Timothei Patriarchae I Epistulae I*. Corpus Scriptorum Christianorum Orientalium Syri 2,67. Paris: J. Gabalda, 1914 (reprint: CSCO 74/75 Syri 30/31, 1953).

Brisman, Shimeon, *A History and Guide to Judaic Dictionaries and Concordances*, Jewish research literature 3. New York: Ktav Publishing House, 2000.

Brock, Sebastian P., "Abraham and the Ravens: A Syriac Counterpart to Jubilees 11–12 and its implications." *Journal for the Study of Judaism* 9 (1978): pp. 135–152.

—, "A Fragment of Enoch in Syriac." *Journal of Theological Studies* 19 (1968): pp. 626–631.

—, "North Mesopotamia in the late seventh century. Book XV of John Bar Penkayē's *Riš Mellē*." Jerusalem Studies in Arabic and Islam 9 (1987): pp. 51–75.

—, *Studies in Syriac Christianity: History, Literature and* Theology, Collected studies series 357. Variorum. Hampshire: Ashgate, 1992.
Brody, Robert, *The Geonim of Babylonia and the Shaping of Medieval Jewish Culture.* New Haven: Yale University Press, 1998.
Brown, Francis, Samuel Rolles Driver and Charles A. Briggs et al., eds., *Hebrew and English Lexicon of the Old Testament.* Boston/ New York: Houghton Mifflin Company, 1906.
Buber, Martin, *The Prophetic Faith,* trans. from the Hebrew by Carlyle Witton Davies. New York: Harper & Row, 1960.
Budge, Ernest A. Wallis, ed., *The History of the Blessed Virgin Mary and The History of the Likeness of Christ Which the Jews of Tiberias Made to Mock at,* 2 vols., Luzac's Semitic Text and Translation Series. London: Luzac and Co., 1899.
Busse, Heribert, *Islamische Erzählungen von Propheten und Gottesmänner: Qiṣaṣ al-anbiyāʾ oder ʿarāʾis al-maǧālis, von Abū Isḥāq Aḥmad b. Muḥammad b. Ibrāhīm aṯ-Ṯaʿlabī,* Diskurse der Arabistik 9. Wiesbaden: Harrassowitz 2005.
Carlyle, Thomas, *On Heroes, Hero Worship, and the Heroic in History.* London: Chapman & Hall, 1872.
Chabot, Jean-Baptiste, ed., *Incerti Auctoris Chronicon Pseudo-Dionysianum vulgo dicto, II.* Corpus Scriptorum Christianorum Orientalium 104. Louvain: Secrětariat du Csco, 1965.
Chapin, Richard Steven, *Mesopotamian Scholasticism: A Comparison of the Jewish and Christian 'Schools'.* Cincinnati: Ph.D. thesis Hebrew Union College, 1990.
Charfi, Ayoub, *Le commentaire de Psaumes 33–60 d'Ibn aṭ-Ṭayyib, reflet de l'exegese syriaque orientale.* PhD Pontificia Università Gregoriana. Rome, 1997.
Charles, Robert Henry, *The Book of Jubilees.* London: Society for Promoting Christian Knowledge, 1917.
—, *The Ethiopic Version of the Hebrew Book of Jubilees.* Oxford: Clarendon Press, 1895.
Charlesworth, James H., ed., *The Old Testament Pseudepigraph,* vol. 2. Garden City, NY: Doubleday, 1985.
Cheikho, Louis, "Mīmar li-Tādurus Abī Qurra fī wujūd al-khāliq wa-l-dīn al-qawīm." *Al-Mashriq* 15 (1912): pp. 825–842.
Chilton, Bruce D., The *Aramaic Bible 11. The Isaiah Targum: Introduction, Translation, Apparatus and Notes.* Edinburgh: T & T Clark, 1987.
Choksy, Jamsheed K., *Conflict and Cooperation: Zoroastrian Subalterns and Muslim Elites in Medieval Iranian Society.* New York: Columbia Univ. Press, 1997.
Ciancaglini, Claudia A., *Iranian Loanwords in Syriac,* Beiträge zur Iranistik 28. Wiesbaden: Harrassowitz, 2008.
Cohen, Boaz, "Quotations from Saadia's Arabic Commentary on the Bible from Two Manuscripts of Abraham ben Solomon." in *Saadia Anniversary Volume,* ed. by Boaz Cohen, pp. 75–139, Texts and studies American Academy for Jewish Research 2. New York: American Academy for Jewish Research, 1943.
Cohen, Martin A., "Anan ben David and Karaite Origins." *Jewish Quarterly Review* 68 (1977/78): pp. 129–145 and 224–234.
Collins, John C., *Introduction to the Hebrew Bible.* Minneapolis: Fortress Press, 2004.
Contini, Riccardo, "Aspects of Linguistic Thought in the Syriac Exegetical Tradition." in *Syriac Encounters. Papers from the Sixth North American Syriac Symposium, Duke University, 26–29 June 2011,* ed. by Maria E. Doerfler et al., pp. 91–117, Eastern Christian studies 20. Leuven: Peeters, 2015.

Cook, David, "New Testament Citations in the Ḥadīth Literature and the Question of Early Gospel Translations into Arabic." in *The Encounter of Eastern Christianity with Early Islam*, ed. by Emmanouela Grypeou, Mark N. Swanson and David Thomas, pp. 185–224, The history of christian-muslim relations 5. Leiden: Brill, 2006.

Costaz, Louis, *Grammaire syriaque*, 3rd edition. Beirut: Imprimerie Catholique, 1994.

Curtis, Edward Lewis, "The Tribes of Israel." in *Biblical and Semitic Studies. Critical and Historical Essays by the Members of the Semitic and Biblical Faculty of Yale University.* New York/London: C. Scribner's Sons, 1901.

Cuypers, Michel, "Semitic Rhetoric as a Key to the Question of the *naẓm* of the Qur'ān." in *Journal of Qur'anic Studies* 13 (2011): pp. 1–24.

—, *The Banquet. A Reading of the Fifth Sura of the Qur'an.* Preface by Muhammad Ali Amir-Moezzi. Miami: Convivium Press, 2009.

—, *The Composition of the Qur'an: Rhetorical Analysis.* Bloomsbury Academic 2010.

Dalman, Gustav, *Sacred Sites and Ways: Studies in the Topography of the Gospels.* Authorised translation by Paul P. Levertoff. New York: The Macmillan Company, 1935.

Dan, Joseph, "Ben Sira, Alphabet of." in *Encyclopaedia Judaica*, vol. 3, pp. 548–549. Jerusalem: Keter, 1974.

D'Ancona, Cristina, "Greek into Arabic: Neoplatonism in Translation." in *The Cambridge Companion to Arabic Philosophy*, ed. by Peter Adamson and Richard C. Taylor, pp. 10–31, Cambridge companions to philosophy. Cambridge: Cambridge University Press, 2005.

Al-Dārimī, ʿAbd Allāh b. ʿAbd al-Rahmān, *Kitāb as-sunan*, ed. by ʿAbdullāh al-Yamanī l-Madanī. 2 vols. Cairo 1386/1966.

De Crussol, Yolande, *La role de la raison dans la réflexion éthique d'Al-Muḥāsibī.* Dissertation Université de Lyon, 2001. Paris: Consep, 2002.

De Lagarde, Paul, *Materialien zur Kritik und Geschichte des Pentateuchs*, 2 vols. Leipzig: B.G. Teubner, 1867.

De Lange, Nicholas R. M., *Origen and the Jews: Studies in Jewish Christian Relations in Third Century Palestine*, University of Cambridge Oriental Publications 25. Cambridge: Cambridge University Press, 1976.

Demiri, Lejla, "Al-Jaʿfarī." in *Christian-Muslim Relations: a Bibliographical History*, vol. 4, ed. by David Thomas and Alex Mallett, pp. 480–485, The history of Christian-Muslim relations 20. Leiden: Brill 2013.

—, *Muslim Exegesis of the Bible in Medieval Cairo: Najm al-Dīn al-Ṭūfī's (d. 716/1316) Commentary on the Christian Scriptures—a Critical Edition and Annotated Translation with an Introduction*, History of Christian-Muslim Relations 19. Leiden: Brill, 2013.

Den Heijer, Johannes, "History of the Patriarchs of Alexandria." in *The Coptic Encyclopedia*, ed. by Aziz Suryal Atiya, vol. 4, pp. 1238–1242. New York and Toronto: Macmillan, 1991.

—, *Mawhūb ibn Manṣūr ibn Mufarriǧ et l'historiographie copto-arabe: Étude sur la composition de l'Histoire des Patriarches d'Alexandrie*, Corpus Scriptorum Christianorum Orientalium 513. Leuven: Peeters, 1989.

—, "*The Martyrdom of Bifām Ibn Baqūra al-Ṣawwāf* by Mawhūb ibn Manṣūr ibn Mufarrij and its Fatimid Background." *Medieval Encounters* 21 (2015):452–484.

De Strycker, Émile, ed., *La forme la plus ancienne du Protévangile de Jacques. Recherches sur le Papyrus Bodmer 5 avec une édition critique du texte grec et une traduction annotée*, Subsidia Hagiographica 33. Bruxelles: Société des Bollandistes, 1961.

Deutsch, Yaacov, "New Evidence of Early Versions of *Toldot Yeshu*." *Tarbiz* 69 (2000): pp. 177–197 (in Hebrew).

De Vaux, Roland, *Ancient Israel: Its Life and Institutions*. English trans. by John McHugh, Biblical resource series. Grand Rapids: W. B. Eerdmans, 1997.

Dietrich, Friedrich, *Arabisch-deutsches Handwörterbuch zum Koran und Thier und Mensch vor dem König der Genien*. Leipzig: F. Hinrichs, 1894.

Diettrich, Gustav, *Išô'dâdh's Stellung in der Auslegungsgeschichte des Alten Testamentes an seinen Commentaren zu Hosea, Joel, Jona, Sacharia 9–14 und einigen angehängten Psalmen veranschaulicht*. Beihefte zur Zeitschrift für die alttestamentliche Wissenschaft 6. Giessen: Ricker'sche Verlagsbuchhandlung, 1902.

Di Matteo, I., "Confutazione contro I Christiani dello Zaydita al-Qāsim b. Ibrāhīm." *Rivista degli Studi Orientali* 9 (1921–23): pp. 301–364.

Dray, Carol A., *Translation and Interpretation in Targum to the Book of Kings*. Studies in the Aramaic Interpretation of Scripture 5. Leiden: Brill, 2006.

Drijvers, Han J.W., "The Gospel of the Twelve Apostles," in *The Byzantine and Early Islamic Near East, 1: Problems in the Literary Source Material*, ed. by Averil Cameron and Lawrence J. Conrad, pp. 189–213, Studies in Late Antiquity and Early Islam 1. Princeton Princeton, NJ: Darwin Press, 1992.

—, "The Testament of our Lord: Jacob of Edessa's response to Islam" in *ARAM* 6 (1994): pp. 104–114.

Driver, Samuel Rolles, *The Book of Genesis*. With Introduction and Notes, 4th edition. London: Methuen, 1905.

Drory, Rina, "Bilingualism and Cultural Images. The Hebrew and the Arabic Introductions of Saadia Gaon's Sefer ha-Egron." in *Language and Culture in the Near East*, ed. by Shlomo Izre'el and Rina Drory, pp. 11–24, Israel Oriental Studies 15. Leiden: Brill, 1995.

—, "Literary Contacts and Where to Find Them: On Arabic Literary Models in Medieval Jewish Literature." *Poetics Today* 14/2 (1993): pp. 277–302.

—, *Models and Contacts: Arabic Literature and Its Impact on Medieval Jewish* Culture, Brill's series in Jewish studies 25. Leiden: Brill, 2000.

Dunlop Gibson, Margaret, ed., "An Arabic Version of the Acts of the Apostles and the Seven Catholic Epistles from an Eighth or Ninth Century MS in the Convent of St Catharine on Mount Sinai, with a Treatise on *The Triune Nature of God*, with Translation, from the Same Codex." *Studia Sinaitica* VII (1899): pp. 1–36 (English), pp. 74–107 (Arabic).

Ebied, Rifaat and David Thomas, *Muslim-Christian Polemic during the Crusades: The Letter from the People of Cyprus and Ibn Abī Ṭālib al-Dimashqī's Response*, History of Christian-Muslim relations 2. Leiden: Brill, 2005.

Ehrman, Bart D., *The New Testament: A Historical Introduction to the Early Christian Writings*, 5th edition. New York and Oxford: Oxford University Press, 2012.

Eid, Hadi, *Lettre du calife Hârûn Al-Rašîd à l'empereur Constantin VI: Texte présenté, commenté et traduit par Hadi Eid, préface de Gérard Troupeau*. Études chrétiennes arabes. Thèse de doctorat Université de Paris, 1990. Paris: Cariscript, 1992.

Eisenman, Robert H. and Michael Wise, *Dead Sea Scrolls Uncovered: The First Complete Translation and Interpretation of 50 Key Documents Withheld for over 35* Years, Penguin Books/History and religion. New York: Penguin Books, 1993.

Elath, Moshe, "Saul at the Apex of his Success and the Beginning of His Decline (The Historiographical Significance of I Samuel 13–14)." *Tarbiz* 63 (1993): pp. 5–25.

Elbaum, Jacob, "Zwischen Midrasch und Mussar-Literatur: Studien zu den Kapiteln 1–6 in Tanna debe Eliyahu." *Jerusalem Studies in Hebrew Literature* 1 (1981): pp. 144–154 (in Hebrew).

Elman, Ya'akov, "Middle Persian Culture and Babylonian Sages: Accommodation and Resistance in the Shaping of Rabbinic Legal Tradition." in *The Cambridge Companion to the Talmud and Rabbinic Literature*, ed. by Charlotte Elisheva Fonrobert and Martin S. Jaffee, pp. 165–197, Cambridge companions to religion. Cambridge: Cambridge University Press, 2007.

Epstein, Isidore, ed., *Babylonian Talmud: Seder Nizikin*, trans. by E. W. Kirzner, Salis Daiches, Maurice Simon et al., Tractate Baba Mezia. London: World Federation for Mental Health, 1935.

Erder, Yoram, "The Doctrine of Abu ʿIsa al-Isfahani and Its Sources." *Jerusalem Studies in Arabic and Islam* 20 (1996): pp. 162–199.

Esack, Farid, *The Qur'an. A User's Guide.* Oxford: Oneworld Publications, 2005.

Eslinger, Lyle, "Inner-Biblical Exegesis and Inner-Biblical Allusion: The Question of Category." *Vetus Testamentum* 42 (1992): pp. 47–58.

Eutychius of Alexandria, *Annals*, ed. by L. Cheikho. Paris: Secretariat Du Corpus SCO, 1906.

Evans, Craig A., *Ancient Texts for New Testament Studies: A Guide to the Background Literature.* Peabody MA: Hendrickson, 2005.

Evetts, Basil Thomas Alfred, ed., *History of the Patriarchs of the Coptic Church of Alexandria*, Patrologia Orientalis 1–10. Paris: Firmin-Didot, 1904–1915.

Farahi, Hamid al-Din and Mahmood Hashmi, trans., *Exordium to Coherence in the Qur'an. An English Translation of Fatihah Nizam al-Qur'an.* Lahore: Al-Mawrid, 2008.

Farber, Walter, "Witchcraft, Magic, and Divination in Ancient Mesopotamia." in *Civilizations of the Ancient Near East*, vol. 3, ed. by Jack M. Sasson, pp. 1895–1909. New York: C. Scribner's Sons, 1995.

Faultless, Julian, "Ibn al-Ṭayyib." in *Christian-Muslim Relations. A Bibliographical History. Volume 2 (900–1050)*, ed. by David Thomas and Alexander Mallett, pp. 667–697, Christian-Muslim Relations. A Bibliographical History. Leiden: Brill, 2010.

Féghali, Paul, "Ibn At-Tayib et son commentaire sur la Genèse." *Parole de l'Orient* 16 (1990–91): pp. 149–162.

Fernández Marcos, Natalio and José R. Busto Saiz, *Theodoreti Cyrensis Quaestiones in Reges et Paralipomena*, Textos y estudios Cardenal Cisneros 32. Madrid: Instituto "Arias Montano", Consejo superior de investigaciones científicas, 1984.

Fields, Weston W., "The Motif 'Night as Danger' associated with Three Biblical Destruction Narratives." in *"Sha'arei Talmon": Studies in the Bible, Qumran and the Ancient Near East presented to S. Talmon*, ed. by Michael A. Fishbane and Emanuel Tov, Shemaryahu Talmon et. al., pp. 17–32. Winona Lake: Eisenbrauns, 1992.

Finkel, Avraham Y., *Pirkei Drebbi Eliezar*, vol. 2. Scranton: Yeshivath Beth Moshe, 2009.

Finkel, Irving L., "Necromancy in Ancient Mesopotamia." *Archiv für Orientforschung* 29/30 (1983/4): pp. 1–17.

Firestone, Reuven, "Abraham's Son as the Intended Sacrifice (*al-Dhabīḥ*, Qur'ān 37:99–113): Issues in Qur'ānic Exegesis." *Journal of Semitic Studies* 34 (1989): pp. 95–131.

—, *Journeys in Holy Lands: The Evolution of the Abraham-Ishmael Legends in Islamic Exegesis.* Albany: State University of New York Press, 1990.

—, "Ṭālūt." in *Encyclopaedia of Islam*, vol. 10, pp. 168–169. Leiden: Brill, 2000.

Fishbane, Michael, *Biblical Interpretation in Ancient Israel.* Oxford: Oxford University Press, 1988.

—, "Inner-Biblical Exegesis." in *Hebrew Bible/Old Testament: The History of Its Interpretation. From the Beginnings to the Middle Ages (Until 1300)*, ed. by Magne Sæbø, vol. 1, pp. 33–48. Göttingen: Vandenhoeck & Ruprecht, 1996.

Fishman, Talya, *Becoming the people of the Talmud: Oral Torah as Written Tradition in Medieval Jewish Cultures, Jewish* culture and context. Philadelphia: University of Pennsylvania Press, 2011.

Förster, Hans, "Die johanneischen Zeichen und Joh 2:11 als möglicher hermeneutischer Schlüssel." *Novum Testamentum* 56 (2014): pp. 1–23.

Frank, Daniel, *Search Scripture Well: Karaite Exegetes and the Origins of the Jewish Bible Commentary in the Islamic East,* Études sur le judaïsme médiéval 29. Leiden: Brill 2004.

—, *The Jews of Medieval Islam: Community, Society, and Identity.* Etudes sur le judaïsme medieval 16. Leiden: Brill, 1995.

Frank, Richard M. and Dimitri Gutas, eds., *Early Islamic Theology: The Muʿtazilites and al-Ashʿari, Texts and Studies on the Development and History of Kalam, Texts and studies on the development and history of* Kalām 2. Aldershot: Ashgate, 2007.

Freytag, Georg Wilhelm, *Lexicon arabico-latinum*, 4 vols. Halle: C.H. Schwetschke et Filium, 1830–1837.

Friedlander, Gerald, transl., *Pirḳê de-Rabbi Eliezer (The chapters of Rabbi Eliezer the Great): according to the text of the manuscript belonging to Abraham Epstein of Vienna*, 4th edition, *The* Judaic studies library 6. New York: Hermon Press, 1981.

—, *Pirḳê de Rabbi Eliezer (The chapters of Rabbi Eliezer, the Great): according to the text of the manuscript belonging to Abraham Epstein of Vienna/translated and annotated with introduction and indices by Gerald Friedlander.* New York: Bloch, 1916.

Fritsch, Erdmann, *Islam und Christentum im Mittelalter: Beiträge zur Geschichte der muslimischen Polemik gegen das Christentum in arabischer Sprache*, Breslauer Studien zur historischen Theologie 17. Breslau: Müller & Seiffert, 1930.

Galbiati, Giovanni, ed., *Iohannis evangelium apocryphum arabice.* Milan: In aedibus Mondadorianis, 1957.

Gallo, Maria, trans., *Palestinese anonimo; Omelia arabo-cristiana dell'VIII secolo.* Roma: Città nuova Editrice, 1994.

Gargano, Innocenzo, "'Lot si rifugió nella grotto con le sue due figlie' (Origene)." *Parola Spirito e Vita* 26 (1992): pp. 215–231.

Garsiel, Moshe, "Torn between Prophet and Necromancer: Saul's Despair (1Sam 28:3–25)." *Beit Mikra: Journal for the Study of the Bible and Its World* 41 (1996): pp. 172–196.

Geiger, Abraham, *Was hat Mohammed aus dem Judenthume aufgenommen*, 2nd edition. Leipzig: Kaufmann, 1902.

Genesis Rabbah: The Judaic Commentary to the Book of Genesis, ed. and trans. by Jacob Neusner, vol. 2. Atlanta: Scholars Press, 1985.

Ghaly, Mohammed, "Evil and Suffering in Islam." in *Philosophy of Religion: Selected Readings, 5th edition*, ed. by Michael Peterson, William Hasker and Bruce Reichenbach, pp. 383–391. Oxford: Oxford University Press, 2014.

Al-Ghazālī, Abū Ḥamīd M., *Deliverance from Error and Mystical Union with the Almighty*, ed. by Muḥammad Abūlaylah and Nurshīf Abdul-Rahīm Rifʿat, Cultural heritage and contemporary change series 2 A, Islam 2 B. Washington: The Council of Research in Values and Philosophy, 2001.

—, *Iljām al-ʿawāmm ʿan ʿilm al-kalām*. Beirut: Dār al-Kitāb al-ʿarabī, 1985.

—, *The Niche of Lights: A Parallel English-Arabic Text*, trans. by David Buchman, Islamic translation series. Provo: Brigham Young University Press, 1998.

Ghersetti, Antonella, "'Like the wick of the lamp, like the silkworm they are': stupid schoolteachers in classical Arabic literary sources." *Journal of Arabic and Islamic Studies* 10 (2010): pp. 75–100.

Gheyturi, Amer and Arsalan Golfam, "The Qurʾān as a Non-Linear Text: Rethinking Coherence." in *The International Journal of Humanities of the Islamic Republic of Iran* 15 (2008): pp. 119–133.

Gil, Moshe, *A History of Palestine, 634*–1099, trans. by Ethel Broido. Cambridge: Cambridge University Press, 1992.

—, *Jews in Islamic countries in the Middle Ages*, trans. by David Strassler, Etudes sur le judaïsme medieval 28. Leiden: Brill, 2004.

—, "The Origins of the Karaites." in *Karaite Judaism: A Guide to its History and Literary Sources,* ed. by M. Polliack, pp. 73–117. Handbook of Oriental Studies, Section 1 – The Near and the Middle East 73. Leiden and Boston: Brill, 2003.

Gilliot, Claude and Pierre Larcher, "Language and style of the Qur'ān." in *Encyclopedia of the Qurʾān vol. 3*, ed. by Jane D. McAuliffe, pp. 108a and 126a. Leiden: Brill, 2003.

Gimaret, Daniel, *Les noms divins en Islam: Exégèse lexicographique et théologique*, Patrimoines: Islam. Paris: Éditions du Cerf, 1988.

Ginzberg, Louis, *The Legends of the Jews*, trans. from the German Manuscript by Henrietta Szold, 6 vols. Philadelphia: The Jewish Publication Society of America, 1909.

—, *The Legends of the Jews*, vol. 6. Philadelphia: Jewish Publication Society, 1926 (new edition: Philadelphia: Jewish Publication Society of America, 2003).

Goldenberg, Robert, "The Destruction of the Jerusalem Temple: its meaning and its consequences." in *Cambridge History of Judaism IV: The Late Roman-Rabbinic Period,* ed. by Steven T. Katz, pp. 191–205. Cambridge: Cambridge University Press, 2006.

Goldstein, Miriam, *Karaite Exegesis in Medieval Jerusalem*, Texts and studies *in medieval* and early modern Judaism 26. Tübingen: Mohr Siebeck, 2011.

Goldziher, Ignaz, *Introduction to Islamic Theology and Law,* trans. by Andras and Ruth Hamori, Modern classics in Near Eastern studies. Princeton: Princeton University Press, 1991.

—, *Die Richtungen der islamischen Koranauslegung*, Publication Stichting De Goeje 6. Brill: Leiden, 1920.

Goodman, Lenn E., *The Book of Theodicy: Translation and Commentary on the Book of Job by Saadia ben Joseph al Fayyumi,* Yale Judaica Series 25. New Haven: Yale University Press, 1988.

Graham, William A., "Basmala." in *Encyclopedia of the Qurʾān* 1, ed. by Jane D. McAuliffe, pp. 207–212. Leiden: Brill, 2001.

—, "Light in the Qurʾān and Other Early Islamic Exegesis." in *God is the Light of the Heavens and the Earth: Light in Islamic Art and Culture*, ed. by Sheila Blair and Jonathan Bloom, pp. 43–60. New Haven: Yale University Press, 2015.

Gramlich, Richard, *Alte Vorbilder des Sufitums*, Veröffentlichungen der Orientalischen Kommission/Akademie der Wissenschaften und der Literatur 42/2. Wiesbaden: Harrassowitz, 1996.

—, *Weltverzicht: Grundlagen und Weisen islamischer Askese*, Veröffentlichungen der Orientalischen Kommission/Akademie der Wissenschaften und der Literatur 43. Wiesbaden: Harrassowitz, 1997.

Gray, John, *I and II Kings*, 2nd rev. edition. Old Testament Library. London: SCM Press, 1970.

Greenspahn, Frederick E., "Sadducees and Karaites: The Rhetoric of Jewish Sectarianism." *Jewish Studies Quarterly* 18/1 (2011): pp. 91–105.

Griffith, Sidney H., "Arguing from Scripture: the Bible in Christian/Muslim Encounter in the Middle Ages." in *Scripture and Pluralism: Reading the Bible in the Religiously Plural Worlds of the Middle Ages and Renaissance*, ed. by Thomas J. Heffernan and Thomas E. Burman, pp. 29–58, Studies in the History of Christian Traditions 123. Leiden: Brill, 2005.

—, "Christians and the Arabic Qurʾān: Prooftexting, Polemics, and Intertwined Scriptures." *Intellectual History of the Islamicate World* 2 (2014): pp. 243–266.

—, "Disclosing the Mystery: The Hermeneutics of Typology in Syriac Exegesis; Jacob of Serūg on Genesis XXII." in *Interpreting Scriptures in Judaism, Christianity and Islam: Overlapping Inquiries*, ed. by Mordechai Z. Cohen and Adele Berlin, pp. 46–64. Cambridge, UK: Cambridge University Press, 2016.

—, "Faith and Reason in Christian *Kalām:* Theodore Abū Qurrah on Discerning the True Religion." in *Christian Arabic Apologetics During the Abbasid Period (750–1258)*, ed. by Samir Khalil Samir and Jørgen S. Nielsen, pp. 1–43, Studies in the history of religions 63. Leiden: Brill, 1994.

—, "From Aramaic to Arabic: The Language of the Monasteries of Palestine in the Byzantine and Early Islamic Periods." in *The Beginnings of Christian Theology in Arabic: Muslim-Christian Encounters in the Early Islamic Period*, ed. by Sidney Griffith, pp. 11–31, Variorum collected studies series 746. Aldershot: Ashgate, 2002.

—, "Paul of Antioch." in *The Orthodox Church in the Arab World 700–1700: An Anthology of Sources*, ed. by Samuel Noble and Alexander Treiger, pp. 216–235, 327–331, Orthodox Christian studies. DeKalb, IL: Northern Illinois University Press, 2014.

—, *The Bible in Arabic: The Scriptures of the 'People of the Book' in the Language of Islam*, Jews, Christians and Muslims from the Ancient to the Modern World. Princeton, NJ: Princeton University Press, 2013.

—, *The Church in the Shadow of the Mosque: Christians and Muslims in the World of Islam*, Jews, Christians, and Muslims from the Ancient to the Modern World. Princeton, NY: Princeton University Press, 2008, 2012.

—, "The Church of Jerusalem and the 'Melkites': The Making of an 'Arab Orthodox' Christian Identity in the World of Islam (750–1050 CE)." in *Christians and Christianity in the Holy Land: From the Origins to the Latin Kingdoms*, ed. by Guy G. Stroumsa and Ora Limor, pp. 175–204, Cultural encounters in late antiquity and the Middle Ages 5. Turnhout: Brepols, 2006.

—, "The Gospel in Arabic: An Inquiry into Its Appearance in the First Abbasid Century." *Oriens Christianus* 69 (1985): pp. 126–167.

—, "The Gospel, the Qur'ān, and the Presentation of Jesus in al-Ya'qūbī's *Ta'rīkh*." in *Bible and Qur'ān: Essays in Scriptural Intertextuality*, ed. by John C. Reeves, pp. 133–160, Symposium Series 24. Atlanta, GA: Society of Biblical Literature, 2003.

—, "The Monks of Palestine and the Growth of Christian Literature in Arabic." *The Muslim World* 78 (1988): pp. 1–28.

—, "The Qur'ān in Arab Christian Texts: The Development of an Apologetical Argument; Abū Qurrah in the Mağlis of al-Ma'mūn." *Parole de l'Orient* 24 (1999): pp. 203–233.

—, "The Poetics of Scriptural Reasoning: Syriac *Mêmrê* at Work." Forthcoming.

—, "The Syriac Letters of Patriarch Timothy I and the Birth of Christian Kalām in the Mu'tazilite Milieu of Baghdad and Baṣrah in Early Islamic Times." in *Syriac Polemics: Studies in Honour of Gerrit Jan Reinink*, ed. by Wout Jac van Bekkum, Jan Willem Drijvers and Alexander Cornelis Klugkist, pp. 103–132, Orientalia Lovaniensia analecta 170. Louvain: Peeters, 2007.

—, "What has Constantinople to do with Jerusalem? Palestine in the Ninth Century: Byzantine Orthodoxy in the World of Islam." in *Byzantium in the Ninth Century: Dead or Alive?*, ed. by Leslie Brubacker, pp. 181–194. Publications/Society for the Promotion of Byzantine Studies 5. Aldershot: Ashgate, 1998.

Grypeou, Emmanouela and Monferrer-Sala, Juan Pedro. ""A tour of the other world": A contribution to the textual and literary criticism of the "Six Books Apocalypse of the Virgin"." *Collectanea Christiana Orientalia* 6 (2009): pp. 115–165, pp. 157–158.

Günther, Sebastian. "Advice for Teachers: The 9th Century Muslim Scholars Ibn Sahnun and al-Jahiz on Pedagogy and Didactics." in *Ideas, Images and Methods of Portrayal: Insights into Classical Arabic Literature and Islam*, ed. by Sebastian Günther, pp. 79–116, Islamic history and civilization 58. Leiden: Brill, 2005.

—, "Al-Jahiz and the Poetics of Teaching: A Ninth Century Muslim Scholar on Intellectual Education." in *Al-Jahiz: A Humanist for Our Time*, ed. by Tarif Khalidi, Arnim Heinemann, John Meloy et al., pp. 17–26, Beiruter Texte und Studien 119. Beirut, Wiesbaden: Orient-Institute of the German Oriental Society, 2009.

—, "Pioneers of Arabic linguistic studies." in *In the Shadow of Arabic: The Centrality of Language to Arabic Culture*, ed. by Bilal Orfali, pp. 197–220. Leiden: Brill, 2011.

Gutas, Dimitri, *Greek Thought, Arabic Culture: the Graeco-Arabic Translation Movement in Baghdad and Early 'Abbasid Society (2nd–4th/8th–10th Centuries)*. Abingdon: Routledge, 2005.

Gwynne, Rosalind Ward, "Patterns of Address." in *Blackwell Companion to the Qur'ān*, ed. by Andrew Rippin, pp. 73–87, Blackwell companions to religion. Malden, MA: Blackwell, 2006.

Haag, Ernst, "Abraham und Lot in Gen 18–19." in *Mélanges bibliques et orientaux en l'honneur de M. Henri Cazelles*, ed. by André Caquot and Mathias Delcor, pp. 173–179, Alter Orient und Altes Testament 212. Kevelaer: Butzon & Bercker, 1981.

Halkin, Abraham S., "Judeo-Arabic Literature." *Encyclopaedia Judaica*, vol. 10, pp. 410–423. Jerusalem: Keter 1972.

Al-Ḥamawī, Yāqūt, *Mu'jam al-buldān*, 5 vols. Beirut: Dār iḥyā' al-Turāth al-'Arabī, 1399.

Hare, Douglas R. A., "The Lives of the Prophets." in *The Old Testament Pseudepigrapha II: Expansions of the Old Testament and Legends, Wisdom and Philosophical Literature, Prayers, Psalms and Odes, Fragments of Lost Judeo-Hellenistic Works*, ed. by James Charlesworth, pp. 379–400. London: Darton and Todd, 1985.

Harrington, Daniel J. and Saldarini, Anthony J., eds. and transl., *Targum Jonathan of the Former Prophets: Introduction, Translation and Notes*, The Aramaic Bible 10. Edinburgh: T & T Clark, 1987.

Harris, James Rendel, ed., *The Gospel of the Twelve Apostles together with the Apocalypses of Each One of them*. Cambridge: Cambridge University Press, 1900.

Hasan-Rokem, Galit, *Tales of the Neighborhood: Jewish Narrative Dialogues in Late Antiquity*, Taubman lectures in Jewish studies 4. London: University of California Press, 2003.

Heemskerk, Margaretha T., *Suffering in the Muʿtazilite Theology: ʿAbd al-Jabbār's Teaching on Pain and Divine Justice*, Islamic philosophy and theology 41. Leiden: Brill, 2000.

Heidet, Louis, "Ségor." in *Dictionnaire de la Bible*, ed. by Fulcran Vigouroux, pp. 1561–1565, V/3, Encyclopédie des sciences ecclésiastiques 1. Paris: Letouzé Et Ané, 1912.

Heimgartner, Martin and Roggema, Barbara, "Timothy I." in *Christian-Muslim Relations: A Bibliographical History*, vol 1, ed. by David Thomas and Barbara Roggema, pp. 515–531, The History of Christian-Muslim Relations. Leiden: Brill, 2009.

—, *Die Briefe 30–39 des ostsyrischen Patriarchen Timotheos I.* Corpus Scriptorum Christianorum Orientalium 662. Leuven: Peeters, 2016 (*Übersetzung und Anmerkungen*. Corpus Scriptorum Christianorum Orientalium. Leuven: Peeters, 2015).

—, *Die Briefe 42–58 des ostsyrischen Patriarchen Timotheos I.* Corpus Scriptorum Christianorum Orientalium 644. Leuven: Peeters, 2012 (*Einleitung, Übersetzung und Anmerkungen*, CSCO 645, Leuven: Peeters, 2012).

—, "Die Disputatio des ostsyrischen Patriarchen Timotheos (780–823) mit dem Kalifen al-Mahdī." in *Christians and Muslims in Dialogue in the Islamic Orient of the Middle Ages: Christlich-muslimische Gespräche im Mittelalter*, ed. by Martin Tamcke, pp. 41–56. Beiruter Texte und Studien 117. Beirut: Ergon Verlag Würzburg in Kommission, 2007.

—, "Neue Fragmente Diodors von Tarsus aus den Schriften 'Gegen Apollinarius','Gegen die Manichäer' und 'Über den heiligen Geist'." in *Apollinarius und seine Folgen*, ed. by Silke-Petra Bergian, Benjamin Gleede and Martin Heimgartner, pp. 211–232, Studien und Texte zu Antike und Christentum 93. Tübingen: Mohr Siebeck, 2015.

—, *Timotheos I., ostsyrischer Patriarch: Disputation mit dem Kalifen al-Mahdī. Textedition*, Corpus Scriptorum Christianorum Orientalium 631 (Scriptores Syri 244). Leuven: Peeters, 2011 (*Einleitung, Übersetzung und Anmerkungen*, Corpus Scriptorum Christianorum Orientalium, 632, Scriptores Syri 245).

Hennecke, Edgar and Schneemelcher, William, eds., *New Testament Apocrypha*, trans. by Robert McLachlan Wilson and Angus John Brockhurst Higgins. Philadelphia: Westminster Press, 1963–1966.

Herr, Moshe David, "Pirke de-Rabbi Eliezer." *Encyclopaedia Judaica*, vol. 13, cols. 558–560. Jerusalem: Keter 1971.

Hezser, Catherine, *The Social Structure of the Rabbinic Movement in Roman Palestine*, Texte und Studien zum Antiken Judentum 66. Tübingen: Mohr Siebeck, 1997.

Himmelfarb, Martha, *Jewish Messiahs in a Christian Empire. A History of the Book of Zerubbabel*. Cambridge, MA: Harvard University Press, 2017.

Hinz, Walther, "Farsakh." in *Encyclopédie de l'Islam*, Nouvelle édition, ed. by H.A.R. Gibb et al., 13 vol. Leyde and Paris: Brill – G.-P. Maisonneuve, 1960–2009.

Hirschfeld, Hartwig, *Qirqisani Studies*, Jews' College London Publication 6. London: Jews' College Publications, 1918 (in Arabic).

Horn, Cornelia, "Arabic Infancy Gospel." in *Encyclopedia of the Bible and Its Reception Vol. 2*, ed. by Hans-Josef Klauck et al., pp. 589–592. Berlin and New York: de Gruyter, 2009.

—, "Children and Violence in Syriac Sources: The *Martyrdom of Mar Ṭalyā' of Cyrrhus* in the Light of Literary and Theological Implications." *Parole de l'Orient* 31 (2006): pp. 309–326.

—, "Editing a Witness to Early Interactions between Christian Literature and the Qur'ān: *Status Quaestionis* and Relevance of the *Arabic Apocryphal Gospel of John*." in *Actes du 8e Congrès International des Études Arabes Chrétiennes (Granada, septembre 2008)*, published in *Parole de l'Orient* 37 (2012): pp. 87–103.

—, "Jesus' Healing Miracles as Proof of Divine Agency and Identity: The Trajectory of Early Syriac Literature." in *The Bible, the Qur'ān, and their Interpretation: Syriac Perspectives*, ed. by Cornelia Horn, pp. 69–97, Eastern Mediterranean Texts and Contexts 1. Warwick, RI: Abelian Academic, 2013.

—, "Syriac and Arabic Perspectives on Structural and Motif Parallels regarding Jesus' Childhood in Christian Apocrypha and Early Islamic Literature: the 'Book of Mary,' the *Arabic Apocryphal Gospel of John*, and the Qur'ān." *Apocrypha* 19 (2008): pp. 267–291.

Hoyland, Robert G., "History, fiction and authorship in the first centuries of Islam." in *Writing and Representation in Medieval Islam*, ed. by Julia Bray, pp. 16–46, Routledge Studies in Middle Eastern Literatures. London: Routledge, 2006.

—, "Jacob and Early Islamic Edessa," in *Jacob of Edessa and the Syriac Culture of his Day*, ed. by Robert Barend ter Haar Romeny, pp. 16–18, Monographs of the Peshiṭta Institute Leiden 18. Leiden: Brill 2008.

—, *Seeing Islam as Others Saw it: A Survey and Evaluation of Christian, Jewish and Zoroastrian Writings on Early* Islam, Studies in Late Antiquity and Early Islam 13. Princeton, NJ: Darwin Press, 1997.

Hughes, Aaron, "The Formative Period of Islam and the Documentary Approach: a Prolegomenon." in *A Legacy of Learning. Essays in Honor of Jacob Neusner*, ed. by Alan J. Avery Peck, Bruce Chilton, William Scott Green et al., pp. 372–385, Brill reference library of Judaism 43. Leiden: Brill, 2014.

Husseini, Sara Leila, *Early Christian-Muslim Debate on the Unity of God: Three Christian Scholars and Their Engagement with Islamic Thought (9th Century C.E.)*, The History of Christian-Muslim Relations. Leiden: Brill, 2014.

Ibn Danān, Sĕʿadyah, *Libro de las raíces: Diccionario de hebreo bíblico*, Introduction, translation and indexes by Milagros Jiménez Sánchez, Textos/Lengua hebrea 2. Granada: Universidad de Granada, 2004.

Ibn al-Jawzī, *Zād al-masīr fī ʿilm al-tafsīr.* Beirut: Al-Maktab al-Islamī, 1404/1984.

Ibn Ḥazm, Abū Muḥammad ʿAlī b. Aḥmad, *Kitāb al-fiṣal fī l-milal wa-l-ahwāʾ wa-n-niḥal*, 3 parts. Cairo: Al-Maṭbaʿa al-Adabiyya, 1900.

Ibn Kammūna, *Ibn Kammunā's Examination of the Three Faiths: A Thirteen-Century Essay in the Comparative Study of Religion*, trans. by Moshe Perlemann. Berkley: University of California Press, 1971.

—, "Risāla fī azaliyyat al-nafs wa-baqāʾihā." in *Azaliyyat al-nafs wa-baqāʾuhā*, ed. by Insīyah Barkhwāh. Tehran: Kitābkhāna, Muza, va Markaz-I Asnād-I Majlis-I Shūrā-Yi Islāmī, 2006.

—, *Sharḥ al-Talwīḥāt al-lawḥiyya wa-l-ʿarshiyya*, ed. by Najaf-Qulī Ḥabībī, vol. 3. Tehran: Mīrāth-i Maktūb, 2008.

Ibn Kathīr, Ismāʿīl, *Tafsīr*, ed. by Muḥammad Bayḍūn. Beirut: Dār al-Kutub al-ʿIlmiyya, 1424/2004.
Ibn al-Qayyim al-Jawziyya, Shams al-Dīn Muḥammad, *Hidāyat al-ḥayāra fī ajwibat al-yahūd wa-l-naṣāra*, ed. by Aḥmad Ḥijāzī s-Saqqā. Egypt: 1980.
Ibn Qutayba, Abū Muḥammad ʿAbdallāh b. Muslim, *Kitāb taʾwīl mukhtalif al-ḥadīth*. Cairo: 1908.
Ibn Sulaymān, Muqātil, *Tafsīr*, ed. by ʿAbdallāh Muḥammad al-Shiḥāta. Beirut: Dār al-Turāth al-ʿArabī, 2002.
—, *Al-Wujūh wa-n-naẓāʾir fī l-qurʾān al-ʿaẓīm*, ed. by Ḥātim Ṣāliḥ aḍ-Ḍāmin. Riyad: Maktabat ar-Rushd Nāshirūn, 2010.
Ibn Taymiyya, Taqī ad-Dīn, *Al-Jawāb aṣ-saḥīḥ li-man baddala dīn al-Masīḥ*. Cairo: 1905.
Ibn al-Ṭayyib, Abū al-Faraj ʿAbd Allāh, *Commentaire sur la Genèse*, ed. and trans. by Joannes Cornelis Josephus Sanders, 2 vols, Corpus scriptorum christianorum orientalium 275 (Scriptores arabici 24). Louvain: Secretariat du Corpus SCO, 1967.
Ikhwān al-Ṣafā, Rasāʾil Ikhwān al-Ṣafāʾ, vol. 4. Beirut 1970 (reprint of: Ikhwān al-Ṣafāʾ, *Rasāʾil Ikhwān al-Ṣafāʾ wa-khullān al-wafāʾ*, vol. 4, ed. by Buṭrus al-Bustānī. Beirut: Dār Ṣādir, 1957).
Isebaert-Cauuet, Isabelle, "Les pères dans les commentaires syriaques." in *Les Pères grecs dans la tradition syriaque*, ed. by Andrea Schmidt and Dominique Gonnet, pp. 77–88, Études syriaques 4. Paris: Geuthner, 2007.
Īshūʿdād of Merv, *Commentaire d'Iṣoʿdad de Merv sur l'Ancient Testament: I. Genèse.*, ed. by Jaques Marie Vosté and Ceslas van den Eynde, trans. by C. van den Eynde. Louvain: SECRÉTARIAT DU CORPUS SCO, 1950, 1955 (I Syriac, II French).
Jacob of Serugh, *Homiliae Selectae Mar-Jacobi Sarugensis*, 5 vol., ed. by Paul Bedjan. Paris: Harrassowitz, 1905–1910.
James, John Courtenay, *The Language of Palestine and Adjacent Regions*. Edinburgh: T. & T. Clark, 1920.
Japhet, Sara, *I & II Chronicles: A Commentary*, Old Testament library. London: SCM Press, 1993.
—, *The Ideology of the Book of Chronicles and its Place in Biblical Thought*, 2nd edition, Beiträge zur Erforschung des Alten Testaments und des antiken Judentums 9. Frankfurt am Main: Peter Lang, 1997.
Jeremias, Joachim, *Die Gleichnisse Jesu*, 11th edition. Göttingen: Vandenhoeck and Ruprecht, 1998.
Jirjis, Murqus, ed., *Aṣ-Ṣafī ibn al-ʿAssāl: Kitāb aṣ-Ṣaḥāʾiḥ fī jawāb an-naṣāʾiḥ & kitāb nahj as-sabīl fī takhjīl muḥarrifī l-injīl*. Cairo: Maṭbaʿat ʿAyn Shams, 1926.
Jones, Alan, *The Qurʾān Translated into English*. Cambridge: Gibb Memorial Trust, 2007.
Josephus, Flavius, *Josephus: in nine volumes with an English translation by Henry St. John Thackeray, Jewish Antiquities. Books V–VIII*, vol. 5, trans. by Henry St. John Thackeray and Ralph Marcus. London: Heinemann, 1934.
—, *Josephus: Jewish Antiquities, books I–IV*, ed. and trans. by Henry St. John Thackeray, Loeb classical library 242. Cambridge, MA: Harvard University Press, 1967.
—, *The Jewish War, IV-VII*. Cambridge, MA: Harvard University Press, 1928 (rep. 1961).
Jüngling, Hans-Winfried, "Das Buch Jesaja." in *Einleitung in das Alte Testament*, ed. by Erich Zenger et al., 5th edition, pp. 427–451, Kohlhammer Studienbücher Theologie 1/1. Stuttgart: Kohlhammer, 2004.

Jullien, Florence, ed., *Le monachisme syriaque*, Études syriaques 7. Paris: Geuthner, 2010.
Al-Juwaynī, Abū l-Maʿālī ʿAbd al-Malik, *Shifāʾ al-ghalīl fī l-tabdīl; lumaʿ fī ʿawāʾid ahl al-sunna: Textes apologétiques de Ǧuwainî (d. 1085)*, ed. and trans. by Michel Allard. Beirut: Dar al-Machreq, 1968.
Juynboll, Gautier H. A., "al-Suddī." in *Encylopaedia of Islam*, vol. 9, pp. 762. Leiden: Brill, 1997.
Kadari, Adiel, "'All drink from the same fountain': The Initial Acceptance of the Halakhot of Eldad Ha-Dani into the Halakhic Discourse." *Review of Rabbinic Judaism* 13, 2 (2010): pp. 211–228.
—, "Talmud Torah in Seder Eliyahu: The Ideological Doctrine in its Socio-historical Context." *Daat* 50–52 (2003): pp. 35–59 (in Hebrew).
Kalimi, Isaac, "The Contribution of the Literary Study of Chronicles to the Solution of its Textual Problems." *Tarbiz* 62 (1992–1993): pp. 471–486.
—, *Zur Geschichtsschreibung des Chronisten – Literarisch-historiographische Abweichungen der Chronik von ihren Paralleltexten in den Samuel-und Koenigsbuecher*, Beihefte zur Zeitschrift für die alttestamentliche Wissenschaft 226. Berlin: de Gruyter, 1994.
Kalmin, Richard L. and Schwartz, Seth, eds., *Jewish Culture and Society Under the Christian Roman Empire*, Interdisciplinary Studies in Ancient Culture and Religion 3. Leuven: Peeters, 2003.
Kannengiesser, Charles, *Handbook of Patristic Exegesis: The Bible in Ancient Christianity: With special contributions by various scholars*, Bible in Ancient Christianity 1. Leiden and Boston: Brill, 2006.
Kaplan, Steven, "Dominance and Diversity: Kingship, Ethnicity, and Christianity in Orthodox Ethiopia." *Church History* 89/1–3 (2009): pp. 291–305.
Kasher, Rimon and Zipor, Moshe, eds., *Studies in Bible and Exegesis Vol. VI: Yehuda Otto Komlosh – In Memoriam*, Bar-Ilan Department Researches. Ramat-Gan: Bar-Ilan University Press, 2002.
Keim, Katharina, *Pirqei de Rabbi Eliezer: Structure, Coherence, Intertextuality*, Ancient Judaism and Early Christianity 96. Leiden: Brill, 2016.
Kennedy, Philip F., "Preface." in *On Fiction and Adab in Medieval Arabic Literature*, ed. by Philip F. Kennedy, pp. i–xii, Studies in Arabic Language and Literature 6. Wiesbaden: Harrassowitz, 2005.
Kermani, Navid "The Aesthetic Reception of the Qurʾān as Reflected in Early Muslim History." in *Literary Structures of Religious Meaning in the Qur'ān*, ed. by Issa J. Boullata, pp. 255–276, Curzon studies in the Qur'an. Richmond Surrey: Curzon Press, 2000.
Kerr, Robert M, "Von der aramäischen Lesekultur zur arabischen Schreibkultur II. Der aramäische Wortschatz des Koran." in *Die Entstehung einer Weltreligion II. Von der koranischen Bewegung zum Frühislam*, ed. by Markus Groß and Karl-Heinz Ohlig, pp. 553–614, Inārah 6. Berlin: Schiler, 2012.
Khalek, Nancy, *Damascus after the Muslim Conquest: Text and Image in Early Islam*. New York: Oxford University Press, 2011.
Khalifé, Ignace-Abdo and Kutsch, Wilhelm, "Ar-Radd ʿalā-n-naṣārā de ʿAlī aṭ-Ṭabarī." *Mélanges de l'Université Saint Joseph* 36 (1959): pp. 3–36.
Khan, Geoffrey, "Abu al-Faraj Harun and the early Karaite grammatical tradition." *The Journal of Jewish Studies* 48 (1997): pp. 314–334.

—, "Al-Qirqisani's opinions concerning the text of the Bible and parallel Muslim attitudes towards the text of the Qur'an." *The Jewish Quarterly Review* 81/1–2 (1990): pp. 59–73.

—, ed., *Exegesis and Grammar in Medieval Karaite* Texts, Journal of Semitic Studies Supplement 13. Oxford: Oxford University Press, 2001.

—, "The Book of Hebrew Grammar by the Karaite Joseph Ben Noah." *Journal of Semitic Studies* 43/2 (1998): pp. 265–286.

—, "The Early Eastern Traditions of Hebrew Grammar." in *Hebrew Scholarship and the Medieval World*, ed. by Nicholas De Lange, pp. 77–91. Cambridge: Cambridge University Press, 2001.

—, *The Early Karaite Tradition of Hebrew Grammatical Thought: Including a Critical Edition, Translation and Analysis of the Diqduq of·Abū Yaʿqūb ibn Nūḥ on the* Hagiographa, Studies in Semitic Languages and Linguistics 32. Leiden: Brill, 2000.

—, "The Karaite Tradition of Hebrew Grammatical Thought" in *Hebrew Study from Ezra to Ben-Yehuda*, ed. by William Horbury, pp. 186–205. Edinburgh: T&T Clark, 1999.

Khoury, Paul, ed. and trans., *Paul d'Antioche, évêque Melkite de Sidon (XIIe. S.)*, Université Saint-Joseph/Institut de lettres orientales, Recherches 24. Beirut: Imprimerie Catholique, 1964.

Khoury, Raif Georges, "Wahb b. Munabbih." in *Encylopaedia of Islam*, vol. 11, 2nd edition, pp. 34–36. Leiden: Brill, 2004.

Kinberg, Leah, "*Muḥkamāt* and *Mutashābihāt* (Koran 3/7): Implication of a Koranic Pair of Terms in Medieval Exegesis." *Arabica* 35 (1988): pp. 143–172.

Al-Kisāʾī, Muḥammad b. ʿAbdillāh, *Qiṣaṣ al-anbiyāʾ*, ed. by Isaac Eisenberg, vol. 2. Leiden: Brill, 1923.

Kitzinger, Ernst, *Byzantine Art in the Making: Main Lines of Stylistic Development in Mediterranean Art, 3rd to 7th Century.* London: Faber And Faber, 1977.

Kohlbacher, Michael, "Wessen Kirche ordnete das Testamentum Domini Nostri Jesu Christi? Anmerkungen zum historischen Kontext von CPG 1743." in *Zu Geschichte, Theologie, Liturgie und Gegenwartslage der syrischen Kirchen. Ausgewählte Vorträge des deutschen Syrologen-Symposiums vom 2.–4. Oktober 1998 in Hermannsburg*, ed. by Martin Tamcke and Andreas Heinz, pp. 55–137. Studien zur Orientalischen Kirchengeschichte 9. Münster, Hamburg and London: LIT, 2000.

Kotter, Bonifatius, ed., *Die Schriften des Johannes von Damaskos*, vol. 4, Patristische Texte und Studien 22. Berlin: de Gruyter 1981.

Kruisheer, Dirk, "Reconstructing Jacob of Edessa's Scholia." in *The Book of Genesis in Jewish and Oriental Christian Interpretation: A Collection of Essays*, ed. by Judith Frishman and Lucas van Rompay, pp. 187–196, Traditio Exegetica Graeca 5. Leuven: Peeters, 1997.

Kulik, Alexander, *Retroverting Slavonic Pseudepigrapha: Toward the Original of the Apocalypse of Abraham*, Text-critical Studies 3. Leiden: Brill, 2004.

Kulp, Joshua, "Review Essay. Reading the Bavli with Monks and Zoroastrians." *Prooftexts* 33, 3 (2013): pp. 381–94.

Lamoreaux, John C., *Theodore Abū Qurrah*, trans. by John C. Lamoreaux, Library of the Christian East 1. Provo: Brigham Young University Press, 2005.

Landau, Brent Christopher, *The Sages and the Star-Child: An Introduction to the Revelation of the Magi, an Ancient Christian Apocryphon.* Ph.D. thesis Harvard University, 2008.

Langerman, Y. Tzvi, "On the Beginnings of Hebrew Scientific Literature and on Studying History through 'Maqbiloṯ' (Parallels)." *Aleph* 2 (2002): pp. 169–189.

Lapin, Hayim, "Aspects of the Rabbinic Movement in Palestine 500–800 C.E." in *Shaping the Middle East: Jews, Christians, and Muslims in an Age of Transition 400–800 C.E.*, ed. by Kenneth G. Holum and Hayim Lapin, pp. 181–194, Studies and Texts in Jewish History and Culture 20. Bethesda: University Press of Maryland, 2011.

—, *Rabbis as Romans: The Rabbinic Movement in Palestine, 100–400 CE.* New York: Oxford University Press, 2012.

Lasine, Stuart, "Guest and Host in Judges 19: Lot's Hospitality in an Inverted World." *Journal for the Study of the Old Testament* 29 (1984): pp. 37–59.

Lassner, Jacob, *Jews, Christians, and the Abode of Islam: Modern Scholarship, Medieval Realities.* Chicago: University of Chicago Press, 2012.

Lazarus-Yafeh, Hava, *Interwined Worlds: Medieval Islam and Bible Criticism.* Princeton: Princeton University Press, 1992 (reprint: Princeton University Press, 2014).

—, "ʿUzayr." in *Encyclopaedia of Islam*, vol. 10, p. 960. Leiden: Brill, 2000.

Leder, Stefan, "The Literary Use of *Khabar.* A Basic Form of Historical Writing." in *The Byzantine and Early Islamic Near East*, ed. by Lawrence Conrad and Averil Cameron, pp. 277–315, Studies in Late Antiquity and Early Islam 1. Princeton: Darwin Press, 1992.

Lehmhaus, Lennart, "*Blessed be He* – Benedictions, Prayers, and Narrative in the Garb of Late Midrashic traditions." in *"It's Better to Hear the Rebuke of the Wise Than the Song of Fools" (Qoh 7:5): Proceedings of the Midrash Section, Society of Biblical Literature,* Volume 6, ed. by Rivka Ulmer and David W. Nelson, pp. 107–151, Judaism in Context 18. Piscataway, NJ: Gorgias Press, 2015.

—, *Derekh Eretz im Tora – Seder Elijahu Zuta als universale, religiöse Ethik für rabbinische und nicht-rabbinische Juden.* Dissertation Martin-Luther-Universität Halle-Wittenberg, 2012/13.

—, "'Es ist mancher scharfsinnig, aber ein Schalk und kann die Sache drehen, wie er es haben will.' (Sirach 19,22) – Intertextuelle Kritik rabbinischer Quellenarbeit im Alphabet des Ben Sira." in *Literatur im Dialog: die Faszination von Talmud und Midrasch*, ed. by Susanne Plietzsch, pp. 127–163. Zürich: Theologischer Verlag, 2007.

—, "Jewish *Adab?* – Midrashic Traditions in Early Islamicate Time Between Narrative and Ethics." (forthcoming).

—, "*Listenwissenschaft* and the Encyclopedic Hermeneutics of Knowledge in Talmud and Midrash." in *In the Wake of the Compendia: Infrastructural Contexts and the Licensing of Empiricism in Ancient and Medieval Mesopotamia,* ed. by John C. Johnson, pp. 59–100, Science, Technology, and Medicine in Ancient Cultures 3. Berlin/New York: de Gruyter, 2015.

—, "Making Others Ourselves(')—Rabbinic Identity Formation in Contexts of Challenge: The Tradition of Seder Eliyahu." (forthcoming)

—, "'The Birth of the Author'—literary strategies and discursive developments in later midrashic texts." (forthcoming)

—, "Ways of Wisdom—The Medieval Ben Sira Tradition as a Bildungsroman in its Historical Context." in *Proceeding of 15th World Congress of Jewish Studies 2009, Jerusalem.* (forthcoming)

—, "'Were not Understanding and Knowledge Given to you from Heaven?' Minimal Judaism and the Unlearned 'Other' in Seder Eliyahu Zuta." *Jewish Studies Quarterly* 19/3 (2012): pp. 230–258.

Leslau, Wolf, *Comparative Dictionary of Geʿez (Classical Ethiopic)*.Wiesbaden: Harrassowitz, 1991.

Le Syrien, Michel, *Chronique de Michel le Syrien, Patriarche jacobite d'Antioche (1166–1199)*, ed. by Jean-Baptiste Chabot, 4 vols. Paris: Ernest Leroux, 1899, 1901, 1905, 1910.

Levenson, Jon, *Inheriting Abraham: The Legacy of the Patriarch in Judaism, Christianity, and Islam*, Library of Jewish ideas. Princeton: Princeton University Press, 2012.

Lewin, Menashe B., ed., *Otzar ha-Geʾonim* 4/2: tractate *Ḥagiga*. Jerusalem: The Hebrew University Press Association, 1931.

Lewis, Bernard, *The Jews of Islam*. Princeton: Princeton University Press, 2014.

Libson, Gideon, "Halakha and Reality in the Geonic Period: Taqqanah, Minhag, Tradition and Consensus: Some Observations." in *The Jews of Medieval Islam: Community, Society and Identity*, ed. by Daniel Frank, pp. 67–99, Etudes sur le judaïsme médiéval 16. Leiden: Brill, 1995.

Lightfoot, Joseph B., "Epistle to Diognetus." in *The Apostolic Fathers: Revised Texts with Short Introductions and English Translations*, ed. and completed by J. R. Harmer, pp. 490–511. London: Macmillan, 1891.

Livne-Kafri, Ofner, "Early Muslim Ascetics and the World of Christian Monasticism." *Jerusalem Studies in Arabic and Islam 20* (1996): pp. 105–129.

Löfgren, Oscar, *Det Apokryfiska Johannesevangeliet: I översättning från den ende kända arabiska handskriften i Ambrosiana*. Stockholm: Natur och Kultur, 1967.

—, "Ein unbeachtetes apokryphes Evangelium." *Orientalische Literaturzeitung* 4 (1943): pp. 153–159.

Lössl, Josef and John W. Watt, eds., *Interpreting the Bible and Aristotle in Late Antiquity: the Alexandria Commentary Tradition between Rome and Baghdad*. Farnham: Ashgate, 2011.

Lord Byron, *The Soul's Pilgrimage: A Poem*. Cambridge: Metcalfe, 1818.

Luisier, Philippe, "De Pilate chez les Coptes." *Orientalia Christiana Periodica* 62 (1996): pp. 411–425.

MacIntyre, Alasdair, *Whose Justice? Which Rationality?* Notre Dame: University of Notre Dame Press, 1988

MacLagan, Eric, "An Early Christian Ivory Relief of the Miracle of Cana." *The Burlington Magazine for Connoisseurs* 38/217 (1921): pp. 178–195.

Al-Majlisī, Muḥammad Bāqir, *Biḥār al-Anwār*. Teheran: Dār al-Kutub al-Islāmiyya, 1992.

Mann, Jacob, *Texts and Studies in Jewish History and Literature*, vol. 2. Cincinnati: Hebrew Union College Press, 1935 (reprint: New York: KTAV, 1972).

Markschies, Christoph and Jens Schröter, eds., in collaboration with Andreas Heiser, *Antike christliche Apokryphen in deutscher Übersetzung. I. Band in zwei Teilbänden: Evangelien und Verwandtes. 7. Auflage der von Edgar Hennecke begründeten und von Wilhelm Schneemelcher fortgeführten Sammlung der neutestamentlichen Apokryphen*. Tübingen: Mohr Siebeck, 2012.

Marogy, Amal Elesha, *The Foundations of Arabic Linguistics: Sībawayhi and Early Arabic Grammatical Theory*, Studies in Semitic Languages and Linguistics 65. Leiden: Brill, 2012.

Marquet, Yves, "Ikhwān al- Ṣafāʾ." in Encyclopaedia of Islam, vol. 3, pp. 1071–7106. Leiden: Brill 1960.

Marshall, David, *God, Muhammad and the Unbelievers: A Qurʾanic Study*. Richmond: Curzon Press, 1999.

Martens, Peter W., *Origen and Scripture: The Contours of the Exegetical Life*, Oxford Early Christian Studies. Oxford: Oxford University Press, 2012.

Marx, Michael, Angelika Neuwirth and Nicolai Sinai, eds., *The Qur'an in Context: Historical and Literary Investigations Into the Qur'anic* Milieu, Texts and Studies on the Qur'ān 6. Leiden: Brill, 2010.

Menn, Esther, "Inner-Biblical Exegesis in the Tanak." in *A History of Biblical Interpretation*, ed. by Alan J. Hauser and Duane F. Watson, pp. 55–79. Grand Rapids: Eerdmans, 2003.

Mescherskaya, Elena N., "'L'Adoration des Mages' dans l'apocryphe syriaque *Histoire de la Vierge Marie*." in *Sur les pas des Araméens chrétiens: mélanges offerts à Alain Desreumaux*, ed. by Françoise Briquel-Chatonnet and Muriel Debié, pp. 95–100, Cahiers d'Études Syriaques 1. Paris: Paul Geuthner, 2010.

Michel, Thomas, *A Muslim Theologian's Response to Christianity: Ibn Taymiyya's Al-Jawāb al-Ṣaḥīḥ*, Studies in Islamic Philosophy and Science. New York: Caravan Books, 1984.

Midrash Bereshit Rabba, ed. by Julius Theodor and Chanoch Albeck. Critical Edition with Notes and Commentary. Jerusalem: Wahrmann Books, 1965.

Midrash Bereschit Rabba, ed. by Julius Theodor and Chanoch Albeck. Jerusalem: Shalem Books, 1996.

Midrash Tanḥuma on the pericope Emor, section 2, ed. by Solomon Buber. Vilna: Verlag Wittwe und Gebrüder Romm, 1885.

Midrash Tanḥuma: Translated into English with Introduction, Indices, and Brief Notes by John T. Townsend. Hoboken, NJ: KTAV Publishing House, 1989.

Midrash Tanḥuma: Translated into English with Introduction, Indices, and Brief Notes (S. Buber Recension), trans. by John T. Townsend, vol. 2. Hoboken, NJ: KTAV, 1997.

Midrash Wayyikra Rabbah, ed. by Mordecai Margulies, 3rd printing. New York: Jewish Theological Seminary of America, 1993.

Mikhail, Maged S. A., *From Byzantine to Islamic Egypt: Religion, Identity and Politics after the Arab Conquest*, Library of Middle East history 45. London and New York: I. B. Tauris, 2014.

Milikowsky, Chaim, "Which Gehenna? Retribution and Eschatology in the Synoptic Gospels and in Early Jewish Texts." *New Testament Studies* 34 (1988): pp. 238–249.

Miller, Stuart S., *Sages and Commoners in Late Antique 'Erez Israel: A Philological Inquiry into Local Traditions in Talmud* Yerushalmi, Texts and Studies in Ancient Judaism 111. Tübingen: Mohr Siebeck, 2006.

Mingana, Alphonse, "Syriac Influence on the Style of the Kur'ān." *Bulletin of The John Rylands Library* 11/1 (1927): pp. 77–98.

—, ed. and trans., *Woodbrooke Studies: Christian Documents in Syriac, Arabic, and Garshūni*, vol. 2. Cambridge: Heffer & Sons, 1928.

—, "Woodbroke Studies: Christian Documents in Syriac, Arabic, and Garshūni." in: *The Early Christian-Muslim Dialogue: A Collection of Documents from the First Three Islamic Centuries (632–900 A.D.). Translations with Commentary*, ed. by N. A. Newman, pp. 169–267. Hatfield: Interdisciplinary Biblical Research Institute, 1993.

Mir, Mustansir, *Coherence in the Qur'an: A study of Iṣāḥī's concept of naẓm in Tadabbur-i Qur'an*. Indianapolis: American Trust Publications, 1986.

—, "The Sūra as Unity. A Twentieth Century Development in Qur'ān-exegesis." in *Approaches to the Qur'ān*, ed. by Gérard R. Hawting and Abdul-Kader A. Shareef, pp. 211–224, Routledge-SOAS Series on Contemporary Politics and Culture in the Middle East. London: Routledge, 1993.

—, "Unity of the Text of the Qurʾan." in *Encyclopedia of the Qur'ān* 5, ed. by Jane D. McAuliffe, pp. 405–406. Leiden: Brill, 2006.

Moawad, Samuel, "Yūḥannā ibn Wahb." in *Christian-Muslim Relations: A Bibliographical History*, vol 4, ed. by David Thomas, Barbara Roggema and Monferrer Sala, pp. 317–319. The history of Christian-Muslim relations 20. Leiden: Brill 2013.

Momigliano, Arnaldo, "Pagan and Christian Historiography in the Fourth Century A.D." in *Conflict between Paganism and Christianity in the Fourth Century*, ed. by Arnaldo Momigliano, pp. 77–99, Oxford-Warburg studies. Oxford: Clarendon Press, 1963.

Monferrer-Sala, Juan Pedro, "An Arabic-Muslim Quotation of a Biblical Text: Ibn Kathīr's *al-Bidāya wa-l-Nihāya* and the Construction of the Ark of the Covenant." in *Studies on the Christian Heritage: In Honour of Father Prof. Dr. Samir Khalil Samir, S.I. at the Occasion of His Sixty-Fifth Birthday*, ed. by Rifaat Ebied and Herman Teule, pp. 263–278, Eastern Christian studies 5. Leuven: Peeters, 2004.

—, "Sacred Readings, Lexicographic Soundings: Cosmology, Men, Asses and Gods in the Semitic Orient." in *Sacred Text: Explorations in Lexicography*, ed. by Juan Pedro Monferrer-Sala and A. Urbán, pp. 201–218. Frankfurt am Main: Peter Lang, 2009.

—, "'Texto', 'subtexto' e 'hipotexto' en el 'Apocalipsis del Pseudo Atanasio' copto-árabe." in *Legendaria Medievalia en honor de Concepción Castillo Castillo*, ed. by Raif Georges Khoury, J.P. Monferrer-Sala and Mª J. Viguera Molins, pp. 427–428, Horizonte de al-Andalus 1. Córdoba: Ediciones el Almendro – Fundación Paradigma Córdoba, 2011.

—, "'The Antichrist is Coming…' The Making of an Apocalyptic *topos* in Arabic (Ps.-Athanasius, Vat. ar. 158 / Par. Ar. 153/32)." in *Bibel, Byzanz und christlicher Orient: Festschrift für Stephen Gerö zum 65. Geburtstag*, ed. by Dmitrij Bumazhnov, Stephen Gero, Emmanouela Grypeou et al., pp. 674–675, Orientalia Lovaniensia analecta 187. Louvain: Peeters, 2011.

Montgomery, James E., *Al-Jahiz: In praise of* books, Edinburgh Studies in Classical Arabic Literature. Edinburgh: Edinburgh University Press, 2013.

Montgomery, Scott L., *Science in Translation: Movements of Knowledge through Cultures and Time*. Chicago: University Of Chicago Press, 2000.

Moraldi, Luigi, *Vangelo Arabo apocrifo dell'Apostolo Giovanni da un Manoscritto della Biblioteca Ambrosiana*, Biblioteca di Cultura Medievale. Milan: Editoriale Jaca Book, 1991.

Morgenthaler, Robert, *Statistik des neutestamentlichen Wortschatzes*. Zürich and Frankfurt a.M.: Gotthelf-Verlag, 1958.

Moulton, William F. and Geden, Alfred S., eds., *A Concordance to the Greek Testament*. Edinburgh: T. & T. Clark, 1897.

Al-Muḥāsibī, al-Ḥārith b. Asad, *al-Waṣāya (al-Qaṣd wa-r-rujūʿ ilā llāh – Badʾ man anāba ilā llāh – Fahm aṣ-Ṣalāh – at-Tawahhum)*, ed. by ʿAbd al-Qādir A. ʿAṭā. Beirut: Dar al-Kutub Al-ʿIlmiyya, 1986.

—, *Kitāb ar-Riʿāya li-ḥuqūq Allāh*, ed. by Margaret Smith. London: Luzac & Co, 1940.

—, *Kitāb Fahm al-Qurʾān*. Beirut: Dār al-Kindī lil-Ṭibāʿa wa-l-Nashr wa-l-Tawzīʿ, 1978.

Munier, Charles, "La femme de Lot dans la literature juive et chrétienne des premiers siècles." in *Figures de l'Ancien Testament chez les Pères*, ed. by Pierre Maraval, pp. 123–142, Cahiers de Biblia patristica 2. Strasbourg: Centre d'analyse et de documentation patristiques, 1989.

Murad, Suleiman Ali, *Early Islam between Myth and History: Al-Ḥaṣan Al-Baṣrī (d. 110H/728CE) and the Formation of His Legacy in Classical Islamic Scholarship*. Leiden: Brill, 2006.

Muraoka, Takamitsu, *A Greek-English Lexicon of the Septuagint*. Louvain: Peeters, 2009.

—, *A Greek-Hebrew/Aramaic Two-way Index to the Septuagint*. Louvain: Peeters, 2010.

Nagel, Tilman, "al-Kisāʾī." in *Encyclopaedia of Islam*, vol. 5, p. 176. Leiden: Brill, 1986.

—, *The History of Islamic Theology: From Muhammad to the Present*, trans. by Thomas Thornton, Princeton Series on the Middle East. Princeton: Markus Wiener Publishers, 2010.

Nassif, Bassam A., "Religious Dialogue in the Eighth Century." *Parole de l'Orient* 30 (2005): pp. 333–340.

Nau, François, "Lettre de Jacques d'Édesse sur la généalogie de la sainte Vierge." *Revue de l'Orient Chrétien* 6 (1901): pp. 512–531.

—, "Traduction des lettres XII et XIII de Jacques d'Édesse (exégèse biblique)." *Revue de l'Orient chrétien* 10 (1905): pp. 197–208, 258–282.

Nautin, Pierre and Marie-Thérèse Nautin, eds., *Origen, Homélies sur Samuel*, Sources chrétiennes 328. Paris: Éditions du Cerf, 1986.

Negev, Avraham and Shimon Gibson, eds., *Archaeological Encyclopedia of the Holy Land*. Nashville: T. Nelson, 1986.

Nemoy, Leon, *Karaite Anthology*, Yale Judaica Series 7. New Haven: Yale University Press, 1952.

Netton, Ian R., *Muslim Neoplatonists: An Introduction to the Thought of the Brethren of Purity (Ikhwān al-Ṣafāʾ)*, Islamic Surveys 19. Edinburgh: Edinburgh University Press, 1991.

Neusner, Jacob, *Chapters in the Formative History of Judaism: Some Current Essays on the History, Literature, and Theology of Judaism*, Studies in Judaism. Lanham, MD: University Press Of America, 2010.

Neuwirth, Angelika, *Der Koran als Text der Spätantike: Ein europäischer Zugang*. Frankfurt a. M.: Verlag der Weltreligionen, 2010.

—, "Form and Structure of the Qur'ān." in *Encyclopedia of the Qurʾān* 2, ed. by Jane D. McAuliffe, pp. 245–266. Leiden: Brill, 2002.

—, "Meccan Texts – Medinan Additions? Politics and the Re-reading of Liturgical Communications." in *Words, Texts, and Concepts Cruising the Mediterranean Sea: Studies on the Sources, Contents, and Influences of Islamic Civilization and Arabic Philosophy and Science*, ed. by Rüdiger Arnzen and Jörn Thielmann, pp. 71–93, Orientalia Lovaniensia Analecta 139. Leuven: Peeters, 2004.

—, "Referentiality and Textuality in *Sūrat al-Ḥijr* (Q. 15): Observations on the Qur'ānic 'Canonical Process' and the Emergence of a Community." in *Scripture, Poetry and the Making of a Community: Reading the Qurʾan as a Literary Text*, ed. by Angelika Neuwirth, pp. 184–215, Qur'anic Studies Series 10. Oxford: Oxford University Press, 2014.

—, ed., *Scripture, Poetry and the Making of a Community: Reading the Qurʾan as a Literary* Text, Qurʾanic Studies Series 10. Oxford: Oxford University Press, 2014.

—, *Studien zur Komposition der mekkanischen Suren: Die literarische Form des Koran-ein Zeugnis seiner Historizität?*, 2nd edition, Studien zur Geschichte und Kultur des islamischen Orients 10 Berlin: de Gruyter, 2007.
—, "Sūra(s)." in *Encyclopedia of the Qur'ān* 5, ed. by Jane D. McAuliffe, pp. 166–177. Leiden: Brill, 2006
—, "Vom Rezitationstext über die Liturgie zum Kanon: Zur Entstehung und Wiederauflosung der Surenkomposition im Verlauf der Entwicklung eines Islamischen Kultus." in *The Qur'an as text*, ed. by Stefan Wild, pp. 69–105, Islamic Philosophy, Theology and Science: Texts and Studies 27. Leiden: Brill, 1996.
New Revised Standard Version Bible. New York: National Council of Churches of Christ in the U.S.A., 1989.
Nickelsburg, George W. E. and James C. VanderKam., eds., *1 Enoch: A New Translation.* Minneapolis: Fortress Press, 2004.
Nihan, Christophe L., "1Samuel 28 and the Condemnation of Necromancy in Persian Yehud." in *Magic in the Biblical World: From the Rod of Aaron to the Ring of Solomon*, ed. by Todd E. Klutz, pp. 23–54, Journal for the Study of the New Testament/Supplement Series 245 London: T & T Clark International, 2003.
Nöldeke, Theodor and Schwally, Friedrich, *Geschichte des Qorāns, vol. 1: Über den Ursprung des Qorāns.* Leipzig: Dieterich'sche Verlagsbuchhandlung, 1909.
Norelli, Enrico, *Ascensio Isaiae: Commentarius*, Corpus Christianorum Series Apocryphorum 8. Turnhout: Brepols, 1995.
Olszowy-Schlanger, Judith, "The Science of Language among Medieval Jews." in *Science in Medieval Jewish Culture*, ed. by Gad Freudenthal, pp. 359–424. Cambridge: Cambridge University Press, 2011.
Orbe, Antonio, "Los hechos de Lot, mujer e hijas vistos por S. Ireneo (Adv. Haer. IV,31,1,15/3.71)." *Gregorianum* 75/1 (1994): pp. 37–64.
Origen, *Contra Celsum*, trans. by Henry Chadwick. Cambridge: Cambridge University Press, 1953, 1980.
—, *Homilies on Jeremiah; Homily on 1 Kings 28*, trans. by John Clark Smith, The Fathers of the Church 97. Washington, D. C.: Catholic University of America Press, 1998.
Palmer, Andrew, *The Seventh Century in the West-Syrian Chronicles*, Translated Texts for Historians 15. Liverpool: Liverpool University Press, 1993.
Paret, Rudi, "Signification coranique de *ḫalīfa* et d'autre dérivés de la racine *ḫalafa*." *Studia Islamica* 31 (1970): pp. 211–217.
Paul, Shalom M., *Amos: A Commentary on the Book of Amos*, ed. by Frank Moore Cross, Hermeneia. Minneapolis: Fortress Press, 1991.
Parker, Kenneth S., "Coptic Language and Identity in Ayyūbid Egypt." *Al-Masāq* 25 (2013): pp. 222–239.
Payne, Richard, *A State of Mixture: Christians, Zoroastrians, and Iranian Political Culture in Late* Antiquity, Transformation of the Classical Heritage 56. Berkeley: University of California Press, 2015.
—, Myles Lavan and John Weisweiler, eds., *Cosmopolitanism and Empire: Universal Rulers, Local Elites, and Cultural Integration in the Ancient Near East and Mediterranean*, Oxford Studies in Early Empires. Oxford: Oxford University Press, 2016.
Peck, Harry Thurston, *Harpers Dictionary of Classical Antiquities.* New York: Harper and Brothers, 1898.

Pelegrini, Silvia, “Das Protevangelium des Jakobus.” in *Antike christliche Apokryphen in deutscher Übersetzung. I. Band in zwei Teilbänden: Evangelien und Verwandtes. 7. Auflage der von Edgar Hennecke begründeten und von Wilhelm Schneemelcher fortgeführten Sammlung der neutestamentlichen Apokryphen*, ed. by Christoph Markschies and Jens Schröter, in collaboration with Andreas Heiser, I.2, pp. 903–929. Tübingen: Mohr Siebeck, 2012.

Penn, Michael P., *Envisioning Islam: Syriac Christians and the Early Muslim World*, Divinations: Rereading Late Ancient Religion. Philadelphia: University of Pennsylvania Press, 2015.

Peters, Francis E., *Aristotle and the Arabs: The Aristotelian Tradition in Islam*, New York University Studies in Near Eastern Civilization 1. New York: New York University Press, London: University of London Press, 1968.

Phillips, David, “The Reception of Peshitta Chronicles: Some Elements for Investigation.” in *The Peshitta: Its Use in Literature and Liturgy. Papers read at the 3rd Peshitta Symposium*, ed. by Robert Barend ter Haar Romeny, pp. 259–295, Monographs of the Peshitta Institute, Leiden 15. Leiden: Brill, 2006.

Phillips, George, *Scholia on Passages of the Old Testament by Mar Jacob Bishop of Edessa*. London: Williams and Norgate, 1864.

Picken, Gavin, “Ibn Ḥanbal and al-Muhasibi: A Study of Early Conflicting Scholarly Methodologies.” *Arabica* 55 (2008): pp. 337–361.

Pickthall, Marmaduke, *The Meaning of the Glorious Qur'an*. New York: Dorset Press, 1930.

Pietruschka, Ute, “Classical Heritage and New Literary Forms: Literary Activities of Christians during the Umayyad Period.” in *Ideas, images, and methods of portrayal: insights into classical Arabic literature and Islam*, ed. by Sebastian Günther, pp. 17–40, Islamic history and civilization 58. Leiden: Brill, 2005.

Pohlmann, Karl-Friedrich, *Die Entstehung des Korans: Neue Erkenntnisse aus Sicht der historisch-kritischen Bibelwissenschaft*. Darmstadt: Wissenschaftliche Buchgesellschaft, 2012.

Polliack, Meira, “Concepts of Scripture among the Jews of the Medieval Islamic World.” in *Jewish Concepts of Scripture: A Comparative Introduction*, ed. by Benjamin Sommer, pp. 80–101. New York: New York University Press, 2012.

—, “Major Trends in Karaite Biblical Exegesis in the Tenth and Eleventh Centuries.” in *Karaite Judaism: A Guide to its History and Literary Sources*, ed. by Meira Polliack, pp. 363–413, Handbook of Oriental Studies, Section 1—The Near and the Middle East 73. Leiden and Boston: Brill, 2003.

—, “Rethinking Karaism: Between Judaism and Islam.” *Association for Jewish Studies Review* 30/1 (2006): pp. 67–93.

—, “Medieval Karaism” in *The Oxford Handbook of Jewish Studies*, ed. by Martin Goodman, pp. 295–326. Oxford: Oxford University Press, 2005.

—, “The ‘Voice’ of the Narrator and the ‘Voice’ of the Characters in the Bible Commentaries of Yefet ben ʿEli.” in *Birkat Shalom: Studies in the Bible, Ancient Near Eastern Literature, and Postbiblical Judaism Presented to Shalom M. Paul*, ed. by Chaim Cohen ans Shalom M. Paul, vol. 2, pp. 891–915. Winona Lake: Eisenbrauns, 2008.

Poonawala, Ismail K., “Muḥammad ʿIzzat Darwaza's Principles of Modern Exegesis: a Contribution Toward Quranic Hermeneutics.” in *Approaches to the Qur'ān*, ed. by Gérard

R. Hawting and Abdul-Kader A. Shareef, pp. 225–246, SOAS Series on Contemporary Politics and Culture in the Middle East. London: Routledge, 1993.

Pourjavady, Reza and Sabine Schmidtke, *A Jewish Philosopher of Baghdad: ʿIzz al-Dawla Ibn Kammūna (d. 683/1284) and His Writings*, Islamic Philosophy, Theology, and Science 65. Leiden: Brill, 2006.

Pregill, Michael, "Methodologies for the Dating of Exegetical Works and Traditions: Can the Lost Tafsir of al-Kalbi Be Recovered from Tafsir Ibn Abbas (also known as *al-Wāḍiḥ*)?" in *Aims, Methods and Contexts of Qurʾanic Exegesis (2nd/8th-9th/15th C.)*, ed. by Karen Bauer, pp. 393–453. Oxford: Oxford University Press, 2013.

Provera, Mario E., ed., *Il vangelo arabo dell'infanzia secondo il ms. Laurenziano Orientale (n. 387)*, Quaderni de "La Terra Santa". Jerusalem: Franciscan Printing Press, 1973.

Prudentius, *Peristephanon 10: 651–845*, ed. by Maurice P. Cunningham, Aurelii Prudentii Clementis Carmina, Corpus Christianorum Series Latina 126. Turnhout: Brepols, 1966

Putman, Hans, *L'église et l'Islam sous Timothée I (780–823)*. Beirut: Dar al-Machreq, 1975.

Al-Qarāfī, Shihāb al-Dīn, *al-Ajwiba al-fākhira ʿan al-asʾila al-fājira*, ed. by Bakr Zakī ʿAwaḍ. Cairo: Maktabat Wahba, 1987.

Al-Qirqisānī, Yaʿqūb, *Al-Anwār wa-l-marāqib*, ed. by Leon Neomy, vol. 6. New York: The Alexander Kohut Memorial Foundation, 1941.

Al-Qurṭubī, Abū ʿAbdillāh Muḥammad, *Al-Jāmiʿ li-aḥkām al-Qurʾān*, ed. by ʿAbd al-Razzāq al-Mahdī. Beirut: Dār al-Kitāb al-ʿarabī 1433/2012.

Quṭb, Sayyid, *Fī ẓilāl al-Qurʾān*, 17th edition. Cairo: Dār al-Shurūq, 1412/1992.

—, *In the Shade of the Qurʾān*, ed. and trans. by A. Salahi. Leicester, UK: Islamic Foundation, 2003.

Rahmani, Ignatius Ephrem II, *Testamentum Domini Nostri Jesu Christi*. Mainz, 1899 (reprint Hildesheim: Olms, 1968).

Razaby, Yehuda, "Muvaot ḥadashot mi-perush R. Saadya la-miqra." *Sinai* 88 (2008): pp. 97–108.

Al-Rāzī, Fakhr al-Dīn, *Mafātīḥ al-ghayb*, ed. by Muḥammad Bayḍūn. Beirut: Dār al-Kutub al-ʿIlmiyya, 1421/2000.

—, *Tafsīr*. Beirut: Dār al-Kutub al-ʿIlmiyya, 2000.

Reeves, John C., "Scriptural Authority in Early Judaism." in *Living Traditions of the Bible: Scripture in Jewish, Christian, and Muslim Practice*, ed. by James E. Bowley, pp. 63–84. St. Louis: Chalice, 1999.

Reinink, Gerard J., *Die syrische Apokalypse des Pseudo-Methodius*, Corpus Scriptorum Christianorum Orientalium 540, 541; 220, 221. Louvain: Peeters, 1993.

—, "Ps.-Methodius: a Concept of History in Response to the Rise of Islam." in *The Byzantine and Early Islamic Near East: I. Problems in the Literary Source Material*, ed. by Averil Cameron and Lawrence I. Conrad, pp. 149–187, Studies in Late Antiquity and Early Islam 1. Princeton, NJ: Darwin Press, 1992.

Reynolds, Gabriel Said, "Noah's Lost Son in the Qurʾān." *Arabica* 64 (2017): pp. 1–20.

—, *The Qurʾān and Its Biblical Subtext*. London/New York: Routledge, 2010.

Ritner, Robert K., "Necromancy in Ancient Egypt." in *Magic and Divination in the Ancient World*, ed. by Leda Ciraolo and Jonathan Seidel, pp. 89–96, Ancient Magic and Divination 2. Leiden: Brill 2002.

Robinson, Neal, *Discovering the Qurʾan: A Contemporary Approach to a Veiled Text*. Washington D.C.: Georgetown University Press, 2003.

Römer, Thomas, "Abraham and the 'Law and the Prophets'." in *The Reception and Remembrance of Abraham*, ed. by Pernille Carstens and Niels Peter Lemche, pp. 87–101, Perspectives on Hebrew Scriptures and its Contexts 13. Piscataway, N.J.: Georgia Press, 2011.

Roggema, Barbara, "A Christian Reading of the Qur'ān: the Legend of Sergius-Baḥīrā and its Use of Qur'ān and Sīra." in *Syrian Christians under Islam: the First Thousand Years*, ed. by David Thomas, pp. 57–74. Leiden: Brill, 2001.

Rosenthal, Franz, *The Classical Heritage in Islam*, trans. by Emile and Jenny Marmorstein, Arabic Thought and Culture. London: Routledge, 1994.

Rosenthal, Judah, *Hiwi al-Balkhi: A Comparative Study*. Philadelphia: Dropsie College, 1949.

Ross, William David, *Aristotelis Metaphysica*, Oxford: Oxford University Press, 1957.

Rousseau, Adelin and Doutreleau, Louis, eds., *Irénée de Lyon. Contre les hérésies. Livre I. Tome II*, Sources chrétiennes 264. Paris: Les Éditions du Cerf, 1979.

Rubenson, Samuel, "Translating the Tradition: Some Remarks on the Arabization of the Patristic Heritage in Egypt." *Medieval Encounters* 2 (1996): pp. 4–14.

Rubenstein, Jeffrey, "From Mythic Motifs to Sustained Myth: The Revision of Rabbinic Traditions in Medieval Midrashim." *Harvard Theological Review* 89/2 (1996): pp. 131–159.

Rustow, Marina, "Karaites Real and Imagined: Three Cases of Jewish Heresy." *Past and Present* 197 (2007): pp. 35–74

—, "Laity versus Leadership in Eleventh-Century Jerusalem: Karaites, Rabbanites, and the Affair of the Ban on the Mount of Olives." in *Rabbinic Culture and Its Critics*, ed. by Daniel Frank and Matt Goldish, pp. 195–248. Detroit: Wayne State University Press, 2008.

Sacks, Steven D., *Midrash and Multiplicity: Pirke de-Rabbi Eliezer and the Renewal of Rabbinic Interpretive* Culture, Studia Judaica: Forschungen zur Wissenschaft des Judentum 48. Berlin: de Gruyter, 2009.

Sadan, Joseph, "In the Eyes of the Christian Writer al-Ḥāriṯ ibn Sinān. Poetics and Eloquence as a Platform of Inter-Cultural Contacts and Contrasts." *Arabica* 56 (2009): pp. 1–26.

Sadeghi, Behnam and Uwe Bergmann, "The Codex of a Companion of the Prophet and the Qur'ān of the Prophet." *Arabica* 57 (2010): pp. 343–436.

Saeed, Abdullah, "The Charge of Distortion of Jewish and Christian Scriptures." *The Muslim World* 92 (2002): pp. 419–436.

Sahas, Daniel, "The Arab Character of the Christian Disputation with Islam: The Case of John of Damascus (ca. 655–ca. 749)." in *Religionsgespräche im Mittelalter*, ed. by Bernard Lewis and Friedrich Niewöhner, pp. 185–206. Wiesbaden: Harrasowitz, 1992.

Saleh, Walid, ed., *In Defense of the Bible: A Critical Edition and an Introduction to al-Biqāʿī's Bible Treatise*, Islamic History and Civilization 73. Leiden: Brill, 2008.

Salvesen, Alison, "Scholarship on the Margins: Biblical and Secular Learning in the Work of Jacob of Edessa." in *Syriac Encounters: Papers from the Sixth North American Syriac Symposium, Duke University, 26–29 June 2011*, ed. by Kyle Smith, Emilio Fiano and Maria Doerfler, pp. 327–344, Eastern Christian Studies 20. Leuven: Peeters, 2015.

—, "'*Tradunt Hebraei*': The Problem of the Function and Reception of Jewish Midrash in Jerome," in *Midrash Unbound: Transformations and Innovations*, ed. by Michael Fishbane and Joanna Weinberg, pp. 57–81, The Littman Library of Jewish Civilization. Oxford: Littmann Library, 2013.

Samir, Samir Khalil, "Christian Arabic Literature in the 'Abbasid Period." in *Religion, Learning and Science in the 'Abbasid Period*, ed. by M. L. Young; J. D. Latham and R. B. Serjeant, pp. 446–460. Cambridge: Cambridge University Press, 1990.

—, "*La place d'Ibn-at-Tayyib dans la pensée arabe.*" *Journal of Eastern Christian Studies* 58/3–4 (2006): pp. 177–193.

—, "La réponse d'al-Ṣafī ibn al-ʿAssāl à la refutation des chrétiens de ʿAlī al-Ṭabarī." *Parole de l'Orient* 11 (1983): pp. 281–328.

—, "Nécessité de l'exégèse scientifique. Texte de ʿAbdallâh Ibn at-Tayyib." *Parole de l'Orient*, 5 (1974): pp. 243–279.

—, "The Earliest Arab Apology for Christianity (c. 750)." in *Christian Arabic Apologetics during the Abbasid Period (750–1258)*, ed. by Samir Khalil Samir and Jørgen S. Nielsen, pp. 57–114, Studies in the History of Religions 63. Leiden: Brill, 1994.

Sands, Kristin Z., *Ṣūfī Commentaries on the Qur'ān in Classical Islam*, Routledge Studies in the Qur'an. London: Routledge, 2006.

Al-Sarrāj, Abū Naṣr ʿAbdallāh b. ʿAlī, *The Kitāb al-Lumaʿ fī t-Taṣawwuf of Abū Naṣr ʿAbdallāh b. ʿAlī as-Sarrāj aṭ-Ṭūsī*, ed. by Reynold Alleyne Nicholson. London: Luzac, 1963.

Sarrio Cucarella, Diego R., "Carta a un amigo musulmán de Sidón de Pablo de Antioquia." *Collectanea Christiana Orientalia* 4 (2007): pp. 189–215.

—, *Muslim-Christian Polemics across the Mediterranean: The Splendid Replies of Shihāb al-Dīn al-Qarāfī (d.684/1285)*, History of Christian-Muslim Relations 23. Leiden: Brill, 2015.

Schäfer, Peter, *Jesus in the Talmud*. Princeton: Princeton University Press, 2007.

Schilling, Alexander Markus, *Die Anbetung der Magier und die Taufe der Sāsāniden: Zur Geistesgeschichte des iranischen Christentums in der Spätantike*, Corpus scriptorum christianorum orientalium: Subsidia 120. Louvain: Peeters, 2008.

Schlossberg, Eliezer, "The Commentary of R. Isaac b. Samuel al-Kanzi on the Story of the Witch of ʿEn Dor." *Studies in Bible and Exegesis* 8 (2008): pp. 193–223 (in Hebrew).

Schmidt, Brian B., *Israel's Beneficent Dead: Ancestor Cult and Necromancy in Ancient Israelite Religion and Tradition*, Forschungen zum Alten Testament 11. Tübingen: Mohr Siebeck, 1994.

Schmidt, Werner H., "Die deuteronomistische Redaktion des Amosbuches." *Zeitschrift für die Alttestamentliche Wissenschaft* 77 (1965): pp. 168–193.

Schmidtke, Sabine and Gregor Schwab, eds., *Jewish and Christian reception(s) of Muslim Theology*, Intellectual History of the Islamicate World 2. Leiden: Brill, 2014.

Schmidtke, Sabine, *The Bible in Arabic among Jews, Christians and Muslims*. Intellectual History of the Islamicate World 1. Leiden: Brill, 2013.

—, "The Muslim Reception of Biblical Materials: Ibn Qutayba and His *Aʿlām al-Nubuwwa*." *Islam and Christian-Muslim Relations* 22 (2011): pp. 249–274.

Schmitz, Bertram, *Der Koran: Sure 2 "Die Kuh". Ein religionshistorischer Kommentar*. Stuttgart: Kohlhammer, 2009.

Schnitzler, Hermann, "Kästchen oder fünfteiliges Buchdeckelpaar?" in *Festschrift für Gert von der Osten*, ed. by Horst Keller, Rainer Budde, Brigitte Klesse et al., pp. 24–32. Köln: DuMont 1970.

Schremer, Adiel, *Brothers Estranged: Heresy, Christianity, and Jewish Identity in Late Antiquity*. Oxford: Oxford University Press, 2009.

Schwartz, Seth, *Imperialism and Jewish Society, 200 B.C.E. to 640 C.E.*, Jews, Christians, and Muslims from the Ancient to the Modern World. Princeton/Oxford: Princeton University Press, 2002.

Scott, James C., *Domination and the Arts of Resistance: Hidden Transcripts*. New Haven: Yale University Press, 1990.

Scurlock, Joann, "Death and the Afterlife in Ancient Mesopotamian Thought." in *Civilizations of the Ancient Near East*, vol. 3, ed. by Jack M. Sasson, pp. 1883–1893. New York: C. Scribner's Sons, 1995.

Sells, Michael, *Approaching the Qur'ān: The Early Revelations*. Oregon: White Cloud Press, 1999.

Seybold, Christian Friedrich, *Severus ben el Moqaffa': Historia patriarcharum Alexandrinorum*, Corpus Scriptorum Christianorum Orientalium 52. Paris: Poussielgue, 1904.

—, *Severus ibn al-Muqaffa': Alexandrinische Patriarchengeschichte von S. Marcus bis Michael I 61–767, nach der ältesten 1266 geschribenen Hamburger Handschrift*. Hamburg: Lucas Gräfe, 1912.

Shafi, Muhammad, *Maariful Qur'an*, trans. by M. Shameen and M.W. Razi. Karachi: Maktaba-e-Darul-Uloom, 1996–2003.

Al-Shahrazūrī, Shams al-Dīn, *Rasā'il al-Shajara al-ilāhīya fī 'ulūm al-ḥaqā'iq al-rabbāniyya*, ed. by Najaf-Qulī Ḥabībī, vol. 3. Tehran: Mu'assasa-yi Pazhūhishī-i Ḥikmat u Falsafa-yi Īrān, 2006.

Sharon, Moshe, *Corpus Inscriptionum Arabicarum Palaestinae*, 5 vols., Handbook of oriental studies Abt.1: Der Nahe und Mittlere Osten 30. Leiden and Boston: Brill, 2003.

Shtober, Shim'on, "Ba'alei ha-'ovot be-ferusho shel rav Saadya Gaon la-torah ve-la-navi." *Sinai* 134 (2004): pp. 3–18.

Sidarus, Adel Y., "Essai sur l'âge d'or de la litterature copte arabe (XIIIe-XIVe siècles)." in *Acts of the Fifth International Congress of Coptic Studies, Washington, 12–15 August 1992*, ed. by David W. Johnson, pp. 443–462, Papers from the Sections, Part 2, vol. 2. Rome: C.I.M., 1993.

—, "La pré-renaissance copte arabe du Moyen Âge (deuxième moitié du XIIe/début du XIIIe siècle)." in *Eastern Crossroads: Essays on Medieval Christian Legacy*, ed. by Juan Pedro Monferrer-Sala, pp. 191–216, Gorgias Eastern Christianity studies 1. Piscataway, NJ: Gorgias Press, 2007.

Sike, Heinrich: *Evangelium infantiae: oder ein so genantes apocryphisches Büchlein, worinnen die Wunder-Geschichte unseres Herrn Jesu Christi, welche sich in seiner Kindheit ... begeben, beschrieben werden. Aus d. Arab. ins Latein durch H. Sike u. nun ins Hochteutsche übersetzet von H. A. v. R. Bi.* 1699.

—, *Evangelium infantiae vel liber apocryphus de infantia Salvatoris*. Trajecti ad Rhenum: Halman, 1697.

Silsilat tārīkh al-ābā' al-baṭārikah, 3 vols., 2nd ed. Wādī l-Naṭrūn: Maktabat Dayr al-Suryān, 2011.

Silverstein, Adam, "Haman's Transition from the Jahiliyya to Islam." *Jerusalem Studies in Arabic and Islam* 34 (2008): pp. 285–308.

—, "The Qur'ānic Pharaoh." in *New Perspectives on the Qur'ān: The Qur'ān in Its Historical Context 2*, ed. by Gabriel Said Reynolds, pp. 467–477, Routledge studies in the Qur'an. London: Routledge, 2011.

Simon, Uriel, "A Balanced Story: The Stern Prophet and the Kind Witch." *Prooftexts* 8 (1988): pp. 159–171.
—, "The Contribution of R. Isaac b. Samuel al-Kanzi to the Spanish School of Biblical Interpretation." *Journal of Jewish Studies* 34 (1983): pp. 171–178.
Sinai, Nicolai, *Die Heilige Schrift des Islams: Die wichtigsten Fakten zum* Koran, Herder Spektrum; 6512. Freiburg: Herder, 2012.
—, *Fortschreibung und Auslegung: Studien zur frühen Koraninterpretation.* Diskurse der Arabistik 16. Wiesbaden: Harrassowitz, 2009.
—, "Inner-Qurʾanic Chronology." in *The Oxford Handbook of Qurʾanic Studies*, ed. by Muhammad Abdel Haleem and Mustafa Shah. Oxford: Oxford University Press (forthcoming).
—, "Processes of Literary Growth and Editorial Expansion in Two Medinan Surahs", forthcoming in *Islam and its Past: Jāhiliyya, Late Antiquity, and the Qurʾan*, ed. by Carol Bakhos and Michael Cook, pp. 69–119, Oxford Studies in the Abrahamic Religions. Oxford: Oxford University Press, 2017.
—, "Qurʾānic Self-Referentiality as a Strategy of Self-Authorization." in *Self-Referentiality in the Qurʾān*, ed. by Stefan Wild, pp. 103–134, Diskurse der Arabistik 11. Wiesbaden: Harrassowitz, 2006.
—, "Religious Poetry from the Quranic Milieu: Umayya b. Abī l-Ṣalt on the Fate of the Thamūd." *Bulletin of the School of Oriental and African Studies* 74 (2011): pp. 397–416.
—, "When Did the Consonantal Skeleton of the Quran Reach Closure?" *Bulletin of the School of Oriental and African Studies* 77 (2014): pp. 273–292 and 509–521.
Ska, Jean-Louis, *Introduction to Reading the Pentateuch*, trans. by Pascale Dominique Winona Lake: Eisenbrauns, 2006.
Sklare, David E., *Samuel ben Hofni Gaon and his Cultural World*, Études sur le Judaïsme Médiéval 18. Leiden: Brill, 1996.
—, "Scriptural Questions: Early Texts in Judaeo-Arabic." in *A Word Fitly Spoken: Studies in Mediaeval Exegesis of the Hebrew Bible and the Qurʾān*, ed. by Meir Bar-Asher et al. Jerusalem: The Ben-Zvi Institute, 2007 (in Hebrew).
—, "The Reception of Mu'tazilism among Jews who were not Professional Theologians." *Intellectual History of the Islamicate World* 2 (2014): pp. 18–36.
Skoss, Solomon L., ed., *The Hebrew-Arabic Dictionary of the Bible known as Kitāb Jāmiʿ al-Alfāẓ of David ben Abraham al-Fāsī the Karaite (tenth century)*, vol. 1. New Haven: Yale University Press, 1936.
Slotki, Judah J., *Midrash Rabbah – Leviticus.* London: Soncino Press, 1939.
Smelik, Klaas A. D., "The Witch of Endor: I Samuel 28 in Rabbinic and Christian Exegesis till 800 A.D." *Vigiliae Christianae* 33 (1979): pp. 160–179.
Smelik, Willem F., "The Aramaic Dialect(s) of the Toledot Yeshu Fragments." *Aramaic Studies* 7 (2009): pp. 39–73.
Smith, Robert Payne, *Thesaurus syriacus: Collegerunt Stephanus M. Quatremere et al.*, 2 vols. Oxford: Clarendon Press, 1879, 1901.
Smith, William Robertson, *The Religion of the Semites.* With a New Introduction by Robert A. Segal. New Brunswisk, NJ: Transaction Publishers, 2002.
Smitmans, Adolf, *Das Weinwunder von Kana. Die Auslegung von Jo 2,1–11 bei den Vätern und heute*, Beiträge zur Geschichte der biblischen Exegese 6. Tübingen: Mohr Siebeck, 1966.

Sokoloff, Michael, "The Date and Provenance of the Aramaic Toledot Yeshu on the Basis of Aramaic Dialectology." in *Toledot Yeshu ("The Life Story of Jesus") Revisited. A Princeton Conference*, ed. by Peter Schäfer, Michael Meerson, and Yaacov Deutsch, pp. 13–26, Texts and Studies in Ancient Judaism 143. Tübingen: Mohr Siebeck, 2011.

Sommer, Benjamin, *A Prophet Reads Scripture: Allusion in Isaiah 40–66*, Contraversions: Jews and Other Differences. Stanford: Stanford University Press, 1998.

—, "Exegesis, Allusion and Intertextuality in the Bible: A Response." *Vetus Testamentum* 46 (1996): pp. 479–489.

Speyer, Heinrich, *Die biblischen Erzählungen im Qoran*. Gräfenhainichen: Schulze, 1931 (reprint: Hildesheim: OLMS, 1961, 1988).

Spitaler, Anton, *Die Verszählung des Koran nach islamischer Überlieferung*. München: Verlag der Bayerischen Akademie der Wissenschaften, 1935.

Spivak, Gayatri Chakravorty, "Can the Subaltern Speak?." in *Marxism and the Interpretation of Culture*, ed. by Cary Nelson and Lawrence Grossberg, pp. 271–313. Urbana, IL: University of Illinois Press, 1988.

Stein, Dina, *Maxims, Magic, Myth: A Folkloristic Perspective on Pirkei de-Rabbi Eliezer*. Dissertation Hebrew University of Jerusalem, 1998. Jerusalem: Magnes, 2004.

—, "Pirke de Rabbi Eliezer and Seder Eliyahu: Preliminary Notes on Poetics and Imagined Landscapes." *Jerusalem Studies in Hebrew Literature* 24 (2011): pp. 73–94 (in Hebrew).

Steinschneider, Moritz, *Die arabische Literatur der Juden*. Frankfurt am Main: Kauffmann, 1902.

Stemberger, Günter, *Einleitung in Talmud und Midrasch*. 8th edition. München: C.H. Beck, 1992.

—, *Introduction to the Talmud and Midrash*, trans. by Markus Bockmuehl. Edinburgh: T & T Clark, 1996.

Stern, David and Mark Mirsky, eds., *Rabbinic fantasies*. New Haven: Yale University Press, 1998.

Stern, David, *Parables in Midrash: Narrative and Exegesis in Rabbinic Literature*. Cambridge, Mass.: Harvard University Press, 1991.

—, "Rhetoric and Midrash: The Case of the Mashal." *Prooftexts* 1 (1981): pp. 261–291.

—, "The 'Alphabet of Ben Sira' and the Early History of Parody in Jewish Literature." in *The Idea of Biblical Interpretation*, ed. by Hindy Najman and Judith H. Newman, pp. 423–448, Supplements to the Journal for the Study of Judaism 83. Leiden and Boston: Brill, 2004.

Stillman, Norman A., *The Jews of Arab Lands*. Philadelphia: Jewish Publication Society, 1991.

Stilt, Kristen, *Islamic Law in Action: Authority, Discretion, and Everyday Experiences in Mamluk Egypt*. Oxford: Oxford University Press, 2011.

Stol, Marten, "Blindness and Night-Blindness in Akkadian." *Journal of Near Eastern Studies* 45 (1986): pp. 295–299.

Strack, Herrman and Günter Stemberger, eds., *Introduction to Talmud and* Midrash, 2nd printing. Minneapolis: Fortress Press, 1996.

Stroumsa, Sarah, *Dawūd ibn Marwān al-Muqammiṣ's Twenty Chapters*, Études sur le judaïsme médiéval 13. Leiden: Brill 1989.

—, "From the Earliest Known Judaeo-Arabic Commentary on Genesis." *Jerusalem Studies in Arabic and Islam* 27 (2002): pp. 375–395.

—, "Soul-searching at the Dawn of Jewish Philosophy: A Hitherto Lost Fragment of al-Muqammas's Twenty Chapters." *Ginzei Qedem* 3 (2007): pp. 137–161.
—, *The Beginnings of the Maimonidean Controversy in the East: Yosef Ibn Shim'on's Silencing Epistle Concerning the Resurrection of the Dead.* Jerusalem: Ben-Zvi Institute, 1999.
—, "The Impact of Syriac Tradition in Early Judaeo-Arabic Exegesis." *Aram* 3 (1991): pp. 83–96.
Al-Subkī, Taqī l-Dīn 'Alī b. 'Abd al-Kāfī, *Ṭabaqāt ash-shāfi'iyya al-kubrā*, ed. by 'Abd al-Fattāḥ Muḥammad al-Ḥilū, vol. 2. Cairo: Dār iḥyā' al-Kutub al-'arabiyya, 1970.
Al-Suyūṭī, Jalāl al-Dīn, *Kitāb at-Taḥadduth bi-ni'mat Allāh*, ed. by Elizabeth M. Sartain. Cambridge: Cambridge University Press, 1975.
Swanson, Mark N., "Apology or its Evasion? Some Ninth-Century Arabic Christian Texts on Discerning the True Religion." *Currents in Theology and Mission* 37 (2010): pp. 389–399, here 391–394.
—, "Beyond Prooftexting: Approaches to the Qur'ān in Some Early Arabic Christian Apologies." *The Muslim World* 88 (1998): pp. 297–319.
—, "John the Deacon." in *Christian-Muslim Relations: A Bibliographical History*, vol. 1, ed. by David Thomas, Barbara Roggema and Monferrer-Sala, pp. 317–321. Leiden: Brill, 2009.
—, "John the Writer." in *Christian-Muslim Relations: A Bibliographical History*, vol. 1, ed. by David Thomas, Barbara Roggema and Juan Pedro Monferrer-Sala, pp. 702–705. Leiden: Brill, 2009.
—, "George the Archdeacon." in *Christian-Muslim Relations: A Bibliographical History*, vol. 1, ed. by David Thomas, Barbara Roggema and Monferrer Sala, pp. 234–238. Leiden: Brill, 2009.
—, "Ibn al-Qulzumī." in *Christian-Muslim Relations: A Bibliographical History*, vol. 3, ed. by David Thomas, Barbara Roggema and Monferrer-Sala, pp. 409–413. Leiden: Brill, 2011.
—, "Ma'ānī ibn Abī l-Makārim." in *Christian-Muslim Relations: A Bibliographical History*, vol 5, ed. by David Thomas, Barbara Roggema and Monferrer Sala, pp. 679–683. Leiden: Brill, 2013.
—, "Marqus ibn Zur'a." in *Christian-Muslim Relations: A Bibliographical History*, vol. 3, ed. by David Thomas, Barbara Roggema and Monferrer-Sala, pp. 643–647. Leiden: Brill, 2011.
—, "Mawhūb ibn Manṣūr ibn Mufarrij." in *Christian-Muslim Relations: A Bibliographical History*, vol. 3, ed. by David Thomas, Barbara Roggema and Monferrer-Sala, pp. 217–222. Leiden: Brill, 2011.
—, "Michael of Damrū." in *Christian-Muslim Relations: A Bibliographical History*, vol. 3, ed. by David Thomas, Barbara Roggema and Juan Pedro Monferrer-Sala, pp. 84–88. Leiden: Brill, 2011.
—, "Reading the Church's Story: The 'Amr–Benjamin paradigm' and its Echoes in the History of the Patriarchs." in *Copts in Contexts: Negotiating Identity, Tradition, and Modernity*, ed. by Nelly van Doorn-Harder. Columbia: University of South Carolina Press, 2017. (Forthcoming)
—, "Sainthood Achieved: Coptic Patriarch Zacharias According to The History of the Patriarchs." in *Writing 'True Stories': Historians and Hagiographers in the Late-Antique and Medieval Near East*, ed. by Arietta Papaconstantinou, Muriel Debié and Hugh

Kennedy, pp. 219 – 230, Cultural Encounters in Late Antiquity and the Middle Ages 9. Turnhout: Brepols, 2010.

—, *The Coptic Papacy in Islamic Egypt (641 – 1517)*, The Popes of Egypt 2. Cairo and New York: American University in Cairo Press, 2010.

—, "The Life of Patriarch Matthew I.." in *Christian-Muslim Relations: A Bibliographical History*, vol 5, ed. by David Thomas, Barbara Roggema and Juan Pedro Monferrer-Sala, pp. 396 – 401. Leiden: Brill, 2013.

Al-Ṭabarī, Abū Jaʿfar Muḥammad, *Taʾrīkh al-rusul wa-l-mulūk*, vol. 1, ed. by Muḥammad Abū ʾl-Faḍl Ibrāhīm. Cairo: Dār al-Maʿārif, 1960.

—, *Jāmiʿ al-bayān ʿan tafsīr āy al-Qurʾān*, vol. 2. Cairo: Muṣṭafā al-Bābī al-Ḥalabī, 1968.

—, *The Children of Israel, The History of al-Tabari*, vol. 3, trans. and annotated by William M. Brinner, SUNY Series in Near Eastern Studies. Albany: State University of New York Press, 1991.

Aṭ-Ṭabarī, ʿAlī, *Kitāb ad-dīn wa-d-dawla*, ed. by Alphonse Mingana. Manchester: Longmans, 1923.

Al-Ṭabaṭabāʾī, Muḥammad Ḥ., *Al-Mizān fī tafsīr al-Qurʾān*. Beirut: Muʾassasat al-ʿālamī li-l-Maṭbūʿāt, 1418/1997.

Talmon-Heller, Daniella, "Charity and Repentance in Medieval Islamic Thought and Practice." in *Charity and Giving in Monotheistic Religions*, ed. by Miriam Frenkel und Yaacov Lev, pp. 265 – 279, Studien zur Geschichte und Kultur des islamischen Orients 22. Boston and Berlin: de Gruyter, 2009.

El-Tahry, Nevin Reda, *Textual Integrity and Coherence in the Qurʾan: Repetition and Narrative Structure in Surat al-Baqara*. Toronto: PhD. Thesis, 2012.

Tanakh, A New Translation of the Holy Scriptures according to the Traditional Hebrew Text. Philadelphia: Jewish Publication Society, 1985.

Tanhum ben Joseph of Jerusalem, *Commentarium arabicum ad librorum Samuelis et Regum*, ed. by Theodor Haarbruecker [with Latin tr.]. Leipzig: F. C. G. Vogel, 1844.

Tanna debe Eliyyahu: The Lore of the School of Elijah, trans. by William G. Braude and Israel J. Kapstein, Philadelphia: Jewish Publication Society, 1981.

Tanwīr al-Miqbās min Tafsīr Ibn ʿAbbās, trans. by M. Guezzou. Amman: Royal aal al-Bayt Institute, 2007.

Ter Haar Romeny, Robert Barend, "Biblical Studies in the Church of the East: The Case of Catholicos Timothy I." in *Papers presented at the Thirteenth International Conference on Patristic Studies held in Oxford 1999: Historica, Biblica, Theologica et Philosophica*, ed. by Maurice F. Wiles and Edward J. Yarnold, pp. 503 – 510, Studia Patristica 34. Leuven: Peeters, 2001.

Tetz, Martin, *Eine Antilogie des Eutherios von Tyana*, Patristische Texte und Studien 1. Berlin: de Gruyter, 1964.

Teule, Herman, "Jacob of Edessa and Canon Law." in *Jacob of Edessa and the Syriac Culture of His Day*, ed. by Robert Barend ter Haar Romeny, pp. 84 – 86, Monographs of the Peshiṭta Institute Leiden 18. Leiden: Brill 2008.

Thackston Jr., Wheeler M., *The Tales of the Prophets of al-Kisāʾi*. Boston: Twayne Publishers, 1978.

Al-Thaʿlabī, Abū Isḥāq Aḥmad b. Muḥammad b. Ibrāhīm, *ʿArāʾis al-majālis fī qiṣaṣ al-anbiyā or "Lives of the Prophets" as Recounted by Abū Isḥāq Aḥmad ibn Muḥammad ibn*

Ibrāhīm al-Thaʿlabī, trans by William M. Brinner, Studies in Arabic Literature, Supplements to the Journal of Arabic Literature 24. Leiden: Brill, 2002.

—, *Qiṣaṣ al-anbiyāʾ*. Beirut: Al-Rāʾid al-ʿarabī, 1985 (reprint of: Cairo: Muṣṭafā Al-Bābī al-Ḥalabī, 1955).

The Apocrypha and Pseudepigrapha of the Old Testament, trans. by Robert H. Charles. Oxford: Clarendon Press, 1913.

The life of Muhammad: A Translation of Isḥāq's Sīrat Rasūl Allāh, with introduction and notes by A. Guillaume. London: Oxford University Press, 1955.

The Midrash Rabbah. I. Genesis: Translated into English with Notes, Glossary and Indices under the Editorship by. H. Freedman and Maurice Simon. Oxford: Oxford University Press, 1977.

Theodor, Julius, "Besprechungen (zu Friedmanns Edition des Seder Eliahu)." *Monatsschrift für Geschichte und Wissenschaft des Judentums* 44/10 (1900): pp. 550–561.

Theodore of Mopsuestia, *Commentary on Psalms 1–81*, trans. by Robert C. Hill, Writings from the Greco-Roman World. Atlanta: Society of Biblical Literature, 2006. .

—, *Le commentaire sur les Psaumes (I–LXXX)*, ed. by Robert Devreese. Città del Vaticano: Biblioteca Apostolica Vaticana, 1939.

The Qurʾān, trans. by Alan Jones. Cambridge: Gibb Memorial Trust, 2007.

The Qurʾan with Phrase-by-Phrase English Translation, ed. and trans. by Ali Quli Qaraʾi. New York: Tahrike Taarsile Qurʾān, 2007.

Thomas, David, "ʿAlī ibn Rabbān al-Ṭabarī: A Convert's Assessment of his Former Faith." in *Christians and Muslims in Dialogue in the Islamic Orient of the Middle Ages*, ed. by Martin Tamcke, pp. 137–156, Beiruter Texte und Studien 117. Würzburg: Ergon Verlag, 2007.

—, "Alī l-Ṭabarī." in *Christian-Muslim Relations: A Bibliographical History*, vol 1, ed. by David Thomas, Barbara Roggema ans Juan Pedro Monferrer-Sala, pp. 669–674. Leiden: Brill, 2009.

—, *Christian Doctrines in Islamic Theology*, History of Christian-Muslim Relations 10. Leiden: Brill, 2008.

—, "Ṭabarī's Book of Religion and Empire." *Bulletin of the John Ryland's Library* 69 (1986): pp. 1–7.

—, Roggema, Barbara, and Monferrer Sala, Juan Pedro, eds., *Christian-Muslim Relations: A Bibliographical History*, vol 1–. Leiden: Brill, 2009–.

—, "The Bible in Early Muslim Anti-Christian Polemic." *Islam and Christian-Muslim Relations* 7 (1996): pp. 29–38.

Thomson, H. J., *Prudentius. Works*, The Loeb Classical Library, 2 vols. Cambridge: Harvard University Press, 1953.

Al-Tirmidhī, al-Ḥakīm, *Al-Jāmiʿ al-ṣaḥīḥ*, ed. by Aḥmad Muḥammad Shākir et al., 5 vols. Cairo: Al-Ḥalabī, 1978.

Tirosh-Becker, Ofra, ed., *Rabbinic Excerpts in Medieval Karaite Literature*, 2 vol. Jerusalem: The Bialik Institute and the Hebrew University, 2011 (in Hebrew).

—, "Karaite Sources for Rabbinic Hebrew (Tannaitic and Amoraic)." in *Encyclopedia of Hebrew Language and Linguistics: Volume 3*, ed. by Geoffrey Khan, pp. 316–319. Leiden: Brill, 2013.

—, "The Use of Rabbinic Sources in Karaite Writings" in *Karaite Judaism: A Guide to Its History and Literary Sources*, ed. by Meira Polliack, pp. 319–338. Handbook of Oriental Studies, Section 1 – The Near and the Middle East 73. Leiden and Boston: Brill, 2003.

Tisserant, Eugène, "Fragments Syriaques du Livre des Jubilés." *Revue Biblique* 30 (1921): pp. 55–86, 206–232.

Toenies Keating, Sandra, *Defending the People of Truth in the Early Islamic Period: The Christian Apologies of Abū Rā'iṭah*, History of Christian-Muslim relations 4. Leiden: Brill, 2006.

Tonneau, Raymond, ed., *Sancti Ephraem Syri in Genesim et in Exodum commentarii.* Corpus scriptorum Christianorum Orientalium 152. Louvain: Secretariat Du Corpus SCO, 1955.

Toorawa, Shawkat M., "Defining *Adab* by (re)defining the *Adib:* Ibn Abi Tahir Tayfu and storytelling." in *On Fiction and Adab in Medieval Arabic Literature*, ed. by Philip F. Kennedy, pp. 287–308, Studies in Arabic Language and Literature 6. Wiesbaden: Harrassowitz, 2005.

Tottoli, Roberto, "Origin and Use of the Term Isrā'īliyyāt in Muslim Literature." *Arabica* 46 (1999): pp. 193–210.

Treiger, Alexander, "The Christology of the 'Letter from the People of Cyprus'." *The Journal of Eastern Christian Studies* 65 (2013): pp. 21–48.

Trombley, Frank R., "Sawīrus ibn al-Muqaffa' and the Christians of Umayyad Egypt: War and Society in Documentary Context." in *Papyrology and the History of Early Islamic Egypt*, ed. by Petra M. Sijpesteijn and Lennart Sundelin, pp. 199–226, Islamic History and Civilization 55. Leiden: Brill, 2004.

Tropper, Josef, *Nekromantie: Totenbefragung im Alten Orient und im Alten Testament*, Alter Orient und Altes Testament 223. Kevelaer: Butzon & Bercker, 1989.

Urbach, Ephraim E., ed., *Sefer Pitron Torah: A Collection of Midrashim and Interpretations.* Jerusalem: Magnes Press, 1978.

—, "On the Question of the Language and the Sources of the Book *Seder Eliyahu.*" *Leshonenu* 21 (1956): pp. 183–197. (in Hebrew)

—, "The Homiletical Interpretations of the Sages and the Expositions of Origen on Canticles and the Jewish-Christian Disputation." *Scripta Hierosolymitana* 22 (1971): pp. 247–275.

Vajda, Georges, "Du prologue de Qirqisani à son commentaire sur la Genèse." in *In Memoriam Paul Kahle*, ed. by Matthew Black and Georg Fohrer, pp. 222–231, Beihefte zur Zeitschrift für die alttestamentliche Wissenschaft 103. Berlin: Töpelmann, 1968.

—, "Études sur Qirqisānī." *Revue des Études Juives* 107 (1946): pp. 52–98.

—, "Judaeo-Arabic Literature." in *Encyclopaedia of Islam*, 2nd edition, pp. 303–307. Leiden: Brill, 1954.

Vanderkam, James C., ed., *The Book of Jubilees: A Critical Text*, 2 vols, I Corpus Scriptorum Christianorum Orientalium, Scriptores Aethiopici 87, II Corpus Scriptorum Christianorum Orientalium 510. Louvain: Peeters, 1989 (I Ethiopic, II English).

Van den Eynde, Ceslas and Vosté, Jacques-Marie, eds., *Commentaire d'Išo'dad de Merv sur l'Ancien Testament: I. Genèse*, traduit par Jacques-Marie Vosté and Ceslas van den Eynde. Corpus Scriptorum Christianorum Orientalium 156 (Syri 75). Louvain: Durbecq, 1955.

Van Ess, Josef, *Die Gedankenwelt des Ḥāriṯ al-Muḥāsibī.* Bonn: Orientalisches Seminar der Universität Bonn, 1961.

——, *The Flowering of Muslim Theology*, trans. by Jane Marie Todd. Cambridge, MAS: Harvard University Press, 2006.
——, *Theologie und Gesellschaft im 2. und 3. Jahrhundert Hidschra: eine Geschichte des religiösen Denkens im frühen Islam*, vol 4. Berlin: de Gruyter, 1997.
Van Gelder, *Close Relationships: Incest and Inbreeding in Classical Arabic Literature*. London and New York: Routledge, 2005.
Van Rompay, Lucas "Development of Biblical Interpretation in the Syrian Churches of the Middle Ages." in *Hebrew Bible/Old Testament: The History of Its Interpretation, Volume I: From the Beginnings to the Middle Ages (Until 1300). Part 2: The Middle Ages*, ed. by Magne Sæbø, pp. 559–577. Göttingen: Vandenhoeck and Ruprecht, 2000.
——, ed., *Commentaire sur Genèse-Exode 9,32 du manuscrit (olim) Diyarbakir 22*, 2 vols. Louvain: Peeters, 1986 (I Syriac, II French).
Varsanyi, Orsolya, "The Role of the Intellect in Theodore Abū Qurrah's *On the True Religion* in Comparison with His Contemporaries' Use of the Term." *Parole de l'Orient* 34 (2009): pp. 51–60.
Versteegh, Cornelis, *Arabic Grammar and Qur'ānic Exegesis in Early* Islam, Studies in Semitic Languages and Linguistics 19. Leiden: Brill, 1993.
Vööbus, Arthur, *The History of the School of Nisibis*, Corpus Scriptorum Christianorum Orientalium 266. Leuven: Peeters, 1965.
——, ed., *The Synodicon in the West Syrian Tradition: vol. I*, Corpus Scriptorum Christianorum Orientalium 367 and 368. Louvain: Secrétariat Du Corpus SCO, 1975.
Wafīq Naṣrī, ed., *Abū Qurrah wal-Ma'mūn: al-Mujādalah*. Beirut: Cedrac, 2010.
Wallis Budge, Ernest A., e. and trans., *The Book of the Bee: The Syriac Text Edited from the Manuscripts in London, Oxford, and Munich with an English Translation by E. A. Wallis Budge*. Oxford: Clarendon Press, 1886.
Wansbrough, John, *Qur'anic Studies: Sources and Methods of Scriptural Interpretation*, London oriental series 31. Oxford: Oxford University Press, 1977.
——, *The Sectarian Milieu: Content and Composition of Islamic Salvation History*, London Oriental Series 34. Oxford: Oxford University Press, 1978.
Wasserstrom, Steven, *Between Muslim and Jew: The Problem of Symbiosis under Early Islam*. Princeton: Princeton University Press, 1995.
Watt, John W., "The Strategy of the Baghdad Philosophers: the Aristotelian Tradition as a Common Motif in Christian and Islamic thought." in *Redefining Christian Identity: Cultural Interaction in the Middle East since the Rise of Islam*, ed. by Jan J. van Ginkel, Heleen L. Murre-van den Berg and Theo M. van Lint, pp. 151–165, Orientalia Lovaniensia analecta 134. Leuven: Peeters, 2005.
Watt, William Montgomery, ed., *Bell's Introduction to the Qur'ān completely revised and enlarged*. Edinburgh: Edinburgh University Press, 1970.
——, "Ḥanīf." Encylopaedia of Islam, vol. 3, pp. 165–166. Leiden: Brill, 1971.
Wenham, Gordon J., "The Old Testament Attitude to Homosexuality." *Expository Times* 102 (1991): pp. 359–363.
Wensinck, Arent Jan, *Concordance et indices de la tradition musulmane: les six livres, le Musnad d'al-Dārimī, le Muwaṭṭa' de Mālik, le Musnad de Aḥmad Ibn Ḥanbal*, 2. edition, 6 Vols. Leiden, Brill: 1992.
Werthmuller, Kurt J., *Coptic Identity and Ayyubid Politics in Egypt, 1218–1250*. Cairo and New York: The American University in Cairo Press, 2010.

Whiston, William, *The genuine works of Flavius Josephus, the Jewish historian*, London, 1737.
Whybray, Norman,"Genesis." in *The Oxford Bible Commentary*, ed. by John Barton and John Muddiman, pp. 38–66. Oxford: Oxford University Press, 2007.
Wild, Stefan, "'Die schauerliche Öde des heiligen Buches'. Westliche Wertungen des koranischen Stils." in *Gott ist schön und Er liebt die Schönheit. God is beautiful and He loves beauty. Festschrift in honour of Annemarie Schimmel presented by students, friends and colleagues on April 7, 1992*, ed. by Alma Giese und Johann C. Bürgel, pp. 429–447. Bern: Peter Lang, 1994.
—, "The Koran as Subtext in Modern Arabic Poetry." in *Representations of the Divine in Arabic Poetry*, ed. by Gert Borg and Ed de Moor, pp. 139–160, Orientations 5. Amsterdam: Rodopi, 2001.
Williamson, Hugh G. M., *The Book Called Isaiah: Deutero-Isaiah's Role in Composition and Redaction*. Oxford: Oxford University Press, 1994.
Witztum, Joseph, *The Syriac Milieu of the Qur'an: The Recasting of Biblical Narratives*. Ph.D. Dissertation. Princeton University, 2011.
—, "Variant Traditions, Relative Chronology, and the Study of Intra-Quranic Parallels." in *Islamic Cultures, Islamic Contexts: Essays in Honor of Professor Patricia Crone*, Islamic History and Civilization: Studies and Texts 114, ed. by Behnam Sadeghi, Asad Q. Ahmed, Adam Silverstein et al., pp. 1–50. Leiden: Brill, 2015.
Wolfson, Harry Austrin, *The Philosophy of the Church Fathers, Structure and Growth of Philosophic Systems from Plato to Spinoza*. Cambridge MA.: Harvard University Press, 1970.
Wright, William, *Lectures on the Comparative Grammar of the Semitic Languages*. Cambridge: Cambridge University Press, 1890.
—, "Two Epistles of Mar Jacob, Bishop of Edessa." *Journal of Sacred Literature and Biblical Record* 10 (1867): pp. 430–460.
Wüstenfeld, Ferdinand, ed., *Das Leben Muhammed's nach Muhammed ibn Ishâk bearbeitet von Abd el-Malik ibn Hischâm: Aus den Handschriften zu Berlin, Leipzig, Gotha und Leyden*, vol. 1. Göttingen: Dieterichs, 1864.
Xella, Paolo, "Death and the Afterlife in Canaanite and Hebrew Thought." in *Civilizations of the Ancient Near East*, vol. 3, ed. by Jack M. Sasson, pp. pp. 2059–2072. New York: C. Scribner's Sons, 1995.
Al-Yaʿqūbī, Aḥmad b. Abī Yaʿqūb b. Jaʿfar b. Wahb b. Wāḍiḥ, *Al-Taʾrīkh*, ed. by Martijn T. Houtsma, 2 vols. Leiden: Brill, 1883 (reprint 1969).
Yassif, Eli, "Pseudo Ben Sira and the 'Wisdom Questions' Traditions in the Middle Ages." *Fabula* 23, 1/2 (1982): pp. 48–63.
—, *The Tales of Ben Sira in the Middle-Ages: A Critical Text and Literary Studies*. Jerusalem: Magnes Press, 1984 (in Hebrew).
Young, F. M., *Judaism and Islam*. Madras: M.D.C.S.P.C.K. Press, 1898.
Yücesoy, Hayrettin, *Messianic Beliefs and Imperial Politics in Medieval Islam: The ʿAbbāsid Caliphate in the Early Ninth* Century, Studies in Comparative Religion. Columbia: University of South Carolina Press, 2009.
—, *Lineaments of Islam: Studies in Honor of Fred McGraw Donner*, ed. by Paul Cobb, pp. 61–84, Islamic History and Civilization 95. Leiden: Brill, 2012.
Zaharopoulos, Dimitri Z., *Theodore of Mopsuestia on the Bible*, Theological Inquiries. New York: Paulist Press, 1989.

Zahn, Molly Marie, *Rethinking Rewritten Scripture: Composition and Exegesis in the 4QReworked Pentateuch* Manuscripts, Studies on the texts of the desert of Judah 95. Leiden and Boston: Brill, 2011.
Zahniser, A. H. Mathias, "Major Transitions and Thematic Borders in Two Long Sūras: al-Baqara and al-Nisāʾ." in *Literary Structures of Religious Meaning in the Qurʾan*, ed. by Issa J. Boullata, pp. 26–55, Routledge Studies in the Qurʾān. Abingdon: Routledge, 2000.
Zalewski, Saul, "The Purpose of the Story of the Death of Saul in 1 Chronicles X." *Vetus Testamentum* 39 (1989): pp. 449–467.
Zaman, Muḥammad Qasim, "al-Yaʿḳūbī." Encyclopaedia of Islam, vol. 11, pp. 257–258. Leiden: Brill, 2002.
Zammit, Martin R., *A Comparative Lexical Study of Qurʾānic Arabic*. Handbook of Oriental Studies. Section 1, The Near and Middle East 61. Leiden, Boston and Köln: Brill, 2002.
Al-Zarkashī, Badr al-Dīn, *Al-Burhān fī ʿulūm al-Qurʾān*. Dār al-Fikr, 1988.
Zawanowska, Marzena, "Was Moses the *mudawwin* of the Torah?: The Question of Authorship of the Pentateuch According to Yefet ben ʿEli." in *Studies in Judeo-Arabic Culture: Proceedings of the Fourteenth Conference of the Society for Judaeo-Arabic Studies*, ed. by Haggai Ben-Shammai, Aron Dothan, Yoram Erder et al., pp. 7–35. Tel Aviv: Tel Aviv University, 2014.
Zeilinger, Franz, *Die sieben Zeichenhandlungen Jesu im Johannesevangelium*. Stuttgart: Kohlhammer, 2011.
Ziakas, Gregorios D., "Islamic Aristotelian Philosophy." in *Two Traditions, One Space: Orthodox Christians and Muslims in Dialogue*, ed. by George C. Papademetriou, pp. 77–108. Boston, MA: Somerset Hall Press, 2011.
Zimmermann, Ruben, in ollaboration with Detlev Dormeyer and Susanne Luther, *Die Wunder Jesu*, Kompendium der frühchristlichen Wundererzählungen 1. Gütersloh: Gütersloher Verlags-Haus, 2013.
Zohar: The Book of Enlightenment, trans. by, Daniel C. Matt. Toronto: Paulist Press, 1983.

Online Resources:

http://www.behindthename.com/names/usage/late-roman
"Behind the Name." Accessed on January 29, 2014.

http://www.earlychristianwritings.com/ text/diognetus-lightfoot.html.
"Early Christian Writings – Diognetus." Accessed on February 4, 2014.

http://faculty.georgetown.edu/jod/augustine/ddc.html
"Philip Schaff, A Select Library of the Nicene and Post-Nicene Fathers of the Christian Church, vol. 2 (St. Augustin's City of God and Christian Doctrine)." Accessed on February 4, 2014.

http://www.pseudepigrapha.com/pseudepigrapha/Apocalypse_of_Abraham.html.
"The Apocalypse of Abraham," last modified September 22, 2015.

Lavee, Moshe, "Seder Eliyahu." in *Encyclopedia of Jews in the Islamic World*. Brill Online, 2015,
<http://referenceworks.brillonline.com/entries/encyclopedia-of-jews-in-the-islamic-world/seder-eliyahu-SIM_0017460> (consulted on16 January 2015).

Swanson, Mark, "Arabic as a Christian Language." on www.copticbook.net/books/20101102002.pdf (consulted on August, 26th, 2013).

Authors

Martin Accad is currently the Chief Academic Officer at the Arab Baptist Theological Seminary in Beirut and the Director of its Institute of Middle East Studies. He has published numerous articles and book chapters in the fields of Islam and Christian-Muslim relations, including *Christian Attitudes toward Islam and Muslims: A Kerygmatic Approach* (2013) and *Loving Neighbor in Word and Deed: What Jesus Meant* (2010).

Najib George Awad is Associate Professor of Christian Theology at Hartford Seminary, Connecticut. His various monographs include *Persons in Relation: An Essay on the Trinity and Ontology* (Fortress Press, 2014) and *Orthodoxy in Arabic Terms: A Study of Theodore Abū Qurrah's Trinitarian and Christological Doctrines in an Islamic Context* (De Gruyter, 2015).

Haggai Ben-Shammai is Professor Emeritus of Arabic Language and Literature at The Hebrew University, Jerusalem. He has published numerous articles and chapters in books on Judaeo-Arabic Bible exegesis and philosophy, history of Jewish communities in Islamic countries and Islamic theology. He is the author of *A Leader's Project: Studies in the Philosophical and exegetical Works of Saadya Gaon* (Hebrew) (Bialik Institute, 2015).

William A. Graham is Murray A. Albertson Professor of Middle Eastern Studies and University Distinguished Service Professor at Harvard University where he currently directs Harvard's Prince Alwaleed bin Talal Islamic Studies Program. Among his publications are *Divine Word and Prophetic Word in Early Islam* (De Gruyter, 1977) and *Islamic and Comparative Religious Studies* (Routledge, 2010).

Sidney Griffith is Professor of Early Christian Studies at the Catholic University of America and an expert on Arab Christianity and Medieval Christian-Muslim encounters. He is the author of *The Church in the Shadow of the Mosque: Christians and Muslims in the World of Islam* (Princeton University Press, 2007) and *The Beginnings of Christian Theology in Arabic Muslim-Christian Encounters in the Early Islamic Period* (Routledge, 2002).

Martin Heimgartner is Professor of Church History at the Martin-Luther-Universität Halle-Wittenberg. He is the author of *Timotheos I., ostsyrischer Patriarch: Disputation mit dem Kalifen al-Mahdi* (Peeters, 2011), *Die Briefe 42–58 des ostsyrischen Patriarchen Timotheos (780–823):* Textedition. Einleitung, Übersetzung und Anmerkungen (Peeters, 2012) and has published numerous articles and book chapters on this topic.

Cornelia Horn is Professor of Languages and Cultures of the Christian Orient at the Martin-Luther-Universität Halle-Wittenberg. She is the author of *Asceticism and Christological Controversy in Fifth-Century Palestine: The Career of Peter the Iberian* (Oxford University Press, 2006) and together with Robert R. Phenix Jr *John Rufus: The* Lives *of Peter the Iberian, Theodosius of Jerusalem, and the Monk Romanus* (Brill, 2008).

Lennart Lehmhaus is a Postdoctoral Research Associate at the Freie Universität. He has co-edited the volumes *Collecting Recipes. Byzantine and Jewish Pharmacology in Dialogue*

(De Gruyter, 2017), and *Defining Jewish Medicine. Transfer of Medical Knowledge in Jewish Cultures and Traditions* (Harrassowitz, 2018).

Berenike Metzler is a Postdoctoral Research Associate at the Chair of Oriental Philology and Islamic Studies at the University of Erlangen-Nuremberg. As specialist in the fields of Qur'ānic studies and Early Islamic theology, she has published a translation and analysis of the *Kitāb Fahm al-Qurʾān* by al-Ḥārith b. Asad al-Muḥāsibī (d. 857) (Harrassowitz, 2016).

Juan Pedro Monferrer-Sala is Full Professor of Arabic and Islamic Studies at the University of Córdoba. Amongst his most recent publications are *Scripta Theologica Arabica Christiana. Andalusi Christian Arabic Fragments Preserved in Ms. 83* (al-Maktabah al-Malikiyyah, 2016) and *The Vision of Theophilus: The Flight of the Holy Family into Egypt* (Gorgias Press, 2015).

Reza Pourjavady is currently professor of Oriental Studies at the Ruhr-Universität Bochum and specialized in Post-Avicennian Philosophy in the Islamic world. His publications include *Philosophy in Early Safavid Iran: Najm al-Dīn Maḥmūd Nayrīzī and His Writings* (Brill, 2011) and, co-authored with Sabine Schmidtke, *A Jewish Philosopher of Baghdad: ʿIzz al-Dawla Ibn Kammūna (d. 683/1284) and His Writings* (Brill, 2006).

Gabriel Said Reynolds is Professor of Islamic Studies and Theology at the University of Notre Dame. He is the editor of *The Qurʾān and Its Biblical Subtext* (Routledge, 2010) and *The Qurʾān and the Bible: Text and Commentary* (Yale University Press, 2018).

Alison Salvesen is Professor of Early Judaism and Christianity at the Oriental Institute, University of Oxford, and Polonsky Fellow in Early Judaism and Christianity at the Oxford Centre for Hebrew and Jewish Studies. She has written numerous articles and book chapters on this field including *A "New Field" for the Twenty-First Century? Rationale for the Hexapla Project, and a Report on Its Progress'* (2017) and *Scholarship on the Margins: Biblical and Secular Learning in the Work of Jacob of Edessa* (2015).

Nicolai Sinai is Professor of Islamic Studies at the University of Oxford and Fellow of Pembroke College. He has published on the Qurʾān, on pre-modern and modern Islamic scriptural exegesis, and on the history of philosophy in the Islamic world. His most recent book is *The Qurʾan: A Historical-Critical Introduction* (Edinburgh University Press, 2017).

Mark N. Swanson is the Harold S. Vogelaar Professor of Christian-Muslim Studies and Interfaith Relations at the Lutheran School of Theology at Chicago. He is the author of *The Coptic Papacy in Islamic Egypt, 641–1517* (AUC Press, 2010) and was the Christian Arabic section editor for *Christian-Muslim Relations: A Bibliographical History*, ed. by David Thomas et al., vols. 1–5 (Brill, 2009–2013).

Stefan Wild is currently Senior Fellow at the Käte Hamburger Kolleg "Law as Culture", Bonn and Professor Emeritus at the University of Bonn. He is the author of *Mensch, Prophet und Gott im Koran: Muslimische Exegeten des 20. Jahrhunderts und das Menschenbild der Moderne* (Rhema, 2001) and editor of *The Qurʾān as Text* (Brill, 1996) and *Self-referentiality in the Qurʾān* (Harrassowitz, 2006).

The Editors

Regina Grundmann is Professor of Jewish Studies at the University of Münster. Since 2012, she is Principial Investigator at the Cluster of Excellence "Religion and Politics in Pre-Modern and Modern Cultures", and since 2015 project leader in the Collaborative Research Center/ SFB 1150 "Cultures of Decision-Making". She has co-edited *Judah Moscato's Sermons: Edition and Translation* (Brill, 2011) and *Martyriumsvorstellungen in Antike und Mittelalter: Leben oder sterben für Gott?* (Brill, 2012).

Assaad Elias Kattan is Professor of Orthodox Theology at the Centre for Religious Studies at the University of Münster. Kattan's publications include *Verleiblichung und Synergie: Grundzüge der Bibelhermeneutik bei Maximus Confessor* (Brill, 2003) and *Jenseits der Tradition: Tradition und Traditionskritik in Judentum, Christentum und Islam* (together with Regina Grundmann, De Gruyter, 2015).

Karl Pinggéra is Professor of Church History at the Philipps-Universität Marburg and specialized in Christian Oriental Studies. He is the author of *All-Erlösung und All-Einheit. Studien zum "Buch des heiligen Hierotheos" und seiner Rezeption in der syrisch-orthodoxen Theologie* (Reichert, 2002) and co-editor of *A Bibliography of Syriac Ascetic and Mystical Literature* (Brill, 2011) and *Die altorientalischen Kirchen. Glaube und Geschichte* (WBG, 2010).

Georges Tamer holds the Chair of Oriental Philology and Islamic Studies at the University of Erlangen-Nuremberg. His numerous publications in the fields of Qur'ānic hermeneutics and Islamic philosophy and theology include *Islamische Philosophie und die Krise der Moderne: Das Verhältnis von Leo Strauss zu Alfarabi, Avicenna und Averroes* (Brill, 2001) and *Zeit und Gott: Hellenistische Zeitvorstellungen in der altarabischen Dichtung und im Koran* (De Gruyter, 2008). He has also edited Vol. I of *Islam and Rationality. The Impact of al-Ghazālī. Papers on the occasion of his 900th anniversary* (Brill, 2015).

Index of used verses from the Bible, the Qur'ān and Apocrypha

I Bible

Gen

Gen 2:8 136
Gen 11 11, 12
Gen 12 20
Gen 12:5–7 20
Gen 12:8 20
Gen 12:10 13
Gen 13 20,
Gen 13:3 13, 14, 20
Gen 13:17–18 20
Gen 14:17 253
Gen 15 12
Gen 15:7 12
Gen 15:13 18
Gen 17:5 10
Gen 18:16–33 130, 135
Gen 18–19 5, 117
Gen 19 5, 117, 118, 124, 129, 140
Gen 19:1–2 130
Gen 19:1–9 130
Gen 19:1–11 117
Gen 19:1–38 140
Gen 19:3–9 131
Gen 19:4–5 131
Gen 19:8 118, 119
Gen 19:11 131, 132
Gen 19:12 118, 122,
Gen 19:12–29 117
Gen 19:17 124
Gen 19:20–22 125
Gen 19:22 123, 124
Gen 19:24 130, 135
Gen 19:25 125, 126
Gen 19:26 128
Gen 19:29 126
Gen 19:30 124
Gen 19:30–38 117
Gen 19:31–38 130, 137
Gen 19:99 135
Gen 20 124
Gen 21 20, 124
Gen 21:33 20
Gen 21:21 22
Gen 22 20, 118, 257, 258
Gen 22:9 20
Gen 22:18 40
Gen 23 15
Gen 25 118
Gen 26:12 174
Gen 28 130, 135
Gen 29 118, 125
Gen 30 125
Gen 38 65
Gen 39:1 295
Gen 40:1 295
Gen 40:5 295

Exod

Exod 1:8–14 67
Exod 1:15 295
Exod 1:18 295
Exod 2:5 288
Exod 2:10 288, 295
Exod 2:12 290
Exod 2:12–15 288
Exod 12:14 290
Exod 3:6 10
Exod 4:19 288
Exod 20:2 12
Exod 20–23 264
Exod 21:23–25 266
Exod 34:29–35 66

Lev

Lev 17–26 264
Lev 18:22 132
Lev 19:13 218

https://doi.org/10.1515/9783110564341-019

Lev 19:31 176, 177
Lev 20:6 176
Lev 20:12 138
Lev 20:13 132
Lev 20:27 176, 191
Lev 25:13 16

Num

Num 20:15 13
Num 24:17 47
Num 33:1 13

Deut

Deut 5:24 261
Deut 10 12
Deut 11 12
Deut 12 311
Deut 12–26 264
Deut 14 311
Deut 16 311
Deut 21 311
Deut 24 311
Deut 26 311
Deut 26:5 11
Deut 28 150
Deut 29 150
Deut 30:1–5 150

Josh

Josh 1:4 12
Josh 10:16 124
Josh 18:13 253
Josh 24:3 10, 12

1 Sam

1 Sam 8:10 172
1 Sam 10:8 172
1 Sam 10:23 188
1 Sam 11–12 180
1 Sam 13:6 124
1 Sam 14:19 172
1 Sam 15:3 189
1 Sam 15–16 179
1 Sam 16:14 179
1 Sam 17:34–37 65
1 Sam 21 180
1 Sam 22:6–19 191
1 Sam 22:9–23 167
1 Sam 25:1 166
1 Sam 28 5, 166, 169, 170, 171, 175, 176, 177, 184, 185
1 Sam 28:3 167
1 Sam 28:7–20 171
1 Sam 28:13 173
1 Sam 28:13–14 166
1 Sam 28:15 192
1 Sam 28:16–19 167
1 Sam 28:19 169
1 Sam 31 170
1 Sam 31:4 189

1 Kgs

1 Kgs 1 65
1 Kgs 2 153
1 Kgs 14 153, 155
1 Kgs 14:21–22 151
1 Kgs 16:29–22 66
1 Kgs 17:17–24 174, 194
1 Kgs 18 146, 148
1 Kgs 18:3–4 148
1 Kgs 19 66
1 Kgs 21:27–29 66

2 Kgs

2 Kgs 1:11–16 147, 148
2 Kgs 1:13 148
2 Kgs 4 147
2 Kgs 4:1 146, 148, 149
2 Kgs 4:8–37 174, 194

1 Chr

1 Chr 8 170
1 Chr 10 171
1 Chr 10:1–7 170
1 Chr 11:4 253
1 Chron 10:13–14 170, 172

2 Chr
2 Chr 12:1–14 153

Ezra
Ezra 9:7 150

Job
Job 16:7 120
Job 18:9 126

Ps
Ps 8 46
Ps 33:6 53
Ps 37:29 266
Ps 39:12 16
Ps 56:10 53
Ps 57:1 125
Ps 65:4 64
Ps 72:1 49
Ps 104:2 163
Ps 105:37 14
Ps 110:1–3 46
Ps 110:1 46, 54
Ps 110:3 46, 51
Ps 118:22 47
Ps 119:19 16

Prov
Prov 9:5 206
Prov 20:13 206, 207
Prov 23:21 207
Prov 32:21 205

Isa
Isa 1:3–9 150
Isa 1:7 150
Isa 1:10 117
Isa 7:14 32, 37
Isa 11:1 47
Isa 21:7 53
Isa 40–66 253, 267
Isa 41:8 19
Isa 42:1 49
Isa 50:4 206
Isa 51:2 10
Isa 52:13 49
Isa 54:3 206
Isa 61:7 206, 207

Jer
Jer 23:14 117
Jer 36 262
Jer 36:32 262

Ezek
Ezek 16:49 117
Ezek 18:18 218
Ezek 36 262
Ezek 36:11 136
Ezek 37 174

Dan
Dan 2:34 47
Dan 6 62
Dan 7:13 49

Joel
Joel 2:25–26 206

Amos
Amos 1:2–2:16 254
Amos 2:4–5 254
Amos 9:8 254, 256, 257
Amos 9:11–15 254
Amos 9:13–15 136

Obad
Obad 1:17 206

Zech
Zech 9:12 206

Mal
Mal 3:20 47

Matt

Matt 1:22 – 23 37
Matt 3:9 258
Matt 3:17 49
Matt 8:22 298
Matt 10:19 41
Matt 10:28 68
Matt 10:37 298
Matt 14:13 – 21 27
Matt 14:22 – 33 27
Matt 17:5 49
Matt 24:7 154
Matt 24:36 88, 89
Matt 26:63 – 64 87
Matt 27 – 31 87
Matt 27:19 68
Matt 27:24 69
Matt 28:1 – 10 27
Matt 28:19 98
Matt 28:19 – 20 98

Mark

Mark 1,11 49
Mark 6:30 – 44 27
Mark 6:45 – 52 27
Mark 9:7 49
Mark 13:9 63
Mark 13:32 88, 89
Mark 16 27

Luke

Luke 1:21 – 22 87, 88
Luke 1:73 10
Luke 2:36 – 38 37
Luke 2:42 – 46 88
Luke 3:22 49
Luke 9:35 49
Luke 9:57 – 60 298
Luke 9:62 127
Luke 10:9 27
Luke 17:32 – 33 118, 127
Luke 18:29 – 30 298
Luke 21:12 63
Luke 22:42 50
Luke 22:67 – 68 87
Luke 24:1 – 12 27
Luke 24:13 – 32 88

John

John 1:5 68
John 1:9 47
John 2:1 – 11 27
John 2:11 30
John 6:38 50
John 8:34 48
John 8:4 48
John 10:7 48
John 11:25 98
John 13:4 49
John 14:6 48
John 14:30 246
John 15:23 – 16:1 83
John 18:4 – 8 87
John 20:1 – 18 27
John 20:21 98
John 21:1 – 8 88

Acts

Acts 2:22 27
Acts 3:1 – 10 27
Acts 7:20 – 23 295
Acts 7:21 288
Acts 8:20 – 23 66
Acts 22:3 66

Rom

Rom 1:26 – 27 132
Rom 4:1 10
Rom 9:6 – 8 15

2 Cor

2 Cor 5:21 48

Gal

Gal 3:7 15
Gal 3:13 48
Gal 4:4 51

1 Tim
1 Tim 2:19 64

Heb
Heb 4:15 48
Heb 5:4 64
Heb 7:1–17 15
Heb 11 15
Heb 11:8–16 15
Heb 11:39–40 16
Heb 13:10–14 17

Jas
Jas 2:21 10

1 Pet
1 Pet 2:11 17

Jude
Jude 1:14–15 145

II Qur'ān

Q 2 261, 281, 308, 309, 310, 311, 312, 313
Q 2:1–286 310
Q 2:21 331
Q 2:21–22 331
Q 2:30 285
Q 2:30–33 285
Q 2:30–34 285
Q 2:30–73 311
Q 2:31 285
Q 2:40 260
Q 2:47 260
Q 2:93 261
Q 2:106 263
Q 2:111 71, 75
Q 2:124 257, 258, 260, 311
Q 2:125 20
Q 2:127 258
Q 2:142–274 311
Q 2:143 310
Q 2:163–283 311
Q 2:164 322
Q 2:170 296
Q 2:212 189
Q 2:217–218 252
Q 2:246–251 188, 189
Q 3:7 318, 319, 321
Q 3:42–51 39
Q 3:45–46 42
Q 3:46 34, 36, 38, 39, 42
Q 3:49 53
Q 3:55 53
Q 3:59 110
Q 3:65 19
Q 3:95 19
Q 3:96 255
Q 4:28 304
Q 4:82 277
Q 4:125 19
Q 4:157 52, 111
Q 4:171 52, 53, 54, 72, 73, 74, 79, 110
Q 5 315
Q 5:3 255
Q 5:32 261, 266
Q 5:45 266
Q 5:104 296
Q 5:109–120 39
Q 5:110 34, 36, 38, 42, 53
Q 5:116–118 45
Q 6:94 74
Q 6:148 296
Q 6:161 19, 36, 38, 42
Q 7 262, 269, 275, 276, 277, 283, 284
Q 7:19 277
Q 7:20 269
Q 7:23 275, 284
Q 7:28 296
Q 7:54 73
Q 7:63 294
Q 7:70–71 296
Q 7:148 277
Q 7:149 284
Q 7:151 284

Q 7:153 284
Q 7:155 284
Q 7:161 284
Q 7:167 284
Q 7:169 284
Q 7:173 296
Q 7:180 73
Q 8 309
Q 8:32–33 330
Q 9 261, 309
Q 9:23 296
Q 9:34 293
Q 9:70 126
Q 9:113–114 265, 297
Q 9:114 263, 264, 265, 266, 267, 268, 275
Q 10:24 322
Q 10:39 331
Q 10:78 296
Q 10:94 75
Q 11:42–47 297
Q 11:82 125
Q 12 297
Q 12:21 298
Q 12:22 298
Q 12:28–29 298
Q 12:33 298
Q 12:35 298
Q 13:8 330
Q 13:19 322
Q 14:32–33 292
Q 14:34 291, 292
Q 14:41 263, 265, 297
Q 15 269, 276, 281, 282, 283, 284, 285
Q 15:1 279
Q 15:26–27 280
Q 15:26–28 282
Q 15:28–31 283
Q 15:28 278, 282, 283
Q 15:29 110
Q 15:31 282
Q 15:33 282
Q 15:36–43 283
Q 15:49 283
Q 15:74 125
Q 15:87 311
Q 16:64 74
Q 16:83 291
Q 16:101 263
Q 17 269, 275, 281, 282
Q 17:5 283
Q 17:17 283
Q 17:30 283
Q 17:53 283
Q 17:61 330
Q 17:61–65 269
Q 17:90 331
Q 17:96 283
Q 18 269, 275, 281
Q 18:50 269
Q 19 40, 42
Q 19:26 41
Q 19:27 39, 40
Q 19:27–36 39
Q 19:28 41
Q 19:29 34
Q 19:30 42, 45
Q 15:40 283
Q 15:42 283
Q 19:47 263, 265, 297
Q 19:48–49 19
Q 20 262, 269, 275, 276, 281, 283, 284
Q 20:40 287, 289
Q 20:88 277
Q 20:115 283
Q 20:116 275
Q 20:117 276, 277
Q 20:120 269
Q 20:121 283
Q 20:122 276, 284
Q 21:5 330
Q 21:51–54 296
Q 21:71 19, 22
Q 21:105 266
Q 22:26–27 23
Q 22:27 23
Q 22:52 263
Q 22:65 292
Q 22:66 291, 292
Q 22:78 10
Q 23:14 333
Q 23:24 296
Q 24:40 334
Q 24:35 21
Q 25:6 330

Q 25:32 302
Q 26 6, 287, 289, 290, 291, 292, 299
Q 26:11 285
Q 26:16–17 298
Q 26:18 289, 290, 291, 299
Q 26:19 289, 291
Q 26:20 291
Q 26:22 293, 299
Q 26:29 292
Q 26:86 263, 265
Q 28 288, 290, 297
Q 28:15 289, 290, 298
Q 28:15–17 290
Q 28 288, 290
Q 28:7–9 287
Q 28:9 295, 298
Q 28:12 288
Q 28:33 290
Q 28:88 335
Q 29:49–50 330
Q 30:1 331
Q 30:27 73
Q 33 289, 307
Q 35:26 310
Q 37 258, 259
Q 37:3–4 259
Q 37:83–98 257
Q 37:99 19, 257
Q 37:101 258
Q 37:99–107 257
Q 37:108–111 257
Q 37:112–113 257, 258, 259, 260, 262
Q 38 269, 275, 276, 281, 282, 283, 284, 285
Q 38:4 294
Q 38:45 283
Q 38:71 278, 283
Q 38:71–74 283
Q 38:72 110
Q 38:74 282
Q 38:75 282
Q 38:76 282
Q 38:79–85 283
Q 38:83 283
Q 39:23 277
Q 40:2 279
Q 41:8 256, 257
Q 42:22 257
Q 42:23 257
Q 42:26 257
Q 42:51 75
Q 43:59 45
Q 44:2 279
Q 44:49 293
Q 45:21 257
Q 45:30 257
Q 46:15–18 297
Q 47:12 257
Q 48:29 257
Q 50:2 294
Q 50:16 321
Q 53:53 126
Q 54:11 74
Q 55:14–15 280, 282
Q 55:15 282
Q 56:42–43 293
Q 56:52–56 293
Q 56:93–94 293
Q 57 310
Q 57:3 333
Q 57:26 258
Q 60:4 263, 265, 266, 267, 268, 297
Q 61:6 53, 245
Q 65:11 257
Q 66:11 298
Q 69 294
Q 73 295
Q 73:2–4 259
Q 73:15–18 276
Q 73:20 295, 262
Q 74:18 310
Q 75:16–19 265
Q 77:4 334
Q 79:15–26 276
Q 80 311
Q 84 256, 257,
Q 84:25 256, 257, 259, 260, 262
Q 85:17–18 276
Q 87:18–19 267
Q 90 257
Q 90:4 74
Q 90:17 257
Q 95:6 257
Q 98:7 257
Q 103:3 257

III Apocrypha

Apocrypha

Book of Jubilees 13, 145, 146
Letter of James 15
Protoevangelium of James 31, 32, 33, 36
Ascension of Isaiah 32
Arabic Infancy Gospel 34, 35
Arabic Apocryphal Gospel of John 35, 36, 38, 44
Revelation of the Magi 43, 44
Book of Enoch 145
Apocalypse of Pseudo-Methodius 151

Index

Abbasid 74, 80, 86, 95 f., 99, 101, 115, 120, 216, 234
'Abd al-'Azīz b. Marwān 66, 68 f., 72
'Abd al-Malik b. Marwān 153, 161
Abraham 3, 5, 9–24, 112, 119 f., 122 f., 128, 137, 141, 148, 163, 176, 179, 181, 200, 206, 259 f., 265–268, 270 f., 277, 290, 298–300, 305
Abrahamic tradition 9, 25
Abraham-Lot narrative cycle 119
Abū Muḥammad 'Abd al-Mālik b. Hishām 55, 80
Adab 224, 233–235, 240 f.
Adam 6, 10, 22, 38, 112, 212, 266, 270–272, 277–288, 296, 324
Adultery 42
Agenda (s) 25, 165, 178, 210, 217, 219, 226–228, 232, 235, 237 f., 244, 249
Ahl al-ḥadīth 218
Ahl al-kitāb 73, 218
Akhbār 230
Aḥmad b. Ḥanbal 334
Alphabeta de-Ben Sira 228 f., 232
Ammonites 139, 154
'Amr b. al-'Āṣ 63, 65, 70, 72
Anagnorisis 131
Anan ben David 220
Anbiyā' 31, 93, 99, 190–193, 227,
Angels 55, 91, 129, 132, 135, 137, 277, 280 f., 283 f., 286 f., 335
Anna 38, 124, 127, 158
Antichrist 128, 156
Apocalypse of Abraham 13, 298
Apocalyptic 128, 156, 160, 241, 298, 310 f.
Apocrypha 13, 28 f., 31, 36, 55, 72, 151
Apocryphal 4, 30, 33–37, 43 f., 120, 147, 151, 159, 161, 225, 229
Apocryphal Infancy Gospel 31 f.
Apologetic 3–5, 53, 73 f., 76, 78–80, 82 f., 93–103, 105 f., 110, 112–114, 117, 177, 243 f., 250, 279, 297, 307
Aqedah 11
'*Aql*, intellect 79, 95, 105, 114
Arabic 4 f., 31, 34–37, 43 f., 49, 53, 61–63, 71, 73–83, 85 f., 93–96, 98–107, 113, 115, 118, 120–137, 139–142, 145, 152, 163–165, 167, 176–179, 181, 183–185, 192–194, 200 f., 205–207, 210–213, 215–217, 222–224, 228–234, 236–238, 240 f., 243 f., 258, 264, 268, 270, 280, 282, 286–288, 304, 308, 312, 315, 317, 336
Arabic Apocryphal Gospel of John 36–39, 44
Arabic Infancy Gospel 34–36
Arabic-Karaite 232
'*Arabiyya* 216
Arabs 5, 10, 55 f., 82, 92, 102, 115, 135, 145, 153 f., 156, 160, 315, 333
Aramaic 42 f., 78, 104, 130, 147, 150, 152, 165, 171, 207, 214 f., 280
Aristotelian 5, 58, 101 f., 142
Aristotle 48 f., 52, 58, 102, 115, 121
Ascension of Isaiah 32
Asceticism 240
Ascetics 227, 235 f., 241, 261, 324
Aṣ-Ṣafā 20, 194 f.
Astren, Fred 200, 203–206, 221, 237, 239
Augustine [of Hippo] 17 f.
Authentic 88, 110, 222, 228, 250
Authenticity 80, 96 f., 103 f., 109, 116 f., 225, 229, 238, 249 f., 331
Author 4, 7, 15 f., 19, 27–29, 39, 42, 57, 62 f., 66, 71, 73–81, 87 f., 102, 111, 116, 120 f., 126, 141–143, 147, 155, 164 f., 167, 171, 176–179, 183, 185, 188, 190–194, 202, 205, 207, 213 f., 222, 224–233, 235 f., 238, 240, 290 f., 296 f., 301, 306, 315, 317, 321 f., 331
Authorial 6, 224 f., 228, 230 f.
Authorship 61, 184, 222, 224, 226, 229 f., 235 f., 291
Avicenna 330
Avraham 'avinu (Abraham our Father) 10, 14
Āyah 134, 313

https://doi.org/10.1515/9783110564341-020

Baby 33 f., 37, 41, 43
Babylonia 13, 163, 201–203, 206, 213, 215, 221, 224, 226
Babylonian Exile 14, 152, 155
Babylonian Talmud 128, 150, 197, 203, 223, 229, 290
Babylonian Yeshivot 203
Bacha, Constantine 99, 103, 115
Baghdad 47, 84, 86, 101 f., 121, 177 f., 184–186, 321, 329 f., 335
Al-Balkhi, Hiwi 189, 217, 240
Bannister, Andrew 253, 268, 270 f., 279–283
Baptism 29–31, 52
Bayt 122, 291
Beaumont, Mark 96–98
Benjamin I, Coptic Orthodox patriarch 64, 87
Ben Sira 155, 206 f., 228–230, 234
Berlin 30 f., 35, 54 f., 58, 78, 138, 164, 172, 210, 212, 222, 230, 236, 258, 323, 325
Bible 4, 6, 27, 35, 49 f., 52–56, 59, 64 f., 67, 71 f., 76–78, 80 f., 84 f., 88, 93, 95–99, 103, 110, 114 f., 119–121, 125, 130–132, 134 f., 141, 150, 152, 155, 163, 170–172, 178–181, 184, 186, 188–191, 193, 195 f., 200 f., 212 f., 215–218, 228, 238, 243 f., 246–249, 253–257, 263–266, 268 f., 271, 288, 290, 292, 296–298, 301, 311, 329
Al-Biqāʿī 293
Birth of Jesus 32 f., 159
Blind 39, 336
Body 4, 33 f., 37, 69, 107, 170, 175, 178, 190, 196 f., 218, 279, 290, 292
Book of Amos 254–256
Book of Enoch 147
Book of Esther 289
Book of Exodus 290
Book of Jubilees 13, 122, 147 f.
Boy 31, 40, 67, 90, 149, 151, 290, 295
Buber, Martin 11, 24, 173
Bughd 135

Caliph 49, 53, 55, 64, 66, 72, 111–113, 155, 247, 319, 321, 334
Canaan 11–14, 18, 20–23
Canonical gospels 27
Cerinthus 29
Chalcedonian Orthodoxy 100, 102–105, 114, 117
Chastity 41, 43
Childhood 30–32, 36, 290, 297 f., 301
Children of Abraham 15
Children of Israel 12–14, 16, 24, 192, 194, 289, 295
Christ 15–18, 29, 34, 36, 38 f., 47, 49–52, 55–58, 65, 69, 71 f., 83, 88 f., 91, 100, 112 f., 129, 145, 147, 154, 156, 158 f.
Christ-child 31
Christian-Arabic 44
Christianity 1–5, 7, 9, 29, 34, 47, 53, 74, 78, 81, 83 f., 87, 96–99, 101, 103 f., 106, 109 f., 114, 117, 120 f., 138, 141 f., 152, 160, 163, 188, 216, 224, 235, 241, 243 f., 246, 253 f., 330 f.
Christians 4–6, 9, 15, 17, 27 f., 47, 53 f., 61, 66, 68, 72 f., 76, 78–83, 85, 90, 93–96, 99–101, 103–105, 114 f., 117 f., 137, 145, 153 f., 156, 158–161, 165, 188 f., 195, 200–202, 204 f., 212, 216, 223, 240 f., 243–250, 331 f.
Christology 29 f., 53, 59, 83 f., 155
Chronicle of Zuqnin 153
Chronography 160 f.
Church 4, 17 f., 29 f., 49, 53, 59, 61–69, 72, 76, 78 f., 82, 95, 104, 117 f., 121, 137, 145, 147, 154–157, 161, 165, 167, 187, 189, 223
Compositional technique 143
Concatenation 39, 281
Contra Celsum 17, 42
Coptic Orthodox Church 61–63
Copto-Arabic 72
Corruption 80, 85, 154, 246–249, 287
Cradle 34 f., 37–39, 42–44, 241
Criticism (*tawbīkh*) 42, 138 f., 305, 329, 337
Crown of the Martyrs 40
Cultural cognates 223

Damascus 54, 84, 104, 145, 205, 227, 237, 241
Dār al-Islām 64

David 55 f., 61 f., 67 f., 78, 81, 83, 86, 89 f., 96 f., 126 f., 130, 147, 153 – 155, 158 f., 163 f., 169, 172, 176, 179 f., 184, 192 f., 202, 207, 218, 222, 229, 255, 281, 308, 336
Day of Judgment 39
Death 11, 40, 51, 68, 71, 82, 86, 88 f., 113, 140, 145, 149 f., 154, 156, 168, 170 – 173, 175, 192 f., 196 f., 249, 262 – 264, 290, 292, 330 f., 335
Depravity 119
Derekh eretz 233, 235
Deuteronomistic 5, 152, 155 f., 160, 171, 256
Dialogue 2, 9, 40, 43, 53 f., 81, 95, 100 f., 115, 117, 168, 188, 205, 207, 218 f., 222, 224 – 229, 231, 237, 239, 243 – 247, 249 f.
Al-Dimashqī, Ibn Abī Ṭālib 83, 114
Diptych 30
Discourse 2 f., 6 f., 10, 34, 40, 42 f., 66, 78, 80, 94, 96 – 99, 107 f., 143, 156, 199, 207, 212 – 217, 221 – 223, 225 f., 228, 231, 233 – 237, 239 f., 243 – 247, 249 – 251, 263, 278 – 280, 295
Discursive 129, 142, 202, 213 f., 216, 218, 223 f., 230, 236
Divinity 27, 38, 49, 79, 83, 88, 91, 136
Dove 29
Drory, Rina 200 – 202, 216 f., 228, 232 f., 236 – 238
Drunken dreams 140

Early authorities on ḥadīth 224
Eastern Christian tradition 121
Eastern Syriac tradition 131, 142
Egypt 12 – 14, 22, 24, 31, 61, 63, 65 f., 68, 72, 84 – 86, 154 f., 172, 179 f., 186, 226, 238, 247, 290, 292, 297, 300 f.
Eldad ha-Dani 215
Elijah 68, 89, 148 f., 151, 176, 196, 206, 220, 225, 227 f., 230, 298
Ephrem the Syrian 120, 127, 139
Epistle to Diognetus 17
Eritrea 29
Ernest A. Wallis Budge 34
Eschatological allegory 138
Ethical 132, 165, 167, 169, 185, 190, 210, 213, 217, 221, 225 f., 231, 233, 235, 238
Ethics 195, 225, 233, 235
Ethiopia 29, 147
Eutychius of Alexandria 120
Examination of the Three Faith (*Tanqīḥ al-abḥāth li-l-milal al-thalāth*) 330, 332, 334 f.
Exodus 10, 12 f., 18, 290, 292, 297
Exodus Rabbah 150, 290, 292

Faith 3, 5, 10 f., 15 – 17, 19, 21 – 25, 27, 33, 41, 59, 65, 68, 71, 74, 77, 81, 96 – 103, 108, 114 – 117, 119, 132, 154, 164, 166, 227, 291, 297 f., 300, 303, 330 f.
Faithful 1, 3, 11, 15 – 17, 24, 65, 137, 149, 191, 268, 286, 299
Farsakh 126
Father 9 – 15, 17 f., 24, 29, 34, 41, 49, 53, 59, 66 – 68, 90 – 92, 96, 100, 115, 118, 137, 140, 148, 151, 154, 159, 165, 167, 187, 189, 203, 221, 229, 260, 265 – 268, 289, 291, 297 – 301, 331
Firdaws 120, 132, 138
Firestone, Reuven 20, 22 f., 190, 260
First Anthropos/First-person 210, 219, 225
Forefather 15, 21 f., 298

Gabriel (Archangel) 6, 23, 34, 53, 89 f., 128 – 130, 200, 289
Galatians 15
Galut 14
Garden of Eden 138
Gehenna 137 f.
Genesis 10 – 16, 18 – 20, 23, 78, 119 f., 122 f., 126, 128, 132, 135, 137, 140, 146 f., 164, 255, 260, 281, 297, 299
Genesis Rabba 165
Geonic 5 f., 163 f., 166 f., 176 – 178, 189, 196, 201, 203 – 205, 211, 214 f., 217 f., 221 f., 224, 228, 230 – 232, 236 – 239
Geonim 163, 165, 177, 203, 206, 215, 222, 238
Al-Ghazālī, Abū Ḥāmid 85, 326, 330, 336
Ghurabā 135
Gnostic 29 f., 204
Gomorrah 127, 135, 141

Gospel 5, 19, 27, 36 f., 39, 67, 73, 75–77, 79–81, 85, 88–96, 99 f., 103–106, 109 f., 112 f., 115, 117, 129 f., 138, 156 f., 160, 245, 247–249, 300, 314
Gospel of Matthew 33, 89, 159, 269
Grammarians 6, 210–212, 240
Greek 4, 28, 53 f., 73, 78, 86, 96, 101, 115, 129 f., 135 f., 145–148, 150 f., 154 f., 157, 160, 171, 175, 187, 213, 237, 243 f., 280, 296
Greek philosophers 42
Griffith, Sidney 4 f., 47, 73, 76–80, 82, 85, 95 f., 98–104, 106, 115–118, 223, 268

Hagar 20, 22–24, 153
Haggadic literature 123
Hagiography 40, 72
Ḥajj 20 f., 23
Al-Ḥākim bi-Amri-llāh 64
Haman 289
Ḥanīf muslim 19
Ḥarrafa 87, 246
Al-Hāshimī, Muḥammad b. ʿAbdillāh 112 f., 115
Ḥatnē 124
Healing 27, 33, 37–39, 109, 116
Hebrew renaissance 214 f., 217, 228
Hebrews, Epistle to the 15, 269
Heilsgeschichte 10, 12, 16
Hell 137 f., 187, 236, 285, 295, 299, 324
Heretics 65, 104 f., 188
Hermeneutical 6, 25, 51, 58 f., 92 f., 117, 131, 189, 199, 211, 243–245, 322
Hermeneutics 9, 47 f., 78, 113, 115, 206, 210, 213, 230, 235, 309, 325
Hijra 24, 254
Historia salutis 143
History of the Patriarchs of Alexandria 4, 61
Holy Spirit 29, 42, 68, 76, 100, 105 f., 114 f., 117, 187

Ibn ʿAbbās 21, 291, 293, 320
Ibn al-Ṭayyib 5, 119–143
Ibn Ḥazm 80, 85, 191, 247 f.
Ibn Kammūna 7, 329–337
Ibn Kathīr 96, 292 f.
Ibn Qayyim al-Jawziyya 247
Ibn Qutayba(h) 97, 250
Ibn Taymiyya(h) 84, 246–248
Identity 5, 27 f., 30, 63, 72, 84, 86, 90, 102–104, 112 f., 137, 147 f., 200, 203 f., 214, 216, 224 f., 237, 239, 260, 265, 269, 316
Ifḥām al-yahūd 331
Iʿjāz al-Qurʾān 304
Impurity 141
Incarnation 29, 38, 52, 57, 74 f., 83, 263
Incest 119, 139 f.
Individual exegesis 218, 221, 235
Infant 20, 27, 31–37, 39–44, 289, 294, 300
Influence 2, 5, 20, 23, 74, 79, 101, 134, 147, 166, 196, 200, 203, 205, 222 f., 239 f., 264, 290, 296, 315, 317
Inner-Biblical interpretation, passim 139, 164, 172, 253, 293, 298
Inqilāb 127 f.
Interreligious crossroads 31
Intertwined Worlds 329
Irenaeus of Lyons 29
ʿĪsā 112
Isaac 10, 15, 19, 22, 66, 112, 172, 180 f., 186, 191, 259 f.
Isaiah 4, 38, 47, 146, 152, 180, 209, 226, 255, 265
Al-Iṣfahānī 205
Ishmael 10, 20, 22–24, 153, 260, 271
Īshūʿdād of Merv 57, 120, 122–127, 129, 137, 139–142
Islam 1–7, 9, 27, 31, 34, 47 f., 50, 54 f., 73, 76, 78, 80–83, 96 f., 101 f., 104 f., 114 f., 126, 138, 142, 147, 152, 156–158, 160, 163–165, 176, 190–194, 200 f., 203–206, 210 f., 216, 219, 221–224, 226 f., 234–237, 241, 243–245, 247–250, 253, 257, 268, 280, 289 f., 292, 306, 308, 311, 319 f., 322, 325 f., 329–332
Islamic 4–6, 9, 20 f., 23, 28, 31, 36 f., 39, 41, 44, 47, 53 f., 61, 63, 65 f., 73, 78–81, 83–86, 88, 90, 93–95, 99–102, 104, 107, 115 f., 126, 143, 147, 156 f., 160, 163, 167, 177, 188–191, 194 f., 200, 202–206, 215, 221–224, 226 f., 231, 233–238, 240, 243–246, 249 f., 253, 257,

263 f., 266, 270, 282, 287, 291, 297, 307, 314, 317 f., 321, 323, 329
Isnād, Chain of tradition 224, 229
Israel 12 f., 15 f., 24, 33, 133, 138, 140, 149, 153 – 155, 163, 168, 171, 178, 181, 184, 186, 205, 208 f., 213, 218 – 220, 253, 255 f.
Isrā'īliyyāt 189 f., 227
Ivory 30 f.
Ivory throne 30
ʿIṣyān 135

Jahannam 137
Al-Jahiz 216, 229, 233 f.
Jamarāt 20
Jerome, Saint 152, 195
Jesus Christ 69 f., 103, 112, 128, 156, 160, 187
Jesus' logion 129
Jewish literature 28, 216 f., 229, 232, 329
Jewish origins 32
Jews 5 f., 9, 15, 17, 34, 41, 56 f., 73, 82, 85, 88 – 90, 96, 99, 104 – 106, 114, 121, 123, 126, 128, 131, 133, 137, 139, 143, 147 f., 150 f., 158 – 161, 164 f., 174, 183, 188, 195, 200 – 205, 210, 212, 214 – 220, 222, 224 f., 229, 233, 237, 239 f., 246, 248, 263, 265, 311, 329, 331 f.
Jirjah (George), Archdeacon 64
Johannine tradition 38
John III of Samannūd 69
Jordan 22, 30, 89
Joseph 13, 32 f., 68, 82, 91, 142, 157, 159, 163, 177, 180, 184 f., 212, 229, 256, 270, 297, 299 – 301
Josephus 34, 135 f., 150, 175
Journey in quest of knowledge (*riḥla fī ṭalab al-ʿilm*) 226
Judaeo-Arabic 6, 163 f., 177 – 185, 189, 193 – 195
Judah ha-Levi 329
Al-Junayd 325 f.
Al-Juwaynī 247, 324

Al-Kaʿba 20, 22 – 24, 312 f.
Kalām 47, 79 f., 96, 98 f., 101 f., 106 f., 114 f., 118, 235, 321, 323, 326
Khalīl („friend") 15, 19, 82, 84, 93, 303
Khatanī 124
Khatanūna 124
King Uzziah 38
Al-Kitāb 139, 293
Kitāb al-bayān wa-l-tabyīn 216
Al-Kitāb al-Khazarī 329
Kitāb al-Ta'rīkh (Eutychius of Alexandria) 120
Kitāb firdaws al-naṣrāniyyah (Ibn al-Ṭayyib) 5, 119 – 143
Ktābā d-Dĕbūrītā (Solomon of Boṣtra) 120
Kutub 93, 95, 114, 190, 291 – 293, 324, 327

Lamoreaux, John 107
Last Judgement 128
Late antiquity 3, 27, 43, 64, 121, 137, 141 f., 157 f., 160, 164, 188, 199, 224, 241, 257, 318
Laurenziana 35
Law of hospitality 133, 135
Lazarus-Yafeh, Hava 329, 332
Leper 39
Linguistics 210 f., 217, 310
Lists 73, 78 f., 85, 210, 248, 320, 323, 325
Löfgren, Oscar 36
Logos 34 f., 49, 56, 58 f., 166
Lot 5, 19, 119 f., 122, 125 – 127, 131 – 137, 139 f., 209
Lot's daughters 119
Lot's house 123, 133 – 135
Lot's refuge 127
Lot's righteousness 133
Lot's wife 122 f., 129 – 131
Luke 10, 27, 38, 51 f., 65, 89 f., 120, 129, 159, 300

Al-Maʿālim 331
Ma'asse Torah 213
Mabīt 133
Maghariyya / Mahgraye 240
Al-Maghribī, Samaw'al 331 f.
Magūshē 139
Mahd 37
Al-Mahdi, Caliph 247
Maimonides 177, 185, 330
Majūs 139
Maṭlaʿ 322 f., 325

Malja' 126
Mamre 137
Manāsik 21, 23
Maqām Ibrāhīm 20
Mar Ṭalyā' of Cyrrhus 40
Mariology 30
Mario Provera 35
Martyrdom 39–41, 44, 53, 63–65, 69–71
Al-Marwa 20
Mary 29, 32–34, 36–38, 41–43, 53, 56, 58, 74–76, 81, 89 f., 139, 158–161, 331
Maryam 99, 112
Masoretes/Masoretic 211, 213, 216, 223, 238, 240
Masoretic Hebrew Text 124
Massacre of the Children at Bethlehem 30
Mathal 130
Mawhūb ibn Manṣūr ibn Mufarrij 4, 62 f.
Maximianus 30
Mayāmir 95, 98 f., 102, 105, 113, 115, 118
Maymar 99–111, 114–117
Mĕʿarat Gazē (Spelunca Thesaurorum) 134, 139
Mecca 11, 20–24, 254, 257–261, 264, 281, 286, 292, 304, 307, 309 f., 313, 315 f., 318
Medina 24, 263, 306, 311, 313
Medinan [revelations] 20, 254, 257 f., 260–264, 268 f., 281, 286, 304, 307, 309–311, 315 f., 318
Melchizedek 15
Melkite 81–83, 88, 100, 102, 104–106, 112, 114, 118, 161
Messiah 27, 37, 74 f., 79, 81, 89–91, 94, 99, 104–106, 112, 129 f., 158–160, 241
Messiahship 104
Messianism/messianic 15, 204 f., 217 f., 225, 228, 231, 240 f.
Meta-dialogue 243, 245–250
Metadiscourse 142
Middle Ages 3, 18, 43, 54, 64, 80 f., 95, 121, 126, 141 f., 165, 189, 197, 207, 212, 253, 317
Midhwad 37
Midrash (classical) 6, 13, 122 f., 126, 128, 130, 140, 143, 148, 152, 165, 173, 199, 201 f., 205 f., 210, 212, 214 f., 218, 223 f., 230–233, 239, 241, 284
Midrashim 14, 201 f., 214, 217, 222, 232 f., 236 f.
Midrash Tanḥuma 123, 173
Midwife 32, 34, 37 f.
Milan 18, 36
Minā 20
Miḥna 321
Minimal Judaism 225, 234
Miracles 27–34, 38 f., 43, 105 f., 110, 114, 184
Moabites 139
Monastery of St. Catherine 36
Moses 1, 6 f., 10, 12–16, 19, 23 f., 67 f., 89, 91, 99 f., 102–105, 114, 117, 125, 148, 175, 180, 184, 186, 190, 220, 226, 277, 289–297, 299–301, 305
Moses and the Law 15
Moshē bar Kēphā 120
Mount Zion 16, 208
Mourners of Zion 204 f.
Ms Diyarbakir 122
Muḥammad 5, 19, 21–24, 31, 54 f., 73, 77, 80–85, 88, 97, 112, 145, 190–193, 247–250, 254, 259, 261–265, 267–269, 278, 280 f., 284, 291–294, 296, 309, 324, 326, 332, 335
Al-Muḥaṣṣal 331
Mujādalah 111–115
Muḥkamāt 320–323
Muqātil b. Sulaymān 21, 291
Muslim 4, 6 f., 9–11, 18–24, 47, 53–56, 62, 64, 66 f., 72–76, 78–88, 92–107, 111–118, 120 f., 139, 154, 157–161, 163, 189–191, 193 f., 200, 202–207, 210 f., 215 f., 219, 221–223, 229, 234–237, 239 f., 243–251, 263, 303–308, 311–314, 317–320, 329–334
Mu'tafikāt 127 f.
Mutakallim 5, 95, 99, 101 f., 112–114, 118
Mutakallimūn 73, 79–81, 95–100, 105, 116–118
Mutashābihāt 319–323, 325
Muʿtazila/Muʿtazalitism/Mutazilite 221 f., 240

Nabī 42
Al-Nahwandi, Benjamin 204, 217, 240
Narratio 142 f., 215 f.
New Testament 4, 10, 15, 18, 24, 27 f., 30, 36, 47, 55, 73, 76, 78–80, 96, 115, 120, 130, 138, 147, 188, 190, 269, 300, 315, 332
Nicaea 106
Nihāyat al-ʿuqūl 331
Noah 10, 212, 277, 298–300
Nocturnal pollution 141
Nomad of faith 11, 24
Numbers 13, 136

On Christian Doctrine 17 f.
Oral Torah (Lore) 22, 201, 203, 206–209, 218 f., 221, 235, 238
Origen of Alexandria 42

Paradise 77, 120, 137 f., 231, 258, 270, 324, 326
Paraphrased translation 142
Para-Scriptural literatures 43
Pardaysā 138
Parting of ways 240
Patterning narrative 10
Paul, the Apostle 15, 35, 68, 82–84, 87, 92 f., 100, 120, 130, 135, 164, 184, 236, 255 f.
Pentateuch 12, 135, 142 f., 146, 164, 184, 201, 218, 220, 266, 313
Pericopes 143, 281, 285 f.
Peristephanon 40
Persian-Arabic 223, 236
Persia/Persian 126, 134, 138 f., 145, 171, 173, 178, 202–204, 210, 212 f., 227, 229, 233, 237, 240 f., 289
Peshīṭtā 124, 128, 133, 140
Pharaoh 6, 13, 68 f., 71, 91, 289–297, 300 f.
Pilgrim 9, 11 f., 15–18, 21
Pirke Avot 221
Pirke de Rabbi Eliezer 232
Pirqoi ben Baboi 205, 221
Pohlmann, Karl-Friedrich 262 f.
Polemic 10, 15, 17, 42, 47, 73, 81–84, 86–88, 93, 95–98, 171, 176, 205, 221, 224, 246, 250, 254, 263, 311
Pontius Pilate 69–72
Potiphar 297, 300
Promised land 11–13, 17, 19, 24, 138
Prooftexting 74 f., 82, 95, 98
Prophet 19, 21–24, 34–36, 38 f., 42–44, 55, 68, 82, 88–90, 97, 145, 148–151, 155, 167–171, 174, 182, 191, 193 f., 225, 227–230, 247, 250, 256 f., 259, 264–266, 269, 293, 295 f., 298–300, 308, 310 f., 314, 318, 322, 332 f., 335
Prophetic 10 f., 21, 24, 34, 42, 51, 80 f., 85, 104, 146, 165, 184, 190, 220, 228, 254, 259, 261–264, 293, 310, 314, 326
Prophets, Lives of the 4, 12, 31, 38 f., 44, 55, 75–77, 94, 99, 103, 148–151, 158 f., 164, 168, 181, 184, 186–188, 190–193, 226 f., 295, 300, 305, 312, 324, 335
Protoevangelium of James 32–34, 36
(Proto-) Karaites 163, 177–179, 204 f., 219 f., 239, 329
Prudentius 40
Psalms 13, 75–77, 94
Pseudepigraphical 32 f., 214
Pseudo Ibn 'Abbās 21
Pseudo-Methodius, Apocalypse of 153

Al-Qarāfī 83–85, 87 f., 90 f., 93, 247
Qarīt 121
Qarītā 121
Al-Qāsim b. Ibrāhīm ar-Rassī 249
Qāṣṣ/quṣṣāṣ 226–228, 235 f.
Al-Qirqisani, Jacob 178, 182–184, 186, 196, 215
Qiṣaṣ al-anbiyāʾ 190–193, 227
Al-Qumisi, Daniel 204, 217, 221, 240
Al-Qummī, 'Alī b. Ibrāhīm 21
Qumrān 143
Qur'ān 81, 95–98, 111, 113, 117, 127 f., 216, 303 f., 307 f., 310, 312, 318

Rabbanite(s) 143, 163–165, 177–179, 202, 204, 217–219, 224, 232, 239, 329
Rabbinic statements 42

Rabbinic tradition 19, 124, 126, 128, 163, 196, 202, 214, 219 f., 230, 236, 263, 287
Rashbam 11
Rationalism 101, 110, 115
Rationalist 79, 101, 114, 183 f., 321
Al-Rāzī, Fakhr al-Dīn 7, 291, 330 – 332, 336
Reason 5, 44, 50 f., 58, 73, 79, 83 f., 93, 95 f., 99, 101 – 103, 105, 109 – 111, 113 – 117, 121, 127, 140 f., 148 f., 151, 153, 155, 169, 172, 179, 182, 184, 195, 235, 248, 250, 258, 263 f., 268, 281, 288 f., 295, 297 f., 320, 323 f.
Redeemer 34
Relief 30 f., 288
Return to Scripture 6, 213 – 216
Rewritten Scriptures 43
Roggema, Barbara 47, 56, 96
Roman Catholic 30
Romans 15, 82, 202, 333
Romanus 40

Sa'adya Gaon 178, 186, 210, 217, 228, 232, 238
Ṣāghār 125
Salome 33 f., 36 – 38
Salvation 15, 34, 38, 52 f., 79, 108, 209, 256, 258
Samir, Samir Khalil 74 f., 87, 96, 101, 120 f.
Sarah 16, 22, 164, 177, 185, 216, 329
Sasanian 35
Satan 20, 38, 270, 277, 283, 292
Saul (King) 126 f., 147, 167 – 177, 180 – 187, 190 – 195
Savior 32
Schmidtke, Sabine 97, 200, 329 f., 335
Science of language 207, 210, 240
Scriptures 1 – 3, 5, 9 f., 24, 39, 43, 75 – 78, 80, 82 – 85, 92 – 103, 105 – 107, 109, 111, 113 – 118, 139, 165, 245 – 249, 268, 312, 314
Second Temple 14, 147, 167
Seder Eliyahu (Zuta and Rabba) 206, 214, 217, 221 f., 225 – 228, 230 – 233, 237
Sefer ha-Egron 217, 228
Severus (Sāwīrus) ibn al-Muqaffa' 61
Sexual appetite/*Shahwah* 136
Sexual intercourse 134
Shahrazūrī, Shams al-Dīn 337
Sharāb 141
Shared discursive space 240
She'iltot 215
Shmuel ben Hofni 232, 236
Shragrāgyātā 134
Sign 7, 27, 30, 33, 39, 82, 100, 123, 134, 145, 202, 236, 258, 323, 332 f.
Sike, Henricus 35
Simon 64, 66 – 68, 122 f., 126, 128, 140, 170, 180
Sinai 6, 10, 12 f., 23, 36, 53, 74, 179, 181, 200, 218, 253 f., 257 f., 260 f., 263 f., 270 f., 278 f., 285 – 287
Sister of Aaron 41
Slave 42, 149 f., 153, 266
Slavery 12, 14, 149 f., 153 – 156
Ṣô'ar 125
Sodom 5, 119, 124, 126 – 128, 131, 134 – 138, 141 f., 255
Sodomites 132 f., 136, 141
Sojourner 3, 9, 11 – 13, 15 f., 18, 20, 22 – 24
Solomon of Boṣtra 120
Son of God 34 f., 38, 47, 76, 112, 154, 158, 331 f.
Son of the First Anthropos 29
Sost Ledat Christology 29
Speech 38 f., 41 – 43, 50 f., 68, 71, 92, 106, 172, 181 f., 210, 216, 225, 268, 278, 297, 304, 318, 334
Spirit 18, 54 – 56, 74 – 76, 81, 106, 112 f., 160, 168, 171, 186, 189, 194, 208, 325
Spiritual 10, 14 f., 18, 24, 47, 66 – 68, 134, 147, 209
Stranger and sojourner/exile 11 – 18, 20, 24, 63, 65
Subaltern 238 – 240
Sufi 7, 235 f., 333
Sufism 326
Al-Suhrawardī, Shihāb al-Dīn 330
Sūra(s) 6 f., 56, 75, 95 f., 254, 258 – 265, 267 f., 270 f., 277 f., 280 f., 285 f., 289, 292, 297, 304 – 318, 321, 333 f.
Sūrat Maryam 41 f.
Swanson, Mark 4, 61 – 66, 69, 74 f., 95 f., 98, 104, 106
Syllogism 158 – 161

Syria 5, 20, 22, 47, 49, 58, 62, 86, 96, 121, 145–147, 151–154, 156 f., 161, 164, 186, 188 f., 205, 237, 241
Syriac-Christian 221 f., 225, 235, 237, 239
Syriac Christian heritage 5, 142
Syriac Life of the Blessed Virgin Mary 34
Syriac Revelation of the Magi 43 f.
Syro-Arabic 28, 34

Al-Ṭabarī, Abū Jaʿfar Muḥammad 192–194
Al-Ṭabarī, ʿAlī b. Rabbān 81, 85, 87, 90 f., 248
Tafsīr 21 f., 24, 120, 125, 135, 176, 185, 190, 192, 211, 263, 291–293, 296, 309, 318–320, 323, 336
Taḥrīf 246–250
Taḥrīf al-lafẓ 246–248
Taḥrīf al-maʿna 246–248
Talmud Babli 128
Al-Talwīḥat 330, 335
Tanakh 10–13, 17 f., 20, 24, 165, 175
Tanna debe Eliyahu 214
Taqallaba 128
Targum 150, 152
Targum Jonathan 150, 152
Ṭawāf 20
Ṭelal 121
Terah 12
Teshuva/repentance 68, 152, 156, 225, 228, 236, 286
Testamentum Domini 156–158, 160
Test of virginity 33, 37
Textual fluidity 35
The City of God 17, 154
The Heavenly Jerusalem 16
The Niche of Lights (*Mishkāt al-anwār*) 336
Theology 5, 29, 58, 80, 86, 88, 104, 106 f., 112 f., 115, 122, 152, 160, 172, 191, 200, 216, 221 f., 234, 246, 314, 321, 329
Theoria 147, 160
The Verses of Challenge (*āyāt al-taḥaddī*) 333
The Way of Abraham (*ṣirāṭ Ibrāhīm*) 19
Thomas, David 12, 30, 47, 56, 61 f., 78, 81, 83 f., 86 f., 95–97, 115, 120, 157, 246, 303
Timothy 47–59, 86, 247 f.
Toldot Ben Sira 228
Toledot Yeshu 42 f.
Torah 1, 5, 10, 12, 19, 39, 73, 76, 91, 94, 104, 160, 177–179, 181, 184, 190, 194, 201, 208 f., 217–221, 231, 235, 313 f.
Treatise on the differences of the Rabbanites and the Karaites 329
Al-Ṭūsī, Naṣīr al-Dīn 330
Al-Tustarī 325
Typology 78, 160

Ṣūfī tafsīr 319
Unctionist Christology 29
Usdūm, Jabal 135
Al-Usdūmāyyīn 134

Verification 95, 99 f., 111, 114, 116
Virgin 29, 32–34, 38, 53, 71, 99, 112, 138, 158 f.
Virginity 33 f., 36 f., 159, 229
Vision of Isaiah 32

Wedding feast at Cana 30 f.
Wine 27, 30, 140 f., 208
Witztum, Joseph 270 f., 278 f., 299
Wonderworker 4, 27–29, 31
Word, God's 1, 4, 7, 14 f., 17, 21, 37, 42, 49, 51 f., 54–57, 66, 68 f., 71, 74–76, 79, 81, 93, 98, 103, 108, 112 f., 122, 124–126, 130 f., 133, 136 f., 142, 146, 148, 153, 159, 166, 170, 173, 175, 183 f., 187 f., 194, 208–210, 212, 221, 231, 239, 244, 246, 255, 260, 264, 278, 282, 285, 287, 308, 313, 318 f., 321 f., 324–326, 332, 334 f.

Al-Yaʿqūbī 191
Yeshivot 203, 206, 222, 239
YHWH 12
Youth 30, 32
Yughdan 240

Zacharias, Coptic Orthodox patriarch 64, 87
Zamzam 20, 22
Zohar 14
Zughar 125

www.ingramcontent.com/pod-product-compliance
Lightning Source LLC
Chambersburg PA
CBHW051247300426
44114CB00011B/921
* 9 7 8 3 1 1 0 5 6 1 4 4 9 *